MW01079407

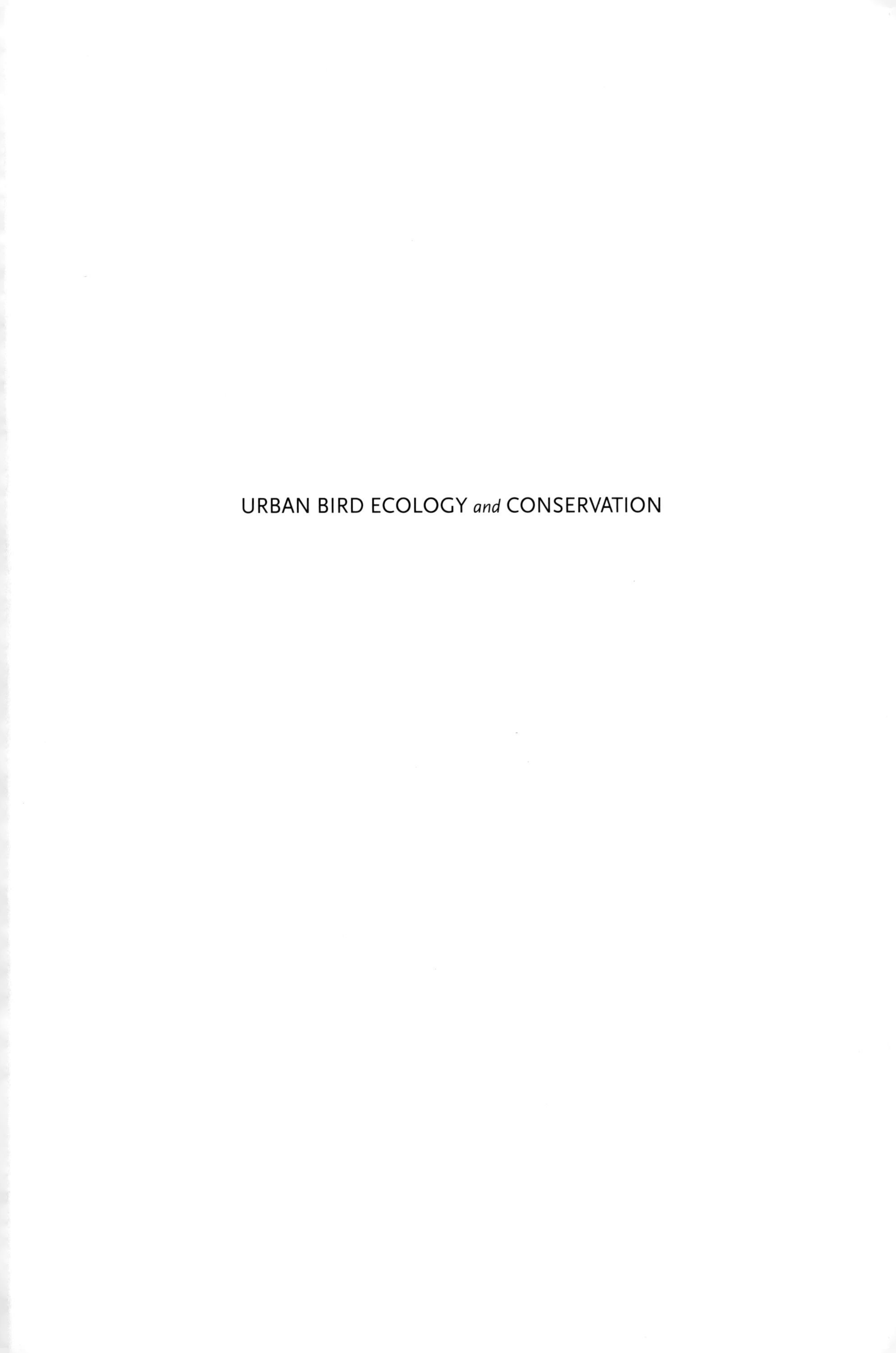

URBAN BIRD ECOLOGY *and* CONSERVATION

STUDIES IN AVIAN BIOLOGY

A Publication of the Cooper Ornithological Society

WWW.UCPRESS.EDU/GO/SAB

Studies in Avian Biology is a series of works published by the Cooper Ornithological Society since 1978. Volumes in the series address current topics in ornithology and can be organized as monographs or multi-authored collections of chapters. Authors are invited to contact the series editor to discuss project proposals and guidelines for preparation of manuscripts.

See complete series list on page 325.

URBAN BIRD ECOLOGY *and* CONSERVATION

Christopher A. Lepczyk and Paige S. Warren, *Editors*

Studies in Avian Biology No. 45

A PUBLICATION OF THE COOPER ORNITHOLOGICAL SOCIETY

University of California Press
Berkeley Los Angeles London

University of California Press, one of the most distinguished university presses in the United States, enriches lives around the world by advancing scholarship in the humanities, social sciences, and natural sciences. Its activities are supported by the UC Press Foundation and by philanthropic contributions from individuals and institutions.
For more information, visit www.ucpress.edu.

Studies in Avian Biology no. 45
For e-book, see the UC Press website.

University of California Press
Berkeley and Los Angeles, California

University of California Press, Ltd.
London, England

Library of Congress Cataloging-in-Publication Data

Urban bird ecology and conservation / Christopher A. Lepczyk and Paige S. Warren, editors.
p. cm.—(Studies in avian biology ; no. 45)
Includes bibliographical references and index.
ISBN 978-0-520-27309-2 (cloth : alk. paper)
1. Birds—Ecology. 2. Birds—Conservation. 3. Urban animals—Ecology.
4. Urban animals—Conservation. I. Lepczyk, Christopher A.
(Christopher Andrew), 1970– II. Warren, Paige S. (Paige Shannon)

QL698.95.U73 2012 598—dc23

Manufactured in the United States of America
19 18 17 16 15 14 13 12
10 9 8 7 6 5 4 3 2 1

The paper used in this publication meets the minimum requirements of ANSI/NISO Z39.48-1992 (R 2002) (*Permanence of Paper*).

Cover illustration: Common Peafowl (*Pavo cristatus*), Kaneohe, Hawaii.
Photo by Christopher A. Lepczyk.

PERMISSION TO COPY

DEDICATION

This volume is dedicated to Carl Marti,
who passed away before seeing the completion of this volume.

CONTENTS

Part III • Human-Avian Interactions and Planning

Part IV • Future Directions

Online Content

This volume of Studies in Avian Biology includes both print and online materials. The publisher is pleased to provide this additional relevant material online as an open-access resource (www.ucpress.edu/go/sab). We encourage you to read the online material in conjunction with the volume.

CONTRIBUTORS

MARINA ALBERTI
College of Built Environments
University of Washington
Box 355726
Seattle, WA 98195-5726, USA
malberti@u.washington.edu

PAUL R. ARMSWORTH
Department of Ecology and Evolutionary Biology
University of Tennessee, Knoxville
569 Dabney Hall
Knoxville, TN 37996-1610, USA
p.armsworth@utk.edu

ANA LAURA BARILLAS-GÓMEZ
Departamento de Ciencias Químico-Biológicas
Universidad de las Américas Puebla
72820 Puebla, México
anabarillas@yahoo.com.mx

LAUREL M. BARNHILL
South Carolina Department of Natural Resources
1000 Assembly Street
Columbia, SC 27202, USA
barnhilll@dnr.sc.gov

SARAH BARTOS SMITH
Environmental Studies Program
Dartmouth College
6182 Steele Hall
Hanover, NH 03755, USA
sbsmith@dartmouth.edu

ANNE M. BARTUSZEVIGE
103 East Simpson Street
Lafayette, CO 80026, USA
anne.bartuszevige@pljv.org

ROBERT B. BLAIR
Department of Fisheries, Wildlife,
and Conservation Biology
University of Minnesota
St. Paul, MN 55108, USA
blairrb@umn.edu

PETER H. BLOOM
13611 Hewes Avenue
Santa Ana, CA 92705, USA
phbloom1@aol.com

CAROLINA BONACHE-REGIDOR
Departamento de Ciencias Químico-Biológicas
Universidad de las Américas Puebla
72820 Puebla, México
carolinabonache@yahoo.es

AMANDA J. BORENS
Ecology and Evolutionary Biology
University of Arizona, LSS 229
Tucson, AZ 85719, USA
jacksona@email.arizona.edu

OWEN D. BOYLE
Wisconsin Department of Natural Resources
101 South Webster Street ER/6
P.O. Box 7921
Madison, WI 53707-7921, USA
owen.boyle@wisconsin.gov

DANIELA BUZO-FRANCO
Departamento de Ciencias Químico-Biológicas
Universidad de las Américas Puebla
72820 Puebla, México
danudla@yahoo.com

SOPHIA N. CHIANG
Western Foundation of Vertebrate Zoology
261 Tangelo
Irvine, CA 92618, USA
snhchiang@hotmail.com

ZOE G. DAVIES
Durrell Institute of Conservation and Ecology
University of Kent
Canterbury, CT2 7NR, UK
z.g.davies@kent.ac.uk

DEAN DEMAREST
U.S. Fish and Wildlife Service
Division of Migratory Birds
1875 Century Boulevard, Suite 240
Atlanta, GA 30345, USA
deand.demarest@fws.gov

STEPHEN DeSTEFANO
USGS Massachusetts Cooperative Fish and
Wildlife Research Unit
Holdsworth Natural Resources Center
University of Massachusetts–Amherst
Amherst, MA 01003, USA
sdestef@nrc.umass.edu

RICHARD A. FULLER
School of Biological Sciences
University of Queensland
St. Lucia, Queensland 4152, Australia
r.a.fuller@dunelm.org.uk

JERÓNIMO GARCÍA-GUZMÁN
Departamento de Ciencias Químico-Biológicas
Universidad de las Américas Puebla
72820 Puebla, México
jeronimo.garcia@udlap.mx

KEVIN J. GASTON
Environment and Sustainability Institute
University of Exeter
Penryn
Cornwall, TR10 9EZ, UK
k.j.gaston@exeter.ac.uk

JOSÉ ANTONIO GONZÁLEZ-OREJA
Departamento de Agroecosistemas y Recursos
Naturales
NEIKER-Tecnalia (Instituto Vasco de Investigación
y Desarrollo Agrario)
Calle Berreaga 1, 48160
Derio, Spain
jgonzalez@neiker.net

JEFFREY HEPINSTALL-CYMERMAN
Warnell School of Forestry and Natural Resources
University of Georgia
Athens, GA 30602-2152, USA
jhepinstall@warnell.uga.edu

LORNA HERNÁNDEZ-SANTÍN
Departamento de Ciencias Químico-Biológicas
Universidad de las Américas Puebla
72820 Puebla, México
lornahs@yahoo.com

MARK HOSTETLER
Department of Wildlife Ecology and Conservation
University of Florida
215 Newins-Ziegler Hall
P.O. Box 110430
Gainesville, FL 32611-0430, USA
hostetm@ufl.edu

ESA HUHTA
Kolari Research Unit
Finnish Forest Research Station
Muoniontie 21, 95900-FI
Kolari, Finland
esa.huhta@metla.fi

KATHERINE N. IRVINE
Institute of Energy and Sustainable Development
De Montfort University
The Gateway
Leicester, LE1 9BH, UK
kirvine@dmu.ac.uk

JUKKA JOKIMÄKI
Arctic Centre
University of Lapland
P.O. Box 122, 96101-FI
Rovaniemi, Finland
jukka.jokimaki@ulapland.fi

MARJA-LIISA KAISANLAHTI-JOKIMÄKI
Arctic Centre
University of Lapland
P.O. Box 122, 96101-FI
Rovaniemi, Finland
marja-liisa.kaisanlahti@ulapland.fi

ANN P. KINZIG
School of Life Sciences
Arizona State University
Tempe, AZ 85287, USA
kinzig@asu.edu

CHRISTOPHER A. LEPCZYK
Department of Natural Resources and Environmental Management
University of Hawai'i at Ma-noa
Honolulu, HI 96822, USA
lepczyk@hawaii.edu

IAN MacGREGOR-FORS
Centro de Investigaciones en Ecosistemas
Universidad Nacional
Autonoma de México
A.P. 27-3, Santa Maria de Guido Morelia
Michoacan, México 5808
ian@oikos.unam.mx

LOUIS MACHABÉE
2675 Cuvillier, Apt. 06
Montreal, Canada
H1W 3A9
louis.machabee@umontreal.ca

JOHN M. MARZLUFF
College of Forest Resources
University of Washington
Box 352100
Seattle, WA 98195, USA
corvid@u.washington.edu

RACHEL E. McCAFFREY
School of Natural Resources
University of Arizona
325 Biological Sciences
East Tucson, AZ 85721-0043, USA
rachmcc@email.arizona.edu

JENNY E. McKAY
Department of Biology
Portland State University
P.O. Box 751
Portland, OR 97207, USA
jenny.e.mckay@gmail.com

ANGELA G. MERTIG
Department of Sociology and Anthropology
Middle Tennessee State University
Murfreesboro, TN 37132, USA
amertig@mtsu.edu

J. MICHAEL MEYERS
Patuxent Wildlife Research Center - Maryland
Warnell School of Forestry and Natural Resources
University of Georgia
180 East Green Street
Athens, Georgia 30602, USA
jmeyers@warnell.uga.edu

JAMES R. MILLER
Department of Natural Resources and Environmental Sciences
University of Illinois
N-407 Turner Hall, MC-047
1102 South Goodwin Avenue
Urbana, IL 61801, USA
jrmillr@illinois.edu

LORENA MORALEZ-PÉREZ
Centro de Investigaciones en Ecosistemas:
Antigua Carretera a Patzcuaro
8701 Col. Ex-Hacienda de San Jose de la Huerta
Morelia 58190
Michoacan, México
lmorales@oikos.unam.mx

WILLIAM P. MUELLER
Wisconsin Society of Ornithology
1242 South 45th Street
Milwaukee, WI 53214, USA
wpmueller1947@gmail.com

MICHAEL T. MURPHY
Department of Biology
Portland State University
P.O. Box 751
Portland, OR 97207, USA
murphym@pdx.edu

DERRIC N. PENNINGTON
World Wildlife Fund-US
1250 24th Street NW
Washington, DC 20037, USA
derric.pennington@wwfus.org

JENNIFER K. RICHARDSON
1326 SE 152nd Place
Portland, OR 97233, USA
jaxely@yahoo.com

SCOTT K. ROBINSON
Florida Museum of Natural History
Dickinson Hall
P.O. Box 117800
Gainesville, FL 32611-7800, USA
srobinson@flmnh.ufl.edu

AMANDA D. RODEWALD
School of Environment and Natural Resources
Ohio State University
2021 Coffey Road
Columbus, OH 43210, USA
rodewald.1@osu.edu

JAMES A. ROTENBERG
Department of Environmental Studies
University of North Carolina–Wilmington
601 South College Road
Wilmington, NC 28403-5949, USA
rotenbergj@uncw.edu

JORGE E. SCHONDUBE
Centro de Investigaciones en Ecosistemas:
Antigua Carretera a Patzcuaro
8701 Col. Ex-Hacienda de San Jose de la Huerta
Morelia 58190
Michoacan, México
chon@oikos.unam.mx

PIRKKO SIIKAMÄKI
Metsähallitus
Natural Heritage Services
P.O. Box 26, FI-93601
Kuusamo, Finland
pirkko.siikamaki@oulu.fi

CHRISTINE M. STRACEY
Department of Biology
Westminster College
1840 South 1300 East
Salt Lake City, UT 84105, USA
cstracey@westminstercollege.edu

SCOTT E. THOMAS
P.O. Box 5447
Irvine, CA 92616, USA
redtail1@cox.net

WILL R. TURNER
Conversation International
2011 Crystal Drive, Suite 500
Arlington, VA 22202, USA
w.turner@conservation.org

TIMOTHY L. VARGO
Urban Ecology Center
1500 East Park Place
Milwaukee, WI 53211, USA
tvargo@urbanecologycenter.org

SARA E. VONDRACHEK
Urban Ecology Center
1500 East Park Place
Milwaukee, WI 53211, USA
sevondra@gmail.com

PAIGE S. WARREN
Department of Environmental Conservation
Holdsworth Hall
University of Massachusetts–Amherst
Amherst, MA 01003, USA
pswarren@eco.umass.edu

CHARLENE M. WEBSTER
4758 East Eastland Street
Tucson, AZ 85711, USA
bckwebster@cox.net

KARA WHITTAKER
College of Forest Resources
University of Washington
10030 40th Avenue SW
Seattle, WA 98146, USA
karaayn@u.washington.edu

PREFACE

Urban bird ecology is not a new field in ornithology. One need look no further than the rich literature from Europe and such classic works as Emlen's 1974 *Condor* article on urban birds in Tucson, Arizona, to recognize the deep roots of study of urban birds. However, interest in urban bird ecology, and urban ecology more generally, waned after the 1970s, with relatively few studies appearing until the late 1990s. The turning point among many of us with interests in urban birds came about with the 2001 publication of *Avian Ecology in an Urbanizing World* (Marzluff et al. 2001). In this volume appeared a number of papers on urban bird distributions, gradient analysis, and scale, to name a few, that demonstrated how understudied, important, and ultimately fascinating the world of urban bird ecology was to science. As many readers of this current volume can surely attest, the volume now stands as a seminal work in the field less than a decade after its publication.

In the spring of 2005, several of the authors from *Avian Ecology in an Urbanizing World* attended the Emerging Issues along Urban/Rural Interfaces: Linking Science and Society conference in Atlanta, Georgia. Over dinner one evening with Roarke Donnelly, Mark Hostetler, Jim Miller, and ourselves, the question arose as to what had been learned since 2001. This dinner conversation led to the idea for a symposium at the IV North American Ornithological Conference to be held in Veracruz, Mexico, during the fall of 2006 that would highlight five years of progress in urban bird ecology. Subsequently, a proposal for the meeting was put forward and accepted, entitled "New Directions in Urban Bird Ecology and Conservation." What we originally proposed as a half-day symposium grew into a full day of 21 talks, with more potential talks submitted to the conference than we had time slots available. The symposium was heavily attended during the meeting, and it was evident to us that interest in urban birds was growing!

Upon seeing the symposium title and talks, Dr. Carl Marti invited us to develop it into a volume for the Studies in Avian Biology series. Following the symposium we set out to produce a volume based not only on the talks, but also involving other noted urban bird ecologists. Thus, our goal was to provide a sequel of sorts to *Avian Ecology in an Urbanizing World*. The volume before you represents both classic approaches and new ideas in urban bird ecology and conservation from around the world. Some authors were recruited with targeted invitations, but we also chose to open the volume to submissions from any participants in the 2006 symposium to provide a mix of both established urban bird ecologists and investigators new to the field. We are pleased that the ultimate body of works includes a wide diversity of topics, locations, and ideas to help move our field forward.

We could not have produced this volume without the significant help of many individuals who served as referees for one or more of the chapters. Thus, we warmly acknowledge the following individuals: C. Bang, K. Belinsky, D. Bolger, A. Budden, P. Clergeau, M. A. Cunningham, D. Curson, S. DeStefano, R. Donnelly, S. Droege, C. Flather, R. A. Fuller, A. Gabela, J. Hepinstall-Cymerman, W. Hochachka, M. Hostetler, M. Katti, D. King, E. Laurent, S. Lerman, I. MacGregor-Fors, P. Marra, J. M. Marzluff, R. E. McCaffrey, N. McIntyre, M. McKinney, J. R. Miller, C. Nilon, D. N. Pennington, C. A. Ribic, A. D. Rodewald, J. A. Rotenberg, T. Roth II, R. Scheller, Ç. Şekercioğlu, E. Shochat, H. Slabbekoorn, S. Bartos Smith, V. St. Louis, C. M. Stracey, E. Strauss, and J. Taylor. In addition, we would like to thank R. Danford for assistance with formatting manuscripts and C. Marti for his patience with us over the course of producing this volume. Unfortunately, Carl passed away before reviewing the final edits of the volume and its publication. We hope he would be proud. Taking over editorial duties for Carl was B. K. Sandercock, who helped shepherd the volume through a new publisher and the many changes this entailed. We would also like to thank each of the volume's authors for sharing their research and ideas. Finally, we would both like to extend a personal thanks to our respective families for their love and support throughout the creation and editing of this volume.

CHRISTOPHER A. LEPCZYK
University of Hawai'i at Manoa
Honolulu, Hawaii

PAIGE S. WARREN
University of Massachusetts–Amherst
Amherst, Massachusetts

LITERATURE CITED

Emlen, J. T. 1974. An urban bird community in Tuscon, Arizona: derivation, structure, regulation. Condor 76:184–197.

Marzluff, J. M., R. Bowman, and R. Donnelly. 2001. Avian ecology in an urbanizing world. Kluwer Academic Press, Boston, MA.

CHAPTER ONE

Beyond the Gradient

INSIGHTS FROM NEW WORK IN THE AVIAN ECOLOGY OF URBANIZING LANDS

Paige S. Warren and Christopher A. Lepczyk

People like birds, and increasingly, people live with birds (Turner et al. 2004, USDI 2006). Many, many people watch, feed, and intentionally attract birds to their homes, yards, and public spaces (Lepczyk et al. 2004, USDI 2006, Davies et al. 2009). Yet not everyone likes all birds, or even some birds, all of the time. Flocks of noisy blackbirds roost in urban street trees and defecate on streets and sidewalks. Pigeons nesting on buildings, geese feeding on golf courses, flocks of birds interfering with airport traffic—the relationships between humans and birds are not always friendly (Conover and Chasko 1985, Coluccy et al. 2001, Sodhi 2002). People actively remove bird habitat, sometimes even for species we admire. Dead branches and snags create real hazards near homes, but they provide nesting habitat for the same woodpeckers that a resident might be feeding from a suet feeder. Birds, in many ways, epitomize the simultaneous bond and tension between humans and nature. As the human population grows and cities continue to sprawl, species conservation becomes increasingly challenging, while at the same time new opportunities for stewardship arise. Science is increasingly needed in support of effective species conservation and environmental stewardship in the places where people live, travel, and work. Because birds can serve as proxies for other biological components of the ecosystem and are easily studied, they represent an ideal model for understanding urban systems. As a result, understanding and conserving birds in urban ecosystems have important repercussions for conservation and management.

In this volume, we explore the complex relationships between humans and birds through three themes in the emerging science and practice of urban bird ecology and conservation: (1) mechanisms structuring bird communities across an urbanization gradient, (2) citizen science and the demography of urban birds, and (3) human-avian interactions and planning. We will use the word "urban" in this introduction for the sake of brevity, but the papers cover a broad range of human-dominated settings, from city centers to suburban residential areas to protected lands interfacing with urban areas to analysis of heterogeneous, urbanizing landscapes. In addition, the papers cover a range of approaches to forming knowledge, including conventional academic approaches (both natural and social science), citizen science partnerships, and what might be considered "clinical studies" of urban ecology, that is,

Warren, P. S., and C. A. Lepczyk. 2012. Beyond the gradient: insights from new work in the avian ecology of urbanizing lands. Pp. 1–6 *in* C. A. Lepczyk and P. S. Warren (editors). Urban bird ecology and conservation. Studies in Avian Biology (no. 45), University of California Press, Berkeley, CA.

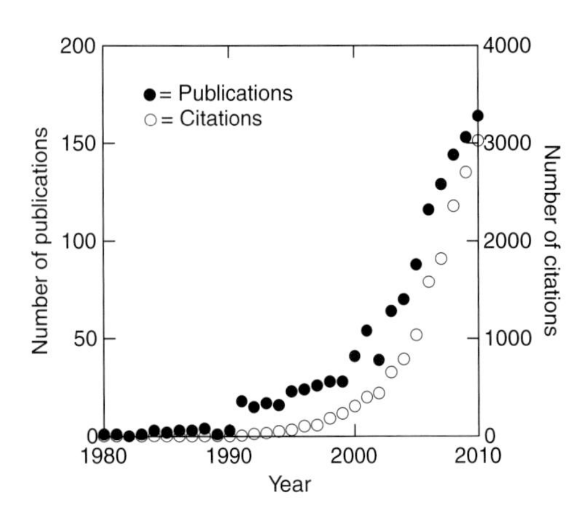

Figure 1.1. The number of urban bird publications and number of citations of urban bird publications by year as identified by ISI Web of Science. Numbers were determined by using the keywords "urban" and "bird."

insights derived from the application of science in practice. A central theme of this volume is that we must employ *all* of these approaches to do effective conservation science in human-dominated lands.

During the past decade, urban ecology has seen a great surge of interest in the United States and abroad, as evident in the establishment of two Long-Term Ecological Research sites (i.e., the Baltimore Ecosystem Study and the Central Arizona-Phoenix LTERs), numerous high profile publications in peer-reviewed journals, and publication of new books and journals (Fig. 1.1). This resurgence of urban ecology is perhaps most notable in the area of urban bird ecology, where the study of birds in urban systems has come to the forefront of ornithological research. One of the sources responsible for promoting urban bird ecology was the 2001 publication of *Avian Ecology and Conservation in an Urbanizing World* (J. M. Marzluff, R. B. Bowman, and R. Donnelly, editors). This book is now considered a seminal publication in urban ecology, with general patterns mentioned in the book having become well established within a decade after publication.

Early work in urban ecology emphasized identifying patterns of altered bird community structure in association with gradients of urbanization. These studies elucidated some general patterns, such as decreased diversity and elevated abundances of remaining species with increasing degree of urbanization along the gradient (Blair 1996, McKinney 2002). In Part I of this volume, we revisit urbanization gradients to illustrate how new work is using this approach to identify potential mechanisms driving the distribution of species along gradients. We begin our visit with a fresh look at the gradient concept by Pennington and Blair (chapter 2, this volume), who provide several comparisons of gradients across the United States as well as a discussion of some common misconceptions regarding the geography of urban gradients. They remind readers that a gradient need not consist of a linear transect from a city center to the countryside. Rather, position on a gradient is defined by the underlying attributes of a given site, and urban areas are more properly considered to be mosaics of habitats and features.

Today, the gradient approach and others are being used to dissect the underlying mechanisms that produce these patterns. According to some studies over the past decade, diversity may peak at intermediate levels of urbanization, a finding attributed by some to Connell's (1978) intermediate disturbance hypothesis (McDonnell and Pickett 1990, Blair 1996, McKinney 2002, Lepczyk et al. 2008). New habitats and resources introduced by suburban development interspersed with remnant wildlands may support both retention of preexisting species and introduction of new species. Patterns, according to this view, are largely driven by the net effects of local extinction and colonization (Rodewald, chapter 5, Marzluff, chapter 18, this volume). Other processes, such as competition, have been invoked to account for changing species composition (Shochat et al. 2006). In addition, debate has arisen over the importance of predation in driving the distribution and abundance of species across an urbanization gradient (Gering and Blair 1999, Roth and Lima 2003, Shochat 2004, Stracey and Robinson, chapter 4, this volume). In an attempt to address this variety of processes, Rodewald (chapter 5, this volume) tests several alternative mechanisms structuring bird communities in riparian forest remnants in Ohio along an urbanization gradient. Her findings suggest the intriguing possibility that a behavioral mechanism, heterospecific attraction, may account for variance in species diversity along the gradient. In addition, both Stracey and Robinson and Kaisanlahti-Jokimäki et al. (chapter 6, this volume) find that predation along a gradient is associated with marked changes in the bird community. For instance, Stracey and Robinson find that small birds (<40 g) were "losers," as they were far less abundant in urban habitats relative to non-urban habitats.

Similarly, Kaisanlahti-Jokimäki et al. observe that increased urbanization is associated with increased ground nest depredation, increased abundance of corvids, and decreased proportion of individuals belonging in ground-nesting species. Finally, both González-Oreja et al. (online material) and MacGregor-Fors et al. (chapter 3, this volume) provide important case studies on bird community patterns along an urbanizing gradient in tropical areas of the world, which have too often been missed in earlier work. As both the González-Oreja et al. and MacGregor-Fors et al. papers demonstrate, the patterns first illustrated by Blair (1996) and others a decade earlier in temperate regions of the world hold fast for the tropics as well.

A key question for avian conservation is whether the species currently found in urban settings can persist under the altered predation, resource, and habitat structure conditions explored in Part I. In order to address this question, long-term demographic data on urban birds is needed. Increasingly, those data are being collected by committed volunteers. Part II considers the relationship between *citizen science*, the enlistment of laypeople in the collection of scientific data, and *demography*, the long-term analysis of data on survival and reproduction of urban birds. The two pursuits are typically considered separately because of their distinct motivations: citizen science for its role in engaging and educating the public about avian conservation, and demography for its critical role in assessing the potential for either persistence or extinction of bird populations. The two are united, however, by a common methodology. Some of the most efficient methods of collecting demographic data are those that involve large groups of volunteers, such as the long-running North American Breeding Bird Survey (Sauer et al. 2003) and numerous banding (i.e., ringing) programs around the world (Greenwood 2007). The papers collected here span a range of topics, from explicitly addressing the successes and challenges of conducting citizen science research (Vargo et al., chapter 7, Rotenberg et al., chapter 8, this volume) to presenting different approaches to incorporating citizen information (McCaffrey and Turner, chapter 9, DeStefano and Webster, chapter 10, this volume) to research that does not enlist citizen scientists, but contributes new insights into demographic processes through studies of particular species (Bartos Smith et al., chapter 11, Whittaker and Marzluff, chapter 12, this volume). In sum, the work in these and other papers in the volume presents a view of the varied responses of species to human alterations of the environment. The question of long-term persistence of species in urban areas remains without a definitive answer. Calls for more work are a common theme among the papers, particularly for studies of the critical post-fledging period. It is clear, however, that not all species show indications of decline, and urban habitats are not always functioning as sinks or ecological traps.

Embedded throughout all the sections of the volume is the theme that humans and their values, perceptions, institutions, and behaviors are perhaps the key defining feature of an urban environment. In Part III, we collect papers that attempt to address human interactions with birds head-on, first by considering the land use planning and prediction perspective (Miller, chapter 13, Hostetler, chapter 14, Hepinstall et al., chapter 15, this volume), and second through the integration of social and natural science approaches (i.e., a human dimensions approach). In terms of land use planning and prediction, there has been considerable change in the past decade as applied ecological perspectives and urban ecology ideas move into the realm of land use planning and landscape architecture. Hostetler (chapter 14, this volume) brings this point home by describing how to consider building with a view for birds. On the other hand, Miller (chapter 13, this volume) reviews how well urban ecological research actually provides useful and/or appropriate management recommendations for birds and other species. Finally, Hepinstall et al. (chapter 15, this volume), demonstrate the value of predicting urban landscape changes and what the implications are for that in the Seattle metropolitan region.

Urban ecologists have struggled at times with integrating social and ecological perspectives into a unified framework (see Grimm et al. 2000 and Pickett et al. 2001 for thoughtful insight and reviews). Ideas for socio-ecological integration were presented in a 2001 benchmark publication (Alberti et al. 2001), but little empirical work in that book integrated social science approaches (Clergeau 2001). Since 2001, a series of papers have begun to identify socioeconomic drivers of spatial structure within cities (Lepczyk et al. 2002,

2004; Hope et al. 2003; Grove et al. 2006a, 2006b). Neighborhoods of higher income and socioeconomic status may support greater species diversity than those of lower income and with greater minority group populations (Kinzig et al. 2005, Melles 2005, but see Loss et al. 2009). Human activities oriented toward birds, such as feeding, are receiving greater scrutiny, with mixed findings on the benefits versus costs of feeding birds (Robb et al. 2008a, 2008b; Chamberlain et al. 2009; Fuller, chapter 16, this volume). Engagement with bird feeding, however, is not uniform, with the amount and kind of bird feeding varying regionally (Lepczyk et al., chapter 17, this volume) and to varying extents with human socioeconomic and demographic factors (Fuller, chapter 16, Lepczyk et al., chapter 17, this volume). The collected papers, together with a growing body of other work, illustrate the varied ways that humans can influence bird community structure and population dynamics.

In the final section of the volume (Part IV), Marzluff takes a further step, suggesting that urbanization and the accompanying array of human interactions with birds can lead to evolutionary changes in the traits of urban-dwelling species. Through a detailed review, he highlights a variety of ways that birds have already adapted to urban areas and suggests that they may continue to do so in the future. He goes further still to suggest that the interactions between humans and birds may have coevolved through the process of cultural evolution. The chapter by Marzluff also highlights an area that is missing from the collected work in this volume: work on physiology and behavior. Recent work has demonstrated that bird songs alter in response to both urban noise and habitat changes associated with urbanization (Slabbekoorn and Peet 2003; Derryberry 2009). Another recent study finds differences in body mass in urban versus rural bird populations, suggesting that anthropogenic food resources (e.g., french fries) may influence growth, reproduction, and survival (Aumen et al. 2008). The question of the capacity for species to adapt to urbanization is clearly relevant to avian conservation, making urban evolutionary ecology a vital new direction for further research.

Our understanding of the ecology of birds in urban areas has grown markedly over the past decade (Fig. 1.1) and shows no signs of slowing. With this growth has come a maturity in the questions that urban avian ecologists are asking and the approaches we are taking. No longer are we simply looking at gradient patterns or bird diversity. Rather, today, we are examining how urban ecosystems may shape avian evolution, how bird physiology and behavior changes or are influenced by urbanization, how urban food webs are structured, and how humans and nature are coupled systems.

Most early work that explicitly set out to address birds in urban areas focused on ecological patterns and processes. The collected papers here show a growing sophistication in our understanding of urban bird ecology. The authors explore potential mechanisms underlying the shifts in community structure along urban gradients (Part I). They consider ways to expand the effectiveness of avian ecological research through engagement of citizen scientists (Part II), and they delve into the urban natural history of increasing numbers of species in ever more diverse locations around the world (all sections). The coupled nature of human-avian interactions is more and more frequently explored (Parts III and IV), and practical suggestions for ways engage humans in conservation are put forth (Part III). Beyond the papers collected here, we have observed new developments during the interim since the 20 papers in this volume were produced: Robb et al. (2008a, 2008b) carried out an extensive meta-analysis of bird feeding, which demonstrated the degree to which it can influence bird population sizes, and a unique study reports the evolution of bird song over time as the landscape changed (Derryberry 2009).

Cities are now growing faster than they ever have in human history, particularly in the developing world. But cities, with their greater efficiencies of scale, also hold the best hope we have of effective conservation of species while supporting the roughly 9.5 billion people the world is projected to hold by 2050. We join the authors of the papers in this volume in calling for continued expansion of urban bird ecology. We expect the field to enjoy a continued expansion from pattern-focused to process-focused research. We hope it will see continued expansions of geographic scope, analytical techniques applied, and degree of interdisciplinary integration. Finally, we hope to see expansion of the practical application of urban ecological science to the conservation of birds and other species.

LITERATURE CITED

Alberti, M., E. Botsford, and A. Cohen. 2001. Quantifying the urban gradient: linking urban planning and ecology. Pp. 89–116 *in* J. M. Marzluff, R. B. Bowman, and R. Donnelly (editors), Avian ecology and conservation in an urbanizing world. Kluwer Academic Publishers, Boston.

Auman, H. J., C. E. Meathrel, and A. Richardson. 2008. Supersize me: does anthropogenic food change the body condition of Silver Gulls? A comparison between urbanized and remote, non-urbanized areas. Waterbirds 31:122–126.

Blair, R. B. 1996. Land use and avian species diversity along an urban gradient. Ecological Applications 6:506–519.

Chamberlain, D. E., A. R. Cannon, M. P. Toms, D. I. Leech, B. J. Hatchwell, and K. J. Gaston. 2009. Avian productivity in urban landscapes: a review and meta analysis. Ibis 151:1–18.

Clergeau, Philippe, G. Mennechez, A. Sauvage, and A. Lemoine. 2001. Human perception and appreciation of birds: a motivation for wildlife conservation in urban environments of France. Pp. 69–88 *in* J. M. Marzluff, R. B. Bowman, and R. Donnelly (editors), Avian ecology and conservation in an urbanizing world. Kluwer Academic Publishers, Boston.

Coluccy, J. M., R. D. Drobney, D. A. Graber, S. L. Sheriff, and D. J. Witter. 2001. Attitudes of central Missouri residents toward local Giant Canada Geese and management alternatives. Wildlife Society Bulletin 29:116–123.

Connell, J. H. 1978. Diversity in tropical rain forests and coral reefs. Science 199:1302–1310.

Conover, M. R., and G. G. Chasko. 1985. Nuisance Canada Goose problems in the eastern United States. Wildlife Society Bulletin 13:228–233.

Davies, Z. G., R. A. Fuller, A. Lorama, K. N. Irvine, V. Sims, and K. J. Gaston. 2009. A national scale inventory of resource provision for biodiversity within domestic gardens. Biological Conservation 142:761–771.

Derryberry, E. P. 2009. Ecology shapes birdsong evolution: variation in morphology and habitat explains variation in White-Crowned Sparrow song. American Naturalist 174:24–33.

Gering, J. C., and R. B. Blair. 1999. Predation on artificial bird nests along an urban gradient: predatory risk or relaxation in urban environments? Ecography 22:532–541.

Greenwood, J. J. D. 2007. Citizens, science and bird conservation. Journal of Ornithology 148 (Suppl. 1): S77–S124.

Grimm, N. B., J. M. Grove, S. T. A. Pickett, and C. L. Redman. 2000. Integrated approaches to long-term studies of urban ecological systems. BioScience 50:571–584.

Grove, J. M., M. L. Cadenasso, W. R. Burch, S. T. A. Pickett, K. Schwarz, J. O'Neil-Dunne, M. Wilson, A. Troy, and C. Boone. 2006a. Data and methods comparing social structure and vegetation structure of urban neighborhoods in Baltimore, Maryland. Society and Natural Resources 19:117–136.

Grove, J. M., A. R. Troy, J. P. M. O'Neil-Dunne, W. R. Burch Jr., M. L. Cadenasso, and S. T. A. Pickett. 2006b. Characterization of households and its implications for the vegetation of urban ecosystems. Ecosystems 9: 578–597.

Hope, D., C. Gries, W. Zhu, W. F. Fagan, C. L. Redman, N. B. Grimm, A. Nelson, C. Martin, and A. Kinzig. 2003. Socio-economics drive urban plant diversity. Proceedings of the National Academy of Sciences of the United States of America 100:8788–8792.

Kinzig, A. P., P. S. Warren, C. Martin, D. Hope, and M. Katti. 2005. The effects of human socioeconomic status and cultural characteristics on urban patterns of biodiversity. Ecology and Society 10:23. <http://www.ecologyandsociety.org/vol10/iss1/art23/>.

Lepczyk, C. A., C. H. Flather, V. C. Radeloff, A. M. Pidgeon, R. B. Hammer, and J. Liu. 2008. Human impacts on regional avian diversity and abundance. Conservation Biology 22:405–446.

Lepczyk, C. A., A. G. Mertig, and J. Liu. 2002. Landowner perceptions and activities related to birds across rural-to-urban landscapes. Pp. 251–259 *in* D. Chamberlain and E. Wilson (editors). Avian landscape ecology. Proceedings of the 2002 Annual UK-IALE Conference.

Lepczyk, C. A., A. G. Mertig, and J. Liu. 2004. Assessing landowner activities that influence birds across rural-to-urban landscapes. Environmental Management 33:110–125.

Loss, S. R., M. O. Ruiz, and J. D. Brawn. 2009. Relationships between avian diversity, neighborhood age, income, and environmental characteristics of an urban landscape. Biological Conservation 142: 2578–2585.

Marzluff, J. M. 2001. Worldwide urbanization and its effects on birds. Pp. 19–47 *in* J. M. Marzluff, R. B. Bowman, and R. Donnelly (editors), Avian ecology and conservation in an urbanizing world. Kluwer Academic Publishers, Boston.

McDonnell, M. J., and S. T. A. Pickett. 1990. Ecosystem structure and function along urban rural gradients: an unexplained opportunity for ecology. Ecological Applications 71:1232–1237.

McKinney, M. L. 2002. Urbanization, biodiversity, and conservation. BioScience 52:883–890.

Melles, S. 2005. Urban bird diversity as an indicator of human social diversity and economic inequality in Vancouver, British Columbia. Urban Habitats 3: 25–48.

Pickett, S. T. A., M. L. Cadenasso, J. M. Grove, C. H. Nilon, R. V. Pouyat, W. C. Zipperer, and R. Costanza. 2001. Urban ecological systems: linking terrestrial ecological, physical, and socioeconomic components of metropolitan areas. Annual Review of Ecology and Systematics 32:127–157.

Robb, G. N., R. A. McDonald, D. E. Chamberlain, and S. Bearhop. 2008a. Frontiers in Ecology and the Environment 6:476–484.

Robb, G. N., R. A. McDonald, D. E. Chamberlain, S. J. Reynolds, T. J. E. Harrison, and S. Bearhop. 2008b. Winter feeding of birds increases productivity in the subsequent breeding season. Biology Letters 4:220–223.

Roth, T. C., and S. L. Lima. 2003. Hunting behavior and diet of Cooper's Hawks: an urban view of the small-bird-in-winter paradigm. Condor 105:474–483.

Sauer, J. R., J. E. Fallon, and R. Johnson. 2003. Use of the North American Breeding Bird Survey data to estimate population change for bird conservation regions. Journal of Wildlife Management 67:317–326.

Shochat, E. 2004. Credit or debit? Resource input changes population dynamics of city-slicker birds. Oikos 106:622–626.

Shochat, E., P. S. Warren, S. H. Faeth, N. E. McIntyre, and D. Hope. 2006. From patterns to emerging processes in mechanistic urban ecology. Trends in Ecology and Evolution 21:186–191.

Slabbekoorn, H., and M. Peet. 2003. Birds sing at a higher pitch in urban noise. Nature 424:267.

Sodhi, N. S. 2002. Competition in the air: birds versus aircraft. Auk 119:587–595.

Turner, W. R., T. Nakamura, and M. Dinetti. 2004. Global urbanization and the separation of humans from nature. BioScience 54:585–590.

U.S. Department of the Interior (USDI), Fish and Wildlife Service, and U.S. Department of Commerce, U.S. Census Bureau. 2006. National survey of fishing, hunting, and wildlife-associated recreation. <http://www.census.gov/prod/2008pubs/fhw06-nat.pdf>.

PART ONE

Mechanisms and Urban-Rural Gradients

CHAPTER TWO

Using Gradient Analysis to Uncover Pattern and Process in Urban Bird Communities

Derric N. Pennington and Robert B. Blair

Abstract. Understanding the processes that create repeatable patterns in urban bird distributions is a challenge that requires investigation at multiple spatial, temporal, and biological scales. Based upon our research of urban systems in California and Ohio, we demonstrate how gradient analysis is a useful tool for understanding both fine- and coarse-scaled processes. We also explore how life-history characteristics permit some species to utilize and even thrive in urban areas while others cease to exist. Finally, we provide recommendations to improve our understanding of urban bird communities using gradient analyses and emphasize the need to derive a common framework that incorporates the ecological and social heterogeneity of urban systems.

Key Words: community, demographics, gradient analysis, migration, scale, stopover, urbanization, urban-rural gradient.

A primary aim of urban ecology is to enhance our understanding of the structure and function of urban systems (McDonnell and Pickett 1990). Urban systems offer a unique ecological situation in which biophysical and socioeconomic factors interact hierarchically across spatial and temporal scales. These complex interactions produce dynamic feedbacks resulting in "emergent properties," such as exacerbated disturbances (e.g., frequent flooding, pollution, climate change, etc.; Pickett et al. 2001), species turnover, and societal impacts (e.g., health problems, gentrification, etc.; Redman and Jones 2005). Insights into these systems allow ecologists to inform policymakers, land use planners, and homeowners regarding land use practices that enhance not only native biodiversity but also the quality of life for the people living and working within these areas by increasing their opportunities to interact with local wildlife (Miller and Hobbs 2002, Turner et al. 2004).

Urbanization can be defined as the "city construction process" (Gottdiener and Hutchinson 2006), which creates a heterogeneous landscape of varying land uses. From an ecological perspective, urbanization represents the process of converting, degrading, and fragmenting natural habitats (McDonnell and Pickett 1993). The direct ecological impact from urbanization involves the replacement of native vegetation structure by city infrastructure (e.g., buildings, roads, utilities). Indirectly, urbanization alters vegetation composition and structure by the process of fragmentation and degradation, which reduces habitat

Pennington, D. N., and R. B. Blair. 2012. Using gradient analysis to uncover pattern and process in urban bird communities. Pp. 9–32 *in* C. A. Lepczyk and P. S. Warren (editors). Urban bird ecology and conservation. Studies in Avian Biology (no. 45), University of California Press, Berkeley, CA.

quality for several native species and increases the quality for early colonizers and nonnative species (McKinney 2002). Not surprisingly, urbanization affects birds in myriad ways (Chace and Walsh 2006).

As a taxonomic group, birds by their conspicuous nature have garnered the greatest attention from urban ecologists. While several vertebrate species fail to persist in urban habitats, birds, through their mobility and plasticity, have been successful at exploiting highly fragmented landscapes. Over the past decade, urban ecologists have applied variations of McDonnell and Pickett's (1990) urban-rural gradient methodology (McKinney 2002, see Chace and Walsh 2006 for review) to examine the patterns and processes structuring urban bird populations and communities. The results from these studies have varied depending upon the diversity (or biological) measures used. In general, urban areas harbor fewer species than more natural areas (McKinney 2002). Many species that do persist are often widespread natives (i.e., cosmopolitan) or nonnatives (Blair 2001). Yet several studies show that certain arrangements of urban land use types can support native species (Marzluff 2005), suggesting the possibility of designing urban landscapes to sustain local indigenous species diversity.

In this paper, we use multiple urban systems to demonstrate how gradient analysis is a useful tool for understanding both fine-scaled and coarse-scaled processes that influence bird community composition. We also demonstrate how life-history characteristics allow some birds to exist and even thrive in urban settings while limiting the existence of others. Finally, we highlight opportunities to improve our understanding of urban bird communities by integrating the ecological and social heterogeneity of urban systems into our studies of urban ecology.

GRADIENT ANALYSIS AS A TOOL FOR STUDYING URBAN BIRDS

The Urban System

Rapid worldwide urbanization presents an immediate challenge to conservationists throughout the world (Vitousek et al. 1997, Czech et al. 2000). The current global population exodus from rural areas has placed a unprecedented majority of the earth's people living within and around urban areas (United Nations 2004), thus severely testing the limits not only of local infrastructure, but of ecosystems worldwide. This growing trend is not limited to developing countries, but it is happening worldwide as complex global economic "incentives" and cultural "preferences" spur development toward ever larger mega-metropolises (Berry 1990, Gottdiener and Hutchinson 2006). The growth of these areas signifies the importance of understanding urban systems. Metropolitan areas represent a kaleidoscopic spatial arrangement of biophysical and socioeconomic patterns and processes within a complex hierarchical system (Pickett et al. 2005). Hence, metropolitan areas provide a unique opportunity to examine and predict ecological effects across multiple spatial and temporal scales, using a multidisciplinary framework that includes sociology, economics, urban planning, and ecology (Alberti 2005). Understandably, then, developing a holistic understanding of the dynamic heterogeneity of urban systems represents the new frontier for urban ecology.

Urban-Rural Gradient Analysis

One goal of urban ecology is to understand how landscape patterns influence ecological process; *how* we describe and define landscape heterogeneity within urban systems is important (Pickett and Cadenasso 1995). Urban ecologists have struggled to define urbanization in a succinct manner (Theobald 2004, Cadenasso et al. 2007), in part because urban systems are exceptionally complex and cannot be categorized simply as either "urban" or "natural" habitats (or landscapes). Instead, urban systems are an intricate arrangement of varying land use types that result in a mosaic of human modification and built structures. Heterogeneity establishes complex gradients extending from dense central cities (or urban cores), through industrial, transportation, residential, and commercial land covers (Cadenasso et al. 2006). To accommodate such heterogeneity, several urban ecologists have adopted the urban-rural gradient paradigm popularized by McDonnell and Pickett (1990), which is based on well-established models of ecological gradients (Whittaker 1967). Specified gradients are one reasonable solution to the problem of multiple concepts of urbanization. Urban-rural gradients differ from traditional gradient analyses in that they explicitly incorporate people as ecological drivers (e.g., built infrastructure). The urban-rural gradient approach has proven useful for studying the complex spatially varying effects of urbanization on forests (McDonnell et al. 1997), terrestrial organisms (Blair 1999), and aquatic systems (Paul and Meyer 2001).

The urban-rural gradient can be quantified in various ways. Traditionally, urbanization has been defined as a "dense, highly developed core, surrounded by irregular rings of diminishing development" (from Dickinson 1966 as cited in McDonnell and Pickett 1990:1233; also similar to von Thünen's location theory; von Thünen 1826). Following this classical definition, the urban-rural gradient paradigm has often been conceived of as a linear transect from the inner city to less-altered landscapes across broad spatial scales. For example, McDonnell and Pickett's (1990) urban-rural gradient was a linear 140-km transect extending north from Central Park in New York City to rural Connecticut. To quantify urbanization in their studies, McDonnell and Pickett included distance from the urban core as one of several landscape factors. However, urban-rural gradients are not necessarily linear transects across the landscape (Austin 1985), especially at smaller spatial scales. Instead, these gradients are "abstract orderings of changes in land cover, land use, human activity, and the fluxes of capital, energy, material, and information in and around cities" (Cadenasso et al. 2006:5). Accordingly, the degree of urbanization can be ordered along conceptual gradients in which scattered sites that can be ordered according to a specific indicator of urbanization, such as human population or vegetation cover. Regardless of the gradient model chosen, it is important to identify the features of the gradient, how they are quantified, and how they change over space for any particular study. To date, no single measure has gained universal approval for quantifying urbanization along gradients. In general, most studies have used multiple measures, such as human population, building, or road density to quantify increasing urbanization (Marzluff 2001).

Scale Considerations When Using Urban-Rural Gradients

Because urbanization operates across a continuum of different spatial scales, gradient analysis is particularly useful as it can be applied at multiple scales. Conducting multiscale studies can improve our understanding of the scale at which ecological processes and patterns uniquely appear (Wiens 1989b). The heterogeneity along a gradient can be distinguished into differing patches at any chosen scale. For example, as the scale of analysis becomes finer, large patches that were internally homogeneous to one another can become a multitude of smaller heterogeneous patches (Cadenasso et al. 2003), thus highlighting the importance of carefully considering the scale(s) in question for a particular study.

Because urban areas are quite heterogeneous, definitions of urbanization will differ depending on the grain (the resolution or detail) and extent (the geographic coverage) of the study (Cadenasso et al. 2007). Most studies have focused on examining large spatial extents (>1 km^2), where urbanization has been classified into broad land cover/land use types—wildland, rural/exurban, suburban, and urban (Marzluff et al. 2001). The reasons for using a coarse-scale approach are two-fold: (1) it is a simple way to document the expansion of urban areas, and (2) it often reflects the best possible resolution (e.g., 30-m^2 resolution Landsat data) due to the limited availability of high-resolution land cover/land use data (<30 m^2 resolution, e.g., IKONOS or Quickbird imagery). Notably, however, as the grain and extent of the study change, so too do ecological relationships. For instance, Pautasso (2007) found that correlations between measures of urbanization and species richness turn from positive to negative below a study grain of <1 km^2 and an extent of $<10{,}000$ km^2. With fine-grain studies, researchers are directly measuring the fine-scale heterogeneity of urban systems (i.e., replacement of vegetation by impervious surfaces); the resultant negative coefficients are caused by the direct loss of habitat by urban infrastructure. In contrast, coarse-grain studies are sampling over a diversity of landscapes (i.e., heterogeneity of landscapes); the positive relationships seen at these broader scales are consistent with the concept of species-area curves (Rosenzweig 1995). In other words, when sampling more area there is a greater likelihood of sampling more species because of the greater diversity of habitats surveyed. In addition, using a relatively coarse grain (>1 km^2) for environmental variables may result in underestimating the heterogeneity or varied patches in a landscape (Kühn et al. 2004, Pennington and Blair 2011).

Increasingly, ecologists are interested in understanding the ecological function of urban areas themselves, which requires fine-grained studies (Grimm et al. 2000). Traditional coarse-grained approaches make it difficult to separate land cover from land use and they often disconnect people from "natural" components. The broad availability of Landsat imagery (U.S. Geological Survey [USGS] NLCD; Vogelman et al. 2001) served urban ecologists well for quantifying the array of landscape factors to establish gradients for coarse-grained studies; however, the 30-m^2 pixel resolution of the data and the coarse land use categories make it difficult to tease out fine-grained heterogeneity or habitat patches (Gottschalk et al. 2005) such as the variety and abundance of vegetation within the "urban" classification (narrow habitat corridors or single trees on parcels; Cadenasso et al. 2007). Given the importance of vegetative cover for native bird diversity (Pennington and Blair 2011), new methods that incorporate vegetative cover within the varying land use types of urban systems would improve our understanding of how birds respond to the fine-scale heterogeneity, and therefore improve conservation and management recommendations for these areas.

New Urban-Rural Gradient Paradigms: Direct Urbanization and Urban Context

For this paper we introduce two paradigms for thinking about and designing urban-rural gradient studies: *direct urbanization* and *urban context*. Under the *direct urbanization paradigm*, the gradient is quantified from the study sites themselves, whereas for the *urban context paradigm*, the gradient is quantified by the surrounding landscape of a site (i.e., landscape context). The direct urbanization paradigm involves explicitly examining sites directly located within various land use types (e.g., sites in wildland, suburban, urban; Fig. 2.1a). The urban context paradigm is similar to "ecology in the city" in that sites are embedded within "natural" features or greenspace (Fig. 2.1b; Grimm et al. 2000).

EXAMPLES FROM CALIFORNIA AND OHIO: THREE STUDIES

To illustrate how both fine-scale and coarse-scale processes influence bird community composition, we present examples from both previously reported and new research on three urban-rural gradients: two studies using the *direct urbanization paradigm* and one study using the *urban context paradigm*. The two *direct urbanization* studies are located in Palo Alto, California, and Oxford, Ohio (see Table 2.1). These two studies represent a broad-scale gradient of land uses ranging from relatively undisturbed to highly developed urban areas to capture a range of human-modified ecosystems based on six sampling sites: a biological preserve, an open-space recreational area, a golf course, single-family detached housing, an industrial-research park (or apartment complex in Oxford), and a business district (see Blair 2004 for details). Each of these sites contained 16 sampling points, which we surveyed multiple times over at least two breeding seasons to obtain estimates of the relative density of all species. For the Oxford study, we additionally examined nest fate and nest location at each site along the gradient (see Reale and Blair 2005 for details). These two gradients are located within two different biomes: the coastal chaparral forest shrub in California and the eastern broadleaf forest in Ohio.

The third study area is located in Cincinnati, Ohio, and is an example of the *urban context paradigm*, where sites of a particular type of "natural habitat" are surrounded by varying types of land use (see Table 2.1). These sites are located within a "natural" riparian area that traverses a spectrum of land use types from relatively undisturbed to highly developed. Many examples of this paradigm have been described across broad spatial scales from "urban" to "natural" (Donnelly and Marzluff 2004). But this fine-scale study differs in that the entire study area is located in what would typically be considered the "urban" land use type of a broad-scale gradient (human population density ~1,498 persons/km^2). Here our objective was to examine the role of fine-scale heterogeneity on riparian bird distributions within the traditional "urban" land use type. The study area is situated within the Mill Creek watershed (42,994 ha) and represents all types of land cover associated with densely built regions: a varying array of parks and residential, industrial, and commercial land uses, including both private and public parcels, all within the Cincinnati metropolitan area (see Pennington et al. 2008 for details). We identified 71 sampling points along the streamside edge of

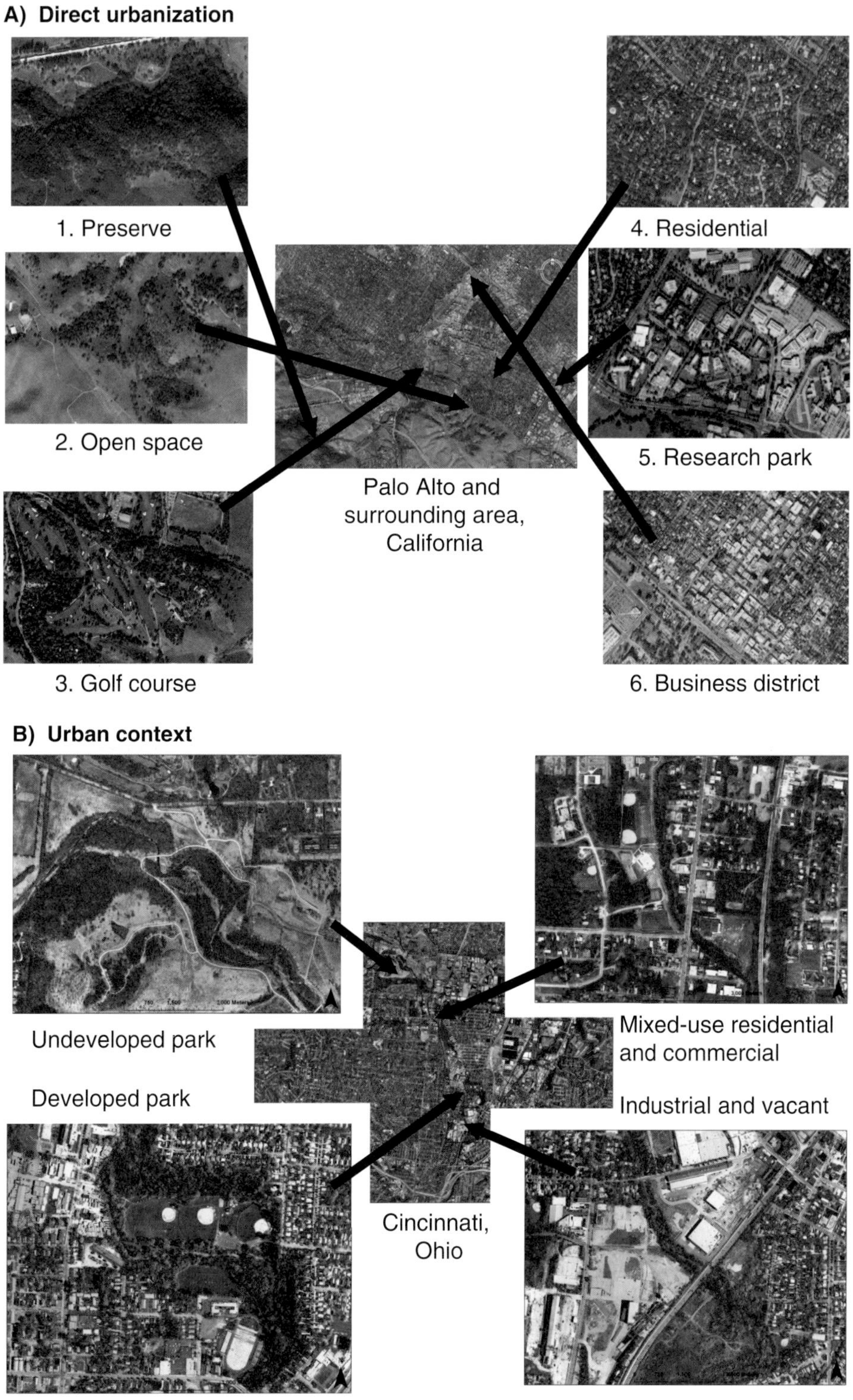

A) Direct urbanization

1. Preserve

2. Open space

3. Golf course

4. Residential

5. Research park

6. Business district

Palo Alto and surrounding area, California

B) Urban context

Undeveloped park

Developed park

Mixed-use residential and commercial

Industrial and vacant

Cincinnati, Ohio

Figure 2.1. Differences in the study design for two urban-rural gradient paradigms: (a) example of direct urbanization gradient paradigm where study sites were located within six land use types (Palo Alto, California, shown; the method was used for the Oxford, Ohio, study); (b) example of the urban context gradient paradigm where sites are embedded within "natural" features or greenspace (represented by forested riparian "corridor") that are surrounded by myriad land use types (Cincinnati, Ohio, shown).

TABLE 2.1

A general description of the three different rural to urban gradient studies.

	Direct urbanization		Surrounding urbanization
Characteristic	Palo Alto, California	Oxford, Ohio	Cincinnati, Ohio
Human population density	955.8/km^2	402.8/km^2	1,498.0/km^2
Extent of areas sampled	~150 ha	~150 ha	~100 ha
No. of plots sampled	6 sites, 16 points per site	6 sites, 16 points per site	71 points; identified systematically from a a random starting point
Amount of the coarse "rural to urban" gradient measured	Urban, office park, residential, golf course, open space, and nature preserve	Urban, office park, residential, golf course, open space, and nature preserve	Predominately urban (with varying surrounding land uses ranging from 'wildland' to 'urban')
Measure of urbanization	Axis 1 scores from PCA of % tree/shrubs area, % grassland and lawn, % pavement, % building	Axis 1 scores from PCA of % tree/shrubs area, % grassland and lawn, % pavement, % building	Building area, tree/ shrubs, and % grassland and lawn cover analyzed as separate variables
Measures of birds	Species richness, relative species density	Species richness, relative species density, nest search, artificial nests	Species richness, relative species density

two Mill Creek tributaries. At each point, we surveyed multiple times during the spring and summer of 2002 to obtain estimates of the relative density of all species. The Cincinnati study area is located in the eastern deciduous forest biome.

For the Palo Alto and Oxford study areas, we ordinated each site along conceptual gradients (based on % tree cover, % grass, % impervious, and % building) using principal component analysis (PCA) to assess how well we captured variation along the gradient (McCune and Grace 2002). We used the first component axis from the PCA as a surrogate measure of urbanization (eigenvalue, percent variance explained, and significance of the first axis based on a Monte Carlo simulation were $l = 2.85$, % = 57.01, $P = 0.001$ for Oxford and $l = 2.38$, % = 47.66, $P = 0.001$ for Palo Alto). For the Cincinnati study, we did not use PCA to measure urbanization, but instead chose to use building area to capture urbanization and examined the biophysical measures of percentage tree cover and percentage grass cover separately.

General Patterns in Bird Communities along Urban-Rural Gradients

Our studies of direct urbanization showed a negative quadratic relationship for species richness and Shannon diversity for Oxford, and reflected a peaking of diversity measures at less developed and more intermediate levels of urbanization (e.g., golf courses and residential areas; Fig. 2.2), which has also been documented for European cities (Clergeau et al. 1998); however, our results for Palo Alto reflected a more linear decline, with increasing levels of urbanization across the gradient suggesting potential geographical differences. This curvilinear trend illustrates local extinction and invasion in that minimal development of a site can create a more heterogeneous habitat (e.g., trees and lawn patches) that supports a higher number of species, but intense development can reduce the numbers of both urban avoider and urban adapter species (Blair 1996).

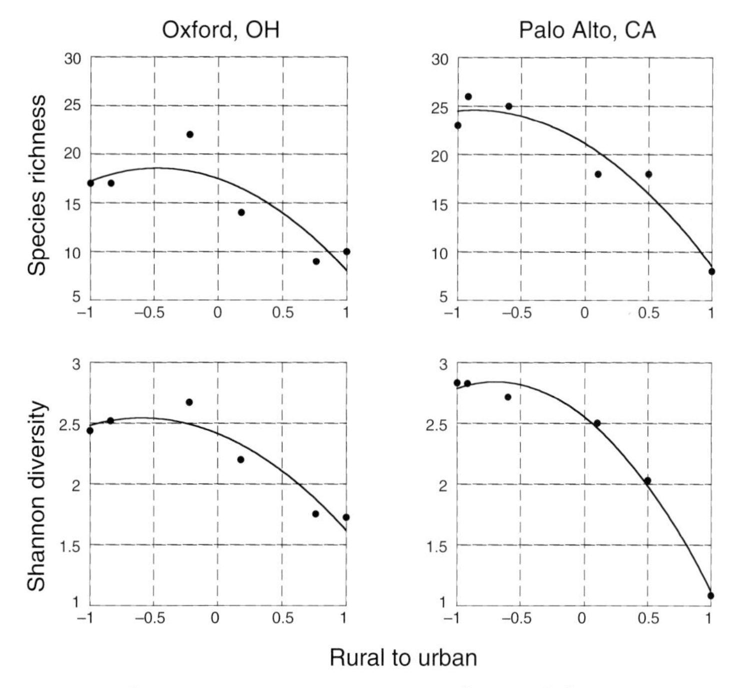

Figure 2.2. Change in community metrics, species richness, and Shannon's diversity, along an intensifying urbanization gradient that was quantified based on the PCA Axis 1 scores (constrained by % tree cover, % grass, % impervious, and % building cover).

Based on the urban context paradigm, Cincinnati riparian bird communities displayed a negative curvilinear response for both density and species richness with surrounding forest cover, with diversity measures peaking at ~70% cover (Fig. 2.3a; also seen in Donnelly and Marzluff 2006). When the proportion of forest cover drops below 30%, a sharp decline in both density and richness occurs; this appears to be a critical threshold in other studies (Donnelly and Marzluff 2004). Both density and richness declined linearly with increasing building area, our surrogate for urbanization (Fig. 2.3b). This trend, which contrasts with what has been generally observed, could be due to a smaller amount of variation captured across the typical urban-rural gradient, as these study sites resided mostly within the "urban" end of the spectrum, and highlights the importance of considering built and natural aspects separately.

As urbanization creates a gradient of disturbance, it can also create a gradient of homogenization (McKinney 2006). The blinking in and out of species densities across the urban gradients implies that some species flourish by exploiting novel resources created from development, while others disappear with increased development (see Table 2.2). At the species level, forest-breeding species, considered the hallmark urban avoiders (Blair 1996), displayed a strong pattern of local extinction as urbanization increased. Many of the species, and also some of the most abundant, displayed the ability to adjust to moderate levels of disturbance (urban adapters), with densities peaking in sites of intermediate development (Northern Cardinal [*Cardinalis cardinalis*] and American Robin [*Turdus migratorius*]; see Table 2.2). It is plausible that these patterns result from the changes in predation and nest survivorship or the presence of fruit-bearing ornamental plants common in the suburban landscape, especially for urban Northern Cardinals and American Robins (Blair 1996). Nonnative species (European Starling [*Sturnus vulgaris*] and House Sparrow [*Passer domesticus*]), representing urban exploiters, had the highest densities at the most urban sites for Palo Alto, Oxford, and Cincinnati. These two exploiters are among the most common species comprising urban bird communities around the world; unlike urban avoiders and adapters, the abundance of these species is not dependent on vegetation, but instead on exploiting

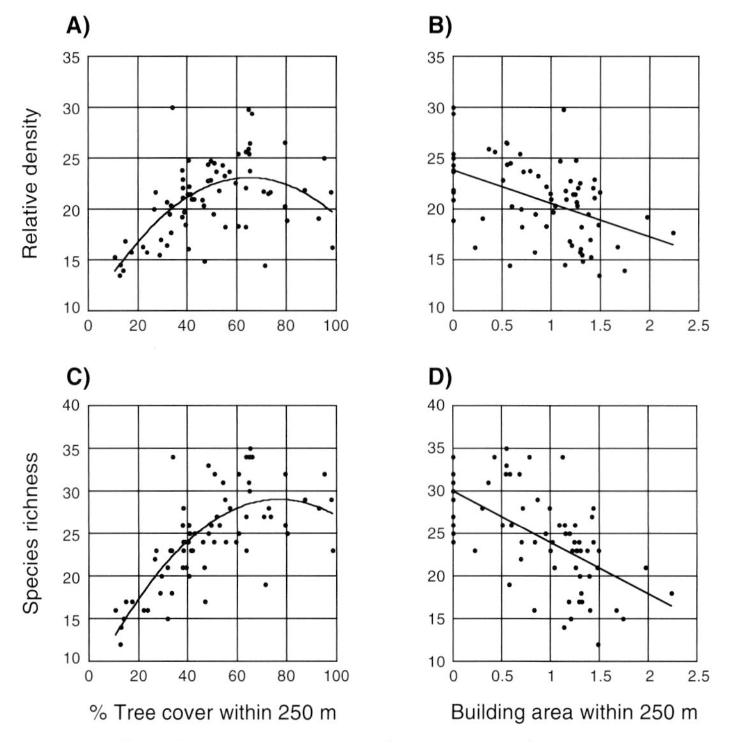

Figure 2.3. Change in community metrics, relative density, and species richness along an intensifying urbanization gradient quantified by examining % tree cover and building area within 250 m separately: (a, c) total bird density and richness displayed strong curvilinear responses to the amount of tree cover surrounding a site; a peak or threshold response for density at 60% tree cover (a) and for richness at 70% tree cover (c); and (b, d) both total bird density and richness declined linearly with increasing building cover.

foods and shelter provided by people (Nilon and VanDruff 1987). Consequently, as urbanization expands globally, biological homogenization will increase as many of the same urban adaptable species become increasingly widespread and locally abundant.

Examples of Fine-scale Processes

Nesting Success and Artificial Nests

Using the broad-scale land use gradient for the Oxford study area, we conducted field experiments investigating fine-scale processes including nesting success and nest predation using artificial nests (Blair 2004, Reale and Blair 2005). We first explored whether increasing levels of urbanization influenced the nesting success of birds by focusing on Northern Cardinals and American Robins because of their ubiquitous distribution across land use types and due to the efficacy of locating their nests (<3 m above ground). Contrary to expectations and previous research (Jokimäki et al. 2005), the probability of nest predation decreased with increasing urbanization ($W = -2.05$, $P = 0.04$; Blair 2004; we used Wald's statistic [W] to identify significant variables [$P < 0.05$]). However, the decrease in predation with increasing urbanization did not increase the nesting success of American Robins or Northern Cardinals ($W = 3.44$, $P = 0.001$; Blair 2004). Since site, and consequently level of predation, was not a significant predictor of nest fate for robins (df = 1, $W = 0.66$, $P = 0.42$; Blair 2004) or cardinals (df = 1, $W = 0.002$, $P = 0.96$; Blair 2004), the reason for nest failure in more urban sites is more likely nest abandonment or food unavailability than predation.

Our findings also failed to support the idea that nesting success influences the density of individual species. Even though the nesting success of robins and cardinals varied, it was neither

TABLE 2.2

Distribution and abundance of summer resident birds in Palo Alto, CA (similar results were found for Oxford and Cincinnati, OH).

Line widths represent ranges of numbers of birds per hectare (*x*). These ranges are displayed graphically to illustrate the ebb and flow of densities across the gradient for each species. Nomenclature follows the American Ornithologists' Union (1998).

	Preserve	Open Space	Residential	Golf	Research	Business
Urban avoiders						
Hutton's Vireo						
Western Wood-Pewee						
Wrentit						
Steller's Jay						
Ash-throated Flycatcher						
Blue-gray Gnatcatcher						
Dark-eyed Junco						
Suburban adapters						
Violet Green Swallow						
Nuttall's Woodpecker						
Cliff Swallow						
European Starling						
Bewick's Wren						
Plain Titmouse						
Mourning Dove						
White-b. Nuthatch						
California Quail						
Red-winged Blackbird						
Acorn Woodpecker						
Rufous-sided Towhee						
Black Phoebe						
Western Bluebird						
Barn Swallow						
Pacific Slope Flycatcher						
Mallard						
California Thrasher						
Scrub Jay						
Brewer's Blackbird						
Northern Mockingbird						
Lesser Goldfinch						
Northern Oriole						
House Finch						
California Towhee						
Anna's Hummingbird						

TABLE 2.2 (*continued*)

TABLE 2.2 (CONTINUED)

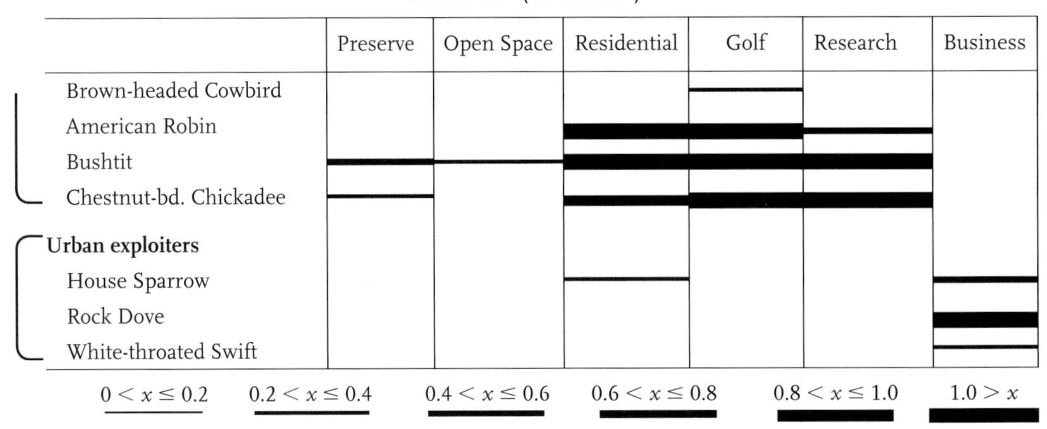

correlated to the density of individuals at a site (for robins: df = 1, $W = 0.08$, $P = 0.67$; for cardinals: df = 1, $W = 0.01$, $P = 0.84$; Reale and Blair 2005) nor clearly related to the level of urbanization (for robins: df = 1, $W = 0.66$, $P = 0.42$; for cardinals: df = 1, $W = 0.002$, $P = 0.96$; Reale and Blair 2005), suggesting that individuals are unable to select nesting sites based on the suitability of the land use in terms of nesting success. This inability to select sites highlights the importance of conducting fine-scale demographic studies in conjunction with diversity measures in assessing habitat quality, especially in urban systems where density may not be a good predictor of nesting success (Bock and Jones 2004).

We also compared predation of both real and artificial nests using two Zebra Finch (*Poephila guttata*) eggs and one plasticine egg within the same nest (Gering and Blair 1999). Artificial nest types had higher probabilities of predation than natural nests ($W = 3.44$, $P = 0.001$; Blair 2004), but both displayed the same general trend of predation decreasing with urbanization ($W = -2.05$, $P = 0.04$; Blair 2004), suggesting that artificial nests can serve as surrogates for real nests (no significant interaction between nest type and site, $W = 0.79$, $P = 0.78$; Blair 2004).

Our findings contradict research that has found that nest predation increases with urbanization (Jokimäki and Huhta 2000). Notably, however, these other studies were conducted using the urban context paradigm that examined greenspaces embedded across an array of surrounding land use intensities. Jokimäki and Huhta's (2000) findings are verified by studies that estimate higher abundances of nest predators in parks found in urban areas than in rural areas (Sorace 2002) and with increasing forest width around greenspaces (Sorace 2002, Sinclair et al. 2005). As an example of the direct urbanization paradigm, our Oxford urban-rural gradient examined sites across land use intensity types and not just patches of "natural" habitat. We expect to see more dramatic responses when examining sites within land use types versus sites within "parks." Caution should be used when drawing conclusions from gradient studies depending on what portion of variation along the urban-rural gradient is being evaluated. For example, Thorington and Bowman (2003) found that predation varied greatly in patches when residential housing density was used to determine the urban gradient.

Nest-site Selection

For the Oxford direct urbanization gradient study, we examined the role of nest-site location and height in influencing nesting outcome. In general, the maturity, and consequently the height, of trees is lower in urban areas than in less developed areas (Porter et al. 2001), plausibly providing fewer nesting options for birds. Nest height significantly influenced nesting outcomes for robins and cardinals (robins: df = 1, $W = 14.157$, $P < 0.01$; cardinals: df = 1, $W = 9.410$, $P < 0.01$; Reale and Blair 2005), with nests higher off the ground being more likely to succeed. Furthermore, mean nest height decreased with urbanization (robins: $r^2 = 0.23$, $P < 0.001$; cardinals: $r^2 = 0.13$, $P = 0.03$;

Reale and Blair 2005), suggesting that birds must select lower-quality nesting sites in more developed areas. In addition, examination of the breeding bird community showed that species that characteristically nest higher in trees dropped out completely at the most urban sites, whereas mid- and low-height nesting species increased (Reale and Blair 2005).

Local Vegetation Effects on Bird Communities

Because local fine-scale vegetation structure and composition can influence the habitat quality for birds (Mills et al. 1989, Donnelly and Marzluff 2006), we explored the role of vegetation using the direct urbanization gradient paradigm in Oxford and the urban context gradient paradigm in Cincinnati. For both the Oxford and Cincinnati study areas, we documented predictable changes in vegetation patterns across the urban gradient (Porter et al. 2001) and how local vegetation influenced riparian bird communities in Cincinnati (Pennington et al. 2008, Pennington and Blair 2011). Along both urban gradients, native bird species were positively correlated with sites having a high proportion of native woody vegetation, while nonnative bird species declined with native woody vegetation. Both bird density and richness were more influenced by vegetation composition, whereas the evenness of the bird community was influenced by vegetation structure (Table 2.3). In other words, both vegetation structure (MacArthur et al. 1966) and composition influence habitat quality (Pennington and Blair 2011). Forest breeding species, in particular neotropical migrants, displayed the greatest correlation with native vegetation across species. A plausible mechanism could be that native vegetation provides a higher abundance of food resources and potentially higher nesting success (Schmidt and Whelan 1999; Reichard et al. 2001). Notably, just as native birds decrease along the urban gradient, so too does the native vegetation community, with nonnative vegetation peaking in the most urban areas for riparian forests in Cincinnati (Pennington et al. 2010). These similarities across the gradient highlight the difficulty of teasing out process from pattern in a system that has coupled factors influencing ecosystem processes. However, these findings also suggest that the loss of native vegetation is a major process through which development affects native and nonnative birds.

Examples of Coarse-scale Processes

Landscape Heterogeneity

As stated earlier, heterogeneity can occur at any scale of investigation, from the individual tree of a forest at a fine scale to a mosaic of forest types across a landscape at coarse scales. For the Oxford study area, we found that landscape heterogeneity (as measured by percent dissimilarity of woody species) varied across the urban gradient (χ^2 = 308.1, df = 5, $P < 0.001$; Porter et al. 2001) and peaked at moderate levels of urbanization (Porter et al. 2001). An intermediate peak corresponds to the pattern we observed for bird community richness and Shannon diversity, suggesting that landscape heterogeneity could be another factor driving community composition. Based on the results of a two-variable regression model, both bird richness and diversity positively correlated with landscape heterogeneity and negatively with the number of canopy patches, implying that bird diversity is highest when there is a mixture of vegetation types in large patches (Blair 2004). One explanation for these patterns is that when woody vegetation is lost due to urbanization, it creates a disturbance that facilitates local invasion by pioneer generalists of both plant and bird species, thus temporarily increasing species richness (Pennington et al. 2010). Over time, this increase is tempered by the loss of species due to habitat loss. Indeed, high loss of woody vegetation associated with development has been shown to facilitate the invasion of nonnative woody species (Hutchinson and Vankat 1997).

Spatial Effects of Surrounding Land Use

Birds respond to habitat heterogeneity at a variety of scales, and the level of response varies for different species (e.g., body size, migrating and foraging habits; Hostetler 1999), yet few studies have addressed this in urban ecology. Such information is important not only for ecologists but for city planners as well (Miller, chapter 13, this volume), since a connection exists between the scales at which birds respond to landscape

TABLE 2.3

Model summary for migratory guild community measures and local vegetation measures from multiple regression analyses[a].

Migratory guild	Vegetation composition			Vegetation structure				
	Native tree frequency	Native shrub/ sapling frequency	Total tree richness	Overstory stem density	Understory stem density	Dead trees	Canopy height	
	Partial r	Partial r	Partial r	Partial r	Partial r	Partial r	Partial r	Adj r^2
Relative density								
AB +	0.2646 −	0.1100 −	0.0340 +	0.1888 −	0.2344 −	0.1494 +	0.0199	0.07
N +	0.4159 +	0.2007 +	0.0275 +	0.2149 −	0.1170 −	0.2100 +	0.1341	0.34
EN +	0.3585 +	0.0637 +	0.0184 +	0.0961 −	0.0472 −	0.1878 +	0.1862	0.24
PBN +	0.3143 +	0.2032 −	0.0618 +	0.2105 −	0.1012 −	0.1719 +	0.0680	0.22
SD +	0.0378 −	0.2872 −	0.0487 −	0.0287 −	0.2346 +	0.0132 −	0.0346	0.05
P −	0.0034 −	0.1149 −	0.0573 +	0.2577 −	0.0814 −	0.1219 −	0.0998	0.00
E −	0.2339 −	0.2893 −	0.1239 −	0.2373 −	0.1589 +	0.2281 −	0.1975	0.35

Species richness								
AB +	0.4241 −	0.0157 +	0.0610 +	0.2733 −	0.2712 −	0.1769 +	0.0719	0.29
N +	0.3974 +	0.2093 +	0.0413 +	0.2576 −	0.1299 −	0.2370 +	0.1234	0.36
EN +	0.3403 +	0.1121 +	0.2245 +	0.1196 −	0.0031 −	0.2163 +	0.1801	0.28
PBN +	0.2932 +	0.1867 −	0.0635 +	0.2454 −	0.1362 −	0.1885 +	0.0565	0.22
SD +	0.2842 −	0.2476 +	0.0511 +	0.0536 −	0.2981 +	0.5850−	0.0441	0.05
P +	0.1154 −	0.1329 +	0.0466 +	0.2516 −	0.1482 −	0.1669 +	0.0162	0.01
Evenness								
AB +	0.3189 +	0.1418 +	0.1278 +	0.1099 −	0.0485 −	0.2269 −	0.0394	0.21
N +	0.0438 +	0.0867 −	0.0467 −	0.4584 −	0.1302 −	0.0003 +	0.2174	0.29
EN +	0.2344 +	0.1085 +	0.1019 +	0.1698 +	0.0097 −	0.1663 +	0.1017	0.15
PBN +	0.1592 +	0.1541 −	0.1296 +	0.3400 −	0.1201 +	0.1267 +	0.1211	0.20
SD −	0.0092 −	0.2108 +	0.1911 +	0.0232 +	0.2543 −	0.2164 −	0.1328	0.12
P +	0.2714 +	0.3162 −	0.0457 +	0.2619 −	0.2178 −	0.2703 −	0.2441	0.32

[a] (+/−) represents effect in relationship; denominator df are 70 for all effects: all birds (AB), neotropical migrants (N), en-route neotropical migrants (EN), potentially breeding neotropical migrants (PBN), short-distance migrants (SD), permanent residents (P), and exotics (E; density only).

structure and the scales at which people drive development (Hostetler 2001). Traditionally, urban habitat studies have arbitrarily selected a scale (e.g., local vegetation as a fine-scale measure and a landscape variable, often within 1 km of the site, as the coarse-scale measure) to correlate bird measures to habitat structure. As a few studies have pointed out, these arbitrary scales may not reflect the most important scale to which birds respond (Hostetler and Holling 2000, Hennings and Edge 2003, Dunford and Freemark 2005). Many of the studies that have examined landscape structure and scale have used broad urban-rural gradients and employed coarse-resolution land use/land cover data (Landsat satellite imagery). Additionally, bird data are often collected at a much finer resolution (e.g., survey point counts or line transects) than the methods used to quantify the urban gradient (e.g., quantifying habitat with 30-m^2 pixel resolution land cover data), creating a scale mismatch between remote sensing and bird data that is rarely addressed (Gottschalk et al. 2005). Depending on the species in question, the scale of interest could vary significantly (Pennington and Blair 2011).

In contrast to previous studies, we used fine-resolution IKONOS satellite imagery (4-m^2 pixel resolution) of the Cincinnati area to classify forest cover and used digitized building footprints from fine-resolution color orthophotos to calculate building area (Pennington et al. 2008). We examined bird diversity and landscape structure at 50, 100, 250, and 500 m radii surrounding each riparian site based on *a priori* knowledge that several breeding bird territory sizes are encompassed within smaller scales and the larger scales might be for migrating species. Our findings show that riparian birds respond most strongly to landscape structure at the 250-m scale (see Fig. 2.4). A strong response implies that, to maintain native birds, we need to avoid activities that diminish tree cover both near and adjacent to urban streams and that riparian buffers of 250 m are sufficient to protect a reasonable amount of bird diversity.

Another aspect to our investigation addressed how migratory guild as a life-history characteristic influenced birds' response to landscape structure. During the spring, neotropical migrant density and richness peaked with increasing tree cover and fewer buildings at the 250-m and 500-m scales (Fig. 2.4). In contrast, exotic species density peaked as building area increased and tree cover decreased at the 250-m scale (Fig. 2.4a–b). Larger areas of forest cover are possibly perceived as having higher habitat quality in terms of food and nesting resources for neotropical migrants. Short-distance migrants and permanent residents displayed the weakest response to landscape structure across scales, with marginal responses to tree cover and grass cover at the 250-m scale and no response to building area (Fig. 2.4). It is possible that these two groups of species have adjusted to the habitat changes because they live in this location most if not all of the year, yet it is also possible that these groups were responding to scales beyond what we measured (Mayer and Cameron 2003b).

Additionally, the grain of the study can affect how much variation is captured and may have limited our ability to discern patterns. Another study of neotropical migrants in Ohio found no relationship between bird abundance and forest cover or urbanization using a 1-km radius and 30-m^2 Landsat imagery (Rodewald and Matthews 2005). The fact that we did find a relationship between migrants and landscape structure at a finer scale suggests that some spatial relationships may begin to break down at larger scales (>500 m). Similar patterns were also seen during the breeding season (Pennington and Blair 2011).

Importance of Considering Life-history Characteristics

Grouping the bird community based on differing life-history guilds can aid in distinguishing process from pattern, as illustrated above in the spatial section for migratory guilds. In this section, we highlight additional examples from the direct and urban context paradigms from Ohio that underscore the importance of considering life-history traits in discerning pattern and process.

For the Oxford study, we partitioned the breeding bird community into brooding strategies (number of broods per year) of single, double, and multiple broods during the breeding season. We then examined whether brooding strategy changed across the urban-rural gradient (Reale and Blair 2005). The proportion of species in the community that rely on multiple brooding sharply increased with urbanization. In contrast, single brooding species were most associated with the least disturbed sites. Interestingly, these patterns are similar to the above patterns for the

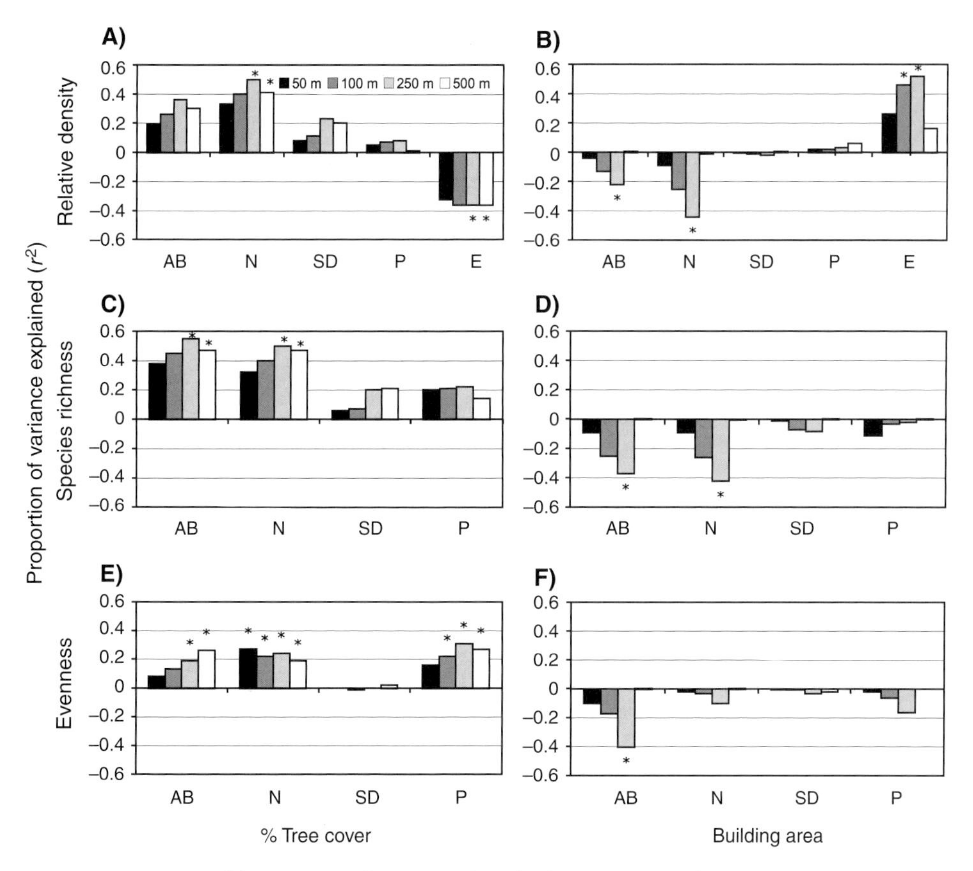

Figure 2.4. Comparison of the proportion of variance explained in linear regressions by % tree cover (a, c, e) and building area (b, d, f) across four scales (50, 100, 250, 500 m radii) for relative density (a, b), species richness (c, d), and evenness (e, f) for all birds (AB), neotropical migrants (N), short-distance migrants (SD), permanent residents (P), and exotic (E; density only) during the spring migratory season for tributaries of Mill Creek, Cincinnati, Ohio; * represents a significant spatial response based on uniqueness index tests ($P < 0.05$). Abundance and richness displayed curvilinear responses (adjusted r^2 values from linear regressions with a quadratic terms; no uniqueness index tests preformed). (a, c, e) In general, native bird density and diversity measures increased with the amount of tree cover around a plot; exotic species responded oppositely. (b, d, f) Total birds and neotropical migrant density and richness decreased sharply with increasing building area; again, exotic species responded oppositely. Total species evenness declined sharply with increasing building area. In general, the 250 m scale was the most important scale across guilds and landscape measures (Pennington et al. 2008).

riparian migratory guilds. The shift in the overall community-level brooding strategy is due to the replacement of single-brooding neotropical migrant species with generalist, multi-brooding species represented by many short-distance and permanent residents of Ohio. Presumably, multiple-brooding species are able to compensate for nesting losses in sites with low nesting success by making multiple attempts, while many neotropical migrants are restricted to one or two broods given the constraints of long-distance migration (Moore et al. 1995). In addition, a change in community-level brooding strategy from low productivity to high productivity at low and highly developed sites suggests a possible change in overall nesting-habitat quality for some species along the gradient. These findings indicate that nesting success over the entire season for the community may provide a better understanding of the distribution of bird species than examining only individual nesting failure alone.

An important aspect of the Cincinnati riparian study is to understand the role of urban areas in providing stopover versus breeding habitat for long-distance migrant species (Pennington et al. 2008, D. N. Pennington, unpubl. data, Vargo et al.,

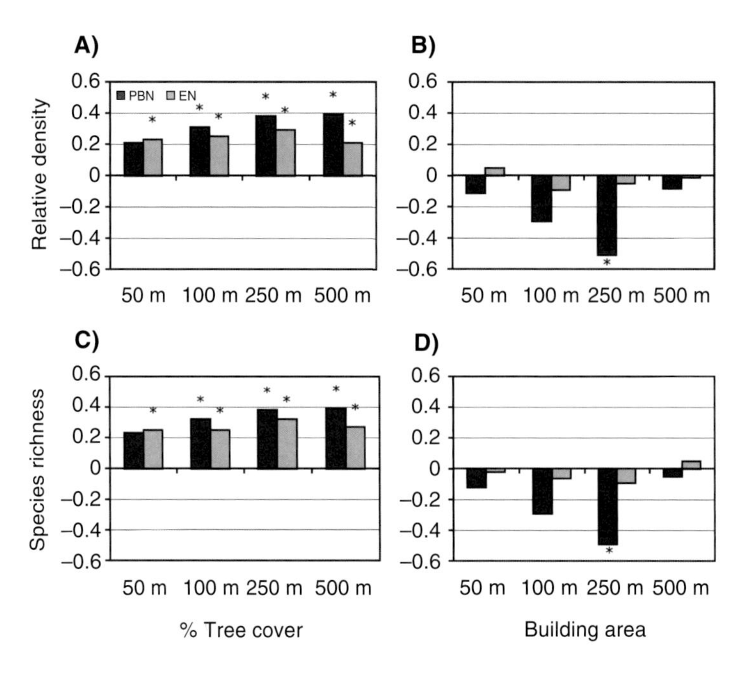

Figure 2.5. Comparison of the proportion of variance explained in linear regressions by % tree cover (a, c) and building area (b, d) across four scales (50, 100, 250, 500 m radii) for relative density (a, b) and species richness (c, d) of potentially breeding (PBN) and en-route (EN) neotropical migrants during the spring migratory season for tributaries of Mill Creek, Cincinnati, Ohio; * represents a significant spatial response based on uniqueness index tests ($P < 0.05$). (a, c) Both potentially breeding and en-route neotropical migrant richness increased with surrounding tree cover; potential breeders displayed a stronger relationship to tree cover >500 m around stream. (b, d) Potential breeding neotropical migrant richness declines sharply with increasing building area surrounding the stream, whereas en-route richness displayed little or no response to building area (Pennington et al. 2008).

chapter 7, this volume). Habitat use during migration has been largely overlooked in the development of conservation strategies (Mehlman et al. 2005). Instead, most research has been conducted during the breeding season. However, different life stages can represent periods when birds respond to the landscape in fundamentally different ways. During the breeding season, birds focus on establishing territories and nesting sites, whereas during migration, birds focus on resting and foraging along the way to and from their wintering grounds (Moore and Kerlinger 1987). Hence, food availability could be the most important criterion for selecting suitable stopover habitat en route (Smith et al. 2007).

We examined riparian bird communities during the spring and summer breeding seasons of 2002 (surveyed six times during 4 April–25 May and six times during 30 May–15 July). To examine the migratory season, we divided the neotropical migrants into en-route (those that are not known to breed within the study area but do utilize the area during stopover) and potentially breeding species (species known to establish breeding territories within the study area). During the migratory season, we found that several neotropical species utilize urban riparian forests. Potentially breeding neotropical migrant density and richness increased with the amount of tree cover and low levels of development surrounding the stream (i.e., less urban sites), while en-route migrant density and richness were highest in areas of high tree cover regardless of the level of development (Fig. 2.5). Effects of tree cover suggest that some urban areas with high tree cover (e.g., a residential neighborhood with mature trees throughout) are perceived as "forest" and suitable stopover sites to long-distance migrants en route, but are perceived as unsuitable habitat by potentially breeding individuals. A comparison of response to landscape structure during spring migration and breeding season further supports this finding. During the spring, neotropical migrants respond most significantly to tree cover, but during the breeding season they respond most strongly to building area surrounding a site (Pennington 2003).

NEW DIRECTIONS FOR APPLYING URBAN-RURAL GRADIANT ANALYSES IN AVIAN RESEARCH

Throughout this paper we have provided examples from our own work on urban-rural gradients to highlight how this methodology can provide insights into complex processes structuring bird communities in urbanizing areas. We recommend the following objectives for future research on birds in urban areas.

Analyze the Community in Total, as Different Migratory Groups, and by Life-history Traits

Researchers should use partitions of the bird community and not limit their analyses to the entire bird community alone. Dividing the total bird community into functional guilds (e.g., migratory guilds, breeding guilds, foraging guilds) can provide key insights into pattern and process (Levin 1992; Fig. 2.4). An important finding from our analyses shows that breeding and migratory guilds respond to urbanization in different ways. A change in species distribution based on brooding strategy from low productivity to high productivity with increased development possibly signifies lower, less predictable, habitat quality and an attempt to overcome this limitation (Reale and Blair 2005). Our analysis of migratory guilds along riparian areas in Cincinnati showed a dramatic decrease in neotropical migrant abundance and diversity with increasing urbanization in contrast to an overall increase in total bird abundance (Pennington et al. 2008). Additionally, studies differentiating the breeding bird community using foraging guilds have shown that urban avoiders tend to be insectivorous leaf gleaners, and urban adapters and urban exploiters are comprised of mainly ground-foraging seedeaters and omnivores (Emlen 1974). Further research investigating food availability (e.g., insect abundance and supplemental feeding) and species competition (e.g., foraging efficiency; Shochat et al. 2004), along urban gradients is needed.

Bird Responses Differ Seasonally

Our examination of riparian bird communities during the spring migratory season highlights the need for more detailed studies on the effects of urbanization on migrant stopover strategies. Conservation strategies have traditionally focused on breeding and overwintering habitat quality, which were often thought to be the primary factors influencing population declines. Consequently, urban areas have been perceived as inhospitable to neotropical migrants and ignored when devising conservation plans. However, recent evidence suggests that factors experienced during a migrant's journey to and from the wintering and breeding grounds could ultimately be driving some neotropical migrant population declines (Sillett and Holmes 2002). Our findings imply that certain levels of development, traditionally considered poor habitat for *breeding* neotropical migrants, are actually important during spring and fall as stopover habitat. Measures of habitat-specific demography and dispersal during migration should also be examined along the urban-rural gradient to further improve conservation recommendations. In addition, studies of the effects of urbanization on migrant wintering habitat are also needed. We caution against the general labeling of neotropical migrants as "urban avoiders" in the future, as this designation depends on whether the bird is breeding, migrating, or overwintering at the time of the study.

Assess Species Richness, Diversity, and Evenness for a Comprehensive View of the Community

There are several indices to measure diversity, yet no single measure includes the number of species and species abundance while revealing how much either contributes to the final value (Hayek and Buzas 1997). Therefore, measures of density, species richness, and evenness are needed to capture different aspects of diversity. Most community-level studies have relied on the community measures of abundance/density or species richness when evaluating bird communities (Marzluff 2001), often ignoring the role of evenness; however, our results from Cincinnati suggest that evenness provides additional information on community processes. Evenness accounts for how evenly partitioned individuals are among species, and is a potentially important measure for urban studies since bird communities are often dominated by a few species. In general, density, species richness, and evenness differed in the magnitude of response to scale and local habitat variables across migratory guilds (Fig. 2.4; Pennington et al. 2008). For example, evenness for the total bird community, and in particular permanent residents, decreases with development, indicating that a few synanthropic species dominate the urban bird community. Additionally, density and species richness are influenced by the local vegetative composition of riparian forests, whereas evenness is influenced by vegetative structure (Pennington et al. 2008).

Incorporate Both Experiments and Field Observations on Demographics to Understand Individual and Community Dynamics Shaping Bird Distributions

As illustrated earlier, both demographic and community-level studies are important for elucidating process from pattern. Community-level studies that quantify relative species abundance have been, and continue to be, essential for understanding the effects of urbanization on bird community structure and composition. However, the insights derived from these observed patterns regarding the processes driving community change are limited and could be potentially misleading. Studies of only species detection/nondetection or relative abundance cannot determine whether detected individuals of a species are experiencing positive or negative survivorship at a site. Consequently, these studies do not differentiate whether a population is a source, sink, or stable (Pulliam 1988). Therefore, field experiments and field observations on population demographics should be incorporated into studies of relative abundance. Given our findings regarding predation, demographic studies should also account for the contribution of food availability, renesting, and multiple brooding to better estimate population growth rate, source-sink dynamics, habitat quality, or population viability (Nagy and Holmes 2004).

Recognize That the Amount of Variation across a Gradient Will Influence the Pattern Observed

It is important to consider the context of the urban-rural gradient for a particular study when drawing conclusions about process from the patterns observed. The amount of variation captured across a gradient can vary from study to study, with some studies examining only a small portion of the entire urban-rural gradient spectrum. In other words, some studies have focused on the entire gradient at broad scales and others on only one or two land use types along the gradient, which may explain why it has been difficult to elucidate generalities regarding bird responses to urbanization (McKinney 2002). For example, the curvilinear responses observed for the Palo Alto and Oxford studies may result from considering an entire array of land use intensity types (Fig. 2.2). Observed linear responses of decreasing richness and diversity with increasing measures of urbanization could be due to only capturing variation along suburban to urban land use types, whereas positive linear responses could be due to only capturing variation along exurban to suburban types (Fig. 2.3). On the other hand, the type of response (i.e., linear vs. curvilinear) may also be due to the urban measure used (Lepczyk et al. 2008).

Recognize That Some Studies Focus on Intensifying Land Use while Others Focus on Intensifying Surrounding Land Use

The two common urban-rural gradient paradigms, direct urbanization and surrounding urbanization, focus on intensifying land use and on intensifying surrounding land use, respectfully. Both are valuable methods for discerning pattern from process at fine and coarse scales. When examining sites within all land use types comprising the spectrum of the urban-rural gradient, species responses will often be more dramatic than when only examining sites located within "greenspaces." Most urban-rural gradients in urban ecology have followed the urban context paradigm. Although this paradigm continues to be important, studies that evaluate sites located within differing intensifying land uses are needed to directly measure how species are responding to various types of development. In addition to increasing the number of studies explicitly examining various land use types, we also need further investigation on the variation within a particular land use type (e.g., not all suburban residential areas are the same and can vary in size, density, age, etc.).

Analyze Land Use/Land Cover at an Array of Scales

Processes determining the richness, composition, and dynamics of communities occur at various scales (Wiens 1989a). Consequently, multiscale analyses should be considered for both demographic and community-level studies along urban-rural gradients (Hostetler 1999). As illustrated by our analyses on migratory stopover, both spatial and temporal scales are important for understanding bird distributions. More studies are needed to address how results vary as a function of scale and to search for consistent patterns in these scale effects (Rahbek 2005). Researchers should also report the spatial grain and extent of their study area, since these two constraints can greatly influence conclusions (Mayer and Cameron 2003a). For example, both coarse- (landscape level) and fine- (vegetation level) scale factors influence stopover habitat selection along riparian areas in Cincinnati (Pennington et al. 2008). In addition, researchers should address the issue of using coarse-scale land use/land cover data with fine-scale bird surveys (Gottschalk et al. 2005). The land use/land cover data classification (e.g., USGS Anderson-derived classifications) used to quantify the gradient influences the amount of fine-scale heterogeneity captured across the urban-rural gradient (Cadenasso et al. 2007).

The intended audience should also be considered when determining the scale of a particular study. For example, an ecologist may want to know only about a forest patch, a homeowner may only be concerned about the parcel, a land manager may want to know about protected lands, and a planner may want to know effects at the zoning or municipal level. Historically, urban ecologists have been satisfied with focusing on the ecology of urban systems (i.e., an organism's response to biophysical variables). As we are now realizing, our current understanding of even the ecology is quite limited in the absence of social and economic factors (Grove et al. 2006a). Recent efforts analyzing socioeconomic and demographic aspects of urban ecology have suggested potential mechanisms by which private landowners influence bird communities (Lepczyk et al. 2004a, 2004b; Kinzig et al. 2005). We need to move toward a more inclusive interdisciplinary approach to studying the effects of urbanization on organisms, as described below.

Incorporate Biophysical and Socioeconomic Factors into Gradient Analysis

A more inclusive approach to studying urban systems—the ecology *of* cities—addresses the entire range of habitats within metropolitan areas and not just the greenspaces that are the focus of the ecology *in* cities described thus far (Pickett et al. 1997a, 1997b; Grimm et al. 2000). In urban systems, people make decisions at multiple scales, from individuals to households to municipalities and agencies, which directly influence the landscape mosaic (Grove et al. 2006b). In this approach, people and their institutions are recognized as a part of the ecosystem and not just outside influences (Cadenasso et al. 2006). Understanding the feedbacks between socioeconomic and biophysical factors, both spatially and temporally, will allow urban ecologists to more explicitly tease apart finer-scale processes driving bird distributions.

CONCLUSIONS

Even though we have highlighted several differences in the ways urban ecologists can, and should, examine patterns and processes using gradient analysis, we want to emphasize that we, as researchers, need to focus on the commonalities among studies using similar methods. It is important to understand how our methodological differences influence the patterns observed, but we also need to derive a common framework so that we can compare our studies and findings. A comment at a recent professional meeting noted that if ecologists were given the task of discovering the Universal Law of Gravitation, they still would be dropping different items from the apple tree to see how differently they fell under different conditions. In our work as urban ecologists and ornithologists, we need to focus on the universal similarities of our systems and not the picayune differences.

ACKNOWLEDGMENTS

We would like to thank M. Cadenasso and two anonymous reviewers whose comments greatly improved our thinking on this topic and structuring of this paper. We would like to thank J. Hansel, T. Arbour, A. Hobday, C. Quinn, J. Mager, J. Minturn, T. Minturn, and B. Forschner for their contributions as field assistants on various facets of the research. We also thank K. Pennington for her careful editing and intellectual spurring. Funding was provided in part by the Hamilton County Park District, Miami University, University of Minnesota, and the U.S. Environmental Protection Agency.

LITERATURE CITED

Alberti, M. 2005. The effects of urban patterns on ecosystem function. International Regional Science Review 28:168–192.

Austin, M. P. 1985. Continuum concept, ordination methods, and niche theory. Annual Review of Ecology and Systematics 16:39–61.

Berry, B. L. 1990. Urbanization. The earth as transformed by human action. Pp. 103–119 *in* B. L. Turner, W. C. Clark, R. W. Kates, J. F. Richards, J. T. Matthews, and W. B. Meyers (editors), The earth as transformed by human actions. Cambridge University Press, Cambridge, UK.

Blair, R. B. 1996. Land use and avian species diversity along an urban gradient. Ecological Applications 6:506–519.

Blair, R. B. 1999. Birds and butterflies along an urban gradient: surrogate taxa for assessing biodiversity? Ecological Applications 9:164–170.

Blair, R. B. 2001. Creating a homogeneous aviafauna, Pp. 459–487 *in* J. M. Marzluff, R. Bowman, and R. Donnelly (editors), Avian ecology and conservation in an urbanizing world. Kluwer Academic Publishers, Norwell, MA.

Blair, R. B. 2004. The effects of urban sprawl on birds at multiple levels of biological organization. Ecology and Society 9:2.

Bock, C. E., and Z. F. Jones. 2004. Avian habitat evaluation: should counting birds count? Frontiers in Ecology and the Environment 2:403–410.

Cadenasso, M. L., S. T. A. Pickett, and M. J. Grove. 2006. Integrative approaches to investigating human-natural systems: the Baltimore ecosystem study. Natures Sciences Sociétés 14:4–14.

Cadenasso, M. L., S. T. A. Pickett, and K. Schwarz. 2007. Spatial heterogeneity in urban ecosystems: reconceptualizing land cover and a framework for classification. Frontiers in Ecology and the Environment 5:80–88.

Cadenasso, M. L., S. T. A. Pickett, K. C. Weathers, and C. G. Jones. 2003. A framework for a theory of ecological boundaries. Bioscience 53:750–758.

Chace, J. F., and J. J. Walsh. 2006. Urban effects on native avifauna: a review. Landscape and Urban Planning 74:46–69.

Clergeau, P., J. P. L. Savard, G. Mennechez, and G. Falardeau. 1998. Bird abundance and diversity along an urban–rural gradient: a comparative study between two cities on different continents. Condor 100:413–425.

Czech, B., P. R. Krausman, and P. K. Devers. 2000. Economic associations among causes of species endangerment in the United States. BioScience 50:593–601.

Dickinson, R. E. 1966. The process of urbanization. Pp. 463–478 *in* F. F. Darling and J. P. Milton, eds. Future Environments of North America. Natural History Press, Garden City, NY.

Donnelly, R., and J. M. Marzluff. 2004. Importance of reserve size and landscape context to urban bird conservation. Conservation Biology 18:733–745.

Donnelly, R., and J. M. Marzluff. 2006. Relative importance of habitat quantity, structure, and spatial pattern to birds in urbanizing environments. Urban Ecosystems 9:99–117.

Dunford, W., and K. Freemark. 2005. Matrix matters: effects of surrounding land uses on forest birds near Ottawa, Canada. Landscape Ecology 20: 497–511.

Emlen, J. T. 1974. An urban bird community in Tucson, Arizona: derivation, structure, regulation. Condor 76:184–197.

Gering, J. C., and R. B. Blair. 1999. Predation on artificial bird nests along an urban gradient: predatory risk or relaxation in urban environments? Ecography 22:532–541.

Gottdiener, M., and R. Hutchinson. 2006. The new urban sociology. 3rd ed. Westview Press, Boulder, CO.

Gottschalk, T. K., F. Huettmann, and M. Ehlers. 2005. Thirty years of analyzing and modeling avian habitat relationships using satellite imagery data: a review. International Journal of Remote Sensing 26:2631–2656.

Grimm, N. B., J. M. Grove, S. T. A. Pickett, and C. L. Redman. 2000. Integrated approaches to long-term studies of urban ecological systems. BioScience 50:571–584.

Grove, J. M., M. L. Cadenasso, W. R. Burch, S. T. A. Pickett, K. Schwarz, J. O'Neil-Dunne, M. Wilson, A. Troy, and C. Boone. 2006a. Data and methods comparing social structure and vegetation structure of urban neighborhoods in Baltimore, Maryland. Society & Natural Resources 19:117–136.

Grove, J. M., A. R. Troy, J. P. M. O'Neil-Dunne, W. R. Burch, M. L. Cadenasso, and S. T. A. Pickett. 2006b. Characterization of households and its implications for the vegetation of urban ecosystems. Ecosystems 9:578–597.

Hayek, L. A. C., and M. A. Buzas. 1997. Surveying natural populations. Columbia University Press, New York, NY.
Hennings, L. A., and W. D. Edge. 2003. Riparian bird community structure in Portland, Oregon: habitat, urbanization, and spatial scale patterns. Condor 105:288–302.
Hostetler, M. 1999. Scale, birds, and human decisions: a potential for integrative research in urban ecosystems. Landscape and Urban Planning 45:15–19.
Hostetler, M., and C. S. Holling. 2000. Detecting the scales at which birds respond to structure in urban landscapes. Urban Ecosystems 4:25–54.
Hostetler, M. E. 2001. The importance of multi-scale analyses in avian habitat selection studies in urban environments. Pp. 139–154 *in* J. M. Marzluff, R. Bowman, and R. Donnelly (editors), Avian ecology and conservation in an urbanizing world. Kluwer Academic Publishers, Norwell, MA.
Hutchinson, T. F., and J. L. Vankat. 1997. Invasibility and effects of Amur honeysuckle in southwestern Ohio forests. Conservation Biology 11:1117–1124.
Jokimäki, J., and E. Huhta. 2000. Artificial nest predation and abundance of birds along an urban gradient. Condor 102:838–847.
Jokimäki, J., M.-L. Kaisanlahti-Jokimäki, A. Sorace, E. Fernández-Juricic, I. Rodriguez-Prieto, and M. D. Jimenez. 2005. Evaluation of the "safe nesting zone" hypothesis across an urban gradient: a multiscale study. Ecography 28:59–70.
Kinzig, A. P., P. S. Warren, C. Martin, D. Hope, and M. Katti. 2005. The effects of human socioeconomic status and cultural characteristics on urban patterns of biodiversity. Ecology and Society 10:23.
Kühn, I., R. Brandl, and S. Klotz. 2004. The flora of German cities is naturally species rich. Evolutionary Ecology Research 6:749–764.
Lepczyk, C. A., A. G. Mertig, and J. Liu. 2004a. Landowners and cat predation across rural-to-urban landscapes. Biological Conservation 115: 191–201.
Lepczyk, C. A., A. G. Mertig, and J. Liu. 2004b. Assessing landowner activities related to birds across rural-to-urban landscapes. Environmental Management 33:110–125.
Lepczyk, C. A., C. H. Flather, V. C. Radeloff, A. M. Pidgeon, R. B. Hammer, and J. Liu. 2008. Human impacts on regional avian diversity and abundance. Conservation Biology 22:405–416.
Levin, S. A. 1992. The problem of pattern and scale in ecology: the Robert H. MacArthur award lecture. Ecology 73:1943–1967.
MacArthur, R., H. Recher, and M. Cody. 1966. On the relation between habitat selection and species diversity. American Naturalist 100:319–332.
Marzluff, J. M. 2001. Worldwide urbanization and its effects on birds. Pp. 19–48 *in* J. M. Marzluff, R. Bowman, and R. Donnelly (editors), Avian ecology and conservation in an urbanizing world. Kluwer Academic Publishing, Norwell, MA.
Marzluff, J. M. 2005. Island biogeography for an urbanizing world: how extinction and colonization may determine biological diversity in human-dominated landscapes. Urban Ecosystems 8: 157–177.
Marzluff, J. M., R. Bowman, and R. Donnelly. 2001. A historical perspective on urban bird research: trends, terms, and approaches. Pp. 1–18 *in* J. M. Marzluff, R. Bowman, and R. Donnelly (editors), Avian ecology and conservation in an urbanizing world. Kluwer Academic Publishers, Norwell, MA.
Mayer, A. L., and G. N. Cameron. 2003a. Consideration of grain and extent in landscape studies of terrestrial vertebrate ecology. Landscape and Urban Planning 65:201–217.
Mayer, A. L., and G. N. Cameron. 2003b. Landscape characteristics, spatial extent, and breeding bird diversity in Ohio, USA. Diversity and Distributions 9:297–311.
McCune, B., and J. B. Grace. 2002. Analysis of ecological communities. MjM Software Design, Gleneden Beach, OR.
McDonnell, M. J., and S. T. A. Pickett. 1990. Ecosystem structure and function along urban–rural gradients: an unexploited opportunity for ecology. Ecology 71:1232–1237.
McDonnell, M. J., and S. T. A. Pickett. 1993. Humans as components of ecosystems: the ecology of subtle human effects and populated areas. Springer-Verlag, New York.
McDonnell, M. J., S. T. A. Pickett, P. Groffman, P. Bohlen, and R. V. Pouyat. 1997. Ecosystem processes along an urban-to-rural gradient. Urban Ecosystems 1:21–36.
McKinney, M. L. 2002. Urbanization, biodiversity, and conservation. BioScience 52:883–890.
McKinney, M. L. 2006. Urbanization as a major cause of biotic homogenization. Biological Conservation 127:247–260.
Mehlman, D. W., S. E. Mabey, D. N. Ewert, C. Duncan, B. Abel, D. Cimprich, R. D. Sutter, and M. Woodrey. 2005. Conserving stopover sites for forest-dwelling migratory landbirds. Auk 122:1281–1290.
Miller, J. R., and R. J. Hobbs. 2002. Conservation where people live and work. Conservation Biology 16:330–337.
Mills, G. S., J. B. Dunning Jr., and J. M. Bates. 1989. Effects of urbanization on breeding bird community structure in southwestern desert habitats. Condor 91:416–428.

Moore, F., and P. Kerlinger. 1987. Stopover and fat deposition by North American wood-warblers (*Parulinae*) following spring migration over the Gulf of Mexico. Oecologia 74:47–54.

Moore, F. R., S. A. Gauthreaux Jr., P. Kerlinger, and T. R. Simons. 1995. Pp. 121–144 *in* T. E. Martin and D. M. Finch (editors), Habitat requirements during migration: important link in conservation. Ecology and management of neotropical migratory birds. Oxford University Press, New York.

Nagy, L. R., and R. T. Holmes. 2004. Factors influencing fecundity in migratory songbirds: is nest predation the most important? Journal of Avian Biology 35:487–491.

Nilon, C. H., and L. W. VanDruff. 1987. Analysis of small mammal community data and applications to management of urban greenspaces. Pp. 53–59 *in* L. W. Adams and D. L. Leedy (editors), Integrating man and nature in the metropolitan environment. National Institute for Urban Wildlife, Columbia, MD.

Paul, M. J., and J. L. Meyer. 2001. Riverine ecosystems in an urban landscape. Annual Review of Ecology and Systematics. 32:333–365.

Pautasso, M. 2007. Scale dependence of the correlation between human population presence and vertebrate and plant species richness. Ecology Letters 10:16–24.

Pennington, D. N. 2003. Land use effects on urban riparian bird communities during the migratory and breeding season in the greater Cincinnati metropolitan area. M.S. thesis, Miami University, Oxford, OH.

Pennington, D. N., and R. B. Blair. 2011. Habitat selection of breeding riparian birds in an urban environment: untangling the relative importance of biophysical elements and spatial scale. Diversity and Distributions 17:506–518.

Pennington, D. N., J. Hansel, and R. B. Blair. 2008. Land cover, scale, and vegetation effects on bird communities along urban riparian areas during spring migration. Biological Conservation 141:1235–1248.

Pickett, S. T. A., W. R. Burch, S. E. Dalton, and T. W. Foresman. 1997a. Integrated urban ecosystem research. Urban Ecosystems 1:183 –184.

Pickett, S. T. A., W. R. Burch, S. E. Dalton, T. W. Foresman, J. M. Grove, and R. Rowntree. 1997b. A conceptual framework for the study of human ecosystems in urban areas. Urban Ecosystems 1:185–199.

Pickett, S. T. A., and M. L. Cadenasso. 1995. Landscape ecology: spatial heterogeneity in ecological-systems. Science 269:331–334.

Pickett, S. T. A., M. L. Cadenasso, and J. M. Grove. 2005. Biocomplexity in coupled natural–human systems: a multidimensional framework. Ecosystems 8:225–232.

Pickett, S. T. A., M. L. Cadenasso, J. M. Grove, C. H. Nilon, R. V. Pouyat, W. C. Zipperer, and R. Costanza. 2001. Urban ecological systems: linking terrestrial ecological, physical, and socioeconomic components of metropolitan areas. Annual Review of Ecology and Systematics 32:127–157.

Porter, E. E., B. R. Forschner, and R. B. Blair. 2001. Woody vegetation and canopy fragmentation along a forest-to-urban gradient. Urban Ecosystems 5:131–151.

Pulliam, H. R. 1988. Sources, sinks, and population regulation. American Naturalist 132:652–661.

Rahbek, C. 2005. The role of spatial scale and the perception of large-scale species-richness patterns. Ecology Letters 8:224–239.

Reale, J. A., and R. B. Blair. 2005. Nesting success and life-history attributes of bird communities along an urbanization gradient. Urban Habitats 3:1–24.

Redman, C. L., and N. S. Jones. 2005. The environmental, social, and health dimensions of urban expansion. Population and Environment 26:505–520.

Reichard, S. H., L. Chalker-Scott, and S. Buchanan. 2001. Interactions among non-native plants and birds. Pp. 179–224 *in* J. M. Marzluff, R. Bowman, and R. Donnelly (editors), Avian ecology and conservation in an urbanizing world. Kluwer Academic Publishers, Norwell, MA.

Rodewald, P. G., and S. N. Matthews. 2005. Landbird use of riparian and upland forest stopover habitats in an urban landscape. Condor 107:259–268.

Rosenzweig, M. L. 1995. Species diversity in space and time. Cambridge University Press, Cambridge, UK.

Schmidt, K. A., and C. J. Whelan. 1999. Effects of exotic *Lonicera* and *Rhamnus* on songbird nest predation. Conservation Biology 13:1502–1506.

Shochat, E., S. B. Lerman, M. Katto, and D. B. Lewis. 2004. Linking optimal foraging behavior to bird community structure in an urban–desert landscape: field experiments with artificial food patches. American Naturalist 164:232–243.

Sillett, T. S., and R. T. Holmes. 2002. Variation in survivorship of a migratory songbird throughout its annual cycle. Journal of Animal Ecology 71:296–308.

Sinclair, K. E., G. R. Hess, C. E. Moorman, and J. H. Mason. 2005. Mammalian nest predators respond to greenway width, landscape context and habitat structure. Landscape and Urban Planning 71:277–293.

Smith, S. B., K. H. McPherson, J. M. Backer, B. J. Pierce, D. W. Podlesak, and S. R. McWilliams. 2007. Fruit quality and consumption by songbirds during autumn migration. Wilson Journal of Ornithology 119:419–428.

Sorace, A. 2002. High density of bird and pest species in urban habitats and the role of predator abundance. Ornis Fennica 79:60–71.

Theobald, D. M. 2004. Placing exurban land-use change in a human modification framework. Frontiers in Ecology and the Environment 2:139–144.

Thorington, K. K., and R. Bowman. 2003. Predation rate on artificial nests increases with human housing density in suburban habitats. Ecography 26:188–196.

Turner, W. R., T. Nakamura, and M. Dinetti. 2004. Global urbanization and the separation of humans from nature. BioScience 54:585–590.

United Nations. 2004. State of world population, 2004. The Cairo consensus at ten: population, reproductive health and the global effort to end poverty. United Nations Population Fund, New York.

Vitousek, P. M., H. A. Mooney, J. Lubchenco, and J. M. Melillo. 1997. Human domination of earth's ecosystems. Science 277:494–499.

Vogelmann, J. E., S. M. Howard, L. Yang, C. R. Larson, B. K. Wylie, and N. van Driel. 2001. Completion of the 1990s National Land Cover data set for the conterminous United States from Landsat Thematic Mapper data and ancillary data sources. Photogrammetric Engineering and Remote Sensing 67:650–662.

Von Thünen, J. H. 1826. Von Thünen's isolated state: an English edition of "Der isolierte Stata" (trans. C. M. Wartenberg; ed. P. Hall). Pergamon, New York.

Whittaker, R. H. 1967. Gradient analysis of vegetation. Biological Review of the Cambridge Philosophical Society 42:207–264.

Wiens, J. A. 1989a. The ecology of bird communities, Vol. 2: Processes and variations. Cambridge University Press, Cambridge, UK.

Wiens, J. A. 1989b. Spatial scaling in ecology. Functional Ecology 3:385–397.

CHAPTER THREE

From Forests to Cities

EFFECTS OF URBANIZATION ON TROPICAL BIRDS

Ian MacGregor-Fors, Lorena Morales-Pérez, and Jorge E. Schondube

Abstract. Urban development modifies natural habitats by replacing their fundamental components with new ones, causing a loss of biodiversity. To understand how urbanization affects bird diversity, we studied the bird communities of forest habitats and the urban system that replaced them in a region of western Mexico. We surveyed resident birds in forest habitats (pine-oak and oak forests) and the city of Morelia. We measured habitat characteristics (vegetation, building height, human activity, population density, and income) in both forests and urban habitats to characterize sampling points. Our results show a clear change in bird diversity between the forest habitats and the urban habitats that replaced them. Bird species richness was negatively related to urbanization, while bird abundance was positively related to it. This trend seems to be explained by the loss of a large number of native species due to natural habitat replacement, and the invasion of the city by two exotic species (House Sparrow [*Passer domesticus*] and Rock Pigeon [*Columba livia*]). Several urban attributes were shown to affect the urban bird diversity. Bird species richness was positively related to tree and herbaceous cover, and negatively affected by human activity. Specifically, bird abundances were positively related to building and herbaceous height. Although we did not find significant relationships between bird diversity and income, residential areas in which we recorded the highest bird species richness corresponded to high-income areas.

Key Words: bird communities, citywide survey, diversity, habitat structure, socioeconomic attributes, species turnover, urban ecology, urbanization gradient.

Urban development modifies natural habitats by replacing their fundamental components with new ones (Vitousek et al. 1997). Specifically, natural habitat structure is replaced by urban elements such as buildings and streets, a process that results in a reduction of plant cover and the loss of native species (Main et al. 1999, Czech et al. 2000, Melles 2005). Hence, urbanization replaces the native plant community structure with human constructions, resulting in fewer species that are often found in high abundances (Emlen 1974, Chace and Walsh 2006). This reduction in local biodiversity allows for the arrival and establishment of exotic species

MacGregor-Fors, I., L. Morales-Pérez, and J. E. Schondube. 2012. From forests to cities: effects of urbanization on tropical birds. Pp. 33–48 *in* C. A. Lepczyk and P. S. Warren (editors). Urban bird ecology and conservation. Studies in Avian Biology (no. 45), University of California Press, Berkeley, CA.

that successfully exploit the urban system (Crooks 2002). With regard to birds, urbanization causes species richness to decline, while the abundance of the remaining species can increase dramatically (Marzluff et al. 2001, Chace and Walsh 2006). However, we know little about how the process of native habitat replacement by urban habitat affects birds in cities, particularly in tropical areas of the world (Fonaroff 1974, Jones 1981, Ruszczyk et al. 1987, Green et al. 1989, Munyenyembe et al. 1989, MacGregor-Fors 2008).

While the relationship between bird communities and vegetation structure has been widely studied in natural systems (Short 1979, Swift et al. 1984, Zimmerman 1992, Wilson and Comet 1996), we know little about the habitat elements that control avian diversity inside urban areas, such as cities (Marzluff et al. 2001, Chace and Walsh 2006, González-Oreja et al., online material, Rodewald, chapter 5, this volume). Cities are ecological systems that are difficult to characterize because they include both natural and social elements. Since the dynamics of urban settlements are principally controlled by socioeconomic variables, biodiversity in cities could be under the control of social factors (Hope et al. 2003, Pickett et al. 2004, Kinzig et al. 2005). However, little information exists on the role that socioeconomic variables play in urban biodiversity (Hope et al. 2003, Melles 2005).

The goal of this study was to investigate the differences between resident bird communities in their natural habitats and the urban system that replaced them in a tropical region of the world. Resident species were used because they have to live with the hazards and resources of the city year-round, unlike migrating birds. We expected that the replacement of forest habitat by urban habitat would reduce bird species richness and increase bird abundances. Based on this expectation, we predicted that within urban habitats, bird species richness would be positively related to the presence and structure of vegetation, and the economic level of the different city neighborhoods. Furthermore, we predicted a negative effect of human activity and population density on bird species richness and abundances. We used two approaches to understand these effects: (1) a before-and-after scenario to describe how the replacement of forests by urban habitats could affect bird communities; and (2) the use of socioeconomic factors, together with human activity, urban elements, and traditional measurements of habitat structure to describe human settlements as birds' habitats. These approaches allowed us to comprehend the magnitude of change in bird communities when forest habitats are substituted by urban ones, and to identify natural, urban, and social factors that should be taken into account in urban management and planning.

METHODS

Study Area

Our study was conducted in and around the city of Morelia, the capital of Michoacán state, México (Fig. 3.1). Morelia City presently covers ~81 km^2 and has a human population of 900,000 inhabitants (Ayuntamiento de Morelia 2002). The city has undergone a rapid and unplanned development, growing 400% from 1960 to 1990 and expanding its area from 10.0 to 50.8 km^2 during this time (López et al. 2001). The city is mainly composed of residential and commercial areas, but also includes three cemeteries, three large parks, and two small industrial areas.

To address our main goal, we studied habitats in the city and adjacent patches of the forest

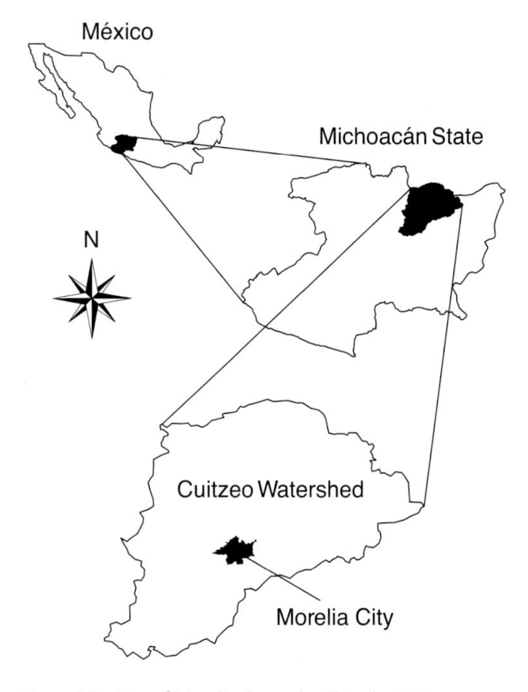

Figure 3.1. City of Morelia, located within the Cuitzeo watershed of Michoacán, México.

habitats that covered the area (pine-oak and oak forests; Madrigal 1997) before the city underwent rapid development (López et al. 2001). Forest habitats still cover a considerable proportion of the area surrounding the city (López et al. 2001). We assumed that bird communities found in the surveyed forest habitats are similar to those that once existed in the area prior to urbanization.

Bird Surveys

Resident birds were surveyed during June and July 2006 from 07:00 to 11:00 CST. For bird surveys, we carried out unlimited radius point counts located at least 250 m from each other to assure survey independence (Ralph et al. 1996, Huff et al. 2000). All birds recorded using the habitat were included in our analyses. We sampled 30 points at each of the forest habitats (pine-oak and oak forests). We also carried out a citywide survey by sampling 204 points within Morelia City limits. Inside the city area, point counts were established at the intersection of a 500 $\times$ 500-m grid randomly set over the city map. This grid allowed us to randomly sample the city's vegetation structure, socioeconomic factors, and human activity. Due to the nonrandom distribution of parks, cemeteries, and industrial areas inside the city, we systematically added 15 sampling points within these areas to assure that all land use categories were well represented in our survey. From the 204 urban sampling points, 104 were located in residential-commercial areas, 60 in residential areas, 15 in parks, 12 in commercial areas, seven in cemeteries, and six in industrial areas. The number of sampling points per land use category was proportional to the total area covered by each category inside the city.

Survey-area Characterization

Because urban habitats are complex and include both natural and human components, we characterized habitats based on four components: (1) vegetation; (2) urban structure; (3) human activity; and (4) socioeconomics. These four components allowed us to describe both forest and urban habitats using a comparable framework. For all sampling points (forest and urban habitats), we measured seven variables to describe their vegetation structure in a 25-m radius area (Ralph et al. 1996): tree cover, tree maximum height, tree abundance, shrub cover, shrub maximum height, herbaceous cover, and herbaceous maximum height. Because tree abundance was highly correlated with tree foliage cover ($r = 0.82$, $P < 0.001$), we only used the latter in our analyses. Additionally, we measured urban structure (building maximum height) and human activity (passing cars/min), and gathered socioeconomic information on population density (inhabitants/km^2) and income (mean monthly income per neighborhood) from published sources (INEGI 2003). Because INEGI's (2003) income data are presented as the number of inhabitants that receive <1, 1–2, 2–5, and >5 official minimum wages (1 official minimum wage $\approx$ 4 U.S. dollars/day), mean income per neighborhood was calculated by adjusting these data to Lopez's (2006) proposed distribution of richness in Mexican medium-sized cities. To calculate human population density, we measured the area of all neighborhoods and divided the number of inhabitants living within each neighborhood by its area.

Data Analysis

Forest-urban Habitat Comparison

To assure that our surveys were representative of the bird communities at the city and forests habitats, we compared rarefaction curves of observed data (mean ±95% CI) with a Chao1 species richness estimator (mean ±95% CI). Rarefaction curves are computed species accumulation curves based on the repeated resampling of all pooled samples. These curves represent the statistical expectation for the corresponding accumulation curves (Gotelli and Colwell 2001). Chao1 is a species richness estimator based on the concept that rare species carry most of the information on the number of missing species in a sample. Chao1 uses singleton and doubleton species to generate a statistical estimate of the total number of expected species (Chao 1987, 2004). Comparing rarefaction statistical expectations with the Chao1 statistical estimations computed from our data allowed us to ensure that our sampling was representative. Both rarefaction and Chao1 were computed on the EstimateS platform (Colwell 2005). Because the coefficient of variance for incidence distributions was >0.50, we used the classic formula for the Chao1 estimator as recommended by Colwell (2005).

To compare bird communities between the surveyed forests and urban habitats, we used a rank/abundance plot approach (= dominance/diversity plot), as recommended by Magurran (2004). Because bird abundance was higher in the urban habitats, abundances are displayed in a log format. To analyze species turnover between the forests and urban habitats, we calculated a sensible species turnover index (β_{sim}; Lennon et al. 2001), as recommended by Koleff et al. (2003). β_{sim} quantifies the relative magnitude of species gains and losses between two samples (Lennon et al. 2001). Since β_{sim} is a relatively new index, we also report Jaccard's index (β_J), expressed as the Colwell-Codington β index ($\beta_{C\text{-}C} = 1 - \beta_J$), a commonly used index to allow data comparison. To contrast species richness among forest and urban habitats, we computed an alpha diversity index not sensible to sample size (Fisher's α index $\pm$SD computed on EstimateS platform; Colwell 2005). To compare bird abundances between forest and urban habitats, we calculated the mean abundance from all of the point counts in each habitat ($\pm$95% confidence intervals). Since the maximum distance at which we recorded birds was 80 m, our abundance estimates represent the mean number of birds per 2 hectares. To evaluate the existence of differences in bird abundances among habitat types (forest, urban) we used a Kruskal-Wallis test.

Land-use Analysis

Because cities are not homogeneous habitats, we characterized the city of Morelia into different land use categories: (1) residential-commercial areas (comprised of houses and commercial buildings); (2) residential areas (encompassing houses only); (3) parks; (4) commercial areas; (5) cemeteries; and (6) industrial areas. To examine if a relationship between land use and bird community diversity existed, we compared the bird communities of each land use category using rarefaction analyses for unequal sample size comparisons (Magurran 2004). We also compared species richness values (Fisher's α index $\pm$SD) using a difference test for means analysis, and bird abundances using a Kruskal-Wallis test among land use categories. For species turnover among land use categories, we calculated β diversity indexes (β_{sim} and $\beta_{C\text{-}C}$).

In addition, we evaluated which of the habitat structures, urban structures, human activities, and socioeconomic variables were related to bird species richness and abundance in the city, using stepwise multiple regression analyses with an $\alpha = 0.05$. Possible correlations among the independent variables were explored, but none were found (all <0.30).

RESULTS

Bird Surveys

In all sampled habitats, rarefaction statistical expectations of the observed data did not differ from the Chao1 species richness statistical estimation (data not shown). For all sampling units (forest habitats, urban habitats, and all urban land use categories), the upper bound of the 95% CI of the observed rarefaction analyses overlapped with the lower bound of the 95% CI of the Chao1 index (95%). This overlap suggests that our sampling effort was enough to represent local species richness in all of our sampling units.

We recorded 64 species of 51 genera in the forest habitats (Appendix 3.1). Of these 64 species 48.4% were insectivores, 29.7% were granivores, 9.4% were omnivores, 6.2% were nectarivores, 4.7% were frugivores, and 1.6% were carnivores. In terms of bird abundances, 56.9% of all birds detected were insectivores, 27.8% were granivores, 9.1% were nectarivores, 6.1% were omnivores, and 0.1% were carnivores. Similarly, in the urban habitats, we recorded 45 species of 42 genera, of which 40% were insectivores, 26.7% were granivores, 15.5% were omnivores, 6.7% were carnivores, 6.7% were nectarivores, and 4.4% were frugivores (Appendix 3.2). In terms of bird abundances, 45.1% of all individuals recorded were omnivores, 31.1% were granivores, 21.3% were insectivores, 1.3% were nectarivores, 1.1% were frugivores, and 0.1% were carnivores.

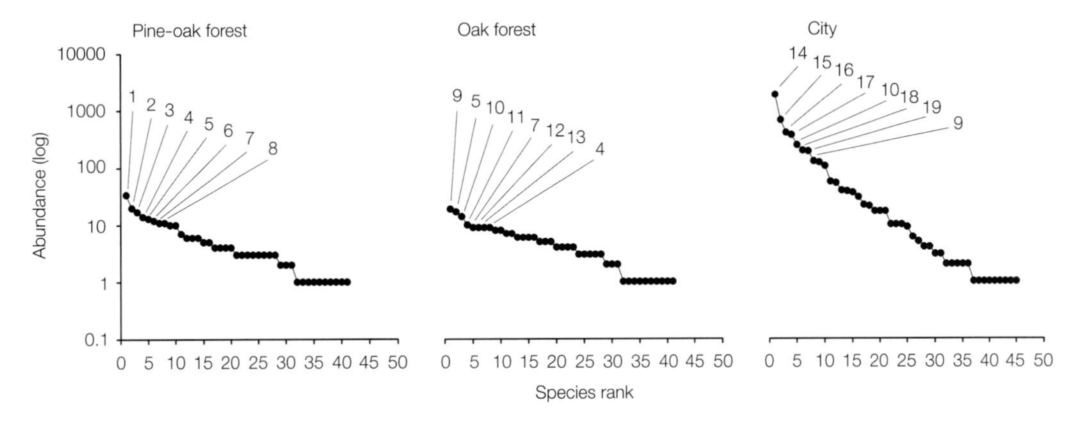

Figure 3.2. Rank/abundance plots representing bird community evenness/dominance for the two forest habitat types (n = 30 each) and the urban habitat that replaced them (n = 204). The bird communities of the forest habitats were highly even, while urban bird communities were dominated by a few species. Numbers represent the eight most common species at each system: 1 = *Myioborus miniatus*, 2 = *Myadestes occidentalis*, 3 = *Contopus pertinax*, 4 = *Piranga flava*, 5 = *Amazilia beryllina*, 6 = *Basileuterus rufifrons*, 7 = *Aphelocoma ultramarina*, 8 = *Pipilo erythrophthalmus*, 9 = *Carduelis psaltria*, 10 = *Pipilo fuscus*, 11 = *Passerina caerulea*, 12 = *Columbina inca*, 13 = *Corvus corax*, 14 = *Passer domesticus*, 15 = *Hirundo rustica*, 16 = *Columba livia*, 17 = *Quiscalus mexicanus*, 18 = *Sporophila torqueola*, 19 = *Molothrus aeneus*.

Forest-urban Habitat Comparison

Rank/abundance plots indicate that both pine-oak and oak forests exhibited highly even bird communities. On the other hand, Morelia's bird community was dominated by a few species (Fig. 3.2). Species composition showed low dissimilarity between the two forests habitats, and between the oak forest and the city (β_{sim} = 0.47, $\beta_{C\text{-}C}$ = 0.64; and β_{sim} = 0.49, $\beta_{C\text{-}C}$ = 0.67, respectively). The pine-oak forest and city bird communities exhibited a higher dissimilarity (β_{sim} = 0.70, $\beta_{C\text{-}C}$ = 0.83). When forest habitats were compared with the different land use categories in the city, the lowest species dissimilarity was for the oak forest/parks cluster (β_{sim} = 0.32, $\beta_{C\text{-}C}$ = 0.64) and the highest was for the residential-commercial/pine-oak forest cluster (β_{sim} = 0.82, $\beta_{C\text{-}C}$ = 0.92; see Table 3.1 for all comparisons).

Species richness was higher in both forest habitats ($\alpha_{Pine\text{-}oak}$ = 14.25 ± 1.53 [SD]; α_{Oak} = 16.01 ± 1.80 [SD]) than in the city (α = 6.4 ± 0.39 [SD]). In contrast, bird abundance per sampling unit was one order of magnitude higher in the city (23.10 ± 2.29 [95% CI]) when compared to both forest habitats (Pine-oak = 8.10 ± 1.23 [95% CI]; Oak = 7.60 ± 1.61 [95% CI]) (Fig. 3.3).

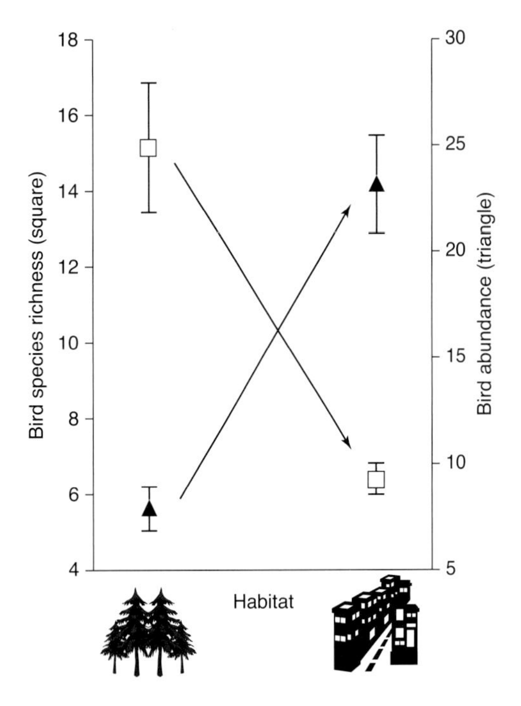

Figure 3.3. Differences in species richness (Fisher's alpha) and abundances (per sampling unit) among the forests and urban bird communities. Species richness was higher at the forest habitats than in the city. Bird abundances were one order of magnitude higher in the city than in the forest habitats. Arrows represent changes in bird species richness and abundance from the forest to the urban habitats.

TABLE 3.1

Species turnover β_{SIM}, ($\beta_{C\text{-}C}$) values for urban land use types and forests.

	Industrial	Res.-Com.	Residential	Cemeteries	Parks	Pine-Oak
Industrial	—					
Res.-Com.	0.31 (0.61)	—				
Residential	0.15 (0.69)	0.14 (0.49)	—			
Cemeteries	0.31 (0.61)	0.47 (0.68)	0.21 (0.61)	—		
Parks	0.31 (0.61)	0.11 (0.29)	0.05 (0.49)	0.21 (0.35)	—	
Pine-Oak	0.77 (0.94)	0.82 (0.92)	0.69 (0.83)	0.79 (0.93)	0.68 (0.89)	—
Oak	0.46 (0.86)	0.41 (0.76)	0.44 (0.68)	0.42 (0.79)	0.32 (0.74)	0.47 (0.64)

TABLE 3.2

Bird species richness and abundance per sampling site along Morelia's land use types.

Land use	Bird species richness (Fisher's $\alpha \pm$ SD)	Bird abundance ($\bar{0} \pm$ SD)
Industrial	3.8 ± 0.7	21.8 ± 5.4
Residential-commercial	3.1 ± 0.6	20.7 ± 8.5
Residential	4.6 ± 0.8	22.9 ± 10.8
Cemeteries	3.9 ± 0.7	26.7 ± 3.9
Parks	4.4 ± 0.8	28.6 ± 19.0

Urban Analysis

Rarefaction analyses showed no difference in bird species richness between urban land use categories (compared to 86 accumulated individuals). We also found no differences between Fisher's α values among land use categories. There were no significant differences in bird abundances among land use categories (H = 8.09, df = 4, 204, P = 0.08; from a Kruskal-Wallis test; Table 3.2). The highest species turnover value (lowest similarity) was for the residential-commercial areas/ cemeteries cluster ($\beta_{sim} = 0.47$, $\beta_{C\text{-}C} = 0.68$), while the lowest turnover (highest similarity) was for the park/residential area cluster ($\beta_{sim} = 0.05$, $\beta_{C\text{-}C} = 0.49$; see Table 3.1 for all comparisons).

Stepwise multiple regression analyses indicated that bird species richness was only related to three of the ten measured habitat variables. Tree and herb cover showed a significant positive effect on bird species richness, while human activity (passing cars/min) exhibited a negative significant effect (Table 3.3). Bird abundances only showed positive significant relationships with herbaceous plant and construction height (Table 3.4).

DISCUSSION

Shifts in Bird Diversity Due to Urbanization

Our results indicate that oak and pine-oak forest habitats differed from urban ones. Species richness of urban bird communities was 60% lower than forest bird communities were, but had abundances that were significantly higher than the forest bird communities (Fig. 3.3). Similar results have been reported for cities located in different latitudes and habitats (desert scrub, closed canopy forests, grasslands, Australian *Eucalyptus* forests, shrublands, subtropical rainforests, coastal sage scrub, and oak woodlands; see Chace and Walsh 2006 and references within), which supports the general pattern described by McKinney (2002). Our results thus add further support for this general pattern holding in tropical latitudes, suggesting that it could be considered a global pattern.

The trend of decreasing richness and increasing abundance from forest to urban habitats (Fig. 3.3) is likely the result of the loss of a large number of native species due to the change of habitat attributes by urbanization processes, and the invasion of the city by the exotic House Sparrow and Rock Pigeon, which are aggressive and/or adapted to take advantage of human subsidies (food and nesting sites). These two exotic species are found in large numbers and dominate the structure of Morelia's bird community. Our results suggest that most of the native species in the area are unable to invade or survive inside the urban habitat, as suggested by Emlen (1974), and generally behave only as suburban adaptable species (as classified by Blair 1996; Appendix 3.2).

Effects of Forest Replacement by Urbanization on the Structure of Bird Communities

Bird community structure changed drastically among forests and urban habitats (Fig. 3.2). Pine-oak and oak forest bird communities were fairly even, where no species doubled the abundance of the next ranked species (Magurran 2004). Morelia's bird community rank/abundance curve (Fig. 3.2), on the other hand, had two different components. The first component was a steep slope segment dominated by a few

TABLE 3.3

Relationship between bird species richness and plant structure, urban structure, human activity, and socioeconomic variables.

General model: $r^2 = 0.23$, $F_{4,146} = 12.71$, $P < 0.001$.

Variable	β	SE	t_{146}	$P\leq$
Intercept	—	—	6.984	0.001
Tree cover	0.382	0.077	4.963	0.001
Herbaceous cover	0.156	0.075	2.079	0.039
No. cars/min.	−0.393	0.077	−5.096	0.001

TABLE 3.4

Relationship between bird abundances and plant structure, urban structure, human activity, and socioeconomic variables.

General model: $r^2 = 0.20$, $F_{4,146} = 18.78$, $P < 0.001$.

Variable	β	SE	t_{146}	$P\leq$
Intercept	—	—	−1.259	0.210
Herbaceous max. height	0.192	0.075	2.564	0.011
Construction max. height	0.445	0.075	5.940	0.001

species classified as urban exploiters (Grussing 1980, Blair 1996, Cupul-Magaña 1996), resembling a logarithmical/geometric distribution common to small, pioneer or stressed animal communities (Boomsma and Van Loon 1982, Hughes 1986, Magurran 2004). The second component was a flat segment composed of a larger number of native species with low abundances that are typically classified as urban/suburban adaptable (*sensu* Blair 1996), and resemble a broken stick distribution, which indicates critical resources being distributed evenly among all community members (MacArthur 1957, Vandermeer and MacArthur 1966, Etennie and Olff 2005).

Species turnover results among forest and urban habitats were surprising. Although urban habitats and pine-oak forests exhibited high dissimilarity values, oak forests showed low dissimilarity values with the urban habitats. This low dissimilarity could be the result of the distribution of pine-oak and oak forest patches around the city. While oak forest patches can still be found surrounding the city, pine-oak forests have been removed. We believe that low turnover rates between oak-forest and urban habitats are the result of geographic proximity.

Within forest habitats, insectivore and granivore trophic guilds dominated the bird communities. However, in Morelia, insectivores (the dominating guild in forests) were replaced by omnivores. This replacement is a common pattern in urban systems, which tend to select for omnivore and granivore species like House Sparrows, Rock Pigeons, and Inca Doves (*Columbina inca*; Emlen 1974, Bessinger and Osborne 1982, Nocedal 1987). All these species are common inside Morelia and represent most of the bird abundance for our study area.

The shift in bird communities that occurred from natural habitats to urban habitats was mainly controlled by two components. First is the introduction of exotic bird species adapted to exploit urban resources. These species can be aggressive competitors for food and for roosting and nesting sites, thereby excluding local species from the urban systems (Lussenhop 1977, Jokimäki 1999). Second is the incapability of a large number of bird species to establish resident populations inside urban habitats due to their hazards (e.g., pollution, passing vehicles), feeding and nesting resources, and habitat structure (Emlen 1974, Bessinger and Osborne 1982, Blair 1996, Stracey and Robinson, chapter 4, this volume).

Urban Attributes Affecting Bird Diversity

Bird diversity (richness and abundance) was similar among all of the city's land use categories. This similarity counters Blair's (1996) findings in Santa Clara County, California (USA), where land use areas with intermediate levels of disturbance had the highest values of species richness and abundance. Differences between bird diversity patterns in the cities of Morelia and Santa Clara could be the result of the unplanned growth that Morelia experienced in the last 40 years (López et al. 2001). Cities that undergo unplanned growth generate complex mixtures of land use categories instead of homogeneous areas devoted to a specific land use (USEPA 2000, Aguilar 2004). We believe that the distribution of land use categories in small patches inside Morelia generated the even bird diversity pattern that we found.

Even though bird diversity in Morelia was not related to land use categories, several attributes of the urban habitat influenced bird diversity. Bird richness was positively related to tree and herbaceous cover, and negatively affected by human activity. Given the importance of plant cover, and specifically trees as roosting, nesting, hiding, and foraging sites, the positive relationships with the vegetation measures are not surprising (Gavareski 1976, Mills et al. 1989, Munyenyembe et al. 1989, MacGregor-Fors 2008). Likewise, human activities have been reported to have negative impacts on bird diversity (Marzluff et al. 2001, Chace and Walsh 2006), which is not surprising if we consider that human activity levels tend to be correlated to noise level and human density, factors that often deter bird presence (Miller et al. 1998).

Bird abundances were positively related to building and herbaceous height. High bird abundances at tall building areas in Morelia are the result of the presence of House Sparrows and Rock Pigeons, which use building façades and roofs as roosting, nesting, and/or foraging areas (Seather et al. 1999, Boren and Hurd 2005). In Morelia, high herbaceous vegetation is mostly found inside unmanaged lots. These lots exhibit habitat structures and resource abundance similar to those found in grassland and shrubland habitats that surround the city (I. MacGregor-Fors et al., unpubl. data). We hypothesize that unmanaged lots act as vegetation islands inside the city, allowing large numbers of both exotic and native bird species to congregate.

Because family income has been shown to influence different elements of urban biodiversity (Hope et al. 2003, Melles 2005), we were surprised not to find a significant positive relation between income and bird diversity in Morelia. A lack of relationship could be due to the heterogeneous distribution of income among the city's neighborhoods. However, residential areas in which we recorded the highest bird species richness corresponded to high-income areas. Because in Mexico and other Latin American countries, high-income areas include large green areas containing most of the city's trees, they act as potential refuges for urban wildlife. Thus, while no statistical relationship was found, there is anecdotal information to suggest a relationship between income and diversity.

The comparison of forest and urban habitat bird communities, representing the before-and-after scenario, revealed five basic patterns that occur when forest habitats are replaced by urban habitats. These patterns are: (1) a decrease of bird species richness, (2) an increase in bird abundances of select species, (3) the loss of community evenness, (4) changes in bird species composition, and (5) shifts in main foraging guild composition. We encourage city planners and managers to consider these variables in order to mitigate the effect that urban habitats have on native bird communities, and therefore maintain and promote native bird community species richness and evenness in subtropical mountain cities.

ACKNOWLEDGMENTS

We would like to thank J. Quesada for statistical analysis support, C. Chávez-Zichinelli for assistance with fieldwork, and Dr. Po for reviewing the manuscript. Research funds were granted to J.E.S. by the Universidad Nacional Autónoma de México (UNAM) through the Megaproyecto–Manejo de Ecosistemas y Desarrollo Humano (SDEI-PTID-02), and PAPIIT project (IN228007). IM-F received a master's scholarship from CONACyT (203142).

LITERATURE CITED

Aguilar, A. D. 2004. La reestructuración del espacio urbano de la ciudad de México. ¿Hacia la metrópoli mutlinodal? Pp. 265–308 *in* A. G. Aguilar (editor), Procesos metropolitanos y grande sciudades: dinámicas recientes en México y otros países. Editorial Porrúa, México, DF.

Ayuntamiento de Morelia, H. 2002. Plan de Desarrollo Municipal de Morelia. H. Ayuntamiento de Morelia, México, DF.

Beissinger, S. R., and D. R. Osborne. 1982. Effects of urbanization on avian community organization. Condor 84:75–83.

Blair, R. B. 1996. Land use and avian species diversity along an urban gradient. Ecological Applications 6:506–519.

Boomsma, J. J., and A. J. Van Loon. 1982. Structure and diversity of ant communities in successive coastal dune valleys. Journal of Animal Ecology 51:957–974.

Boren, J., and B. J. Hurd. 2005. Controlling nuisance birds in New Mexico. Guide L-212. Cooperative Extension Service, College of Agriculture and Home Economics, New Mexico State University.

Chace, J. F., and J. J. Walsh. 2006. Urban effects on native avifauna: a review. Landscape and Urban Planning 74:46–69.

Chao, A. 1987. Estimating the population size for capture-recapture data with unequal catchability. Biometrics 43:783–791.

Chao, A. 2004. Species richness estimation. Pp. 1–23 *in* C. B. Read, B. Vidakovich, and N. Balakrishnan (editors), Encyclopedia of statistical sciences. Wiley-Interscience, New York.

Colwell, R. K. 2005. EstimateS: Statistical estimation of species richness and shared species from samples, version 7.5. <http://purl.oclc.org/estimates>.

Crooks, J. A. 2002. Characterizing ecosystem-level consequences of biological invasions: the role of ecosystem engineers. Oikos 97:153–166.

Cupul-Magaña, F. G. 1996. Incidencia de avifauna en un parque urbano de Los Mochis, Sinaloa, México. Ciencia Ergo Sum 3:193–200.

Czech, B., P. R. Krausman, and P. K. Devers. 2000. Economic associations among causes of species endangerment in the United States. Bioscience 50:593–601.

Emlen, J. T. 1974. An urban bird community in Tucson, Arizona: derivation, structure, regulation. Condor 76:184–197.

Etienne, R. S., and H. Olff. 2005. Confronting different models of community structure to species-abundance data: a Bayesian model comparison. Ecology Letters 8:493–504.

Fonaroff, L. S. 1974. Urbanization, birds and ecological change in northwestern Trinidad. Biological Conservation 6:258–262.

Gavareski, C. A. 1976. Relation of park size and vegetation to urban bird populations in Seattle, Washington. Condor 78:375–382.

Gotelli, N. J., and R. K. Colwell. 2001. Quantifying biodiversity: procedures and pitfalls in the measurement and comparison of species richness. Ecology Letters 4:379–391.

Green, R. J., C. P. Catterall, and D. N. Jones. 1989. Foraging and other behaviour of birds in subtropical and temperate suburban habitats. Emu 89:216–222.

Grussing, D. 1980. How to control House Sparrows. Roseville Publishing House, Roseville, MN.

Hope, D., C. Gries, W. Zhu, W. F. Fagan, C. L. Redman, N. B. Grimm, A. L. Nelson, C. Martin, and A. Kinzig. 2003. Socioeconomics drive urban plant diversity. Proceedings of the National Academy of Sciences of the USA 100: 8788–8792.

Huff, M. H., K. A. Bettinger, H. L. Ferguson, M. J. Brown, and B. Altman. 2000. A habitat-based point-count protocol for terrestrial birds emphasizing Washington and Oregon. General Technical Report PNW-GTR-501. U.S. Department of Agriculture, Forest Service, Pacific Northwest Research Station, Portland, OR.

Hughes, R. G. 1986. Theories and models of species abundance. American Naturalist 128:879–899.

Instituto Nacional de Estadística, Geografía e Informática (INEGI). 2003. SCINCE por colonias – Michoacán de Ocampo – XII Censo General de Población y Vivienda 2000. INEGI, Aguascalientes, México.

Jokimäki, J. 1999. Occurrence of breeding bird species in urban parks: effects of park structure and broad-scale variables. Urban Ecosystems 3:21–34.

Jones, D. N. 1981. Temporal changes in the suburban avifauna of an inland city. Australian Wildlife Research 8:109–119.

Kinzig, A. P., P. S. Warren, C. Martin, D. Hope, and M. Katti. 2005. The effects of human socioeconomic status and cultural characteristics on urban patterns of biodiversity. Ecology and Society 10:23.

Koleff, P., K. J. Gaston, and J. J. Lennon. 2003. Measuring beta diversity for presence–absence data. Journal of Animal Ecology 72:367–382.

Lennon, J. J., P. Koleff, J. J. D. Greenwood, and K. J. Gaston. 2001. The geographic structure of British bird distributions: diversity, spatial turnover and scale. Journal of Animal Ecology 70:966–979.

López, E., G. Bocco, M. Mendoza, and E. Duhau. 2001. Predicting land-cover and land-use change in the urban fringe: a case in Morelia City, México. Landscape and Urban Planning 55:271–285.

López, H. 2006. Avances AMAI: distribución de niveles socio-económicos en el México urbano. Datos, Diagnósticos, Tendencia 6:5–11.
Lussenhop, J. 1977. Urban cemeteries as bird refuges. Condor 79:456–461.
MacArthur, R. H. 1957. On the relative abundance of bird species. Proceedings of the National Academy of Sciences of the USA 43:293–295.
MacGregor-Fors, I. 2008. Relation between habitat attributes and bird richness in a western Mexico suburb. Landscape and Urban Planning 84:92–98.
Madrigal, X. 1997. Ubicación fisiográfica de la vegetación en Michoacán, México. Ciencia Nicolaita 15:65–75.
Magurran, A. E. 2004. Measuring biological diversity. Blackwell Publishing, Oxford.
Main, M. B., F. M. Roka, and R. F. Noss. 1999. Evaluating costs of conservation. Conservation Biology 13:1262–1272.
Marzluff, J. M., R. Bowman, and R. Donnely. 2001. A historical perspective on urban bird research: trends, terms, and approaches. Pp. 1–17 *in* J. M. Marzluff, R. Bowman, and R. Donnelly (editors), Avian conservation in an urbanizing world. Kluwer Academic Publishers, New York.
McKinney, M. L. 2002. Urbanization, biodiversity and conservation. BioScience 52:883–890.
Melles, S. J. 2005. Urban bird diversity as an indicator of human social diversity and economic inequality in Vancouver, British Columbia. Urban Habitats 3: 25–48.
Mills, G. S., J. B. Dunning, and J. M. Bates. 1989. Effects of urbanization on breeding bird community structure in Southwestern desert habitats. Condor 91:416–428.
Miller, S. G., R. L. Knight, and C. K. Miller. 1998. Influence of recreational trails on breeding bird communities. Ecological Applications 8:162–169.
Moreno, C. E. 2001. Métodos para medir la biodiversidad. M&T–Manuales y Tesis SEA, Vol. 1. Zaragoza, Spain.
Munyenyembe, F., J. Harris, and J. Hone. 1989. Determinants of bird populations in an urban area. Australian Journal of Ecology 14:549–557.
Nocedal, J. 1987. Las comunidades de pájaros y su relación con la urbanización en la ciudad de México. Pp. 73–109 *in* E. Rapoport and I. R. López-Moreno (editors), Aportes a la ecología urbana de la ciudad de México. Limusa, México, DF.
Pickett, S. T. A., M. L. Cadenasso, and J. M. Grove. 2004. Resilient cities: meaning, models, and metaphor for integrating the ecological, socio-economic, and planning realms. Landscape and Urban Planning 69:369–384.
Ralph, C. J., G. R. Geupel, P. Pyle, T. E. Martin, D. F. DeSante, and B. Milá. 1996. Manual de métodos de campo para el monitoreo de aves terrestres. General Technical Report PSW-GTR-159. Pacific Southwest Research Station, Forest Service, U.S. Department of Agriculture, Albany, CA.
Ruszczyk, A. J., J. S. Rodrigues, T. M. T. Roberts, M. M. A. Bendati, R. S. Del Pino, J. C. V. Marques, and M. T. Q. Melo. 1987. Distribution patterns of eight bird species in the urbanization gradient of Porto Alegre, Brazil. Ciência e Cultura 39:14–19.
Seather, B. E., T. H. Ringsby, Ø. Bakke, and E. J. Solberg. 1999. Spatial and temporal variation in demography of a house sparrow population. Journal of Animal Ecology 68:628–637.
Short, J. J. 1979. Patterns of alpha-diversity and abundance in breeding bird communities across North America. Condor 81:21–27.
Swift, B. L., J. S. Larson, and R. M. DeGraff. 1984. Relationship of breeding bird density and diversity to habitat variables in forested wetlands. Wilson Bulletin 96:48–59.
U.S. Environmental Protection Agency (USEPA). 2000. Projecting land-use change: a summary of models for assessing the effects of community growth and change on land-use patterns. EPA/600/R-00/098. U.S. Environmental Protection Agency, Office of Research and Development, Cincinnati, OH.
Vandermeer, J. H., and R. H. MacArthur. 1966. A reformulation of alternative (b) of the broken stick model of species abundance. Ecology 47: 139–140.
Vitousek, P. M., H. A. Mooney, J. Lubchenco, and J. M. Melillo. 1997. Human domination of earth's ecosystems. Science 277:494–499.
Wilson, M. F., and T. A. Comet. 1996. Bird communities of northern forests: patterns of diversity and abundance. Condor 98:337–349.
Zimmerman, J. L. 1992. Density-independent factors affecting the avian diversity of the tallgrass prairie community. Wilson Bulletin 104:85–94.

APPENDIX 3.1

Bird species recorded in forest habitats (pine-oak and oak forests).

Scientific name	Main trophic guild	Total abundance	Frequency (records/ total point counts)
Accipiter striatus	C	1	1/60
Zenaida macroura	G	1	1/60
Columbina inca	G	12	5/60
Columbina passerina	G	1	1/60
Leptotila verreauxi	G	9	9/60
Cynanthus latirostris	N	7	6/60
Hylocharis leucotis	N	3	3/60
Amazilia beryllina	N	30	21/60
Eugenes fulgens	N	1	1/60
Trogon elegans	F	1	1/60
Melanerpes formicivorus	G	15	13/60
Melanerpes aurifrons	O	1	1/60
Lepidocolaptes leucogaster	I	7	7/60
Contopus pertinax	I	21	14/60
Empidonax affinis	I	1	1/60
Empidonax occidentalis	I	6	6/60
Attila spadiceus	I	1	1/60
Myiarchus tuberculifer	I	3	2/60
Myiarchus nuttingi	I	1	1/60
Myiodynastes luteiventris	I	1	1/60
Tyrannus melancholicus	I	3	2/60
Tyrannus vociferans	I	8	6/60

Scientific name	Main trophic guild	Total abundance	Frequency (records/ total point counts)
Vireo gilvus	I	4	2/60
Cyanocitta stelleri	O	7	2/60
Aphelocoma ultramarina	O	20	7/60
Corvus corax	O	10	6/60
Hirundo rustica	I	3	1/60
Campylorhynchus gularis	I	11	10/60
Catherpes mexicanus	I	3	3/60
Thryomanes bewickii	I	15	12/60
Troglodytes aedon	I	10	10/60
Polioptila caerulea	I	18	14/60
Sialia sialis	I	8	5/60
Myadestes occidentalis	I	22	18/60
Catharus occidentalis	I	3	2/60
Turdus migratorius	F	1	1/60
Toxostoma curvirostre	I	1	1/60
Melanotis caerulescens	I	6	6/60
Peucedramus taeniatus	I	1	1/60
Oreothlypis superciliosa	I	6	6/60
Cardellina rubra	I	1	1/60
Myioborus pictus	I	6	2/60
Myioborus miniatus	I	34	20/60
Basileuterus rufifrons	I	19	14/60
Icteria virens	I	3	2/60
Piranga flava	I	23	18/60
Piranga rubra	I	1	1/60
Sporophila torqueola	G	3	3/60
Atlapetes pileatus	G	3	3/60
Pipilo ocai	G	2	1/60
Pipilo erythrophthalmus	G	11	10/60
Melozone fusca	G	17	13/60
Aimophila ruficeps	G	6	3/60
Spizella passerina	G	4	4/60
Spizella atrogularis	G	6	4/60
Junco phaeonotus	G	1	1/60
Pheucticus melanocephalus	G	6	6/60
Passerina caerulea	G	10	7/60
Icterus pustulatus	O	7	4/60

APPENDIX 3.1 (*continued*)

APPENDIX 3.1 (CONTINUED)

Scientific name	Main trophic guild	Total abundance	Frequency (records/ total point counts)
Icterus bullockii	O	1	1/60
Euphonia elegantissima	F	1	1/60
Carpodacus mexicanus	G	2	2/60
Spinus pinus	G	4	1/60
Spinus psaltria	G	24	16/60

Main trophic guilds are abbreviated as I = insectivore, G = granivore, O = omnivore, N = nectarivore, F = frugivore, C = carnivore.

APPENDIX 3.2

Bird species recorded within the Morelia urban area.

Scientific name	Main trophic guild	Total abundance	Frequency (records/ total point counts)
Elanus leucurus	C	1	1/204
Accipiter striatus	C	1	1/204
Buteo jamaicensis	C	1	1/204
Columba livia	G	403	22/204
Columbina inca	G	367	124/204
Crotophaga sulcirostris	I	21	11/204
Chaetura vauxi	I	22	2/ 204
Cynanthus latirostris	N	53	40/204
Amazilia beryllina	N	4	3/204
Amazilia violiceps	N	3	1/204
Melanerpes aurifrons	O	56	35/204
Contopus pertinax	I	1	1/ 204
Pyrocephalus rubinus	I	36	29/204
Tyrannus vociferans	I	17	10/204
Lanius ludovicianus	I	1	1/204
Corvus corax	O	9	2/204
Stelgidopteryx serripennis	I	17	5/204
Hirundo rustica	I	671	181/204
Campylorhynchus gularis	I	1	1/ 204
Catherpes mexicanus	I	105	84/204
Thryomanes bewickii	I	17	13/204
Troglodytes aedon	I	30	11/204
Catharus aurantiirostris	I	2	1/ 204

APPENDIX 3.2 (*continued*)

APPENDIX 3.2 (CONTINUED)

Scientific name	Main trophic guild	Total abundance	Frequency (records/ total point counts)
Turdus rufopalliatus	F	38	16/204
Toxostoma curvirostre	I	10	7/204
Melanotis caerulescens	I	4	2/204
Ptilogonys cinereus	F	10	5/204
Geothlypis speciosa	I	2	2/204
Geothlypis poliocephala	I	2	1/204
Volatinia jacarina	G	5	4/204
Sporophila torqueola	G	191	104/204
Melozone fusca	G	244	115/204
Spizella passerina	G	1	1/204
Pheucticus melanocephalus	G	2	1/204
Passerina caerulea	G	10	7/204
Agelaius phoeniceus	I	2	1/204
Sturnella magna	O	3	3/204
Quiscalus mexicanus	O	196	50/204
Molothrus aeneus	G	122	27/204
Icterus wagleri	O	1	1/204
Icterus bullockii	O	1	1/204
Carpodacus mexicanus	G	39	21/204
Loxia curvirostra	G	6	1/204
Spinus psaltria	G	128	59/204
Passer domesticus	O	1,858	198/204

Main trophic guilds are abbreviated as I = insectivore, G = granivore, O = omnivore, N = nectarivore, F = frugivore, C = carnivore.

CHAPTER FOUR

Does Nest Predation Shape Urban Bird Communities?

Christine M. Stracey and Scott K. Robinson

Abstract. Urban bird communities differ fundamentally from those in non-urban habitats. We provide a preliminary test of the hypothesis that altered communities of nest predators in urban habitats affect which species do and do not thrive. To address this hypothesis, we censused urban and non-urban bird communities (including species that regularly act as nest predators) and measured nest predation rates in a variety of communities in Florida, USA. We predicted that (1) avian nest predators would be more abundant in urban habitats, (2) species with life-history traits that provide effective defenses against avian predation (large body size, aggressive nest defense, protected nest sites) would be more abundant in urban habitats relative to non-urban habitats, and (3) urban adapters would experience lower nest predation rates in urban habitats relative to non-urban habitats. As predicted, avian nest predators were more abundant in urban habitats relative to non-urban habitats, and the life-history traits of birds also differed considerably between urban and non-urban habitats. Small-bodied individuals were far more abundant in non-urban habitats, whereas large-bodied birds, many of which are facultative nest predators and most of which mob avian nest predators, were far more abundant in urban habitats. Most (90%) of the small-bodied (<40 g) birds that were detected in urban habitats nest in enclosed sites. Small-bodied open-cup nesters were almost entirely absent from urban areas, but made up about 50% of the detections in non-urban habitats. These data are consistent with our hypothesis that urban adapters are species that can protect themselves against avian nest predators. In fact, for two urban adapters and two successful introduced species, nest predation rates were significantly lower in urban areas, further supporting the urban refuge hypothesis. Our results are therefore consistent with our hypothesis that nest predation at least partly determines which species are able to survive in urban habitats and which species cannot invade or sustain populations in urban habitats. These results have parallels in small fragmented forests, which are also home to altered nest-predator assemblages and have communities dominated by larger species and those that nest in protected sites.

Key Words: body size, community composition, nest predation, urbanization.

Stracey, C. M., and S. K. Robinson. 2012. Does nest predation shape urban bird communities? Pp. 49–70 *in* C. A. Lepczyk and P. S. Warren (editors). Urban bird ecology and conservation. Studies in Avian Biology (no. 45), University of California Press, Berkeley, CA.

The ways in which bird communities change in response to urbanization have been well documented (Marzluff et al. 2001, Shochat 2004, MacGregor-Fors et al., chapter 3, this volume). Four general patterns have emerged: (1) a reduction in species richness, (2) a decrease in evenness (urban bird communities have fewer but more abundant species), (3) an increase in abundance of large species, and (4) an increase in abundance of introduced species. While these patterns have been documented in various urban areas, few studies have investigated the underlying mechanisms causing such patterns. Results to date emphasize two competing, but not mutually exclusive mechanisms—food availability and decreased predation pressure in urban habitats (Shochat et al. 2004, 2006; Bartos Smith et al., chapter 11, this volume). Here we explore the role of nest predation in structuring urban bird communities. Our working hypothesis, the urban nest predator hypothesis, is that altered communities of nest predators in urban habitats affect which species do and do not thrive. We refer to three groups of species: urban exploiters (species that are essentially human commensals), urban adapters (species that are abundant in urban areas but also maintain populations in more natural habitats), and urban avoiders (species that reach their highest densities in natural habitats; Blair 1996, Shochat et al. 2006). Specifically, we predict that (1) urban adapters have lower rates of nest loss to predators in urban habitats than those same species in non-urban habitats, and (2) species missing from urban communities will have life-history traits (nest defense and placement) that make them more vulnerable to the kinds of nest predators that thrive in urban areas. Unfortunately, we are not able to test the reciprocal version of our first prediction, that urban avoiders have higher rates of nest predation in urban areas because we did not locate enough nests of urban avoiders in urban habitat.

Nest predation has been hypothesized to be a major force structuring bird communities (Martin 1993) and a critical determinant of the viability of populations in fragmented habitats (Robinson et al. 1995). It is now widely accepted that human-caused alterations of communities of nest predators can have cascading effects on community structure in remnant patches of natural habitat in human-dominated landscapes (Heske et al. 1999, Chalfoun et al. 2002). Indeed, one of the primary changes in fragmented habitats is the loss of top predators and the resulting increase in generalist "mesopredators," many of which incidentally eat bird eggs and nestlings (Soulé et al. 1988, Crooks and Soulé 1999). Nesting success of many birds in fragmented habitats is so low that they have been hypothesized to be population sinks (Robinson et al. 1995) or even ecological traps that attract nesting birds but that produce too few young to compensate for adult mortality (Heske et al. 1999, Weldon and Haddad 2005). As a result, bird communities in human-altered landscapes (e.g., forest fragments within an agricultural matrix) are dominated by medium-sized species that can defend their nests against predators and species that nest in sites that are inaccessible to all but a small subset of nest predators (e.g., cavities, Whitcomb et al. 1981, Brown and Sullivan 2005). If there are comparable differences in nest-predator communities in relation to urbanization, then it is possible that those differences in predation pressure could be driving the distribution and abundance of avian species.

To date, however, studies of urban nest predation have produced mixed results. For example, Gering and Blair (1999) found that rates of nest predation on artificial nests decreased along an urban gradient and hypothesized that urban areas may actually offer a refuge from nest predation, at least for some species. Because urban bird communities often have high populations of avian nest predators, this result has been called a "predator paradox" (Shochat et al. 2006). Other studies have found no relationship between urbanization and nest predation rates (Melampy et al. 1999, Haskell et al. 2001), whereas still others have found an increase in nest predation along the urban gradient (Jokimäki and Huhta 2000, Thorington and Bowman 2003, Kaisanlahti-Jokimäki et al., chapter 6, this volume). Such increased nest predation rates may be especially prevalent in species that live in remnant patches of native forest within urban areas (Borgmann and Rodewald 2004).

The goals of this study were to compare nesting success and composition of avian bird communities, including species that regularly act as nest predators (e.g., corvids) and introduced species, in various urban and non-urban habitats. We predicted that (1) avian species that are nest predators would be more abundant in urban habitats (Marzluff et al. 2001, Sorace 2002, Shochat et al. 2006), (2) species with life-history traits that provide effective defenses (e.g., large body size, aggressive nest defense, protected nest sites) against avian predation would be more abundant in urban habitats, and (3) urban adapters would experience lower nest predation rates in urban habitats.

METHODS

Surveys

To address the hypothesis that nest predation rates account for many of the changes in urban bird communities and test our predictions, we censused urban and non-urban bird communities and measured nest predation rates in a variety of habitats in Florida, USA. The goal of our surveying was to characterize the bird communities of habitats believed to be representative of northern and central Florida. Surveys were conducted in the spring and summer of 2004. We used 5-minute, 100-m fixed-radius point counts during which we recorded species and number of individuals within the 100-m radius. We chose 100 m rather than the traditional 50 m (Ralph et al. 1995) because it increased sample sizes for habitats for which we had relatively few census points and because detectability was generally high in all habitats, with the exception of some broadleaf forests. The comparatively dense vegetation of hammock and floodplain forests may have decreased detectability, especially of small species. However, this bias against detecting small species in forested habitat makes our results showing an increase in small-bodied birds in these habitats (see below) a conservative test of our hypotheses.

All surveys were conducted after 1 May, by which time all breeding species had returned. Surveys began at sunrise and ended no later than 10:30 EST. Within each survey area, points were located at least 250 m apart. Most points were located within 100 km of Gainesville, but additional surveys were conducted in the Archibald Biological Station and the MacArthur Ranch, in coastal scrub near St. Augustine and Jacksonville, and in urban settings in Jacksonville, St. Augustine, and Sarasota. Points were grouped into the following habitat categories (number of points in parentheses): parking lot (42), residential neighborhood (16), urban forest fragment, as defined by intact canopy and understory (11), pasture (15), non-urban scrub/secondary growth (52), and non-urban forest (48). We calculated the average number of detections per point in the different habitats for each species for which we had nesting data. The average number of avian nest predators detected per point was also calculated. We considered the following species nest predators: Fish Crow (*Corvus ossifragus*), American Crow (*Corvus brachyrhynchos*), Common Grackle (*Quiscalus quiscula*), Boat-tailed Grackle (*Quiscalus major*), Blue Jay (*Cyanocitta cristata*), and Red-shouldered Hawk (*Buteo lineatus*). Estimates of the abundance of avian nest predators are conservative because we counted flocks of the same species as one detection in the analyses. Avian nest predators were surveyed at the same time as the rest of the species. While surveying nest predators at the same time might result in an underestimate of their true abundance, due to their large home ranges, we feel this approach is justified because we are comparing the relative abundance of predators across different habitats and not directly comparing their abundance with that of other species. We also calculated the average number of detections per point of introduced species. We included the following species in this category: House Sparrow (*Passer domesticus*), House Finch (*Carpodacus mexicanus*), Eurasian Collared-Dove (*Streptopelia decaocto*), European Starling (*Sturnus vulgaris*), Monk Parakeet (*Myiopsitta monachus*), and Rock Pigeon (*Columba livia*). To test for an effect of habitat on the average number of detections of introduced species, avian nest predators, and the different nesting species, we performed Kruskal-Wallis tests because our data were not normally distributed. If there was a significant effect of habitat on the average number of detections, we then conducted post hoc comparisons of the habitats using Tukey's pairwise comparisons with alpha levels adjusted for multiple comparisons.

To examine how the abundance of birds of different body masses was affected by urbanization, we calculated the average number of individuals detected per point in the following size classes: 0–14.9 g, 15–39.9 g, 40–99.9 g, 100–199.9 g, 200+ g. These size classes were chosen based on natural breaks in the data and in an effort to generate a normal distribution for body mass. Body mass estimates for each species were taken from Dunning (1993). In addition, we examined how nesting guild (as defined by either open cup or enclosed nest) differed according to habitat by calculating the average number of enclosed nesting individuals detected and the average number of open-cup nesting individuals detected in each size class. To analyze the effect of habitat, mass, and nesting guild on the average number of detections, we performed a quasi-Poisson regression (SAS 9.1, SAS Institute Inc., Cary, NC). Because there were a few treatments where we observed no individuals (e.g., we did not detect any individuals of mass 100–199.9 g that were cavity nesters in non-urban forests), we added 0.001 to each data point to allow the model to converge. We checked the effect of adding 0.001 by performing the same analysis with the addition of 0.01 and comparing the model parameters for each. In both cases we obtained very similar *F*-statistics and the same *P*-values; thus the model did not seem to be sensitive to this small addition. Because we are also interested in whether any differences in abundance reflect the urban versus non-urban land use dichotomy, we tested the effect of land use category (urban, urban forest fragment, and non-urban) in the same manner.

Nest Monitoring

The nesting study was conducted from 2004 to 2006 in areas in and around Gainesville, Florida (Fig. 4.1). The following study sites were used, followed by the years for which we have data from each site: the University of Florida campus (2004), downtown Gainesville (2004–2005), urban forest fragments within Gainesville (2004), two wildlife preserves (Ordway Swisher Biological Station, 2005–2006, and Payne's Prairie State Park, 2005), three residential neighborhoods (2005–2006), two parking lots (2005–2006), and three pastures (2006).

We searched each study site for nests of all common species. At least 15 nests of the following species were located to allow for analysis (Table 4.1): Northern Mockingbird (*Mimus polyglottos*), Brown Thrasher (*Toxostoma rufum*), Northern Cardinal (*Cardinalis cardinalis*), Mourning Dove (*Zenaida macroura*), House Finch (*Carpodacus mexicanus*), Eurasian Collared-Dove (*Streptopelia decaocto*), and Loggerhead Shrike (*Lanius ludovicianus*). Nests were monitored approximately every third day following discovery. Daily survival rates were calculated using Shaffer's logistic regression model (Shaffer 2004, SAS 9.1, SAS Institute Inc., Cary, NC). Because we are interested only in predation rates, we only considered nests that failed due to predation as "unsuccessful." Therefore, the daily survival rates we report are not a measure of overall nest survival, but rather the probability of a nest escaping predation. For some species we had enough data to model survival based on habitat type (e.g., parking lot, residential, etc.), but for other species for which we found fewer nests we combined habitat types and compared urban land use, including parking lot, residential, town, and campus, to non-urban land use, including pasture and wildlife preserve.

Data for the Mourning Dove and Loggerhead Shrike, for which we have relatively small sample sizes, were pooled across all three years to compare survival rates between urban and non-urban land use. To test for an effect of land use on daily survival, we modeled survival based on urban versus non-urban land use using the logistic regression model. Because House Finch and Eurasian Collared-Dove nests were not located at any of the non-urban study sites, we calculated the daily survival rate in urban land use (parking lots, residential neighborhoods, downtown), and because of low sample size we used data pooled from 2004–2006.

For the Northern Mockingbird and Brown Thrasher, we had at least 15 nests from 2005–2006 combined in each habitat, allowing us to model daily survival rates as a function of urban versus non-urban land use and also as a function of habitat type (e.g., parking lot, residential, etc.). In 2004 we only had data from urban habitat types, predominantly from campus, and so we report 2004 daily survival rates separately. For both mockingbirds and thrashers we used information-theoretic methods to evaluate candidate models (Burnham and Anderson 2002). We compared models that included the following variables (with and without interactions): year, nest stage (incubation, nestling), and habitat type or land use, as well as a model that assumed constant survival Table 4.2).

For the Northern Mockingbird we proceeded to compare the confidence intervals for estimates of daily survival between the five habitat types for each nest stage and then for urban and non-urban land use at each nest stage. For the Brown Thrasher we compared confidence intervals of daily survival rates between habitat types and land use, but not between nest stages. We examined confidence intervals for 2005 and 2006 separately. We did not include parking lot or town in our comparisons for thrashers because sample sizes

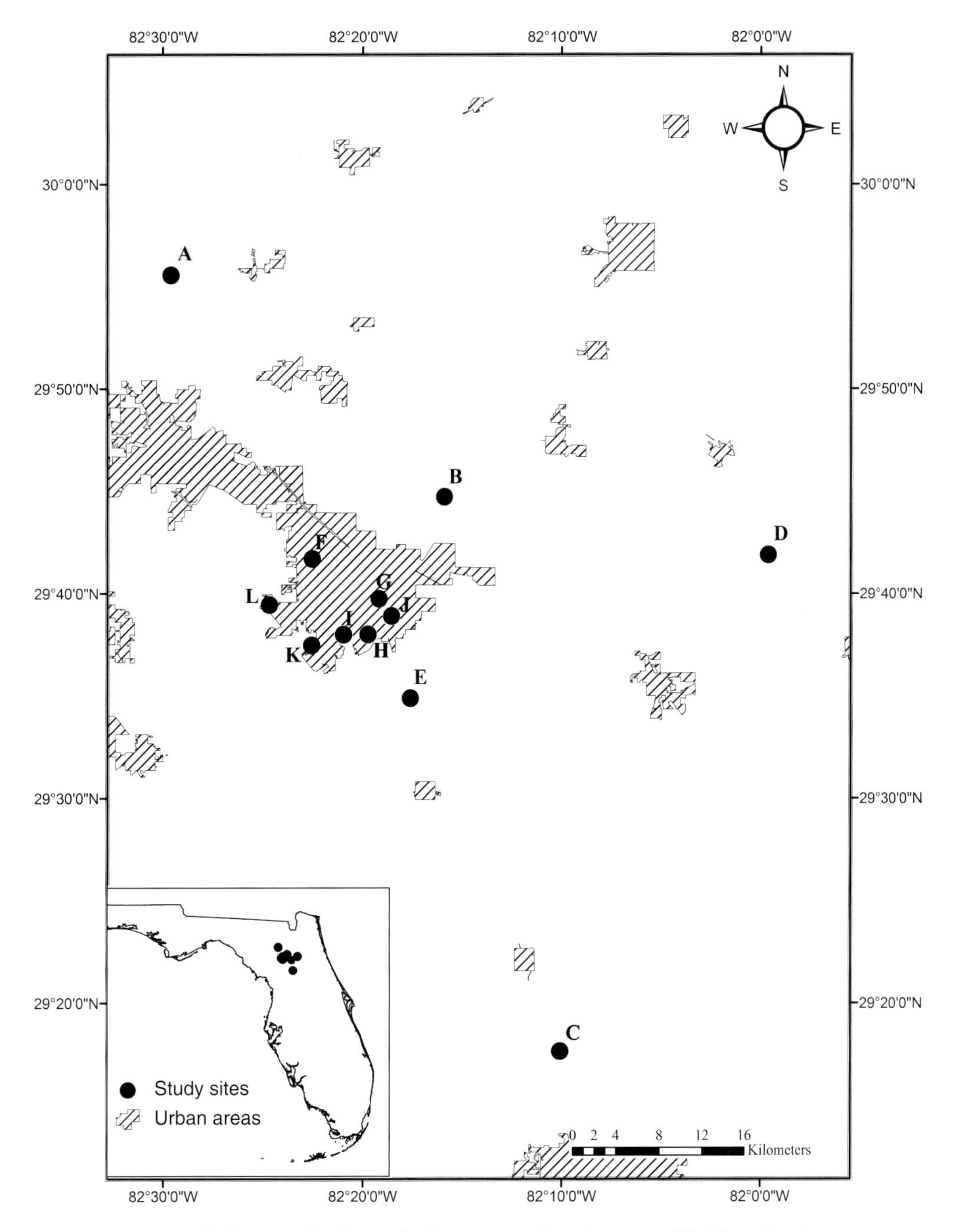

Figure 4.1. The location of study sites for the nesting study on images modified from Google Earth™ mapping service. A–C: pastures; D–E: wildlife preserves; F–H: residential neighborhoods; I: University of Florida campus; J: downtown Gainesville; K–L: parking lots.

for these two categories were limited to three and five nests, respectively.

The Northern Cardinal was the only species for which we had data from urban forest fragments (2004 only), which allowed us to compare daily survival in urban forest fragments with urban and non-urban land use. We compared candidate models that included land use, year, and nest stage (with and without interactions) and a model that assumed constant survival.

TABLE 4.1

Daily survival rate percentages (95% CL) in the years indicated, number of nests monitored, and the statistical significance of habitat on daily survival rates.

Species (years)	Habitat	Daily survival rate (DSR)	*N*	Chi-square	df	*P*≤
MODO (2004–2006)	Urban	96.27 (94.23–97.61)	44			
	Non-urban	92.00 (84.76–95.93)	11	3.43	1	0.064
LOSH (2004–2006)	Urban	97.80 (94.80–99.08)	11			
	Non-urban	94.95 (85.45–98.36)	4	1.32	1	0.25
HOFI (2004–2006)	Urban	99.68 (99.16–99.98)	60			
	Non-urban	n/a	0	n/a		
ECDO (2004–2006)	Urban	96.89 (90.74–97.60)	15			
	Non-urban	n/a	0	n/a		
NOCA (2004–2006)	Urban	97.1 (95.63–98.08)	65			
	UFF	93.5 (86.96–96.87)	11			
	Non-urban	94.13 (92.22–95.59)	73	n/a		
NOMO (2004)	Urban	95.50 (94.30–96.46)	124			
	Non-urban	n/a	0	n/a		
NOMO (2005–2006)	Urban		314			
	Incubation	96.94 (96.19–97.55)				
	Nestling	96.81 (95.94–97.49)				
	Non-urban		147			
	Incubation	94.53 (93.07–95.70)				
	Nestling	90.75 (87.75–93.07)				
BRTH (2004)	Urban	95.12 (92.29–96.95)	31			
	Non-urban	n/a	0	n/a		
BRTH (2005–2006)	Urban		50			
	2005	97.41 (94.67–98.76)				
	2006	96.31 (93.62–97.90)				
	Non-urban		82	n/a		
	2005	89.70 (86.95–91.93)				
	2006	92.43 (89.03–94.84)				

UFF: urban forest fragment, MODO: Mourning Dove, LOSH: Loggerhead Shrike, HOFI: House Finch, ECDO: Eurasian Collared-Dove, NOCA: Northern Cardinal, NOMO: Northern Mockingbird, BRTH: Brown Thrasher.

RESULTS

Avian Nest-Predator Abundance

Avian nest predators were detected far more often in parking lots and residential neighborhoods than in urban forest fragments, pastures, non-urban scrub, and non-urban forests (Fig. 4.2; $\chi^2 = 73.68$, df $= 5$, $P < 0.001$). Because we recorded flocks as one detection, the relative abundance of avian predators is even greater, given that American Crows typically traveled in groups of 3–6 individuals, Fish Crows typically occurred in groups of 10–30 individuals, and both species of grackles traveled in flocks of up to 30 individuals. However, the six species of predators were not equally responsible for the pattern we saw (Fig. 4.2). There was

TABLE 4.2

Model-selection results for the top six logistic-exposure models of daily survival for the Northern Mockingbird (2005–2006), Brown Thrasher (2005–2006), and Northern Cardinal (2004–2006).

"Stage" = stage of the nesting cycle (incubation or nestling), "X" = interaction terms included in model, "+" = no interaction terms included in model.

Model	Log(*L*)	*K*	AIC_c	ΔAIC_c	w_i
		Northern Mockingbird			
Habitat type					
Habitat × stage	−730.00	10	1,480.04	0	0.584
Habitat × stage + year	−729.05	12	1,482.16	2.13	0.202
Habitat × year × stage	−725.29	16	1,482.68	2.64	0.156
Habitat + stage	−737.15	6	1,486.32	6.29	0.025
Habitat + year + stage	−737.02	7	1,488.05	8.01	0.011
Habitat	−739.19	5	1,488.39	8.36	0.009
Land use category					
Category × year × stage	−735.43	8	1,486.88	0	0.342
Category × year + stage	−738.10	6	1,488.22	1.34	0.175
Category + year + stage	−738.10	6	1,488.22	1.34	0.175
Category + year × stage	−738.46	6	1,488.94	2.06	0.122
Category + stage	−741.82	3	1,489.64	2.76	0.086
Category × year	−741.15	4	1,490.31	3.43	0.062
		Brown Thrasher			
Habitat typc					
Habitat	−234.43	5	478.90	0	0.432
Habitat × year	−231.93	8	479.95	1.05	0.255
Habitat + year	−234.15	6	480.36	1.46	0.208
Habitat + stage	−234.33	7	482.73	3.83	0.064
Habitat + year + stage	−234.04	8	484.18	5.28	0.031
Habitat × year + stage	−231.20	12	486.62	7.72	0.009
Land use category					
Land use × year	−232.85	4	473.73	0	0.479
Land use	−234.98	2	473.97	0.24	0.426
Land use + stage	−234.88	4	477.78	4.05	0.063
Land use × year + stage	−232.34	8	480.78	7.05	0.014
Land use + year + stage	−232.34	8	480.78	7.05	0.014
Land use + year × stage	−233.33	9	484.79	11.05	0.002

TABLE 4.2 *(continued)*

TABLE 4.2 (CONTINUED)

Model	Log(L)	K	AIC_c	ΔAIC_c	w_i
		Northern Cardinal			
Land use	−208.97	3	423.96	0	0.540
Land use + year	−208.43	5	426.91	2.95	0.124
Land use + stage	−208.61	5	427.26	3.30	0.104
Land use × year	−207.61	6	427.28	3.12	0.103
Constant	−213.25	1	428.51	4.55	0.056
Land use + year + stage	−208.04	7	430.16	6.20	0.024

no difference in the average number of detections of American Crows ($\chi^2 = 9.17$, df = 5, $P = 0.10$) or Red-shouldered Hawks ($\chi^2 = 4.14$, df = 5, $P = 0.53$) in different habitats. There was a significant effect of habitat on the average number of detections of Blue Jays ($\chi^2 = 12.16$, df = 5, $P = 0.03$); however, none of the Tukey's pairwise comparisons were significant at the corrected alpha level, so this trend is weak at best. The remaining species—Fish Crow ($\chi^2 = 62.31$, df = 5, $P < 0.001$), Common Grackle ($\chi^2 = 38.23$, df = 5, $P < 0.001$), and Boat-tailed Grackle ($\chi^2 = 61.03$, df = 5, $P < 0.001$)— all show the same pattern: they were most abundant in parking lots and residential neighborhoods.

Community Composition

Urban and non-urban communities differed greatly in the relative abundance of most species (Appendix 4.1). The effect of habitat on introduced birds is especially pronounced (Fig. 4.3; $\chi^2 = 111.12$, df = 5, $P < 0.001$), with these species restricted to parking lots and residential areas, with the exception of the Eurasian Collared-Dove, which also occurred around barns in pastures. Even the House Sparrow (*Passer domesticus*) and European Starling were not recorded in agricultural areas (Fig. 4.3). Other urban adapters included the Northern Mockingbird (Fig. 4.4a; $\chi^2 = 93.66$, df = 5, $P < 0.001$) and Mourning Dove (Fig. 4.4b; $\chi^2 = 49.59$, df = 5, $P < 0.001$). Brown Thrashers and Loggerhead Shrikes were also detected in urban habitats, but their detection probabilities were so low, as a result of their extremely infrequent singing, that we could not test for any patterns. Northern Cardinals (Fig. 4.4c; $\chi^2 = 40.59$, df = 5, $P < 0.001$) and Carolina Wrens (*Thryothorus ludovicianus*; Fig. 4.4d; $\chi^2 = 27.25$, df = 5, $P < 0.001$) tended to be slightly more common in non-urban habitats; however, both species were abundant in all habitats with any woody vegetation.

The life-history traits of urban and non-urban birds also differed considerably (Fig. 4.5; also see Pennington and Blair, chapter 2, this volume). There was a significant three-way interaction between habitat, mass, and nesting guild ($F = 3.39$, df = 20, $P < 0.001$) and between land use, mass, and nesting guild ($F = 6.34$, df = 8, $P < 0.001$). Small-bodied cup-nesting individuals were far more abundant in non-urban habitats (Fig. 4.5). Conversely, large-bodied individuals, many of which are also nest predators, were far more abundant in urban habitats (Fig. 4.5). In the smallest size class (0–14.9 g), open-cup nesters made up roughly half of the detections, whereas in non-urban habitats, nearly 100% were open-cup nesters (Fig. 4.5). Interestingly, the urban forest fragments appear more similar to the two non-urban habitats than they do to the urban habitats. The same pattern holds for the next size class (15–39.9 g)—there were more cavity-nesting individuals detected in urban areas than in non-urban areas, with urban forest fragments falling roughly in between. Indeed, the only abundant urban birds of <40 g were the House Sparrow and Carolina Wren, both of which place most of their nests in human structures in urban areas (S. K. Robinson and C. M. Stracey, unpubl. data).

Nesting Success

A total of 1,044 nests were monitored during the three years of the study. For both the Mourning

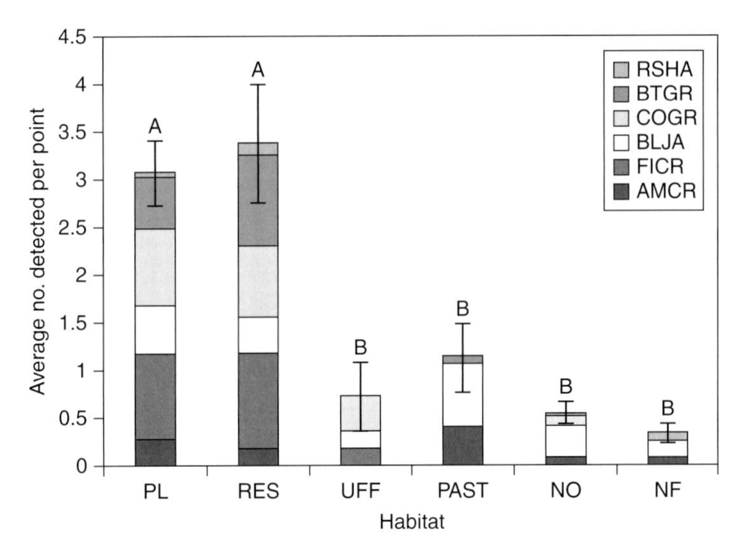

Figure 4.2. Average (±SE) number of total avian nest predators detected per survey point (5 minute, 100 m fixed radius, $\chi^2 = 73.68$, df = 5, $P < 0.001$) and the average number of each species detected. Letters refer to significant post-hoc comparisons of total number of predators. The following species were considered avian nest predators: AMCR: American Crow, FICR: Fish Crow, COGR: Common Grackle, BTGR: Boat-tailed Grackle, BLJA: Blue Jay, and RSHA: Red-shouldered Hawk. PL: parking lot, RES: residential neighborhood, UFF: urban forest fragment, PAST: pasture, NO: non-urban scrub, NF: non-urban forest.

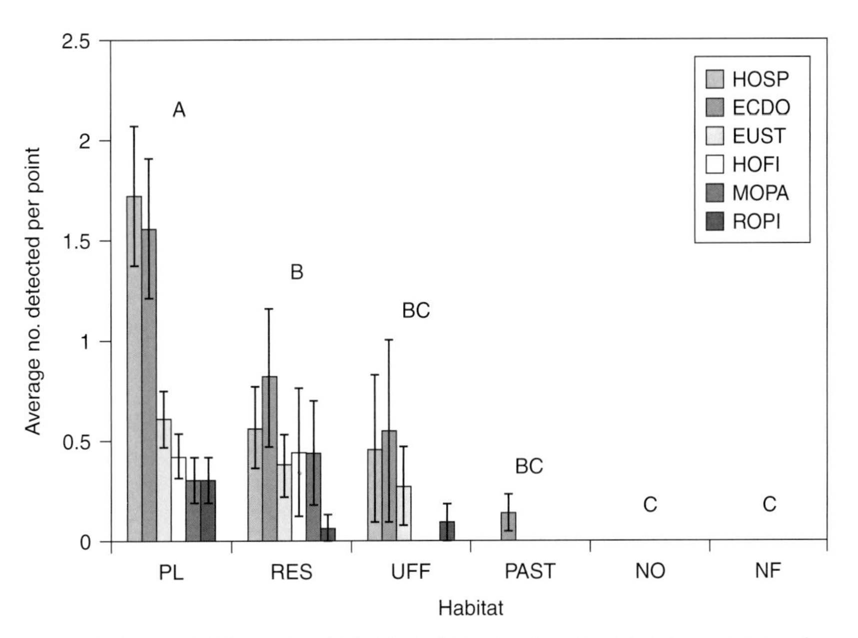

Figure 4.3. Average (±SE) number of individuals of introduced species detected per point in each habitat type based on 5-minute, 100-m fixed radius points ($\chi^2 = 111.12$, df = 5, $P < 0.001$). Letters refer to significant post-hoc comparisons. HOSP: House Sparrow, ECDO: Eurasian Collared-Dove, EUST: European Starling, HOFI: House Finch, MOPA: Monk Parakeet, ROPI: Rock Pigeon. PL: parking lot, RES: residential neighborhood, UFF: urban forest fragment, PAST: pasture, NO: non-urban scrub, NF: non-urban forest.

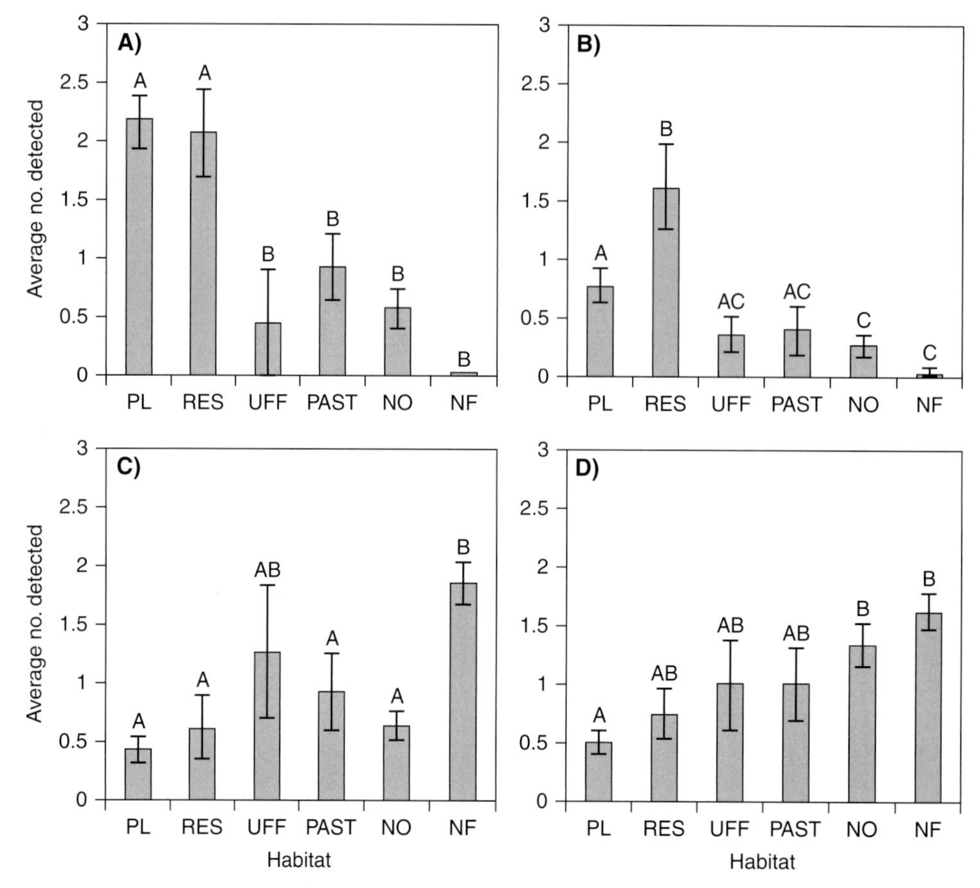

Figure 4.4. Average (±SE) number of individuals detected per point in each habitat type based on 5-minute, 100-m fixed radius points. (a) Northern Mockingbird ($\chi^2 = 93.66$, df = 5, $P < 0.001$). (b) Mourning Dove ($\chi^2 = 46.59$, df = 5, $P < 0.001$). (c) Northern Cardinal ($\chi^2 = 40.59$, df = 5, $P < 0.001$). (d) Carolina Wren ($\chi^2 = 27.25$, df = 5, $P < 0.001$). Letters refer to significant post-hoc comparisons. PL: parking lot, RES: residential neighborhood, UFF: urban forest fragment, PAST: pasture, NO: non-urban scrub, NF: non-urban forest.

Dove ($\chi^2 = 3.43$, df = 1, $P = 0.06$) and Loggerhead Shrike ($\chi^2 = 1.32$, df = 1, $P = 0.25$), the daily survival rate was higher in urban land use (parking lots, residential, campus, town), yet neither of these differences was significant (Table 4.1). Among the introduced species restricted to urban land use, the daily survival rate of House Finches (99.68%) was very high (Table 4.1). The survival rate of the Eurasian Collared-Dove was similar to the survival rate of the Mourning Dove in urban land use (Table 4.1). We have no data for the small-bodied open-cup nesters in either urban or non-urban land use.

We monitored 461 Northern Mockingbird nests and 144 Brown Thrasher nests, for an effective sample of 5,863 and 1,447, respectively. For the mockingbird at the land use level, the three models with $\Delta AIC_c < 2$ all included land use, year, and nest stage and differed in the interactions included between these three variables (Table 4.2). In urban land use there were no differences in survival rates between the incubation and nestling stage in either 2005 or 2006 (Table 4.1). However, in 2006 in non-urban land use, the nestling stage had lower rates of nest survival than the incubation stage. Survival rates were consistently lower in non-urban land use relative to urban land use across years and nest stage, although in some cases the 95% confidence intervals of the daily survival rates overlapped slightly. At the habitat level, the model that included habitat and nest stage and their interaction was the only model with a $\Delta AIC_c < 2$ (Table 4.2). Nests in pastures and wildlife preserves tended to have lower survival rates in the nestling stage; however, the confidence intervals for these survival

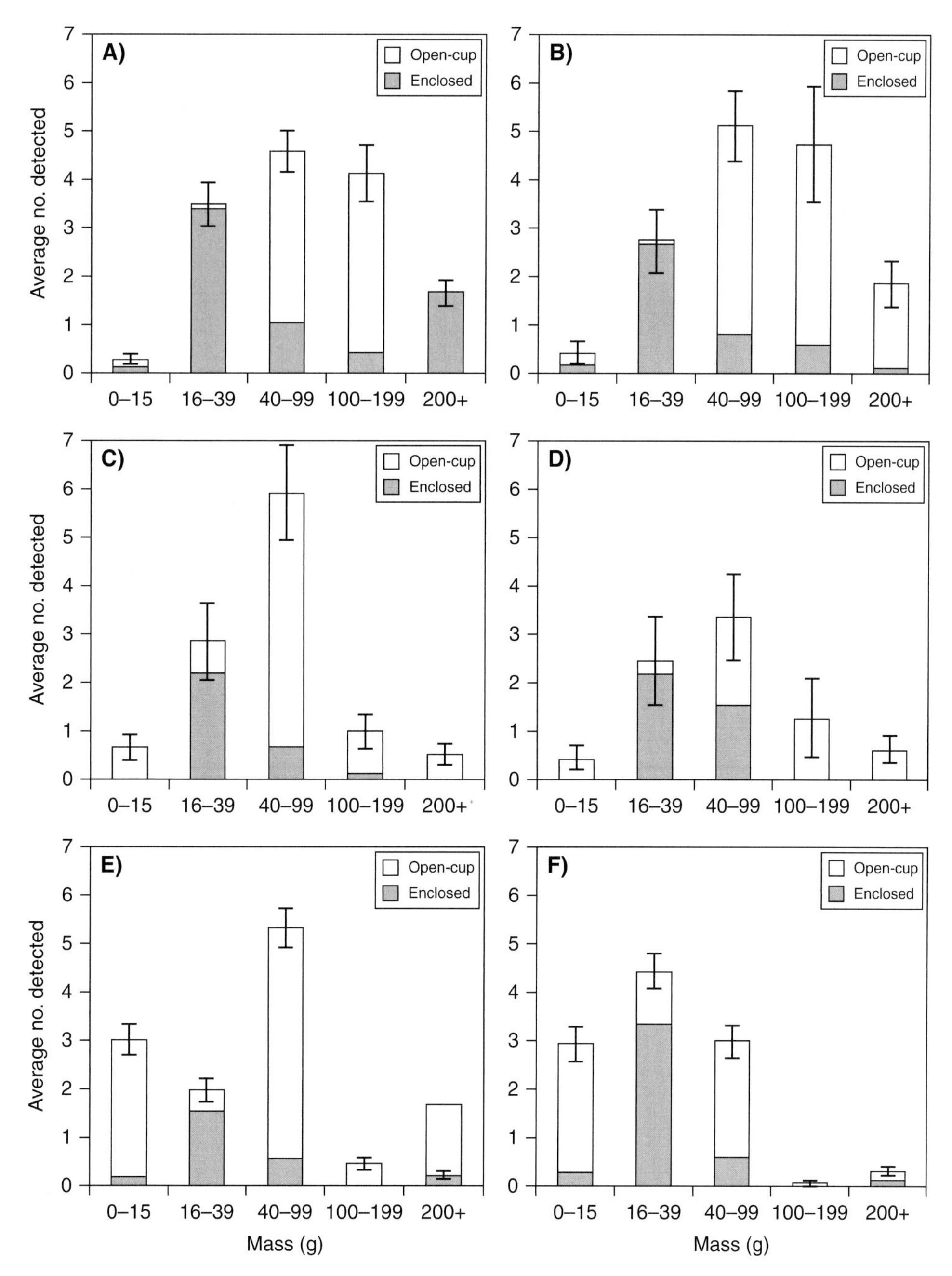

Figure 4.5. Average (±SE) number per survey point (5 minute, 100 m fixed radius) of individuals of all species combined detected in each habitat based on mass and nesting guild: (a) parking lot, (b) residential neighborhood, (c) pasture, (d) urban forest fragment, (e) non-urban scrub, (f) non-urban forest.

rates are much wider and overlap with those for the incubation period (Fig. 4.6). In parking lot, town, and residential areas, the estimates of daily survival rates for incubation and nestling stages were much more similar to each other (Fig. 4.6). There was also a trend for parking lot and town to have higher rates of survival than pastures and wildlife preserves in both the incubation and nestling stages (Fig. 4.6). Survival rates in residential habitat tended to be intermediate between those in parking lot and town and those in pastures and wildlife preserves (Fig. 4.6).

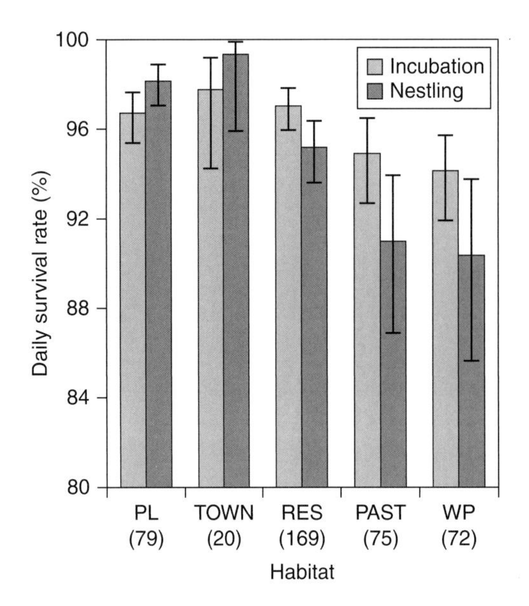

Figure 4.6. Daily survival rate (±95% CL) of the Northern Mockingbird at the incubation and nestling stage in each habitat type, with number of nests indicated below each habitat in parentheses. PL: parking lot, TOWN: downtown Gainesville, RES: residential neighborhood, PAST: pasture, WP: wildlife preserve.

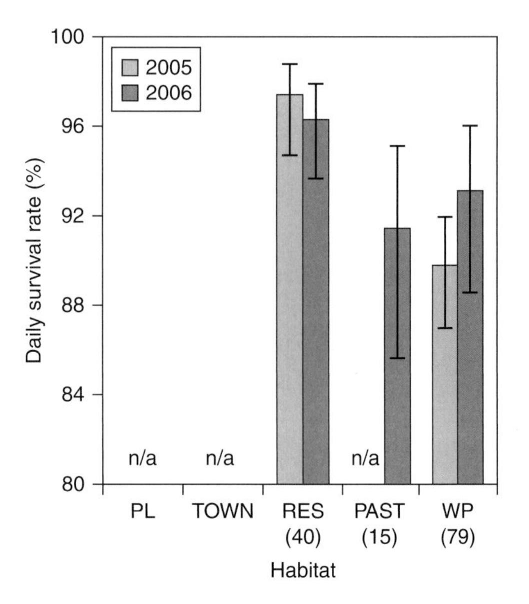

Figure 4.7. Daily survival rate (±95% CL) of the Brown Thrasher in 2005 and 2006 in each habitat type, with number of nests indicated below each habitat in parentheses. PL: parking lot, TOWN: downtown Gainesville, RES: residential neighborhood, PAST: pasture, WP: wildlife preserve.

For the Brown Thrasher, the top model included land use, year, and their interaction (Table 4.2). The next best model ($\Delta AIC_c = 0.235$) included only land use. In both 2005 and 2006, survival rates were lower in non-urban land use than in urban land use; however, in 2006 the difference was greatly reduced because survival rates in non-urban land use were higher than those in 2005 (Table 4.1). At the habitat level, there were three models with ΔAIC_c values less than 2, all of which included habitat as a parameter (Table 4.2). Year was included in two of the three models. In 2005 residential nests had higher survival rates than wildlife preserves (Fig. 4.7). In 2006 the 95% confidence intervals for all three habitat types (residential, pasture, and wildlife preserve) overlapped, although there was a trend for higher survival rates in residential habitat than in pastures and wildlife preserves (Fig. 4.7).

For the Northern Cardinal (effective sample size = 1,534), the model with land use was the only model with $\Delta AIC_c < 2$ (Table 4.2). Daily survival rates in urban land use were higher than those in non-urban land use (Table 4.1). Daily survival rates in urban forest fragments were similar to those in non-urban land use and tended to be lower than survival rates in urban land use (Table 4.1).

DISCUSSION

Our results are generally consistent with the hypothesis that differences in abundances of nest predators drive many of the changes in bird community composition often documented in urban habitats. We found much higher abundances of avian nest predators in urban habitats (Fig. 4.2). Urban bird communities consisted mostly of large-bodied birds (Fig. 4.5), many of which aggressively defend their nests against predators (C. M. Stracey, pers. obs.), and smaller-bodied birds that nest in inaccessible sites. The near absence of small-bodied open-cup nesters in urban areas may reflect the vulnerability of these species to avian predators. As predicted, urban adapters experienced lower nest predation rates in urban habitats relative to non-urban habitats (Table 4.1).

Nest-Predator Abundance

There is little dispute that urban areas are host to altered predator communities. Avian nest predators are more abundant in the urban communities we surveyed, which is similar to other studies where comparable data exist (Gregory and Marchant 1996, Jokimäki and Huhta 2000,

Marzluff et al. 2001, Sorace 2002). This greater abundance may relate to a loss of top predators (Soulé et al. 1988), but may also result from increased anthropogenic food resources and decreased home range size (Marzluff and Neatherlin 2006). In addition, these increased food resources are typically concentrated at point sources that are highly predictable, which larger birds may be more effective at exploiting (Daily and Ehrlich 1994, Shelley et al. 2004, French and Smith 2005, MacNally and Timewell 2005). Such large species often incidentally take eggs and nestlings (Marzluff et al. 2001). We do not, however, know the relative importance of these species as nest predators and whether that importance varies by habitat. We have evidence that the predator communities likely differ: in urban areas there was no difference in predation rates at different nest stages, but there was a difference in non-urban areas. These data suggest that different predators were responsible for the majority of predation events in different habitats.

We have no data from mammalian predators or snakes, and the majority of previous studies concerning the abundance of mammalian nest predators relate to their abundance in habitat fragments within either an agricultural or an urban matrix, and rarely to their abundance in the urban areas themselves (Chalfoun et al. 2002). Within urban forest fragments, rats, foxes, cats, dogs, and opossums have been shown to occur at higher densities than in non-fragmented areas (Crooks 2002, Sorace 2002). Raccoons (*Procyon lotor*) are positively affected by urbanization and reach higher densities in both urban forest fragments and the urban environment itself (Prange et al. 2003, Randa and Yunger 2006). The effect of urbanization on snakes is likewise not well known. Snakes were more abundant in fragmented habitat surrounded by agriculture (Weatherhead and Charland 1985, Durner and Gates 1993). However, Patten and Bolger (2003) found that within urban fragments of coastal scrub, snake abundance was lower. In urban areas themselves, it is likely that snake abundance is further reduced owing to persecution by people. If there are fewer snakes in urban areas, then the high nest predation rates in rural areas may be largely a result of increased abundance of snakes, which are potentially very important nest predators (Weatherhead and Blouin-Demers 2004). Schmidt et al. (2001) found that weak predators (species that only occasionally depredate nests) did not fully compensate for strong predators (species that drive patterns of nest predation) when the strong predator was removed. Therefore, increases in avian predators in urban areas may be more than offset by a decrease in snakes. However, snake abundance and diversity varies from one geographic region to the next, and we do not know how this relates to the effects of urbanization on snakes. Future studies should explicitly consider the effects of urbanization on snakes and the importance of snake predation on nests.

The way in which these changes in the predator community translate into changes in nest predation rates is not likely to be this straightforward because different species are likely to be differentially affected depending on their life histories. For example, nesting guild is an important determinant of the susceptibility of nests to different predators (Patten and Bolger 2003, Schmidt et al. 2006) and thus differences in predator community composition in urban areas might favor the persistence of certain species over others. To fully understand the role of nest predation in structuring urban bird communities we need to determine how the abundances of the entire predator community change, not simply avian predators. Moreover, we need to explore how these changes in abundance translate into changes in nest predation rates, paying particular attention to their relative effect on species with varying life histories.

Urban Bird Community Composition

Several patterns in bird community composition in Florida were consistent with the hypothesis that nest predation is shaping urban communities. The reduced abundance of small-bodied birds in urban environments (Fig. 4.5) may at least partly reflect their vulnerability to avian nest predators. Some of the most successful urban species defend their nests very aggressively against nest predators. Mockingbirds, in particular, mob nest predators when they are still far from their nests (C. M. Stracey, pers. obs.). In addition, because of higher mockingbird abundance in urban habitats, urban mockingbirds may be more successful at deterring avian predators before they reach the nest due to group mobbing behavior (Robinson 1985, Wiklund and Anderson 1994, Picman et al.

2002). Indeed Breitwisch (1988) found that mockingbirds that mobbed more aggressively were also less likely to lose their nests to predators. Brown Thrashers and Loggerhead Shrikes were also extremely aggressive when predators or nest checkers approached the nest. If successful nest defense is necessary to offset increases in avian nest predators, then it is possible that nest predation could explain the pattern of increased average body size in urban areas.

Among small-bodied individuals that do thrive in urban environments, the vast majority nest in cavities or other enclosed places (Fig. 4.5). Even open-cup nesters, such as the House Finch, usually place their nests in enclosed places in human structures in urban environments. Thus, small-bodied birds can thrive in urban environments, but only if they place their nests in sites that are inaccessible to large-bodied avian nest predators. This same pattern also holds for forest fragments, which tend to retain their small-bodied cavity nesters but lose many or even most open-cup nesters (Whitcomb et al. 1981, Brown and Sullivan 2005).

An alternative explanation for the increased abundance of large-bodied birds in urban areas is what we term the interspecific dominance hypothesis. Larger species are better able to exploit point sources of food (Daily and Ehrlich 1994, Shelley et al. 2004, French and Smith 2005, MacNally and Timewell 2005) that are typically found in urbanized settings (e.g., bird feeders, garbage bins, fruiting ornamental shrubs and trees), in addition to being better competitors for limited nest sites (Ingold 1994). In edges of Australian urban forest fragments, the hyper-aggressive Noisy Miner (*Manorina melanocephala*) is correlated with a decrease in species richness of smaller-bodied birds (Piper and Catterall 2003). Thus, the interspecific dominance hypothesis is consistent with Brown and Sullivan's (2005) meta-analysis of the differential response of birds of varying body sizes to fragmentation. They found that large birds (excluding the largest species) were more abundant in small woodlots and that small species were more common in large tracts of forest. They attributed this pattern to the competitive dominance of large species. The interspecific dominance hypothesis expands Shochat's (2004) hypothesis that the predictability of food resources in time leads to changes in population structure. The interspecific dominance hypothesis and the urban nest predator hypothesis are not mutually exclusive and may even act synergistically to favor large-bodied species in urban areas.

Some of the differences we observed in abundance could be due to the structural suitability of the habitat or habitat selection. We do not evaluate these possibilities here, although we attempted to control for them on a gross level by restricting our comparisons to urban settings that most closely resemble non-urban counterparts (e.g., parking lots compared with non-urban open habitat and residential areas compared with forests). Future studies should examine the role of habitat structure explicitly.

Patterns of Nest Predation in Urban Habitats

For most species for which we were able to compare nesting success in urban and non-urban habitats, rates of nest predation were lower in urban habitats, although this difference was not always significant, likely due to small sample sizes (Table 4.1). Urban areas appear to offer a refuge from predation for many of these species, and their increased abundance and status as urban adapters may result from their high nesting success. An exception to this generalization is the cardinal, which was slightly more abundant in non-urban areas in spite of having higher nesting success in urban habitats (Fig. 4.4d; Table 4.1). Differences in nest predation rates for the cardinal tended to be less extreme than for the other species for which we have data. Leston and Rodewald (2006) found slightly higher rates of nest predation in urban forest fragments relative to non-urban habitats, but the differences were insufficient to cause a change in season-long productivity. Perhaps cardinals, which are abundant in both urban and non-urban habitats, may nest successfully in both settings. We do not have data on nesting success of Carolina Wrens, which show a pattern similar to cardinals in terms of their relative abundance in urban and non-urban areas (Fig. 4.4c). Unfortunately, Carolina Wrens hide their nests so well in non-urban settings that even our most skilled nest searchers were unable to find more than a few. Nesting data for Carolina Wrens would be particularly useful in further testing the link between nest predation, abundance of the species, and habitat.

Nest Predation and the Success of Introduced Species

Nest predation pressure may also help explain why introduced species in Florida tend to be restricted to urban areas (Fig. 4.3) and why some introduced species succeed where others fail to become established in the urban areas where they are usually released. To our knowledge, none of the open-cup nesting species that have been released in Florida have succeeded and spread. The Blue-gray Tanager (*Thraupis episcopus*), for example, became extinct after a brief period of apparent success, and the Red-whiskered Bulbul (*Pycnonotus jocosus*) has not spread beyond a restricted area in south Florida (Robertson and Woolfenden 1992, Stevenson and Anderson 1994). The birds that have succeeded are overwhelmingly enclosed nesters such as parrots, House Sparrows, and starlings. For at least one of these species, the House Finch, human structures such as parking garages provide nest sites that are essentially invulnerable to predation (Table 4.1). The House Finch is usually considered an open-cup nesting species; however, nearly all of the nests (55 of 60) that we found and monitored were located in spaces that were functionally cavities. For example, many nests were located in small (approximately 15 cm × 6.2 cm) openings in the concrete ceiling of parking garages and in narrow spaces above lights in breezeways. Of all the species we studied, the House Finch had the highest daily nest survival rate (99.68%). The availability of such safe nest sites has likely contributed as much to their successful invasion of eastern North America as their ability to exploit bird feeders. The lack of such enclosed nest sites in non-urban areas may explain why they continue to be mostly restricted to human settlements in their introduced range.

There are, however, two open-cup nesters that have been successful invasives, the Eurasian Collared-Dove (*Streptopelia decaocto*) and the Rock Pigeon (*Columba livia*). Both species are relatively large and defend their nests to some extent against predators. The Rock Pigeon nests on buildings in loose colonies where access by predators is limited, but the Eurasian Collared-Dove nests in places similar to the native Mourning Dove, including trees in parking lots and residential neighborhoods. Although the Eurasian Collared-Dove does lose nests to predators, their nesting success is comparable to urban Mourning Doves and is quite high compared to Mourning Doves nesting in rural areas. Thus, the relatively high nesting success of Eurasian Collared-Doves may be fueling their phenomenal population increase and range expansion.

Alternative Hypotheses

An alternative explanation for the reduction in nest predation observed in urban areas is the predator satiation hypothesis. According to the predator satiation hypothesis, the increase in abundance of avian predators does not increase nest predation rates because the avian predators are satiated by alternative food sources provided by humans in urban areas. An increase in food, however, often results in an increase in foragers, which typically leads to increased competition over that resource (Sol et al. 1998, Shochat 2004). While most avian nest predators are generalists and no doubt take advantage of alternative food sources, it is possible that incidental predation (*sensu* Vickery et al. 1992) would actually increase, especially over time, as generalists become more abundant; this hypothesis remains to be tested.

Another possible explanation for the reduction in nest predation rates is the suitability of the nest site. Nest sites may be inherently safer in urban settings as a result of differences in the structure of the nest site. Many of the nests in town were located in thick, dense shrubs that were maintained with constant trimming. Such sites might be inherently safer than those available in non-urban settings.

CONCLUSIONS

One of the goals of our study was to determine the mechanisms underlying the "urban nest predator paradox," which is the phenomenon of reduced rates of nest predation in urban areas in spite of conspicuously higher abundances of many nest predators, especially avian predators (Shochat 2004). Indeed, for species that are more abundant in urban habitats, we did find lower rates of nest predation in more urban areas. For at least this subset of species, urban areas may offer a refuge from nest predation, as hypothesized by Gering and Blair (1999). As we have argued above, however, this result may only hold for species that have effective defenses against the kinds of predators that thrive in urban areas. For those species that cannot defend themselves effectively against avian predators, such as most small-bodied open-cup nesters, their absence from urban areas may reflect an inability to nest successfully in the face of abundant flocks of crows and grackles. These open-cup nesters were too rare to study in urban areas, which prevented us from making any comparisons. Thus, there may not be an "urban nest predator paradox"; urban areas may only be a refuge for species that are less vulnerable to the kinds of predators that are also urban adapters. Conversely, the scarcity of many urban adapters in rural areas may reflect their vulnerability to predators such as snakes, which are likely more abundant in non-urban areas.

An additional test of the effects of nest predators on urban bird communities could come from studies of species nesting in small urban fragments (Leston and Rodewald 2006). These fragments may have the worst of both worlds—high populations of both avian nest predators and of other predators such as snakes, which may persist in small urban woodlots. The highest rates of nest predation on nests of the cardinal were indeed in urban fragments, but we lack data from small-bodied species that sometimes persist in these small fragments to make the comparison.

The generality of our results will depend to a large extent on how different predator communities in different geographic regions respond to urbanization. There is some evidence that avian nest predators thrive in urban habitats worldwide (Gregory and Marchant 1996, Jokimäki and Huhta 2000, Marzluff et al. 2001, Sorace 2002), but the responses of mammals to urbanization may vary much more among regions. Clearly, much work remains to be done in our efforts to understand the role of predation in structuring urban bird communities. It will be important to design experiments that distinguish between the above explanations of the predator paradox. In addition, it is necessary to experimentally assess the relationship between body size, nest defense, nesting guild, and nesting success in urban areas. Also, incorporating the interaction between food and predation, both of which differ between urban and non-urban areas, will be critical for advancing our understanding of urban ecology. One such interaction may be between food and nest defense. Increased food resources have been shown to lead to increased levels of nest defense (Komdeur and Kats 1999, Duncan Rastogi et al. 2006). Therefore, birds in urban areas, which are thought to have increased food resources, may show increased nest defense, resulting in reduced nest predation rates. Until we are able to understand the mechanistic role of predation in urban areas, we will be left with numerous, and often conflicting, case studies on predation rates.

ACKNOWLEDGMENTS

We would like to thank our field assistants, T. Richard, S. Daniels, A. Savage, and R. Sandidge. We would also like to thank the numerous homeowners who gave us permission to work on their property, as well as the Oaks Mall and Butler Plaza for allowing us to work in their parking lots. We would also like to thank M. Christman for statistical help with the logistic regression model and B. Bolker for statistical help with the census data, and D. Levey for comments on the manuscript. This research was funded in part by the Florida Ornithological Society.

LITERATURE CITED

Blair, R. B. 1996. Land use and avian species diversity along an urban gradient. Ecological Applications 6:506–519.

Borgmann, K. L., and A. D. Rodewald. 2004. Nest predation in an urbanizing landscape: the role of exotic shrubs. Ecological Applications 14:1757–1765.

Breitwisch, R. 1988. Sex differences in defense of eggs and nestlings by Northern Mockingbirds, *Mimus polyglottos*. Animal Behaviour 36:62–72.

Brown, W. P., and P. J. Sullivan. 2005. Avian community composition in isolated forest fragments: A conceptual revision. Oikos 111:1–8.

Burnham, K. P., and D. R. Anderson. 2002. Model selection and multimodel inferences: a practical information-theoretic approach. 2nd ed. Springer, New York.

Chalfoun, A. D., M. J. Ratnaswamy, and F. R. Thompson III. 2002. Songbird nest predators in forest-pasture edge and forest interior in a fragmented landscape. Ecological Applications 12:858–867.

Crooks, K. R. 2002. Relative sensitivities of mammalian carnivores to habitat fragmentation. Conservation Biology 16:488–502.

Crooks, K. R., and M. E. Soulé. 1999. Mesopredator release and avifaunal extinctions in a fragmented system. Nature 400:563–566.

Daily, G. C., and P. R. Ehrlich. 1994. Influence of social-status on individual foraging and community structure in a bird guild. Oecologia 100:153–165.

Duncan Rastogi, A., L. Zanette, and M. Clinchy. 2006. Food availability affects diurnal nest predation and adult antipredator behavior in Song Sparrows, *Melospiza melodia*. Animal Behaviour 72:933–940.

Dunning, J. B. 1993. CRC handbook of avian body masses. CRC Press, Boca Raton, FL.

Durner, G. M., and J. E. Gates. 1993. Spatial ecology of black rat snakes on Remington Farms, Maryland. Journal of Wildlife Management 57:812–826.

Ehrlich, P. R., D. Wheye, and D. Dobkin. 1988. Birder's handbook. Simon and Schuster, New York.

French, A. G., and T. B. Smith. 2005. Importance of body size in determining dominance hierarchies among diverse tropical frugivores. Biotropica 37:96–101.

Gering, J. C., and R. B. Blair. 1999. Predation on artificial bird nests along an urban gradient: predatory risk or relaxation in urban environments? Ecography 22:532–541.

Gregory, R. D., and J. H. Marchant. 1996. Population trends of jays, magpies, Jackdaws, and Carrion Crows in the United Kingdom. Bird Study 43:28–37.

Haskell, D. G., A. M. Knupp, and M. C. Schneider. 2001. Nest predator abundance and urbanization. Pp. 243–258 *in* J. M. Marzluff, R. Bowman, and R. Donnelly (editors), Avian ecology and conservation in an urbanizing world. Kluwer Academic Publishers, Boston, MA.

Heske, E. J., S. K. Robinson, and J. D. Brawn. 1999. Predator activity and predation on songbird nests on forest-edges in east-central Illinois. Landscape Ecology 14:345–354.

Ingold, D. J. 1994. Influence of nest-site competition between European Starlings and woodpeckers. Wilson Bulletin 106:227–241.

Jokimäki, J., and E. Huhta. 2000. Artificial nest predation and abundance of birds along an urban gradient. Condor 102:838–847.

Komdeur, J., and R. K. H. Kats. 1999. Predation risk affects trade-off between nest guarding and for aging in Seychelles warblers. Behavioral Ecology 10:648–658.

Leston, L. F. V., and A. D. Rodewald. 2006. Are urban forests ecological traps for understory birds? An examination using Northern Cardinals. Biological Conservation 131:566–574.

MacNally, R., and C. A. R. Timewell. 2005. Resource availability controls bird-assemblage composition through interspecific aggression. Auk 122:1097–1111.

Martin, T. E. 1993. Nest predation and nest sites, new perspectives on old patterns. Bioscience 43:523–532.

Marzluff, J. M., and E. Neatherlin. 2006. Corvid response to human settlements and campgrounds: causes, consequences, and challenges for conservation. Biological Conservation 130:301–314.

Marzluff, J. M., R. Bowman, and R. Donnelly. 2001. Avian ecology and conservation in an urbanizing world. Kluwer Academic Publishers, Boston, MA.

Melampy, M. N., E. L. Kershner, and M. A. Jones. 1999. Nest predation in suburban and rural woodlots of northern Ohio. American Midland Naturalist 141:284–292.

Patten, M. A., and D. T. Bolger. 2003. Variation in top-down control of avian reproductive success across a fragmentation gradient. Oikos 101:479–488.

Picman, J., S. Pribil, and A. Isabelle. 2002. Antipredation value of colonial nesting in Yellow-headed Blackbirds. Auk 119:461–472.

Piper, S. D., and C. P. Catterall. 2003. A particular case and a general pattern: hyperaggressive behaviour by one species may mediate avifaunal decreases in fragmented Australian forests. Oikos 101:602–614.

Prange, S., S. D. Gehrt, and E. P. Wiggers. 2003. Demographic factors contributing to high raccoon densities in urban landscapes. Journal of Wildlife Management 67:324–333.

Ralph, C. J., S. Droege, and J. R. Sauer. 1995. Managing and monitoring birds using point counts: standards and applications. Pp. 161–169 *in* C. J. Ralph, S. Droege, and J. R. Sauer (editors), Monitoring bird populations by point counts. USDA Forest Service General Technical Report PSW-GTR-149. USDA Forest Service, Pacific Southwest Research Station, Albany, CA.

Randa, L. A., and J. A. Yunger. 2006. Carnivore occurrence along an urban-rural gradient: a landscape-level analysis. Journal of Mammalogy 87: 1154–1164.

Robertson, W. B., and G. E. Woolfenden. 1992. Florida bird species: an annotated list. Florida Ornithological Society, Gainesville, FL.

Robinson, S. K. 1985. Coloniality in the Yellow-rumped Cacique as a defense against nest predators. Auk 102:506–519.

Robinson, S. K., F. R. Thompson III, T. M. Donovan, D. R. Whitehead, and J. Faaborg. 1995. Regional forest fragmentation and the nesting success of migratory birds. Science 267:1987–1990.

Schmidt, K. A., J. R. Goheen, R. Naumann, R. S. Ostfeld, E. M. Schauber, and A. Berkowitz. 2001. Experimental removal of strong and weak predators: mice and chipmunks preying on songbird nests. Ecology 82:2927–2936.

Schmidt, K. A., R. S. Ostfeld, and K. N. Smyth. 2006. Spatial heterogeneity in predator activity, nest survivorship, and nest-site selection in two forest thrushes. Oecologia 148:22–29.

Shaffer, T. L. 2004. A unified approach to analyzing nest success. Auk 121:26–540.

Shelley, E. L., M. Y. U. Tanaka, A. R. Ratnathicam, and D. T. Blumstein. 2004. Can Lanchester's laws help explain interspecific dominance in birds? Condor 106:395–400.

Shochat, E. 2004. Credit or debit? Resource input changes population dynamics of city-slicker birds. Oikos 106:622–626.

Shochat, E., S. B. Lerman, M. Katti, and D. B. Lewis. 2004. Linking optimal foraging behavior to bird community structure in an urban-desert landscape: field experiments with artificial food patches. American Naturalist 164:232–243.

Shochat, E., P. S. Warren, S. H. Faeth, N. E. McIntyre, and D. Hope. 2006. From patterns to emerging processes in mechanistic urban ecology. Trends in Ecology and Evolution 21:186–191.

Sol, D., D. M. Santos, J. Garcia, M. Cuadrado. 1998. Competition for food in urban pigeons: the cost of being juvenile. Condor 100:298–304.

Sorace, A. 2002. High density of bird and pest species in urban habitats and the role of predator abundance. Ornis Fennica 79:60–71.

Soulé, M. E., D. T. Bolger, A. C. Alberts, J. Wright, M. Sorice, and S. Hill. 1988. Reconstructed dynamics of rapid extinctions of chaparral-requiring birds in urban habitat islands. Conservation Biology 2:75–92.

Stevenson, H. M., and B. H. Anderson. 1994. The birdlife of Florida. University Press of Florida, Gainesville, FL.

Thorington, K. K., and R. Bowman. 2003. Predation rate on artificial nests increases with human housing density in suburban habitats. Ecography 26:188–196.

Vickery, P. D., M. L. Hunter, and J. V. Wells. 1992. Evidence of incidental nest predation and its effects on nests of threatened grassland birds. Oikos 63:281–288.

Weatherhead, P. J., and G. Blouin-Demers. 2004. Understanding avian nest predation: why ornithologists should study snakes. Journal of Avian Biology 35:185–190.

Weatherhead, P. J., and M. B. Charland. 1985. Habitat selection in an Ontario population of the snake, *Elaphe obsoleta*. Journal of Herpetology 19:12–19.

Weldon, A. J., and N. M. Haddad. 2005. The effects of patch shape on Indigo Buntings: evidence for an ecological trap. Ecology 86:1422–1431.

Whitcomb, R. F., C. S. Robbins, J. F. Lynch, B. L. Whitcomb, M. K. Klimkiewicz, and D. Bystrak. 1981. Effects of forest fragmentation on avifauna of the eastern deciduous forest. Pp. 125–205 *in* R. L. Burgess and D. M. Sharpe (editors), Forest island dynamics in man-dominated landscapes. Springer-Verlag, New York.

Wiklund, C. G., and M. Anderson. 1994. Natural selection of colony size in a passerine bird. Journal of Animal Ecology 63:765–774.

APPENDIX 4.1

Total number of individuals detected in each habitat during 100-m, fixed-radius point counts of each of 75 species along with their mass and nesting guild.

The number of points per habitat is indicated in parentheses.

Species	Mass (g)	Nest type	Parking lot (43)	Residential neighborhood (16)	Forest fragment (11)	Pasture (15)	Non-urban scrub (52)	Non-urban forest (48)	Grand total (185)
Laughing Gull	325	O	5	1	1	0	0	0	7
Cattle Egret	338	O	0	0	0	1	0	0	1
Eurasian Collared-Dove	149	O	67	13	6	2	0	0	88
Mourning Dove	119	O	33	26	4	6	14	2	85
White-winged Dove	153	O	0	0	0	0	1	0	1
Common Ground-Dove	30.1	O	0	0	0	0	6	0	6
Rock Pigeon	354.5	O	13	1	1	0	0	0	15
Turkey Vulture	1,467	O	0	0	0	0	1	0	1
Mississippi Kite	278	O	0	3	0	0	0	0	3
Wild Turkey	5,811	O	0	0	0	1	0	1	2
Northern Bobwhite	178	O	0	0	0	4	2	1	7
Cooper's Hawk	439	O	0	1	0	0	0	0	1

APPENDIX 4.1 *(continued)*

Species	Mass (g)	Nest type	Parking lot (43)	Residential neighborhood (16)	Forest fragment (11)	Pasture (15)	Non-urban scrub (52)	Non-urban forest (48)	Grand total (185)
Yellow-throated Warbler	10.02	O	0	1	0	0	0	0	1
Red-tailed Hawk	1,126	O	0	0	1	0	1	0	2
Red-shouldered Hawk	559	O	2	2	0	0	2	4	10
Osprey	1,485.5	O	1	1	2	0	0	0	4
Barred Owl	731	C	0	0	0	0	0	1	1
Burrowing Owl	155	C	0	0	0	1	0	0	1
Black-hooded Parakeet	110	C	2	1	0	0	0	0	3
Monk Parakeet	101	C	13	7	0	0	0	0	20
Yellow-billed Cuckoo	64	O	0	0	0	0	2	3	5
Hairy Woodpecker	66.3	C	0	0	0	0	2	0	2
Downy Woodpecker	27	C	14	5	0	5	18	20	62
Pileated Woodpecker	287	C	0	2	0	0	4	4	10
Red-headed Woodpecker	71.6	C	0	0	0	0	2	0	2
Red-bellied Woodpecker	61.7	C	19	7	14	10	24	27	101
Northern Flicker (yellow-shafted)	132	C	4	2	0	1	2	0	9
Common Nighthawk	61.5	O	0	0	0	0	1	0	1
Chimney Swift	23.6	C	1	0	0	0	0	0	1
Ruby-throated Hummingbird	3.15	O	0	1	1	0	1	0	3
Eastern Kingbird	43.6	O	0	0	0	3	0	0	3
Gray Kingbird	43.8	O	3	4	0	0	0	0	7
Great-crested Flycatcher	33.5	C	10	6	2	6	10	16	50
Eastern Wood-Pewee	14.1	O	0	0	0	0	0	1	1

Species	Mass (g)	Nest type	Parking lot (43)	Residential neighborhood (16)	Forest fragment (11)	Pasture (15)	Non-urban scrub (52)	Non-urban forest (48)	Grand total (185)
Acadian Flycatcher	12.9	O	0	0	0	0	0	8	8
Blue Jay	86.8	O	22	6	2	10	17	8	65
Florida Scrub-Jay	80.2	O	0	0	0	0	7	0	7
American Crow	448	O	12	3	0	6	4	4	29
Fish Crow	285	O	38	16	2	0	0	0	56
European Starling	82.3	O	26	6	3	0	0	0	35
Brown-headed Cowbird	43.9	N/A	2	1	0	2	8	12	25
Red-winged Blackbird	52.6	O	3	8	1	14	2	2	30
Eastern Meadowlark	89.0	O	0	0	0	16	1	0	17
Orchard Oriole	19.6	O	0	0	0	3	0	0	3
Common Grackle	113.5	O	35	12	4	0	5	0	56
Boat-tailed Grackle	166.5	O	23	15	0	1	0	0	39
House Finch	21.4	C	18	7	0	0	0	0	25
Bachman's Sparrow	19.7	O	0	0	0	0	4	0	4
Eastern Towhee	40.5	O	0	0	0	0	105	21	126
Northern Cardinal	44.7	O	24	12	11	15	70	78	210
Blue Grosbeak	34.4	O	0	1	0	5	5	0	11
Indigo Bunting	14.5	O	0	0	0	1	3	1	5
Summer Tanager	28.2	O	2	0	1	1	4	20	28
Purple Martin	49.4	C	0	0	0	0	1	0	1
Loggerhead Shrike	47.4	O	6	4	0	1	0	0	11
Red-eyed Vireo	16.7	O	0	0	1	0	1	27	29
Yellow-throated Vireo	18	O	2	0	1	1	2	6	12
White-eyed Vireo	11.4	O	0	0	3	3	39	32	77

APPENDIX 4.1 (*continued*)

Species	Mass (g)	Nest type	Parking lot (43)	Residential neighborhood (16)	Forest fragment (11)	Pasture (15)	Non-urban scrub (52)	Non-urban forest (48)	Grand total (185)
Northern Parula	8.6	O	1	2	2	2	6	36	49
Pine Warbler	11.9	O	3	1	0	0	25	11	40
Common Yellowthroat	10.1	O	0	0	0	3	39	7	49
Yellow-breasted Chat	25.3	O	0	0	0	0	1	0	1
Hooded Warbler	10.5	O	0	0	0	0	0	8	8
Painted Bunting	15.6	O	2	0	0	0	21	9	32
House Sparrow	27.7	C	74	9	5	0	0	0	88
Northern Mockingbird	48.5	O	93	33	5	14	30	1	176
Brown Thrasher	68.8	O	1	2	1	6	13	3	26
Carolina Wren	18.7	C	19	10	14	14	33	89	179
Brown-headed Nuthatch	10.2	C	0	0	0	0	3	0	3
Eastern Tufted Titmouse	21.6	C	9	6	3	5	12	35	70
Carolina Chickadee	10.2	C	5	3	0	0	6	12	26
Blue-gray Gnatcatcher	6	O	2	0	0	1	14	15	32
Eastern Bluebird	31.6	C	1	0	0	3	7	0	11
Grand total			610	241	91	167	581	525	2215

NOTE: Body mass taken from Dunning (1993) and nesting guild from Ehrlich et al. (1988). Nest: O = open cup, C = cavity nest.

CHAPTER FIVE

Evaluating Factors That Influence Avian Community Response to Urbanization

Amanda D. Rodewald

Abstract. Although numerous studies have documented shifts in avian community structure in urbanizing landscapes, the ecological mechanisms that prompt these changes are poorly understood. I evaluated evidence for five alternate hypotheses to explain the commonly reported negative responses of many neotropical migratory species and positive responses of many residents and temperate migrants to urbanization. Specifically, I examined the extent to which local vegetation, winter microclimate, supplemental food resources, predators and brood parasites, and heterospecific attraction were related to abundances of seven species of urban adapters and six species of urban avoiders as well as the productivity of two focal species. From 2001 to 2004, birds and nest predators were surveyed and habitat and microclimate characteristics were measured at 33 riparian forest sites located along a rural-urban gradient in central Ohio. Annual productivity was estimated for 163 pairs of Acadian Flycatcher (*Empidonax virescens*) from 2001 to 2006 and for 294 pairs of Northern Cardinal (*Cardinalis cardinalis*) from 2003 to 2006. An information-theoretic approach showed that the two hypotheses receiving the most support in explaining abundance of urban avoiders were heterospecific attraction and dominance of understory by Amur honeysuckle (*Lonicera maackii*). In contrast, variation in abundances of urban adapters was explained by local vegetation, winter microclimate, supplemental food, and heterospecific attraction, with all hypotheses receiving roughly equivalent support. For Acadian Flycatcher, productivity was best explained by and negatively related to both abundance of potential nest predators and dominance by honeysuckle. All models were similarly ranked in terms of ability to explain variation in productivity of Northern Cardinals. Results show that species-specific responses to urbanization are complex and may not be amenable to broad generalizations. Interestingly, findings suggest that distribution of certain sensitive species may be partly a consequence of heterospecific attraction, specifically the presence of other insectivores settling within the landscapes, as well as the extent to which habitats are dominated by exotic plants. Future research that experimentally manipulates these potential causative factors would greatly advance our understanding of the mechanisms that govern avian responses to urbanization.

Key Words: exotic plants, feeders, heterospecific attraction, honeysuckle, mechanisms, predators, temperature, urban.

Rodewald, A. D. 2012. Evaluating factors that influence avian community response to urbanization. Pp. 71–92 *in* C. A. Lepczyk and P. S. Warren (editors). Urban bird ecology and conservation. Studies in Avian Biology (no. 45), University of California Press, Berkeley, CA.

Research in urban systems has consistently demonstrated that the distribution and abundance of birds can be strongly affected by urban development at multiple spatial scales (Beissinger and Osborne 1982, Mills et al. 1989, Blair 1996, Germaine et al. 1998, Marzluff et al. 2001, Pickett et al. 2001). Still, there remain several conspicuously large gaps in our understanding of avian ecology in urban areas, particularly with respect to empirical evidence of the ecological mechanisms responsible for relationships between urbanization and avian communities (Faeth et al. 2005, Shochat et al. 2006). Urban development can profoundly affect bird communities in remnant forest systems by modifying local habitat characteristics (Rottenborn 1999, Hennings and Edge 2003, Miller et al. 2003, Borgmann and Rodewald 2005), microclimates (Wachob 1996, Botkin and Beveridge 1997, Atchison and Rodewald 2006, Leston and Rodewald 2006), food resources (Wilson 1994, Atchison and Rodewald 2006, Leston and Rodewald 2006), and communities of predators and brood parasites (Haskell et al. 2001, Odell and Knight 2001, Sorace 2002, Sinclair et al. 2005, Smith and Wachob 2006, Stracey and Robinson, chapter 4, this volume). Although these various factors are recognized as potential causative agents, their roles in prompting urban-associated changes in bird communities are not well studied. Avian ecologists have a particularly poor understanding of the relative importance of resource-based changes versus altered biotic interactions. Understanding the local ecological mechanisms that generate landscape patterns in urban systems is critical to both predict and ameliorate impacts of large-scale urbanization.

In this paper, I evaluate evidence for five alternate hypotheses to explain commonly reported changes in bird communities in urbanizing landscapes patterns. I focus on avian communities of riparian forests in the urbanizing landscapes of Ohio. Rodewald and Bakermans (2006) showed that the amount of urban development within the landscape matrix surrounding these riparian forests was strongly associated with avian community structure. In particular, they found that most neotropical migratory species were negatively related to the amount of urbanization within landscapes, whereas many residents and temperate migrants were positively associated with urbanization. Taking their work one step further, I examined the extent to which the following hypotheses explained patterns in avian community structure and productivity of two focal species:

1. Vegetation. Urban development prompts changes in the structure and floristic composition of vegetation, which influences species according to their autecological requirements.
2. Winter microclimate. Because urban systems can act as heat islands, producing warmer microclimates, urbanization improves conditions for resident and short-distance migratory species and thereby promotes higher densities of their populations.
3. Supplemental food. Artificial food resources provided by bird feeders in urban systems support greater numbers of omnivorous species.
4. Predators and brood parasites. Interspecific interactions in urban systems are altered due to changes in abundance or behavior of key predators and brood parasites, which exert strong influence on certain sensitive species.
5. Heterospecific attraction. Species use the presence of individuals with similar feeding strategies to indicate habitat quality, which can magnify patterns of urban avoidance or attraction.

METHODS

Bird communities were studied on 33 mature riparian forest sites within and surrounding Columbus, Ohio, USA (Rodewald and Bakermans 2006; Fig. 5.1). Study sites were located in mature riparian forests along rivers, and mean forest width was 164 m ± 90 SD (range = 69–565 m; Table 5.1). Riparian forests in the study area had relatively high connectivity along rivers and occurred within landscapes dominated by varying amounts of agricultural (e.g., row crop, pasture, fallow fields) and urban (e.g., roads, parking lots, buildings) disturbances. Common trees included sycamore (*Plantanus occidentalis*), boxelder (*Acer negundo*), sugar maple (*A. saccharum*), black walnut (*Juglans nigra*), white ash (*Fraxinus americana*), cottonwood (*Populus deltoides*), and American hackberry (*Celtis occidentalis*). Common

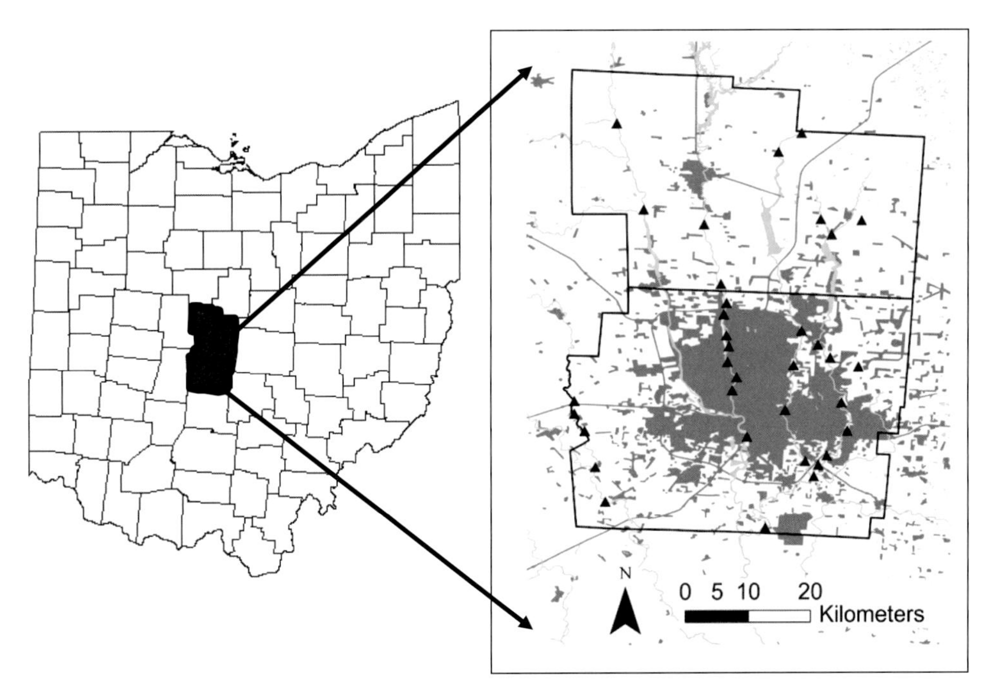

Figure 5.1. Location of study sites (black triangles) in central Ohio. Dark gray tones indicate urban land. Figure modified from Shustack and Rodewald (2010).

woody understory plants included Amur honeysuckle (*Lonicera maackii*), common spicebush (*Lindera benzoin*), tall paw paw (*Asimina triloba*), Ohio buckeye (*Aesculus octandra*), and saplings.

Landscape Analysis

A 1-km-radius area centered on each study site was used for landscape delineation because this scale is strongly associated with bird communities (Tewksbury et al. 1998, Saab 1999; Rodewald and Yahner 2001, Rodewald and Bakermans 2006), is commonly used in and easily applied to conservation efforts, and far exceeds average territory size of most common breeders at study sites. Landscape composition (percent built land, agricultural land, forest, shrubland, wetland, and open water) was quantified using recent (2002–2004) digital orthophotos (Table 5.1). Numbers of buildings were derived from county data and subsequently verified by visually inspecting orthophotos. To reduce the number of landscape variables used in analyses, a principal components analysis was performed on disturbance metrics. The first principal component explained nearly 80% of the variation among sites, and the first factor, hereafter termed the "urban index," loaded strongly positively (>0.9) for number of buildings, percent cover by roads, pavement, and lawn, but loaded negatively for percent cover by agriculture (−0.8). Because the urban index was not correlated with width of riparian forest tract ($r = -0.12$, $P = 0.52$, $n = 33$), the amount of forest habitat available to breeding birds at sites was not confounded with urbanization.

Avian Surveys

Bird and predator communities were surveyed at each forest site three times each June in 2001–2004 between 0545 and 1030 on days without rain or strong wind. Birds and potential nest predators were sampled along a 40-m-wide × 250-m-long belt transect located within the forest and adjacent to the river's edge. A trained observer slowly walked (ca. 10–15 m/min) along the mid-line of transects (i.e., 20 m from the river) and recorded all birds, mammals, and snakes seen or heard. Because transect width was narrow and habitat structure was similar among sites, differences in detection probability were not initially expected across sites. Indeed Rodewald and Bakermans (2006) found that average distance at which an observer detected a bird was not significantly

TABLE 5.1

Landscape composition surrounding 33 riparian forest study sites in Ohio.
The urban index is the first factor from a principal components analysis of disturbance metrics.

Site	Forest width (m)	Proportion of 1-km radius area covered by: Forest	Agriculture	Mowed	Paved	Road	Number of buildings	Urban index
bexley	133	0.14	0	0.50	0.14	0.08	1,692	1.23
bigwal	115	0.17	0	0.45	0.16	0.08	2,233	1.31
campmary	565	0.46	0	0.34	0.07	0.04	681	0.21
casto	202	0.19	0	0.42	0.20	0.08	1,776	1.25
chapman	87	0.54	0.24	0.16	0.02	0.02	153	−0.87
cherry	165	0.22	0.02	0.36	0.16	0.07	997	0.76
creeks	133	0.53	0.10	0.10	0.04	0.02	92	−0.71
elkrun	167	0.22	0.31	0.27	0.06	0.05	812	−0.16
galena	277	0.35	0.15	0.22	0.04	0.02	360	−0.48
gardner	125	0.55	0.20	0.13	0.02	0.01	248	−0.87
girlcamp	200	0.50	0.23	0.15	0.02	0.01	377	−0.82
heisel	144	0.35	0.06	0.27	0.17	0.06	603	0.36
highbank	235	0.48	0.06	0.28	0.04	0.03	262	−0.30
innis	69	0.32	0.02	0.46	0.06	0.03	959	0.33
kilbourn	106	0.46	0.30	0.16	0.02	0.02	202	−0.87
klondike	88	0.25	0.53	0.12	0.03	0.02	199	−1.16
lock	256	0.24	0.39	0.12	0.02	0.01	333	−1.05
lou	156	0.11	0.00	0.28	0.23	0.08	2,272	1.26
ngalena	135	0.54	0.36	0.05	0.01	0.01	34	−1.27
olentan	102	0.20	0.00	0.50	0.15	0.12	1,373	1.46
osuwet	87	0.11	0.00	0.35	0.29	0.09	2,886	1.75
prairie	148	0.29	0.47	0.12	0.03	0.02	58	−1.12
prindle	158	0.11	0.81	0.03	0.01	0.01	57	−1.73
pubhunt	194	0.50	0.32	0.08	0.01	0.01	210	−1.15
rocky	150	0.56	0.17	0.22	0.02	0.01	266	−0.68
rushrun	150	0.32	0.00	0.41	0.09	0.06	1,611	0.75
sgalena	163	0.43	0.14	0.30	0.02	0.01	185	−0.57
smith	144	0.10	0.35	0.33	0.09	0.03	729	−0.28
sunbury	129	0.29	0.30	0.26	0.05	0.03	733	−0.42
tnc	292	0.38	0.41	0.11	0.03	0.03	340	−0.96
whetston	154	0.20	0.00	0.46	0.16	0.08	2,017	1.31
whitehall	106	0.18	0.00	0.41	0.20	0.05	545	0.68
woodside	104	0.25	0.11	0.40	0.07	0.05	1,227	0.32

correlated with either forest width ($r = -0.13$, $P = 0.52$) or percent urban within 1 km ($r = -0.17$, $P = 0.40$), suggesting that there was no systematic detection bias.

Vegetation Measurements

Composition and structure of vegetation were measured at four systematically located 11.3-m-radius plots in July–August 2001 and again in 2004. At each plot, field teams estimated canopy height and then counted the number of small [diameter breast height (dbh = 8–23 cm), medium-sized (dbh = 23.1–38 cm), and large trees (dbh > 38 cm) identified to species, as well as the amount of woody debris (i.e., numbers of logs and stumps) and numbers of snags (>12 cm dbh). Canopy cover and ground cover were estimated using ocular tubes at 20 points located at 2-m intervals along two perpendicular 20-m-long transects. To describe the vertical structure of the stands, we recorded the number of contacts made by woody stems within 0.5-m height intervals on a 3-m-tall vegetation pole at 2-m intervals along transects. The extent to which honeysuckle dominated plots was described by calculating the proportion of plots where it was within the top three most abundant woody plants in the understory. Measurements were averaged over plots within each site and over the two years.

Non-breeding Season Variables

Winter temperatures were measured in December–February 2001–2002 and 2002–2003 as part of a complementary study of wintering bird communities (Atchison and Rodewald 2006). During monthly avian surveys (three visits per year) that were conducted between 08:00 and 12:00 EST on days without strong wind or precipitation, ambient temperature was recorded (° C thermometer). Prior analysis showed that ambient temperature was strongly correlated with both minimum ($r = 0.67$, $P < 0.001$, $n = 57$) and maximum ($r = 0.63$, $P < 0.001$, $n = 57$) temperatures during the entire month of January 2002 and 2003 (Atchison and Rodewald 2006). In February 2002, numbers of bird feeders <300 m of transects (15 ha area) were counted. This distance was chosen using the winter flock territory range of the chickadee as a guide, which ranges from 9.5 ha (Glase 1973) to 14.6 ha (Odum 1942).

Nest Monitoring and Productivity

Annual productivity (i.e., the number of young fledged over the entire breeding season) of Acadian Flycatchers was studied from 2001 to 2006 ($n = 163$ pairs at 18 sites) and of Northern Cardinals from 2003 to 2006 ($n = 294$ pairs at 14 sites). To facilitate determining the cumulative number of fledglings produced over all nest attempts for breeding pairs, birds were individually marked with a USGS aluminum band and a unique combination of color bands. Study sites were visited at 1–4-day intervals from late March through September for cardinals ($n = 14$ sites) and May through August for flycatchers ($n = 18$ sites). Pairs were observed to assess the breeding activity and to monitor nests. Once located, nests were checked at 1–3-day intervals to determine status (i.e., abandoned, failed, or active). Numbers of young were determined by counting the number of nestlings immediately prior to fledging and/or by observing parents and young for extended periods within 1–2 days of fledging. Following either failure or success of a given nesting attempt, field teams continued to observe the breeding pair in order to determine if and when they re-nested. Individual cardinals at sites generally made 1–5 nesting attempts each breeding season, whereas flycatchers made 1–3 nesting attempts. For each breeding pair, all nesting records over the season were compiled to determine the total number of young fledged during the season.

Statistical Analyses

My basic approach was to first establish the pattern of association between urbanization and variables describing habitat, microclimate, food, predators, brood parasites, and abundance of heterospecifics to identify which variables had the potential to guide avian community structure. A canonical correlation analysis (CCA) was used to determine the strength of association between urbanization and local ecological variables. CCA is a multivariate technique used to describe the relationship between multiple sets of variables and identify the underlying gradients of variation and degree to which they are correlated (McGarigal et al. 2000). The following 17 variables were examined: canopy height, number of snags, numbers of trees in three size classes (dbh 8–23 cm, 23–38 cm, and >38 cm), percent ground cover by bare soil, tree

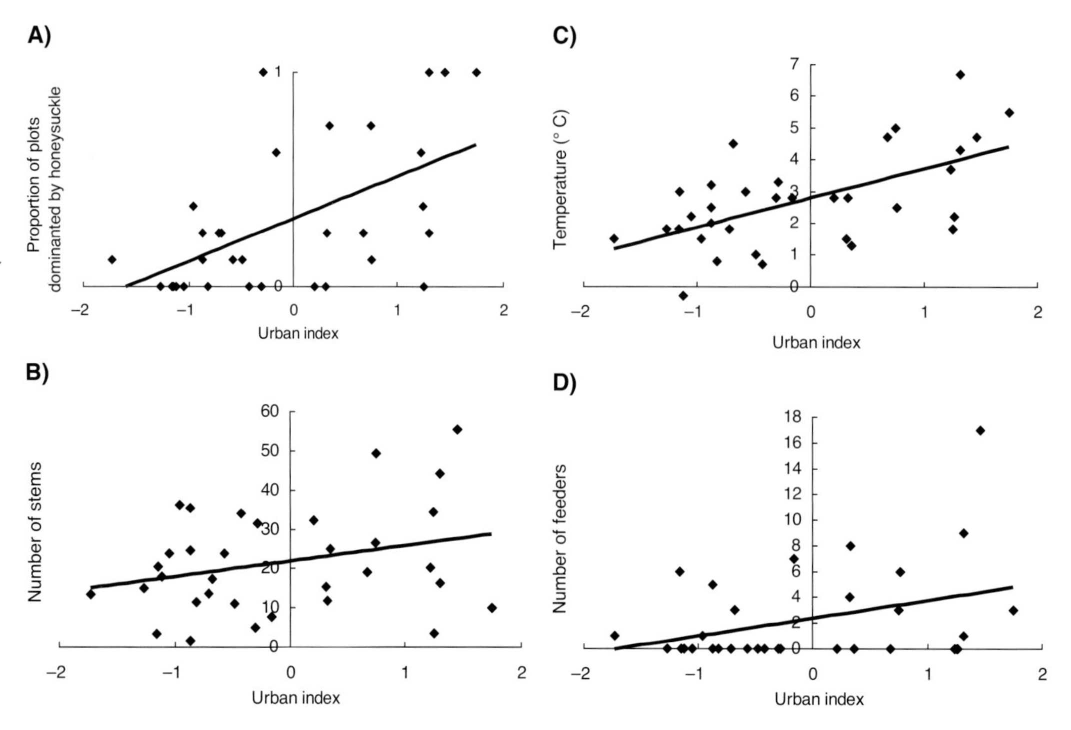

Figure 5.2. Relationship between urbanization in landscapes surrounding 33 riparian forests and (a) proportion of plots dominated by *Lonicera,* (b) understory stem density, (c) ambient winter temperature, and (d) number of bird feeders.

species richness, numbers of understory stems (≤3 m in height), percentage of plots dominated by honeysuckle in the understory, numbers of feeders, winter temperature, avian predators, mammalian predators, Brown-headed Cowbirds (*Molothrus ater*), abundance of omnivorous birds, and abundance of insectivorous birds. Local variables were strongly correlated with urbanization (Wilks' lambda $F_{16,16} = 5.02$, $P = 0.001$). Specifically, dominance by honeysuckle ($F_{1,31} = 14.49$, $P < 0.001$, $r^2 = 0.32$), numbers of feeders ($F_{1,31} = 4.54$, $P = 0.041$, $r^2 = 0.13$), winter temperature ($F_{1,31} = 16.77$, $P < 0.001$, $r^2 = 0.35$), numbers of avian predators ($F_{1,31} = 9.97$, $P = 0.004$, $r^2 = 0.24$), numbers of mammalian predators ($F_{1,31} = 16.44$, $P < 0.001$, $r^2 = 0.35$), and abundances of omnivores ($F_{1,31} = 24.42$, $P < 0.001$, $r^2 = 0.44$) increased with increasing urban development (Figs. 5.2 and 5.3), whereas numbers of insectivores ($F_{1,31} = 15.21$, $P = 0.001$, $r^2 = 0.33$) were negatively related to urbanization (Fig. 5.2). Number of understory stems also was marginally positively related to urbanization ($F_{1,31} = 2.85$, $P = 0.101$, $r^2 = 0.08$) and was included in subsequent analyses. The following variables were not significantly associated with urbanization: canopy height ($F_{1,31} = 0.03$, $P = 0.86$), number of snags ($F_{1,31} = 0.21$, $P = 0.65$), percent ground cover by soil ($F_{1,31} = 0.00$, $P = 0.97$), tree species richness ($F_{1,31} = 0.51$, $P = 0.48$), numbers of small trees ($F_{1,31} = 0.93$, $P = 0.34$), medium trees ($F_{1,31} = 0.12$, $P = 0.73$), and large trees ($F_{1,31} = 0.45$, $P = 0.51$), and numbers of Brown-headed Cowbirds ($F_{1,31} = 0.14$, $P = 0.71$).

An information-theoretic approach was used to rank the relative importance of potential local ecological factors that might produce observed patterns in avian abundance among study sites. An information-theoretic approach allows the support for multiple hypotheses to be compared by ranking their representative models (Burnham and Anderson 1998). I used a generalized linear model with an appropriate specified distribution (e.g., negative binomial, normal) to calculate log-likelihood estimates for each model (PROC GENMOD, SAS Institute, Cary, NC, 1990) and then calculated Akaike's Information Criterion corrected for bias due to small sample size (AIC_c). The model with lowest AIC_c value was considered best, and the difference between this best model and all others was represented by ΔAIC_c (the difference between the AIC_c of a given model and the best

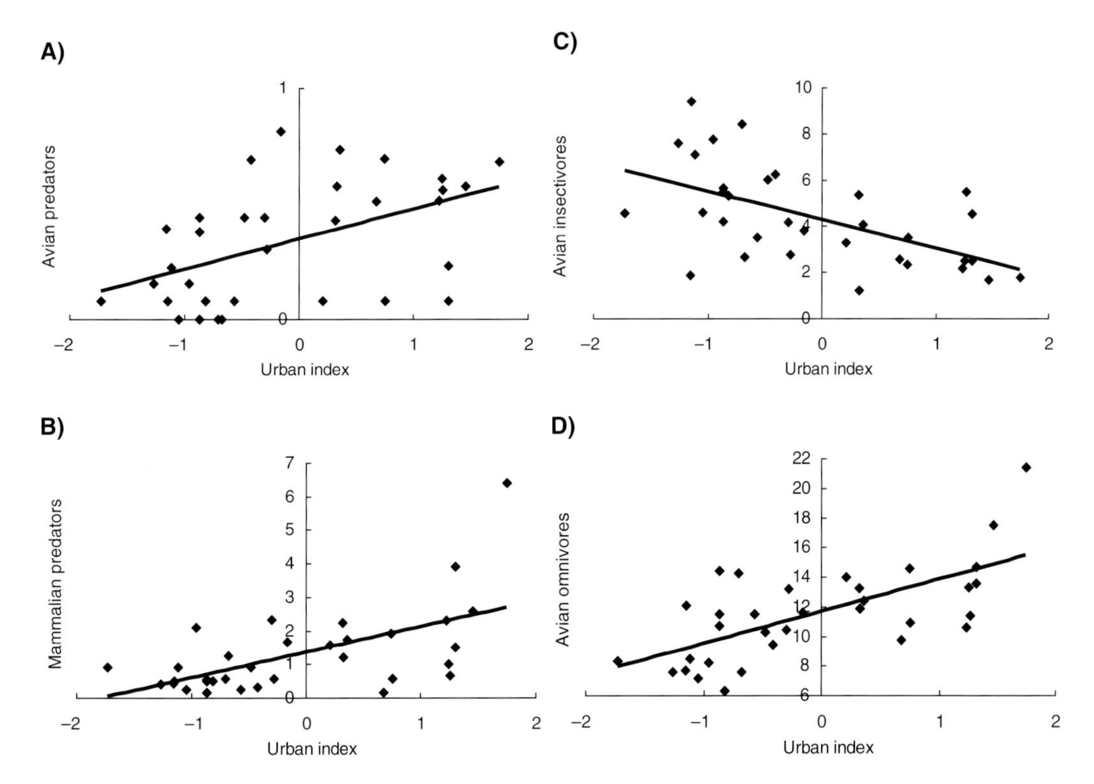

Figure 5.3. Relationship between urbanization in landscapes surrounding 33 riparian forests and relative abundance of (a) avian predators, (b) mammalian predators, (c) avian insectivores, and (d) avian omnivores detected during surveys along 40 × 250-m transects, 2001–2004.

model). Models with a $\Delta_i < 2$ were considered equally plausible given the data (Burnham and Anderson 1998). Akaike weights (w_i, weight of evidence for each model) indicated the relative support for each model and represented the likelihood that any given model was the true best model.

Based on Rodewald and Bakermans (2006), the following focal species were negatively related to urbanization and therefore were considered to be urban avoiders in analyses: Acadian Flycatcher (*Empidonax virescens*), Eastern Wood-Pewee (*Contopus virens*), Red-eyed Vireo (*Vireo olivaceus*), Blue-gray Gnatcatcher (*Polioptila caerulea*), Northern Parula (*Setophaga americana*), and Indigo Bunting (*Passerina cyanea*). The following species were positively related to urbanization and were considered to be urban adapters: Downy Woodpecker (*Picoides pubescens*), Red-bellied Woodpecker (*Melanerpes carolinus*), Carolina Chickadee (*Poecile carolinensis*), American Robin (*Turdus migratorius*), Cedar Waxwing (*Bombycilla cedrorum*), Northern Cardinal (*Cardinalis cardinalis*), and American Goldfinch (*Carduelis tristis*).

I used abundance averaged over all visits and years per site (2001–2004) as the response variable and considered study site as the replicate. Only variables showing a significant association with urbanization were presumed to be potential drivers of urban-associated changes in bird communities, and therefore models were restricted to include only these variables. Numbers of avian and mammalian predators were combined to simplify analysis both because they were strongly positively correlated and because I had no *a priori* reason for expecting one predator group to be more or less important than another. For each species, I compared only univariate models because I was interested in identifying the relative importance of various ecological factors rather than building a model *per se*. Because numbers of feeders and winter temperatures were not expected to affect distribution of insectivorous neotropical migrants, they were not included in models for this group of species. The following models were considered for urban avoiders:

1. number of understory stems
2. dominance by honeysuckle

3. numbers of predators
4. abundance of insectivores minus the abundance of the focal species (e.g., for the Acadian Flycatcher, numbers of insectivores does not include numbers of Acadian Flycatchers)

The following models were considered for urban adapters:

1. understory stems
2. dominance by honeysuckle
3. winter temperatures
4. supplemental food (number of birdfeeders within 300 m)
5. abundance of omnivores minus the abundance of the focal species

Number of potential predators was not used in models for urban adapters given that there was no ecological rationale for expecting birds to respond positively to numbers of predators, which were strongly positively associated with urbanization.

For two focal species, the Northern Cardinal and Acadian Flycatcher, I also examined the concordance between ecological factors that best explained abundance and those that best explained variation in productivity among sites. A similar information-theoretic approach was used as was applied to the abundance data.

RESULTS

Abundance of Urban Adapters

Local vegetation conditions were included within the top model set for three of seven urban adapters (American Goldfinch, American Robin, and Cedar Waxwing; Table 5.2), and in all cases abundance increased as the understory vegetation became more dense and dominated by honeysuckle. Supplemental food was included within the top models for two of seven species (American Goldfinch and Carolina Chickadee), with both showing positive associations with numbers of feeders (Table 5.2). Warmer winter temperatures were correlated with numbers of Cedar Waxwing and Northern Cardinal. Abundance of other omnivores was the top model for the two woodpecker species (Red-bellied Woodpecker and Downy Woodpecker; Table 5.2).

Abundance of Urban Avoiders

Number of other insectivores was most frequently included in the top model set explaining abundance of urban avoiders. Numbers of five of six species (Acadian Flycatcher, Eastern Wood-Pewee, Indigo Bunting, Northern Parula, and Red-eyed Vireo) were positively related to insectivore abundance (minus their own abundance; Fig. 5.4; Table 5.3). Even for Blue-gray Gnatcatcher, which had dominance by honeysuckle as the top model, its abundance was still positively related to numbers of insectivores (Fig. 5.4). Proportion of plots dominated by honeysuckle was in the top model set for three of six species (Blue-gray Gnatcatcher, Eastern Wood-Pewee, and Indigo Bunting), and in all cases abundances were negatively related to honeysuckle dominance.

Productivity

Productivity of the two model species was highly variable among sites. Average numbers of fledglings produced by an Acadian Flycatcher pair over a breeding season ranged from 0 to 2.5 per site. Despite abundance of other insectivores being the most important explanatory variable for abundances of Acadian Flycatchers, the heterospecific attraction hypothesis received little support for its ability to explain differences in productivity among sites. Instead, the abundance of potential nest predators was the most highly ranked model ($\Delta AIC = 0.00$, $w_i = 0.57$), followed closely by dominance of understory by honeysuckle ($\Delta AIC = 0.75$, $w_i = 0.39$; Table 5.4). Annual productivity of flycatchers was strongly negatively associated with both abundance of predators (β estimate $= -0.51 \pm 0.16$ SE) and honeysuckle (β estimate $= -1.708 \pm 0.550$ SE; Fig. 5.5). Average number of fledglings produced by cardinals over a season ranged from 1.42 to 2.8 per site. Productivity of Northern Cardinals, in contrast, was not particularly well explained by any single variable, and all univariate models were closely ranked ($\Delta AIC \leq 1$) and with low Akaike weights ($w_i < 0.21$; Table 5.4).

Correlations among Explanatory Variables

As expected, several explanatory variables were moderately correlated. The extent to which a site was dominated by *Lonicera* proved to be most

TABLE 5.2

Comparison of a priori *models explaining variation in abundance of urban adapters among 33 riparian forest sites in Ohio, 2001–2004.*

Bold text indicates the top-ranked model.

	Log-likelihood	K	AIC_c	ΔAIC_c	w_i	Estimate	SE	Lower 95% CI	Upper 95% CI
American Goldfinch									
Understory stems	**−20.37**	**3**	**48.24**	**1.85**	**0.14**	**0.023**	**0.016**	**−0.008**	**0.053**
Dominance by *Lonicera*	−20.60	3	48.69	2.30	0.11	0.879	0.557	−0.212	1.970
Winter temperature	−20.47	3	48.44	2.05	0.13	0.203	0.155	−0.100	0.506
Birdfeeder numbers	**−19.45**	**3**	**46.39**	**0.00**	**0.36**	**0.080**	**0.020**	**−0.041**	**0.119**
Omnivore abundance	−20.86	3	49.23	2.84	0.09	0.078	0.054	−0.028	0.183
Stems + *Lonicera*	−19.75	4	50.16	3.77	0.05				
All habitat[a]	−19.00	6	56.46	10.07	0.00				
Full model	−18.78	7	60.90	14.51	0.00				
Null model	−21.93	2	48.57	2.18	0.12				
American Robin									
Understory stems	−49.13	3	105.76	9.66	0.01	0.021	0.014	−0.007	0.049
Dominance by *Lonicera*	**−44.30**	**3**	**96.10**	**0.00**	**0.79**	**1.800**	**0.479**	**0.862**	**2.738**
Winter temperature	−49.51	3	106.53	10.43	0.00	0.147	0.125	−0.099	0.393
Birdfeeder numbers	−49.41	3	106.31	10.21	0.00	0.064	0.050	−0.035	0.163
Omnivore abundance	−49.17	3	105.84	9.74	0.01	0.101	0.070	−0.036	0.238
Stems + *Lonicera*	−44.24	4	99.14	3.04	0.17				
All habitat[a]	−43.39	6	105.23	9.13	0.01				
Full model	−43.34	7	110.02	13.92	0.00				
Null model	−50.19	2	105.08	8.98	0.01				

TABLE 5.2 *(continued)*

TABLE 5.2 (CONTINUED)

	Log-likelihood	K	AIC_c	ΔAIC_c	w_i	Estimate	SE	Lower 95% CI	Upper 95% CI
			Carolina Chickadee						
Understory stems	−30.01	3	67.53	4.55	0.05	0.011	0.008	−0.005	0.027
Dominance by *Lonicera*	−30.67	3	68.85	5.87	0.03	0.241	0.317	−0.380	0.862
Winter temperature	**−28.51**	**3**	**64.52**	**1.54**	**0.24**	0.153	0.066	0.023	0.283
Birdfeeder numbers	**−27.74**	**3**	**62.98**	**0.00**	**0.52**	**0.070**	**0.026**	**0.019**	**0.121**
Omnivore abundance	−30.35	3	68.21	5.23	0.04	0.040	0.036	−0.030	0.110
Stems + *Lonicera*	−29.97	4	70.61	7.63	0.01				
All habitat	−25.34	6	69.14	6.16	0.02				
Full model	−25.08	7	73.50	10.52	0.00				
Null model	−30.96	2	66.63	3.65	0.08				
			Cedar Waxwing						
Understory stems	−23.64	3	54.78	6.90	0.01	0.011	0.016	−0.021	0.043
Dominance by *Lonicera*	**−20.19**	**3**	**47.88**	**0.00**	**0.46**	**1.257**	**0.481**	**0.314**	**2.200**
Winter temperature	**−21.02**	**3**	**49.53**	**1.65**	**0.20**	**0.260**	**0.089**	**0.085**	**0.435**
Birdfeeder numbers	−22.33	3	52.17	4.29	0.05	0.072	0.066	−0.057	0.200
Omnivore abundance	−21.33	3	50.15	2.27	0.15	0.122	0.040	0.043	0.200
Stems + *Lonicera*	−20.32	4	51.30	3.42	0.08				
All habitat[a]	−19.51	6	57.47	9.59	0.00				
Full model	−19.29	7	61.92	14.04	0.00				
Null model	−23.97	2	52.65	4.77	0.04				
			Downy Woodpecker						
Understory stems	−12.43	3	32.35	17.33	0.00	0.000	0.005	−0.009	0.009
Dominance by *Lonicera*	−10.28	3	28.05	13.03	0.00	0.366	0.171	0.032	0.700
Winter temperature	−9.67	3	26.83	11.81	0.00	0.092	0.038	0.019	0.165

Birdfeeder numbers	−11.37	3	30.24	15.22	0.00	0.024	0.016	−0.008	0.055
Omnivore abundance	**−3.76**	**3**	**15.02**	**0.00**	**0.99**	**0.078**	**0.016**	**0.046**	**0.110**
Stems + *Lonicera*	−9.88	4	30.43	15.41	0.00				
All habitat[a]	−8.71	6	35.87	20.85	0.00				
Full model	−2.89	7	29.12	14.10	0.00				
Null model	−12.43	2	29.56	14.54	0.00				
			Northern Cardinal						
Understory stems	−41.16	3	89.81	8.96	0.01	0.020	0.011	−0.002	0.042
Dominance by *Lonicera*	−39.95	3	87.39	6.54	0.03	1.017	0.419	0.195	1.839
Winter temperature	**−36.67**	**3**	**80.85**	**0.00**	**0.81**	**0.323**	**0.085**	**0.156**	**0.489**
Birdfeeder numbers	−38.76	3	85.03	4.18	0.10	0.108	0.037	0.037	0.180
Omnivore abundance	−42.02	3	91.54	10.69	0.00	0.060	0.053	−0.044	0.164
Stems + *Lonicera*	−39.42	4	89.51	8.66	0.01				
All habitat[a]	−34.82	6	88.10	7.25	0.02				
Full model	−34.74	7	92.82	11.97	0.00				
Null model	−42.65	2	90.01	9.16	0.01				
			Red-bellied Woodpecker						
Understory stems	6.83	3	−6.16	6.85	0.03	0.002	0.003	−0.003	0.008
Dominance by *Lonicera*	6.57	3	−5.63	7.38	0.02	0.054	0.102	−0.147	0.254
Winter temperature	6.55	3	−5.60	7.41	0.02	0.011	0.023	−0.034	0.056
Birdfeeder numbers	6.55	3	−5.61	7.40	0.02	−0.005	0.009	−0.023	0.014
Omnivore abundance	**10.26**	**3**	**−13.01**	**0.00**	**0.82**	**0.030**	**0.010**	**0.010**	**0.050**
Stems + *Lonicera*	6.85	4	−3.04	9.97	0.01				
All habitat[a]	7.43	6	3.61	16.62	0.00				
Full model	14.17	7	−5.01	8.00	0.01				
Null model	6.43	2	−8.15	4.86	0.07				

[a] The all-habitat model includes understory stems, dominance by *Lonicera*, winter temperatures, and birdfeeder numbers.

TABLE 5.3

Comparison of a priori *models explaining variation in abundance of urban avoiders among 33 riparian forest sites in Ohio, 2001–2004.*

Bold text indicates the top-ranked model.

	Log-likelihood	K	AIC_c	ΔAIC_c	w_i	Estimate	SE	Lower 95% CI	Upper 95% CI
Acadian Flycatcher									
Understory stems	−29.59	3	66.68	14.91	0.00	−0.001	0.017	−0.035	0.033
Dominance by *Lonicera*	−26.54	3	60.58	8.81	0.01	−2.147	1.041	−4.188	−0.107
Predator abundance	−27.18	3	61.87	10.10	0.01	−0.511	0.271	−1.043	0.021
Insectivore abundance	**−22.14**	**3**	**51.77**	**0.00**	**0.81**	**0.374**	**0.110**	**0.158**	**0.590**
Stems + *Lonicera*	−26.22	4	63.10	11.33	0.00				
Predators + insectivores	−22.16	4	54.99	3.22	0.16				
Full model	−22.42	6	63.30	11.53	0.00				
Null model	−29.59	2	63.89	12.12	0.00				
Blue-gray Gnatcatcher									
Understory stems	−7.83	3	23.16	17.16	0.00	−0.007	0.004	−0.015	0.002
Dominance by *Lonicera*	**0.75**	**3**	**6.00**	**0.00**	**0.52**	**−0.633**	**0.122**	**−0.872**	**−0.393**
Predator abundance	−4.05	3	15.61	9.61	0.00	−0.120	0.035	−0.188	−0.051
Insectivore abundance	−1.67	3	10.85	4.85	0.05	0.118	0.027	0.065	0.172
Stems + *Lonicera*	0.76	4	9.14	3.14	0.11				
Predators + insectivores	1.02	4	8.63	2.63	0.14				
Full model	5.17	6	8.12	2.12	0.18				
Null model	−9.06	2	22.83	16.83	0.00				
Eastern Wood-Pewee									
Understory stems	**18.84**	**3**	**−30.17**	**0.20**	**0.23**	**−0.004**	**0.002**	**−0.008**	**0.000**
Dominance by *Lonicera*	**18.94**	**3**	**−30.37**	**0.00**	**0.25**	**−0.157**	**0.070**	**−0.295**	**−0.019**

Predator abundance	17.67	3	−27.84	2.53	0.07	−0.027	0.018	−0.062	0.009
Insectivore abundance	**18.37**	**3**	**−29.23**	**1.14**	**0.14**	**0.139**	**0.017**	**0.109**	**0.177**
Stems + *Lonicera*	**20.04**	**4**	**−29.42**	**0.95**	**0.16**				
Predators + insectivores	18.68	4	−26.68	3.69	0.04				
Full model	20.81	6	−23.15	7.22	0.01				
Null model	**16.63**	**2**	**−28.55**	**1.82**	**0.10**				
Indigo Bunting									
Understory stems	0.79	3	5.92	14.47	0.00	0.006	0.003	−0.001	0.012
Dominance by *Lonicera*	2.08	3	3.33	11.88	0.00	−0.289	0.117	−0.519	−0.059
Predator abundance	1.84	3	3.82	12.37	0.00	−0.069	0.029	−0.126	−0.011
Insectivore abundance	1.31	3	4.89	13.44	0.00	0.044	0.021	0.002	0.085
Stems + *Lonicera*	**7.12**	**3**	**−6.73**	**1.82**	**0.28**				
Predators + insectivores	2.71	3	2.08	10.63	0.00				
Full model	**8.03**	**3**	**−8.55**	**0.00**	**0.71**				
Null model	−0.69	3	8.89	17.44	0.00				
Northern Parula									
Understory stems	−11.35	3	30.20	6.90	0.02	0.000	0.018	−0.036	0.036
Dominance by *Lonicera*	−11.26	3	30.01	6.71	0.02	−0.194	0.842	−1.844	1.455
Predator abundance	−11.26	3	30.01	6.71	0.02	−0.043	0.206	−0.447	0.362
Insectivore abundance	**−7.90**	**3**	**23.30**	**0.00**	**0.66**	**0.216**	**0.196**	**−0.168**	**0.600**
Stems + *Lonicera*	−9.42	4	29.51	6.21	0.03				
Predators + insectivores	−7.82	4	26.30	3.00	0.15				
Full model	−8.64	6	35.74	12.44	0.00				
Null model	−11.24	2	27.18	3.88	0.09				

TABLE 5.3 (*continued*)

TABLE 5.3 (CONTINUED)

	Log-likelihood	K	AIC_c	ΔAIC_c	w_i	Estimate	SE	Lower 95% CI	Upper 95% CI
				Red-eyed Vireo					
Understory stems	1.27	3	4.97	6.85	0.02	0.001	0.003	−0.005	0.007
Dominance by *Lonicera*	2.65	3	2.21	4.09	0.08	−0.197	0.115	−0.423	0.029
Predator abundance	2.02	3	3.46	5.34	0.04	−0.037	0.029	−0.094	0.020
Insectivore abundance	**4.69**	**3**	**−1.88**	**0.00**	**0.62**	**0.054**	**0.020**	**0.016**	**0.092**
Stems + *Lonicera*	3.07	4	4.53	6.41	0.03				
Predators + insectivores	4.71	4	1.24	3.12	0.13				
Full model	5.23	6	8.00	9.88	0.00				
Null model	1.24	2	2.22	4.10	0.08				

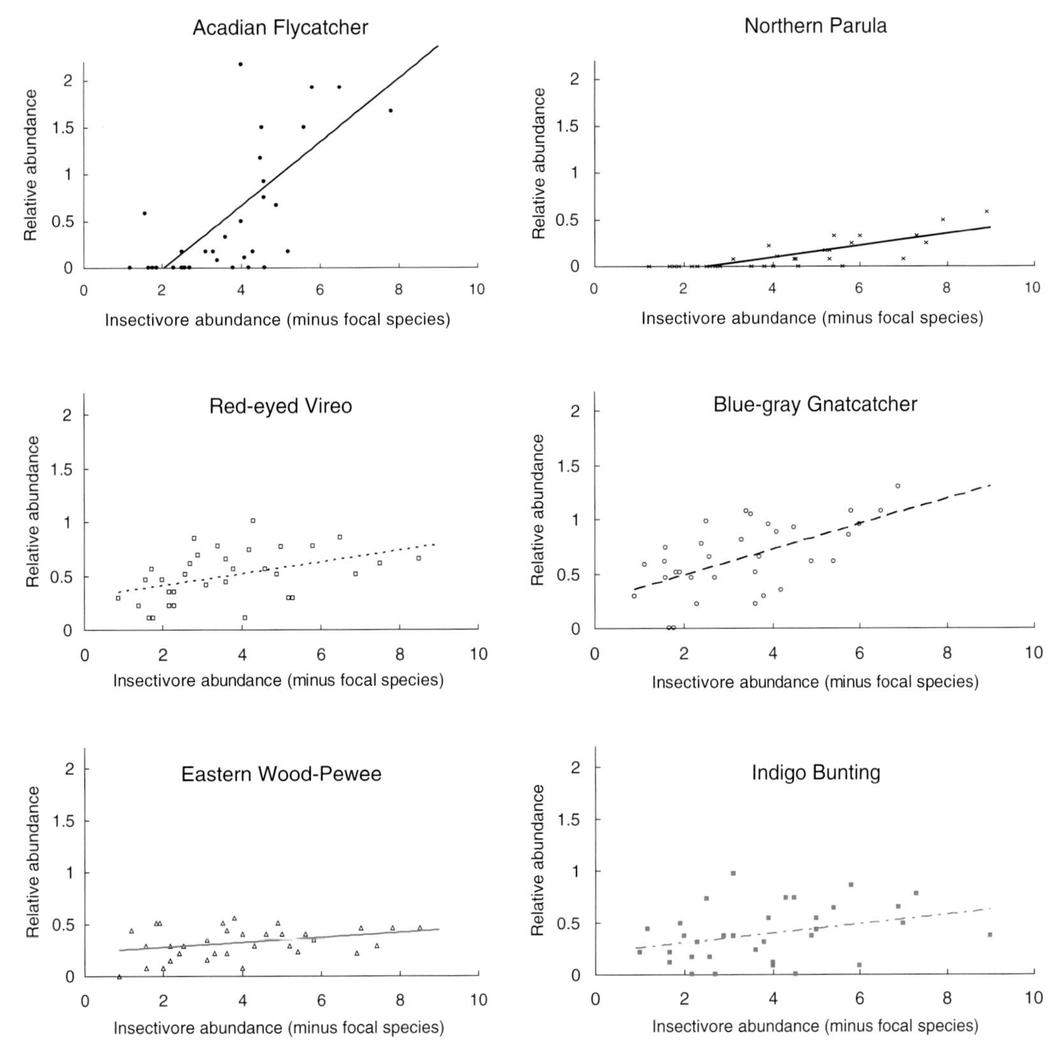

Figure 5.4. Association between relative abundance of urban avoiders and insectivore abundance along 40 × 250-m transects at 33 study sites, 2001–2004.

strongly correlated with other variables, particularly winter temperatures (r = 0.63), predator abundance (r = 0.66), and omnivore abundance (mean r = 0.58, range: 0.50–0.64; Table 5.5).

Discussion

Responses of individual avian species to urbanization were explained by a variety of local ecological mechanisms, making it difficult to cast broad generalizations across multiple species. This was especially true for urban adapters, as results provided support for the local vegetation, winter microclimate, supplemental food, and heterospecific attraction hypotheses. Examination of urban avoiders proved clearer, and two factors, abundance of other insectivores and dominance of understory by honeysuckle, were most frequently included in top model sets. Thus, many species that are sensitive to and avoid urbanizing landscapes may respond strongly to the presence of other insectivores settling within the landscapes and to the degree of invasion by exotic plants.

As a whole, findings regarding the responses of urban adapters are consistent with previous research showing that local vegetation, food, and microclimate changes promote greater abundance of certain human commensal species (reviewed in Marzluff et al. 2001). In addition to changes in local vegetation induced by urbanization, urban areas may provide new and/or abundant food

TABLE 5.4

Comparison of a priori *models explaining variation in reproductive productivity of Acadian Flycatcher and Northern Cardinal in riparian forests of Ohio, 2001–2006.*

Bold text indicates the top-ranked models.

	Log-likelihood	K	AIC_c	ΔAIC_c	w_i	Estimate	SE	Lower 95% CI	Upper 95% CI
			Acadian Flycatcher productivity						
Understory stems	−17.02	3	41.53	7.53	0.01	0.013	0.013	−0.013	0.038
Dominance by *Lonicera*	**−13.62**	**3**	**34.75**	**0.75**	**0.39**	**−1.708**	**0.550**	**−2.786**	**−0.630**
Predator abundance	**−13.25**	**3**	**34.00**	**0.00**	**0.57**	**−0.511**	**0.155**	**−0.816**	**−0.206**
Insectivorous species	−16.49	3	40.48	6.48	0.02	0.146	0.100	−0.051	0.342
Stems + *Lonicera*	−20.73	4	52.12	18.12	0.00				
Predators + insectivores	−17.11	4	44.89	10.89	0.00				
Full model	−17.00	6	52.46	18.46	0.00				
Null model	−21.50	2	47.70	13.70	0.00				
			Northern Cardinal productivity						
Understory stems	**−6.97**	**3**	**21.44**	**0.46**	**0.17**	**−0.008**	**0.010**	**−0.029**	**0.012**
Dominance by *Lonicera*	**−6.81**	**3**	**21.12**	**0.14**	**0.19**	**−0.488**	**0.499**	**−1.465**	**0.490**
Winter temperature	**−6.74**	**3**	**20.98**	**0.00**	**0.21**	**−0.105**	**0.099**	**−0.300**	**0.090**
Birdfeeder numbers	**−7.10**	**3**	**21.71**	**0.73**	**0.14**	**0.029**	**0.050**	**−0.069**	**0.127**
Predator abundance	**−7.17**	**3**	**21.85**	**0.87**	**0.14**	**0.069**	**0.155**	**−0.235**	**0.373**
Omnivore abundance	**−7.24**	**3**	**21.97**	**0.99**	**0.13**	**−0.019**	**0.072**	**−0.123**	**0.161**
Stems + *Lonicera*	−7.47	4	25.61	4.63	0.02				
All habitat	−5.32	6	29.10	8.12	0.00				
Full model	−5.31	7	33.95	12.97	0.00				
Null model	−13.81	2	32.32	11.34	0.00				

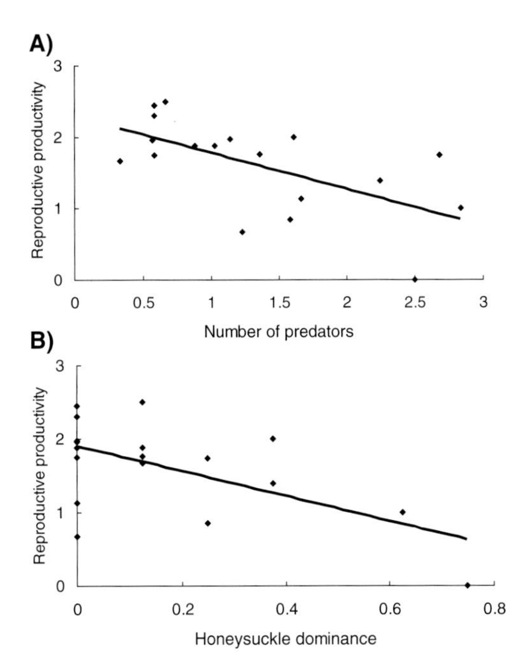

Figure 5.5. Relationship between productivity of Acadian Flycatchers (mean number of fledglings per pair per site) and (a) number of predators and (b) dominance of *Lonicera* at 18 riparian forests in Ohio, 2001–2006.

sources. Supplemental food provided at feeding stations can affect distribution, abundance, habitat selection, annual survivorship, sociality, and foraging behavior of wintering birds (Brittingham and Temple 1988; Desrochers et al. 1988; Wilson 1994, 2001; Doherty and Grubb 2002; Atchison and Rodewald 2006). Similarly, wintering birds in urban areas are expected to benefit from elevated temperatures due to the combined effects of increased anthropogenic heating, decreased vegetated surfaces, and increased concrete and pavement, which collectively can result in a 1–2° C increase in ambient temperatures during winter months (Botkin and Beveridge 1997). Warmer winter temperatures can improve the energetic environment for resident birds, especially for small-bodied species (Grubb 1975, Wachob 1996, Dolby and Grubb 1999). In this way, the combination of supplemental food and altered microclimate may create habitats that are beneficial to wintering birds in harsh winter climates.

One of the most interesting findings of this study was that positive social interactions may contribute to the pattern of urban avoidance displayed by most neotropical migratory birds in my study area. Social attraction to conspecifics or heterospecifics is thought to be the consequence of individuals using public information about habitat quality (Kiester 1979, Stamps 1988, Mönkkönen et al. 1990). In this study, numbers of all six urban avoiders but only two urban adapters were positively related to numbers of heterospecifics in the same foraging guild. Interestingly, one of the latest migrants to the study area, the Acadian Flycatcher, showed the strongest association with numbers of other insectivores. Arriving late in the spring presumably would facilitate the use of heterospecifics as cues to habitat selection because other breeding birds would have already settled sites.

Most urban ecology studies have emphasized the importance of negative species interactions (e.g., predation and brood parasitism), and positive biotic interactions have generally been overlooked in ecological field studies (Dickman 1992, Stachowicz 2001). However, birds may use nonhabitat cues and other types of "public information" (e.g., abundance of conspecifics, number of juveniles, reproductive success of neighbors) to guide decisions about habitat selection (Doligez et al. 2003, Danchin et al. 2004). Although conspecific attraction has received the most attention in recent decades (Kiester 1979, Stamps 1988, Muller et al. 1997, Fletcher 2006), individuals also may use cues derived from other species with similar requirements, often termed "heterospecific attraction" (Mönkkönen et al. 1990, 1997, 1999; Forsman et al. 1998, 2002). Others have experimentally demonstrated that the abundance of arboreal insectivores is positively related to the abundance of heterospecifics with similar ecological requirements and species with previous or more extensive knowledge about sites, such as resident species and species in foliage-gleaning guilds (Mönkkönen et al. 1997, Forsman et al. 1998, Thomson et al. 2003, Fletcher 2007). If heterospecific attraction is a contributing factor to avian distribution and abundance, this may partly explain why the development of mechanistic explanations of responses of bird communities to urbanization has proven elusive. Heterospecific attraction could lead to large changes in avian community structure in cases where only a few key species responded directly to changes in local resources or even predator/brood parasite pressures. From a conservation perspective, heterospecific attraction also has the potential to either impede (when the key species are unidentified or difficult to manage) or facilitate (when key species

TABLE 5.5

Pearson correlation coefficients (and P-values) among explanatory variables included in models.

Variable	Dominance by *Lonicera*	Understory stems	Winter temperature	Birdfeeder abundance	Predator abundance
Dominance by *Lonicera*		0.368 (0.035)	0.628 (<0.001)	0.457 (0.008)	0.660 (<0.001)
Understory stems			0.255 (0.15)	0.314 (0.08)	0.059 (0.74)
Winter temperature				0.422 (0.014)	0.526 (0.002)
Birdfeeder numbers					0.380 (0.029)
Predator abundance					
			Omnivore abundance[a]		
American Goldfinch	0.605 (<0.001)	0.132 (0.47)	0.518 (0.002)	0.315 (0.07)	—
American Robin	0.501 (0.003)	0.100 (0.58)	0.547 (0.001)	0.343 (0.051)	—
Carolina Chickadee	0.631 (<0.001)	0.134 (0.46)	0.492 (0.004)	0.302 (0.09)	—
Cedar Waxwing	0.578 (<0.001)	0.161 (0.37)	0.497 (0.003)	0.343 (0.051)	—
Downy Woodpecker	0.629 (<0.001)	0.188 (0.29)	0.534 (0.001)	0.368 (0.035)	—
Northern Cardinal	0.564 (0.001)	0.100 (0.58)	0.422 (0.015)	0.264 (0.14)	—
Red-bellied Woodpecker	0.635 (<0.001)	0.169 (0.35)	0.549 (0.001)	0.385 (0.027)	—
			Insectivore abundance[a]		
Acadian Flycatcher	−0.530 (0.002)	−0.180 (0.32)	—	—	−0.425 (0.014)
Blue-gray Gnatcatcher	−0.424 (0.014)	−0.072 (0.69)	—	—	−0.373 (0.033)
Eastern Wood-Pewee	−0.498 (0.003)	−0.102 (0.57)	—	—	−0.409 (0.018)
Indigo Bunting	−0.480 (0.005)	−0.198 (0.26)	—	—	−0.380 (0.029)
Northern Parula	−0.517 (0.002)	−0.136 (0.45)	—	—	−0.423 (0.014)
Red-eyed Vireo	−0.514 (0.002)	−0.151 (0.40)	—	—	−0.424 (0.014)

[a] Omnivore and insectivore abundance reflects abundance of omnivores minus the focal species listed.

are known and amenable to management) restoration and/or conservation efforts.

An important cautionary note regarding the apparent support for the heterospecific attraction hypothesis is, of course, that correlations do not necessarily indicate causation. For example, insectivore abundance was moderately correlated with other local ecological variables, particularly dominance by honeysuckle. Furthermore, all insectivores may have responded similarly to a habitat variable that was overlooked in this study. One obvious gap is the lack of arthropod data in my analyses. However, three pieces of evidence, each with its own weakness, suggest that arthropods may not be driving the pattern. First, previous sampling of aerial arthropods used by Acadian Flycatchers at sites indicated that arthropod biomass was positively related to the amount of urbanization within the landscape (Bakermans 2003). Second, previous studies of nestling provisioning at sites showed that provisioning rates were higher in forests within urban rather than rural landscapes (Leston 2005). Third, although urban Acadian Flycatchers were smaller, body condition (i.e., fatness or thinness determined by deviation from expected weight-frame relationship) was not significantly related to urbanization (Rodewald and Shustack 2008a), suggesting that urban birds were not more food-deprived than their rural counterparts. While it is premature to invoke heterospecific attraction as the cause of distribution patterns, results do make a case for the need to give positive species interactions consideration in future studies.

Another important finding in this study was that invasion of urban forests by exotic honeysuckle may play a particularly important role in determining the response of avian communities to urbanization. Exotic plants are well known to disrupt ecosystem processes (Vitousek 1990, Wilcove et al. 1998, Ehrenfeld 2003, Levine et al. 2003), shift floristic and faunal composition (Luken and Thieret 1996, Hutchison and Vankat 1997), and alter species interactions at higher trophic levels (Schmidt and Whelan 1999, Remes 2003, Borgmann and Rodewald 2004, Schmidt et al. 2005). Honeysuckle is one of the most problematic exotic and invasive plants in my study area as well as elsewhere in the U.S. (Luken and Thieret 1996, Hutchison and Vankat 1997), and invasion by honeysuckle seems to be promoted by urban development (Borgmann and Rodewald 2005). Although the full range of consequences of honeysuckle to higher trophic levels are poorly understood, honeysuckle clearly changes the structure and floristic composition of habitat, and thereby alters food and nesting resources available to birds (Mills et al. 1989, Reichard et al. 2001). Indeed, honeysuckle provided the vast majority (>90%) of available fruits at sites during late fall and winter (Leston 2005) and was largely responsible for the pattern of greater fruit availability in urban compared to rural sites (Leston and Rodewald 2006). Honeysuckle also promoted a dense understory layer in riparian forests and, not surprisingly, was actively selected as a nesting substrate by certain understory birds, such as the Northern Cardinal (Leston and Rodewald 2006). On the other hand, dense understory growth may have dissuaded other species from settling at sites, especially species such as the Acadian Flycatcher, which prefers an open understory (Bakermans and Rodewald 2006). However, the fact that understory stem density received little support in analyses suggests that the effects of honeysuckle on forest ecosystems result from more than its tendency to alter understory habitat structure.

My analysis failed to show strong concordance between factors that performed well in explaining variation in abundance among sites and those that explained variation in productivity. Previous work has indicated that annual reproductive productivity of Acadian Flycatchers (Rodewald and Shustack 2008a), but not Northern Cardinals (Rodewald and Shustack 2008b), declined with increasing levels of urbanization. Identifying the factors that underlie these patterns has proven more challenging. In this study, although abundance of Acadian Flycatchers was most tightly linked with insectivore abundance and dominance by honeysuckle, of the two variables, only dominance by honeysuckle was included in the top model set explaining productivity. The negative association between flycatcher productivity and dominance by honeysuckle is consistent with recent studies showing that exotic plants can increase vulnerability to nest depredation (Schmidt and Whelan 1999, Remes 2003, Borgmann and Rodewald 2004, Schmidt et al. 2005). Moreover, birds that selected honeysuckle as a nesting substrate early in the breeding season produce 20% fewer young over the course of the season compared to birds using other substrates (Rodewald et al. 2010).

Honeysuckle cover also best explained, and was positively associated with, brood parasitism of flycatcher nests at these same study sites (Rodewald 2009). Because parasitized flycatcher nests in my system seldom fledge host young, brood parasitism can strongly reduce productivity. Number of nest predators also was identified as a top model, and productivity of flycatchers was negatively related to numbers of predators. This association with nest predators is curious in my study system, given that flycatchers do not experience comparatively lower nest success in urban than rural landscapes (Rodewald and Shustack 2008a; A. D. Rodewald, unpubl. data, 2001–2010). In fact, recent work demonstrates that relationships between breeding birds and nest predators can become completely decoupled when synanthropic predators are subsidized by anthropogenic resources (Rodewald et al. 2011). All ecological variables were similarly ranked in terms of ability to explain variation in cardinal productivity.

CONCLUSION

This study did not point to a single ecological mechanism that is likely responsible for most urban-associated changes in bird communities. Nevertheless, the data make an important contribution to the growing literature on urban ecology by showing that a variety of local ecological factors, particularly positive social interactions and plant invasions, may be important in shaping avian communities in urbanizing landscapes. Fortunately, investigating the influence of exotic substrates and social attraction are particularly amenable to experimental manipulation, which could lead to large advances in our understanding of animal responses to urbanization. Identifying the local ecological mechanisms that generate landscape patterns in urban systems is ultimately critical to both predict and ameliorate impacts of large-scale urbanization.

ACKNOWLEDGMENTS

Funding for this research was provided by the National Science Foundation (DEB-0340879 and DEB-0639429), Ohio Division of Wildlife, Ohio Agricultural Research and Development Center, and The Ohio State University Swank Program in Rural-Urban Policy. I am most grateful to M. H. Bakermans, K. L. Borgmann, F. V. L. Leston, L. J. Kearns, D. P. Shustack, J. R. Smith-Castro, and N. Sundell-Turner for their
the
M. (
Haz
S.
D.
O'
in
of
kl
P
a

LITERATURE CITED

Atchison, K. A., and A. D. Rodewald. 2006. The value of urban forests to wintering birds. Natural Areas Journal 26:280–288.

Bakermans, M. H. 2003. Hierarchical habitat selection in the Acadian Flycatcher: implications for conservation of riparian forests. M.S. thesis, The Ohio State University, Columbus, OH.

Bakermans, M. H., and A. D. Rodewald. 2006. Scale-dependent habitat use of Acadian Flycatcher (*Empidonax virescens*) in central Ohio. Auk 123:368–382.

Beissinger, S. R., and D. R. Osborne. 1982. Effects of urbanization on avian community organization. Condor 84:75–83.

Blair, R. B. 1996. Land use and avian species diversity along an urban gradient. Ecological Applications 6:506–519.

Borgmann, K. L., and A. D. Rodewald. 2004. Nest predation in an urbanizing landscape: the role of exotic shrubs. Ecological Applications 14:1757–1765.

Borgmann, K. L., and A. D. Rodewald. 2005. Forest restoration in urbanizing landscapes: interactions between land uses and exotic shrubs. Restoration Ecology 13:334–340.

Botkin, D. B., and C. E. Beveridge. 1997. Cities as environments. Urban Ecosystems 1:3–19.

Brittingham, M. C., and S. A. Temple. 1988. Impacts of supplemental feeding on survival rates of Black-capped Chickadees. Ecology 69:581–589.

Burnham, K. P., and D. R. Anderson. 1998. Model selection and inference: a practical information-theoretic approach. 1st ed. Springer-Verlag, New York.

Danchin, E., L. A. Giraldeau, T. J. Valone, and R. H. Wagner. 2004. Public information: from nosy neighbors to cultural evolution. Science 305:487–491.

Desrochers, A., S. J. Hannon, and K. E. Nordin. 1988. Winter survival and territory acquisition in a northern population of Black-capped Chickadees. Auk 105:727–736.

Dickman, C. R. 1992. Commensal and mutualistic interactions among terrestrial vertebrates. Trends in Ecology & Evolution 7:194–197.

Doherty, P. F., and T. C. Grubb. 2002. Survivorship of permanent-resident birds in a fragmented forested landscape. Ecology 83:844–857.

Dolby, A. S., and T. C. Grubb. 1999. Effects of winter weather on horizontal vertical use of isolated forest fragments by bark-foraging birds. Condor 101:408–412.

Doligez, B., C. Cadet, E. Danchin, and T. Boulinier. 2003. When to use public information for breeding habitat selection? The role of environmental predictability and density dependence. Animal Behaviour 66:973–988.

Ehrenfeld, J. G. 2003. Effects of exotic plant invasions on soil nutrient cycling processes. Ecosystems 6:503–523.

Faeth, S. H., P. S. Warren, E. Shochat, and W. A. Marussich. 2005. Trophic dynamics in urban communities. Bioscience 55:399–407.

Fletcher, R. J. 2006. Emergent properties of conspecific attraction in fragmented landscapes. American Naturalist 168:207–219.

Fletcher, R. J. 2007. Species interactions and population density mediate the use of social cues for habitat selection. Journal of Animal Ecology 76:598–606.

Forsman, J. T., M. Mönkkönen, P. Helle, and J. Inkeroinen. 1998. Heterospecific attraction and food resources in migrants' breeding patch selection in northern boreal forest. Oecologia 115:278–286.

Forsman, J. T., J. T. Seppänen, and M. Mönkkönen. 2002. Positive fitness consequences of interspecific interaction with a potential competitor. Proceedings of the Royal Society of London Series B: Biological Sciences 269:1619–1623.

Germaine, S. S., S. S. Rosenstock, R. E. Schweinsburg, and W. S. Richardson. 1998. Relationships among breeding birds, habitat, and residential development in Greater Tucson, Arizona. Ecological Applications 8:680–691.

Glase, J. C. 1973. Ecology of social organization in the Black-capped Chickadee. Living Bird 12:235–267.

Grubb, T. C. 1975. Weather-dependent foraging behavior of some birds wintering in a deciduous woodland. Condor 77:175–182.

Haskell, D. G., A. M. Knupp, and M. C. Schneider. 2001. Nest predator abundance and urbanization. Pp. 243–258 *in* J. M. Marzluff, R. Bowman, and R. Donnelly (editors), Avian ecology and conservation in an urbanizing world. Kluwer Academic Publishers, Boston, MA.

Hennings, L. A., and W. D. Edge. 2003. Riparian bird community structure in Portland, Oregon: habitat, urbanization, and spatial scale patterns. Condor 105:288–302.

Hutchinson, T. F., and J. L. Vankat. 1997. Invasibility and effects of Amur honeysuckle in southwestern Ohio forests. Conservation Biology 11:1117–1124.

Kiester, A. R. 1979. Conspecifics as cues: a mechanism for habitat selection in the Panamanian grass anole (*Anolis auratus*). Behavioral Ecology and Sociobiology 5:323–330.

Leston, L. F. V. 2005. Are urban riparian forests ecological traps for understory birds? Habitat selection by Northern Cardinals (*Cardinalis cardinalis*) in urbanizing landscapes. M.Sc. thesis, Ohio State University, Columbus, OH.

Leston, L. F. V., and A. D. Rodewald. 2006. Are urban forests ecological traps for understory birds? An examination using Northern Cardinals. Biological Conservation 131:566–574.

Levine, J. M., M. Vila, C. M. D'Antonio, J. S. Dukes, K. Grigulis, and S. Lavorel. 2003. Mechanisms underlying the impacts of exotic plant invasions. Proceedings of the Royal Society of London Series B: Biological Sciences 270:775–781.

Luken, J. O., and J. W. Thieret. 1996. Amur honeysuckle, its fall from grace. Bioscience 46:18–24.

Marzluff, J. M., R. Bowman, and R. E. Donnelly. 2001. Avian Ecology and Conservation in an Urbanizing World. Kluwer Academic Publishers, Boston, MA.

McGarigal, K., S. Cushman, and S. Stafford. 2000. Multivariate statistics for wildlife and ecology research. Springer-Verlag, New York.

Miller, J. R., J. A. Wiens, N. T. Hobbs, and D. M. Theobald. 2003. Effects of human settlement on bird communities in lowland riparian areas of Colorado (USA). Ecological Applications 13:1041–1059.

Mills, G. S., J. B. Dunning, and J. M. Bates. 1989. Effects of urbanization on breeding bird community structure in southwestern desert habitats. Condor 91:416–428.

Mönkkönen, M., R. Hardling, J. T. Forsman, and J. Tuomi. 1999. Evolution of heterospecific attraction: using other species as cues in habitat selection. Evolutionary Ecology 13:91–104.

Mönkkönen, M., P. Helle, G. J. Niemi, and K. Montgomery. 1997. Heterospecific attraction affects community structure and migrant abundances in northern breeding bird communities. Canadian Journal of Zoology–Revue Canadienne de Zoologie 75:2077–2083.

Mönkkönen, M., P. Helle, and K. Soppela. 1990. Numerical and behavioral responses of migrant passerines to experimental manipulation of resident tits (*Parus* spp.): heterospecific attraction in northern breeding bird communities. Oecologia 85:218–225.

Muller, K. L., J. A. Stamps, V. V. Krishnan, and N. H. Willits. 1997. The effects of conspecific attraction and habitat quality on habitat selection in territorial birds (*Troglodytes aedon*). American Naturalist 150:650–661.

Odell, E. A., and R. L. Knight. 2001. Songbird and medium-sized mammal communities associated with exurban development in Pitkin County, Colorado. Conservation Biology 15:1143–1150.

Odum, E. P. 1942. Annual cycle of the Black-capped Chickadee. Auk 59:499–531.

Pickett, S. T. A., M. L. Cadenasso, J. M. Grove, C. H. Nilon, R. V. Pouyat, W. C. Zipperer, and R. Costanza. 2001. Urban ecological systems: linking terrestrial ecological, physical, and socioeconomic components of metropolitan areas. Annual Review of Ecology and Systematics 32:127–157.

Reichard, S. H., L. Chalker-Scott, and S. Buchanan. 2001. Interactions among non-native plants and birds. Pp. 179–223 in J. M. Marzluff, R. Bowman, and R. Donnelly (editors), Avian ecology and conservation in an urbanizing world. Kluwer Academic Publishers, Boston, MA.

Remes, V. 2003. Effects of exotic habitat on nesting success, territory density, and settlement patterns in the Blackcap (*Sylvia atricapilla*). Conservation Biology 17:1127–1133.

Rodewald, A. D. 2009. Urban-associated habitat alteration promotes brood parasitism of Acadian Flycatcher. Journal of Field Ornithology 80:234–241.

Rodewald, A. D., and M. H. Bakermans. 2006. What is the appropriate paradigm for riparian forest conservation? Biological Conservation 128:193–200.

Rodewald, A. D., L. E. Hitchcock, and D. P. Shustack. 2010. Exotic shrubs as ephemeral ecological traps for nesting birds. Biological Invasions 12:33–39.

Rodewald, A. D., L. J. Kearns, and D. P. Shustack. 2011. Anthropogenic resources decouple predator–prey relationships. Ecological Applications 20:2047–2057.

Rodewald, A. D., and D. P. Shustack. 2008a. Urban flight: understanding individual and population-level responses of nearctic–neotropical migratory birds to urbanization. Journal of Animal Ecology 77:83–91.

Rodewald, A. D., and D. P. Shustack. 2008b. Consumer resource-matching in urbanizing landscapes: are synanthropic species over-matching? Ecology 89: 515–521.

Rodewald, A. D., and R. H. Yahner. 2001. Influence of landscape composition on avian community structure and associated mechanisms. Ecology 82:3493–3504.

Rottenborn, S. C. 1999. Predicting the impacts of urbanization on riparian bird communities. Biological Conservation 88:289–299.

Saab, V. 1999. Importance of spatial scale to habitat use by breeding birds in riparian forests: a hierarchical analysis. Ecological Applications 9:135–151.

Schmidt, K. A., L. C. Nelis, N. Briggs, and R. S. Ostfeld. 2005. Invasive shrubs and songbird nesting success: effects of climate variability and predator abundance. Ecological Applications 15:258–265.

Schmidt, K. A., and C. J. Whelan. 1999. Effects of exotic *Lonicera* and *Rhamnus* on songbird nest predation. Conservation Biology 13:1502–1506.

Shochat, E., P. S. Warren, S. H. Faeth, N. E. McIntyre, and D. Hope. 2006. From patterns to emerging processes in mechanistic urban ecology. Trends in Ecology & Evolution 21:186–191.

Shustack, D. P., and A. D. Rodewald. 2010. Attenuated nesting season of the Acadian Flycatcher (*Empidonax virescens*) in urban forests. Auk 127:421–429.

Sinclair, K. E., G. R. Hess, C. E. Moorman, and J. H. Mason. 2005. Mammalian nest predators respond to greenway width, landscape context and habitat structure. Landscape and Urban Planning 71:277–293.

Smith, C. M., and D. G. Wachob. 2006. Trends associated with residential development in riparian breeding bird habitat along the Snake River in Jackson Hole, WY, USA: implications for conservation planning. Biological Conservation 128:431–446.

Sorace, A. 2002. High density of bird and pest species in urban habitats and the role of predator abundance. Ornis Fennica 79:60–71.

Stachowicz, J. J. 2001. Mutualism, facilitation, and the structure of ecological communities. Bioscience 51:235–246.

Stamps, J. A. 1988. Conspecific attraction and aggregation in territorial species. American Naturalist 131:329–347.

Tewksbury, J. J., S. J. Hejl, and T. E. Martin. 1998. Breeding productivity does not decline with increasing fragmentation in a western landscape. Ecology 79:2890–2903.

Thomson, R. L., J. T. Forsman, and M. Mönkkönen. 2003. Positive interactions between migrant and resident birds: testing the heterospecific attraction hypothesis. Oecologia 134: 431–438.

Vitousek, P. M. 1990. Biological invasions and ecosystem processes: towards an integration of population biology and ecosystem studies. Oikos 57:7–13.

Wachob, D. G. 1996. The effect of thermal microclimate on foraging site selection by wintering Mountain Chickadees. Condor 98:114–122.

Wilcove, D. S., D. Rothstein, J. Dubow, A. Phillips, and E. Losos. 1998. Quantifying threats to imperiled species in the United States. Bioscience 48:607–615.

Wilson, W. H. 1994. The distribution of wintering birds in central Maine: the interactive effects of landscape and bird feeders. Journal of Field Ornithology 65:512–519.

Wilson, W. H. 2001. The effects of supplemental feeding on wintering Black-capped Chickadees (*Poecile atricapilla*) in central Maine: population and individual responses. Wilson Bulletin 113:65–72.

CHAPTER SIX

Impacts of Seasonal Small-scale Urbanization on Nest Predation and Bird Assemblages at Tourist Destinations

Marja-Liisa Kaisanlahti-Jokimäki, Jukka Jokimäki, Esa Huhta, and Pirkko Siikamäki

Abstract. Urbanization and urban sprawl may have a fundamental effect on bird assemblages. Northern tourist destinations are nowadays like towns with high numbers of people during the peak seasons and the urban structure of the nearby landscape. In this study, our aim was to investigate the effects of tourist destinations on nest predation risk, potential nest predator abundance, and the assemblages of birds across an urban gradient from uninhabited forests via tourist destinations to towns in northern Finland. The impact of the relatively small-scale seasonal tourist destinations (ski resorts in our case) in areas close to the wilderness end of the urban-rural-wildland gradient has not received attention before. Our data include all the main ski resorts located in northern Finland and their surrounding forest areas as well as the biggest towns located within the study area. We evaluated nest predation risk by means of an artificial ground nest predation experiment. The abundance of potential nest predators as well as bird community assemblages were evaluated using the point count method. Our results showed that artificial ground nests preyed upon and the abundance of corvids increased with increasing urbanization, and the proportional abundance of ground-nesting species decreased with increasing urbanization. Nest losses, the abundance of corvids, and the proportional abundance of ground-nesting species were intermediate in the tourist destinations when compared to the towns and forests. Our results indicated that seasonal ski resorts may have negative impacts on bird assemblages, and more attention should be paid to area planning in tourist destinations and their surroundings. These results have significant implications for improving land use planning of northern ski resorts and conserving birds at seasonal tourist towns.

Key Words: avian nest predators, birds, gradient, indicators, nest predation, ski resorts, tourism, tourist destinations, urbanization, urban sprawl.

Urbanization is a large-scale process that fragments the landscape and significantly alters the distribution and abundance of many native species and assemblages of bird communities (Vale and Vale 1976, Beissinger and Osborne 1982, Bezzel 1985, Blair 1996, Fernández-Juricic and Jokimäki 2001, Marzluff 2001, McKinney 2002, Chace and Walsh 2004). In general, urbanization decreases bird species diversity and richness, but increases the total density

Kaisanlahti-Jokimäki, M.-L., J. Jokimäki, E. Huhta, and P. Siikamäki. 2012. Impacts of seasonal small-scale urbanization on nest predation and bird assemblages at tourist destinations. Pp. 93–110 *in* C. A. Lepczyk and P. S. Warren (editors). Urban bird ecology and conservation. Studies in Avian Biology (no. 45), University of California Press, Berkeley, CA.

of birds (Batten 1972, Bezzel 1985, Jokimäki and Suhonen 1993, Marzluff 2001, McKinney 2002). Urbanization favors omnivores (Beissinger and Osborne 1982, Bezzel 1985, Diamond 1986, Clergeau et al. 1998, Jokimäki and Suhonen 1998), and partly for this reason corvids have expanded their ranges and increased their abundance in urban habitats (Jokimäki et al. 1996, Jokimäki and Suhonen 1998, Sorace 2002). This, in turn, may be disadvantageous for ground-nesting birds (Jokimäki 1999, Jokimäki and Huhta 2000). Cavity nesters breeding in buildings or nest boxes have been reported to benefit from urbanization (Suhonen and Jokimäki 1988). Some species may be considered as urban exploiters, others as urban avoiders or suburban adapters (Blair 1996, McKinney 2002). Urban avoiders fail to successfully reproduce and may become locally extinct in urban landscapes, whereas urban exploiters reproduce successfully in urban habitats and colonize new locations (Blair 2004). Invasions by these ubiquitous species and decrease in the numbers of ground-nesting species may cause general homogenization of biota (Blair 2004, Clergeau et al. 2006). Therefore, it is important to know where urban growth will occur and what the biological impacts of the urban sprawl are (Allen 2006).

The impacts of urban sprawl on bird communities can be studied by means of urban gradient analyses. These gradient studies have indicated that bird species richness and diversity peak at intermediate levels of urbanization, whereas density of individual species peaks at different levels of urbanization (Jokimäki and Suhonen 1993, Blair 1996). Urban gradient studies have dealt with gradients changing either from relatively undisturbed sites to highly developed sites, or alternatively from small villages to large towns. The increase in people's leisure time has indirectly increased the extent of urban sprawl. New summer cottages and tourist destinations have been established in areas formerly undisturbed, and areas with old cottages and ski resorts have encroached upon wilderness areas in northern Finland. Cottages may change the bird community structure and the occurrence and abundance of nest predators (Nilon et al. 1995, Lehtilä et al. 1996). In general, this kind of urban sprawl may be especially harmful to the ecosystem because it is directed toward wilderness areas. More information about the effects of the urban sprawl caused by tourist destinations is needed and would be valuable for planners involved with tourist destination areas.

Only limited studies of environmental impacts of recreation and tourism on nature have been published (Sun and Walsh 1998). Earlier tourism-related studies have mainly analyzed the effects of recreational trails (van der Zande et al. 1984, Miller et al. 1998, Miller and Hobs 2000, Chettri et al. 2001, Watson and Moss 2004, Mallord et al. 2007), as well as ski runs or lifts on birds (Watson 1979, Zeitler 2000, Laiolo and Rolando 2005, Rolando et al. 2007). The results of these studies have shown that bird species composition has been altered adjacent to recreational trails, and generalist species are more abundant near trails. An influx of crows following the development of ski resorts increased the nest losses of Rock Ptarmigan (*Lagopus muta*) at a Scottish ski resort area (Watson and Moss 2004). Bird species richness and diversity are reported to be lower at high-altitude ski runs than at more natural areas at similar latitudes (Laiolo and Rolando 2005, Rolando et al. 2007). The impact of the relatively small-scale seasonal tourist destinations or towns (ski resorts in our case) in areas close to the wilderness end of the urban-rural-wildland gradient has not received attention before. We do not know of any studies examining bird assemblages or nest predation of the "urban" parts (i.e., densely built-up areas with hotels, shops, spas, etc.) of tourist destinations in comparison with the surrounding wilderness areas and towns. We need to know how severe the disturbance caused by ski resorts on birds is and whether this type of urban sprawl is comparable to traditional urbanization processes in towns.

A typical feature of ski resorts is the seasonality in the numbers of visitors. The numbers of tourists may increase five-fold during the peak skiing season as compared to the summer season in northern Finland (Regional Council of Lapland 2003). Another feature that sets ski resorts apart from towns is the low number of local residents—only some hundreds in the tourist destinations in northern Finland (also known as Lapland). During the peak season, the ratio of tourists to locals is more than 100:1. Tourists need services (e.g., hotels, shops, fireplaces, camping places, etc.), and these services must be planned based on the peak season figures. The seasonality of the visitor flows offers interesting possibilities for ecologists to study "experimentally" the effects of infrastructure and its peaking seasonal use on the ecosystem.

Tourism-related urban sprawl is increasing markedly in northern Finland. The annual growth rate of the number of registered guest nights in accommodation facilities is about 2.7% (Regional Council of Lapland 2003). During 2004, the total number of registered overnight stays in northern Finland was 1.9 million (Statistics Finland). However, registered overnights show only a part of the truth, as many overnights go unregistered. Based on indirect calculations with the help of water consumption, it has been estimated that there are as many unregistered as there are registered overnights in northern Finland (Regional Council of Lapland 2003). In addition, day trips, which have also increased in northern Finland, are not included in accommodation statistics. At the same time, the number of visitors has almost tripled in national parks located in northern Finland during 1992–2000 (Regional Council of Lapland 2003). The increasing numbers of visitors require more space, infrastructure, and other facilities.

Our aim in this study was to explore the intensity of nest predation pressure on birds and patterns of nest predation risk and avian assemblage organization along the urban gradient from wilderness forest areas and tourist destinations to towns in northern Finland. In addition, we compared nest predation risk and assemblage organization between tourist destinations of different sizes. Because vegetation characteristics and the abundance of potential nest predators may influence nest predation, we also evaluated the role of vegetation structure and predator abundance on nest losses. We hypothesized that the nest predation risk, the abundance of nest predators, and the avian assemblage organization should change along the urban gradient. We predicted that both the nest predation risk and the number of nest predators should increase with urbanization, and that correspondingly the proportion of ground-nesting birds in the assemblages should decrease.

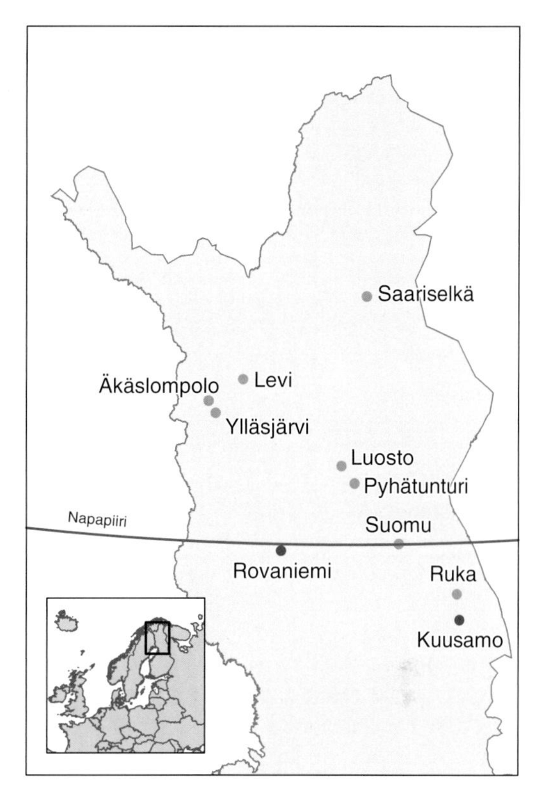

Figure 6.1. Location of the study sites included in this study. Black dots = cities.

MATERIAL AND METHODS

Study Sites

The study was conducted at eight tourist destinations, their surrounding forests, and two towns in northern Finland (Fig. 6.1). The study sites include all the main ski resorts ("tourist towns") and towns located in northern Finland. All of the tourist destinations are located within the north-boreal zone (Ahti et al. 1968). The height of the hills, meters above sea level, used for downhill skiing at these tourist destinations varies from 401 m (Suomu) to 718 m (Äkäslompolo-Ylläsjärvi). Taiga forests, composed mainly of Scots pine (*Pinus sylvestris*, L.), dominate the landscape. In addition, open mires are a significant feature of the landscape.

The average length of the growing season (days with an average temperature of +5° C or higher) is about 100–140 days. Snow covers the ground for about 6–7 months.

The size of the tourist destination is depicted by the number of beds, registered overnight stays by visitors, and the area of the destination (Table 6.1). Information on the accommodation facilities was gathered from the available literature. The areas (ha) of the tourist destinations were measured from topographic maps (1:50,000) using GIS tools. The area variable includes all urban structures as well as ski runs. If the urban

TABLE 6.1

The basic features of the tourist destinations included in this study.

Tourist destinations	Number of beds in 2003	Area (ha)	Number of registered overnight stays	
			March 2003	June 2003
Large				
Levi	18,000	393	39,756	6,220
Ruka	16,000	230	41,954	19,584
Saariselkä	11,000	280	33,726	16,216
Äkäslompolo	11,700	596	30,500	2,400
Small				
Ylläsjärvi	6,300	180	16,423	1,079
Pyhätunturi	3,500	248	8,172	3,194
Luosto	3,500	175	8,172	3,194
Suomu	1,500	110	—	—

core of the ski resort and the ski runs were disjunct, we measured each subunit separately and summed them. The largest tourist destinations with the highest registered overnight stay numbers are Levi, Ruka, Saariselkä, and Äkäslompolo. These tourist destinations have already been transformed into small leisure-time towns and are located relatively peripheral to other settlements. All the studied tourist destinations are ski resorts offering a diversity of outdoor and indoor activities. The busiest time is the winter with its slalom season, but nature-based tourism is increasing and spreading into uninhabited areas in the summer as well. The average population density of the tourist destinations included in the study is approximately 20 inhabitants/km^2, in their surroundings approximately 1.9 inhabitants/km^2, in the town of Rovaniemi 1,000 inhabitants/km^2, and in the town of Kuusamo 500 inhabitants/km^2. Tourism has been growing since the beginning of the 1980s in northern Finland. In the 1990s, northern Finland became a tourist-oriented center, thus becoming a typical peripheral tourist region, with several ski resorts located in otherwise sparsely populated areas (Regional Council of Lapland 2003). The number of permanent inhabitants at tourist destinations in northern Finland is low, normally in the hundreds.

On examining each of the eight tourist destinations, we distinguished the centers from the surroundings of the destinations: the centers consist of buildings (hotels, spas, cottages, ski runs, etc.), while the surroundings are mainly uninhabited forest areas. Uninhabited areas did not have any settlements or tourism-related infrastructure, and uninhabited areas were located at least 2 km from the outer border of the ski resorts or other settled areas. Forests, tourist destinations, and towns formed an urban gradient based on the number of inhabitants. Tourist destinations of different sizes can be considered as another gradient based on the scale of available accommodation facilities.

Potential avian nest predators in our study area, northern Finland, are the Eurasian Magpie (*Pica pica*), Hooded Crow (*Corvus corone cornix*), Jackdaw (*Corvus monedula*), Common Raven (*Corvus corone corax*), Eurasian Jay (*Garrulus glandarius*), and Siberian Jay (*Perisoreus infaustus*). Potential mammalian nest predators in the area are the red squirrel (*Sciurus vulgarius*), red fox (*Vulpes vulpes*), pine marten (*Martes martes*), stoat (*Mustela erminea*), and least weasel (*Mustela nivalis*).

An artificial nest predation experiment, point counts, and vegetation measurements were made at the exact same sites.

Artificial Nest Experiment

We established 20 artificial ground nests at each tourist destination and 15–20 artificial ground nests in the surrounding forests after conducting a 5-min bird survey at the each point. A total of

155 nests were established at the tourist destinations and 123 nests in the surrounding forests. Twenty artificial nests were established in each of the towns of Rovaniemi and Kuusamo. The nests were located at least 400 m from each other. Nests were put on leaf litter, directly on the ground, under a small tree or shrub, which covered the nest directly from above and exposed it in the other directions. In that way, we could standardize the quality of nest site, which may affect the encounter rate of nests by predators. Standardized sampling with dummy nests provides reasonable information on the potential risk of nest predation in different habitats (Wilcove 1985). Further, in this study we used the same method at every site, and thus the effect of the experimental procedure is equivalent and did not affect predation risk differentially. The nests in the towns and tourist destinations were placed in the most urbanized areas, with hotels, restaurants, and shops, that is, in the middle of the tourist town. Because of the patchy distribution of vegetation, random placement of nests was not possible in the ski resorts and towns. In the forests, the nests were placed randomly at least 200 m from the road leading into the forest. A nest was a hand-made cup in the soil without any particular structures. To reduce human scent around the nests we wore rubber boots and gloves when setting the nests. One egg of the Japanese Quail (*Coturnix coturnix*) was placed in each nest. The nest sites mimicked typical nest sites of the ground-nesting species in the study area. We did not use any nest markers.

The artificial nests were established during the period 5–17 June 2005, and the status of the nests was determined after two weeks of exposure. Two weeks is a typical incubation time for many ground-nesting passerines in Finland (Solonen 1985). We scored the nests as having been preyed upon if the egg had disappeared or had been broken. In this study, we did not measure the predators responsible for the nest losses. The experiment was conducted only once during the breeding season because birds rarely nest twice per season in northern latitudes.

Results provided by our earlier studies have indicated that artificial ground-nest predation risk is fairly consistent from year to year in the towns of northern Finland (Jokimäki and Huhta 2000). However, no corresponding data were available from the tourist destinations and forests of northern Finland. Therefore, we used multi-year data on artificial nest predation experiments from one small tourist destination (Pyhätunturi; data from 1996, 2001, 2002, and 2005) and one large tourist destination (Saariselkä; data from 1996 and 2005) located in northern Finland to test whether there is between-year variability in nest-predation risk. Multi-year data (1996, 2001, 2002, and 2005) collected from the surrounding forests of the Pyhätunturi tourist destination were used to test whether there was between-year variability in the nest-predation risk in forest areas. The study methods and the study sites providing these comparison data were the same as in this study.

Bird Surveys and Nest-predator Surveys

We determined bird abundances using the single-visit point-count method at each artificial nest site (Koskimies and Väisänen 1988). The single-visit survey detects about 60% of the breeding pairs and 90% of the species in forested areas (Järvinen and Lokki 1978, Järvinen et al. 1978). Because of the short breeding season and simple habitat structure, survey efficiency may be even greater in the north (Järvinen et al. 1978). We recorded each bird seen or heard regardless of the observation distance at the site during a 5-min count between 03:00 and 09:00 a.m. on weekdays. Over-flying birds that did not land within the study site and obvious feeding visitors that do not breed in the area were excluded (e.g., gulls feeding in towns). Whenever we were sure that a bird had already been observed, it was not included in the results for the second time. Bird surveys were conducted on the same day as placement of the artificial nests, with the survey conducted first and the artificial nest established right afterward at every site. All surveys were conducted during the period 5–17 June 2005 during good weather conditions.

The bird species recorded during the surveys were grouped according to their breeding habits into four nesting guilds: ground nesters, shrub nesters, tree nesters, and hole nesters (Harrison 1975; Appendix 6.1). We included the species nesting in buildings in the hole-nesters' group. There might be some variation in detectability of birds between the habitats but not within a habitat. By using the proportion of individuals belonging in the different groups, we avoided problems associated with possible differences in detectability between habitats.

We surveyed the potential avian nest predators (Eurasian Magpie, Hooded Crow, Common Raven, Siberian Jay, Eurasian Jay, and Jackdaw) and red squirrels during the 5-min bird surveys using the method used in ordinary bird surveys.

Vegetation Measurements

We collected information on the vegetation characteristics of the tourist destinations and their surrounding forests after the end of the artificial nest predation experiment during July–August 2005. The vegetation measurements were made using circular plots 3.99 m in radius. We located five circular plots at each bird survey site, the central plot being the location of the artificial nest, that is, the bird survey station. The other circles were located 25 m to the north, east, south, and west from the bird survey station.

We measured the tree-stem frequency distribution series for pine, spruce, and deciduous trees by height classes (1–3 m, >3–5 m, >5–10 m, >10–15 m, and >15 m), and the cover (%) of pine, spruce, juniper, and deciduous saplings or shrubs (<2 m) within each circular plot 3.99 m in radius was visually estimated. We also counted the total number of pines, spruces, and deciduous trees, the total number of trees belonging to different height categories, and the total number of shrubs within 3.99 m radius. Number of nest boxes was calculated within a 50-m-radius circle.

We visually estimated the area covered (%) by dwarf shrubs, herbs, grasses, mosses, lichens, bare ground, and dead cover (asphalt, parking areas, and buildings) from squares measuring 1 m × 1 m. The squares were located at the midpoints of every vegetation measurement plot.

A total number of 298 vegetation measurement plots (155 in tourist destinations, 123 in forests, and 20 in towns) were surveyed during the summer of 2005. The five measurements of the variables determined at each bird survey site were later averaged and used in statistical testing.

Statistical Methods

We made an arcsin transformation of the percentage vegetation and bird variables before any statistical testing. Because the bird variables did not have normal distributions even after arcsin transformation, nonparametric tests were used in comparing nest predator abundance and relative bird abundance in different nesting guilds. We used a nonparametric Tukey-type posterior test for pairwise comparisons (towns vs. tourist destinations, towns vs. forests, and tourist destinations vs. forests, Zar 1984). A *G*-test was used to compare nest predation risk between the towns, the tourist destinations, and the forest areas as well as between tourist destinations of different sizes. When using multiple tests, the sequential Bonferroni correction was made to minimize table-wise errors (Rice 1989).

The effects of vegetation characteristics and potential nest predators on nest losses were analyzed using stepwise logistic regression analysis (Hosmer and Lemeshow 1989, Trexler and Travis 1993). Separate analyses were made for both the vegetation variables and potential nest predators. The variables used in vegetation analysis were selected after analyzing the correlation matrix of all possible variables. The following variables that were not intercorrelated or that showed only weak intercorrelations ($P \geq 0.05$) were selected for the analysis: the total number of trees of the height class 1–3 m, the total number of trees of the height class 5–10 m, the total number of deciduous trees, the total number of pines, the total number of shrubs, and the cover of dwarf shrubs. We did not use approaches like principal component analysis that combine information from several variables into multivariate factors because we wanted to get information about variables that are more easily applicable for planning purposes at ski resorts. The independent variables used in the logistic regression analysis of nest predators were the most common avian nest predators (Eurasian Magpie and Hooded Crow) and red squirrel. The significance level required for each variable to be entered in and removed from the analysis was set at 0.05. The significance of the variables included in the model was evaluated by the Wald statistic. The values reported below in the Results section are mean ± SD unless otherwise stated.

RESULTS

Vegetation Differences between Towns, Tourist Destinations, and Forests

The total numbers of trees, trees 3–5 m and 5–10 m in height, and herb cover were less in towns and tourist destinations than in the forests (Table 6.2). There were also fewer spruces and trees 1–3 m

TABLE 6.2

Mean (±SD) of the vegetation characteristics in a town (Kuusamo) at the tourist destinations, and in the forest areas surrounding the tourist destinations.

Vegetation	Town ($n = 20$)	Tourist destinations ($n = 155$)	Forests ($n = 123$)	*K-W* test $P \leq$	Paired comparisons
Tree layer (number of trees)					
Scots pine	3.1 ± 4.5	8.8 ± 13.8	11.6 ± 14.2	0.05	F > T, D > T
Norway spruce	3.0 ± 3.7	3.3 ± 4.0	8.3 ± 7.8	0.05	F > T, F > D
Deciduous trees	5.7 ± 6.1	8.5 ± 12.7	8.6 ± 7.5	ns	
Tree layer 1–3 m	2.7 ± 3.7	6.8 ± 9.5	6.8 ± 6.1	0.05	F > T
Tree layer > 3–5 m	1.5 ± 2.0	4.0 ± 5.9	6.2 ± 4.7	0.05	F > T, F > D
Tree layer > 5–10 m	1.9 ± 3.4	5.6 ± 7.2	9.5 ± 10.1	0.05	F > T, F > D, D > T
Tree layer > 10–15 m	3.8 ± 5.0	3.4 ± 4.7	3.9 ± 3.1	ns	
Tree layer > 15 m	2.0 ± 3.3	0.9 ± 1.6	2.2 ± 3.4	ns	
Total number of trees	11.8 ± 7.4	20.6 ± 20.6	28.6 ± 13.4	0.05	F > T, F > D
Shrub layer					
Shrub cover (%)	10.3 ± 8.9	5.9 ± 10.4	1.1 ± 2.4	0.05	T > F, D > F
Field layer					
Dwarf shrubs (%)	7.5 ± 14.3	23.8 ± 14.1	44.6 ± 14.6	0.05	F > T, F > D, D > T
Asphalt (%)	32.3 ± 17.7	34.6 ± 20.8	0.5 ± 4.1	0.05	D > F, T > F

NOTE: The statistical differences tested by Kruskal-Wallis test and *P*-values indicate significance at the table-wide 0.05 level after sequential Bonferroni correction. Paired comparisons with a Tukey-type nonparametric test at $P < 0.05$ level. Statistically significant differences between habitats are indicated by letters: T = town, D = tourist destination, F = forest.

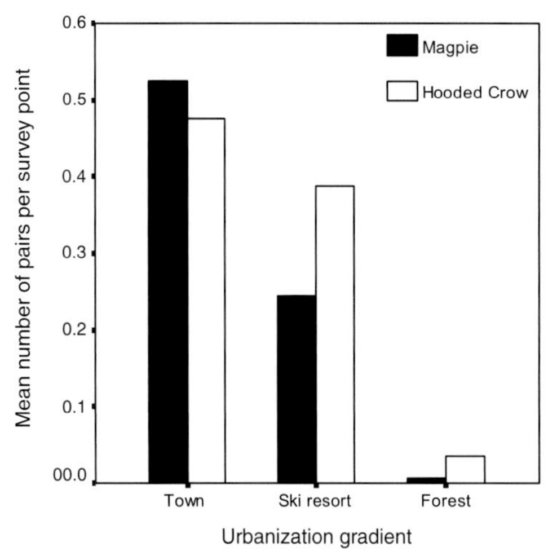

Figure 6.2. Mean number of Magpies and Hooded Crows in towns, tourist destinations, and forests.

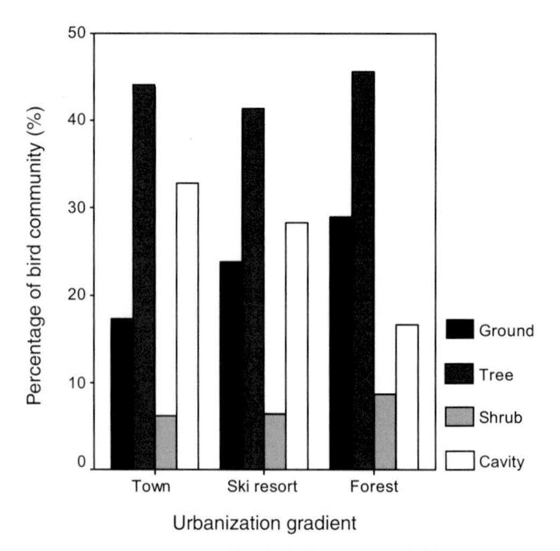

Figure 6.3. Percentage of birds belonging to different nesting guilds in towns, tourist destinations, and forests.

in height in towns than in the forests. The tourist destinations had more pines, trees 5–10 m in height, and dwarf shrubs than towns. Towns and tourist destinations had greater shrub and herb cover than the forest areas.

Nest Predation, Predators, and Bird Assemblages in Towns, Tourist Destinations, and Forests

The percentage of nests preyed upon differed between the towns, the tourist destinations, and the forest areas and was highest in the towns (57.5%, $n = 40$), intermediate at the tourist destinations (11.0%, $n = 155$), and lowest in the forests (4.7%, $n = 123$; $G_2 = 62.8$, $P < 0.001$). The percentage of nests preyed upon was higher in the towns than in the tourist destinations (paired comparison; $G_1 = 36.2$, $P < 0.001$) and in the forests (paired comparison; $G_1 = 25.7$, $P < 0.001$). The percentage of nests preyed upon in the tourist destinations was significantly higher than in the forests (paired comparison; $G_1 = 3.9$, $P = 0.049$).

Nests preyed upon did not differ between study years at the Pyhätunturi tourist destination (1996: 5%, 2001: 16%, 2002: 20%, and 2005: 5%; $G_3 = 3.4$, $P = 0.33$), at the Saariselkä tourist destination (1996: 25% and 2005: 40%; $G_1 = 0.4$, $P = 0.50$), and in the surrounding forests of the Pyhätunturi tourist destination (1996: 0%, 2001: 0%, 2002: 6%, and 2005: 0%; $G_3 = 2.8$, $P = 0.43$).

The abundance (pairs per survey station) of Eurasian Magpie and Hooded Crow differed between the towns, the tourist destinations, and the forests ($\chi^2 = 45.1$, df $= 2$, $P < 0.001$ and $\chi^2 = 42.7$, df $= 2$, $P < 0.001$, respectively; Fig. 6.2). The abundance of Eurasian Magpie increased from the forests (0.01 ± 0.08, $n = 123$) to tourist destinations (0.24 ± 0.54, $n = 155$) and to towns (0.53 ± 0.78, $n = 40$). In a similar manner, the abundance of Hooded Crow increased from the forests (0.03 ± 0.18, $n = 123$) to tourist destinations (0.39 ± 0.65, $n = 155$) and to towns (0.48 ± 0.71, $n = 40$). All paired comparisons of Eurasian Magpie abundance between habitats were significant (Tukey-type nonparametric test; $P < 0.05$); towns and tourist destinations had more magpies than forests; and towns had more magpies than tourist destinations. For the Hooded Crow, the paired comparisons between town and forests and between tourist destination and forests were significant (Tukey-type nonparametric test; $P < 0.05$); towns and tourist destinations had more crows than forests. However, the abundance of the Siberian Jay, Eurasian Jay, and Common Raven did not differ between the towns, the tourist destinations, and the forests ($P > 0.05$ in all cases). We observed Jackdaw only in the town of Rovaniemi, Eurasian Jay only at the tourist destinations, and Siberian Jay both at the tourist destinations and in the forests.

According to stepwise logistic regression analysis with predator species as independent variables, the high abundance of the magpie was negatively correlated with nest survival (Table 6.3). Other possible nest predator species did not correlate with nest losses.

TABLE 6.3

Parameter estimates and statistics from the logistic regression model on nest predation (dependent variable) and nest predators (independent variables).

Variable	β	SE	Wald χ^2	df	$P\leq$
Constant	2.180	0.204	114.278	1	0.001
Magpie	−0.767	0.272	7.957	1	0.005
Hooded Crow	−0.421	0.255	2.729	1	0.100
Red Squirrel	−0.625	0.831	0.566	1	0.450

A total of 71 bird species were observed in surveys (Appendix 3.1). Thirty-two species were observed in towns, 53 species were observed in ski resorts, and 47 species were observed in forests (Appendix 3.1). The proportional abundance of ground nesters and cavity nesters in bird assemblages differed between the towns, the tourist destinations, and the forests ($\chi^2 = 22.6$, df = 2, $P < 0.001$ and $\chi^2 = 50.1$, df = 2, $P < 0.001$, respectively; Fig. 6.3). The proportional abundance of ground nesters decreased from the forests (29.0% ± 14.6, $n = 123$) to tourist destinations (23.8% ± 13.4, $n = 155$) and to towns (17.3% ± 17.1, $n = 40$). The proportional abundance of cavity nesters increased from the forests (16.9% ± 11.8, $n = 123$) to the tourist destinations (28.4% ± 16.4, $n = 155$) and to thc towns (32.8% ± 19.9, $n = 40$). All paired comparisons of the ground-nester guild between habitats were significant (Tukey-type nonparametric test; $P < 0.05$); the proportional abundance of ground nesters was higher in the forests than in the towns or tourist destinations, and their proportion was higher at the tourist destinations than in the towns. For the cavity-nester guild, the paired comparisons between towns and forests and between tourist destinations and forests were significant (Tukey-type nonparametric test; $P < 0.05$); the proportional abundance of cavity nesters was lower in the forest than at the tourist destinations and towns. The proportional abundance of tree or shrub nesters did not differ between the towns, the tourist destinations, and the forests ($P > 0.01$ in both cases).

Nest Predation, Predators, and Bird Assemblages at Tourist Destinations

We found that nests preyed upon differed among the eight studied tourist destinations ($G_7 = 16.0$, $P < 0.025$). The highest nest losses were observed at the Saariselkä (40.0%) and Äkäslompolo (15.5%) tourist destinations. Nests preyed upon differed statistically between the tourist destinations and their surrounding forests at only one location, Saariselkä (Table 6.4).

The abundance of Hooded Crow differed among the tourist destinations ($\chi^2 = 19.3$, df = 7, $P < 0.007$) and was higher at Äkäslompolo (on average, 0.86 pairs per survey station) than in other areas ($P < 0.05$ in all 7 paired comparisons). Crow abundance was between 0.25 and 0.45 pairs per survey point at the other tourist destinations. At the smallest ski resort, Suomu, no magpies or Hooded Crows were observed. The abundance of the other nest predators did not differ among the tourist destinations.

According to the results of stepwise logistic regression analysis with vegetation variables as independent variables, nest survival increased with increasing dwarf shrub cover (Table 6.5).

The proportional abundance of tree nesters and cavity nesters differed among the tourist destinations ($\chi^2 = 32.3$, df = 7, $P < 0.001$ and $\chi^2 = 28.7$, df = 7, $P < 0.001$, respectively). The proportional abundance of tree nesters was highest at the Äkäslompolo (53.8%) tourist destination. The proportion of tree nesters at Äkäslompolo was higher than at the other six destinations (paired comparisons; $P < 0.05$). The proportional abundance of cavity nesters was higher at Pyhätunturi (36.6%) than at the other destinations ($P < 0.05$ in all seven paired comparisons). The proportional abundance of ground or shrub nesters did not differ between the tourist destinations ($P > 0.05$ in both cases). The proportional abundance of ground nesters varied between 21% and 26% of all breeding bird pairs at all tourist destinations.

TABLE 6.4

The percentage of artificial ground nests preyed upon at the different tourist destinations and in their surrounding forests and statistical differences between tourist destinations and their surroundings.

Study site	Levi	Ruka	Saariselkä	Äkäslompolo	Ylläsjärvi	Pyhätunturi	Luosto	Suomu
Tourist destination	10.0	0.0	40.0	15.50	5.0	5.0	10.0	10.0
Forest	13.30	0.0	6.70	5.30	4.80	0.0	0.0	5.0
G_1	0.09	—	5.060	1.05	0.01	1.36	2.77	0.37
$P\leq$	0.76	—	0.025	0.31	0.97	0.24	0.10	0.55

NOTE: In the case of Fuka, *G*-test was not conducted because nest losses were constant.

TABLE 6.5

Parameter estimates and statistics from the logistic regression model on nest predation (dependent variable) and vegetation variables (independent variables).

Variable	β	SE	Wald χ^2	df	$P\leq$
Constant	0.131	0.578	0.051	1	0.82
Dwarf shrubs cover	0.062	0.026	7.700	1	0.006
Trees 1–3 m	−0.400	0.117	0.118	1	0.73
Trees 5–10 m	0.059	0.118	0.247	1	0.62
Deciduous trees	0.148	0.101	2.151	1	0.14
Pines	0.053	0.075	0.509	1	0.48
Shrubs	−0.024	0.025	0.891	1	0.35

The size of the tourist destination (based on accommodation facilities) was not associated with the number of nests preyed upon, the abundance of magpies and Hooded Crows, or the proportional abundance of any of the nesting guilds (Spearman rank correlation; all tests $P \geq 0.05$, $n = 8$). The proportional abundance of the different nesting guilds did not differ between the large and small tourist destinations (Mann–Whitney *U*-test; $P > 0.05$ in all tests). The pooled abundance of the Eurasian Magpie and the Hooded Crow was higher at the large tourist destinations (0.79 ± 1.08) than at the small tourist destinations (0.47 ± 0.81; $U = 2596$, df $= 1$, $P = 0.020$).

DISCUSSION

Tourist Destinations Increase Nest-predation Risk

We found that the artificial ground-nest predation rate was highest in towns, intermediate at tourist destinations, and lowest in the surrounding forest areas. This observation indicates that towns and tourist destinations have some negative impacts on breeding success, at least with respect to ground-nesting bird species. According to the results of our earlier artificial nest predation experiments conducted in northern Finland, nests preyed upon in towns varied within the range of 69–80%, and nests preyed upon in forests varied between 0% and 4%, with the difference statistically significant (Jokimäki and Huhta 2000; Huhta et al. 1996, 1998). In this study, predation rate was about 5–15% at most of the tourist destinations. This corresponds well with the artificial ground-nest predation rate (about 20%) observed in small country villages (of 350 inhabitants) located in northern Finland (Jokimäki et al. 2005). In regard to the Saariselkä tourist destination, we found higher (40%) nest losses than at the other tourist destinations. One reason for this could be that the tourist season continues until June at Saariselkä, while at the other sites the peak period ends in April or early May. Predation level at the large-sized Saariselkä tourist destination approaches the predation levels observed in towns and corresponds well with the predation level observed in villages of approximately 1,000 inhabitants in northern Finland (Jokimäki and Huhta 2000, Jokimäki et al. 2005). Our results agree with the view that ground-nest predation increases with increasing urbanization (see Gering and Blair 1999 for opposite results, and Stracey and Robinson, chapter 4, this volume for similar results). We also found that the predation patterns at the tourist destinations were not significantly variable from year to year. At the individual tourist destination level, nest losses were significantly higher only at one tourist destination (Saariselkä) than in its surrounding forests, although the general trend was that nest losses were higher at the tourist destinations than in their surroundings. This might be partly explained by quite small sample sizes (20 nests per individual tourist destination and 20 nests in the surroundings).

During the past decade, corvids have expanded their distributions into urban environments (Gregory and Marchant 1996). Avian predators have been identified as the main nest predators in urban environments (Groom 1993, Major et al. 1996, Matthews et al. 1999, Jokimäki and Huhta 2000, Thorington and Bowman 2003). Our results suggest that a high abundance of Eurasian Magpies may decrease nest survival. As we did not use plasticine eggs or camera monitoring in the nests, we were unable to identify the predator species causing the nest losses. However, our earlier studies with plasticine eggs in Finland have shown that avian predators are the main predators in both urban and agricultural areas (Jokimäki and Huhta 2000).

In our study area, magpies and Hooded Crows were more abundant at the tourist destinations than in their surroundings. However, we did not observe either species at the Suomu tourist destination. The reason for this could be in the low visitor numbers and the very small size of the destination. Corvids probably benefit from increased food availability at larger tourist destinations. Waste and winter-feeding attracts magpies to settle in the vicinity of tourist destinations, and in this way they may impact on the breeding success of other birds.

In addition to the abundance of potential nest predators, the vegetation structure and the amount of covering vegetation can be important factors influencing nest losses. Our results highlighted the importance of the dwarf shrub cover. Nest survival increased with increasing dwarf shrub cover, and dwarf shrub cover decreased with increasing urbanization.

Tourist Destinations Affect Breeding-assemblage Structure

Birds may avoid breeding areas characterized by high nest predation risk (Suhonen et al. 1994). Open-cup nesters and ground-nesting birds are more vulnerable, especially to avian nest predation, than are, for example, cavity nesters (Huhta et al. 1998). Therefore, avian nest predation may influence the bird assemblages by changing the breeding success of birds with different breeding habits (see also Stracey and Robinson, chapter 4, this volume). Our results support this view. The proportional abundance of ground nesters was lowest in towns where the nest losses were the highest. The proportion of ground nesters was at its highest in forests where predation rate was at its lowest. The opposite trend was observed for cavity nesters: their proportion of the bird assemblages was highest in towns and lowest in forests. Tourist destinations manifested intermediate nest losses, and the proportional abundances of both ground- and cavity-nesting birds were intermediate. Reduced vegetation cover and a high avian nest-predation rate in urban habitats may impair nesting possibilities and the success of the species nesting within the vegetation. However, cavity nesters find suitable and safe nesting sites in buildings and nest boxes in towns and at tourist destinations. In our case, the exceptionally high amount of cavity nesters at the Pyhätunturi and Saariselkä tourist destinations can be explained by the great number of nest boxes in these areas (Kaisanlahti-Jokimäki et al., unpubl. data). Our results agree with the results of Stracey and Robinson (chapter 4, this volume) that urban winners are species that can protect themselves against avian nest predators.

Earlier studies have indicated that the total number of nest predators (Haskell et al. 2001) and nest predation increased with housing density (Thorington and Bowman 2003). In our study, the artificial ground nests preyed upon were found to differ among the eight tourist destinations. However, this could not be explained by the size of the tourist destinations when we used the number of beds as a surrogate for the size of the ski resort. The lowest predation rate was observed at the second-largest ski resort, Ruka, and the highest rate was observed at the third-largest destination, Saariselkä. Perhaps the size variation between the tourist destinations or the scale of our study were too small to measure the appropriate vegetation or landscape structure factors that influence, for example, corvid density (Hostetler and Holling 2000). The high nest losses at Äkäslompolo can be at least partly explained by the high abundance of the Hooded Crow at this tourist destination.

CONCLUSIONS

Urbanization as a process impacting on bird communities is clearly demonstrated in our data: the gradient of change from natural habitats to ski resorts to urban environments gradually molds the bird assemblages. The characteristic features of urban bird assemblages were already observed at the ski resorts, even though these are relatively new. The reasons for the observed changes are not necessarily straightforward. Partly they are caused by direct impacts such as human-caused disturbance, but there is also the mediating influence of modified landscapes. Habitat destruction and fragmentation with greater proportions of edge habitats and habitat modification and further attraction of certain generalists and urban exploiters can all modify bird assemblages. However, as nature-based tourism is nowadays a rapidly growing activity, it is also important for tourism operators to promote high bird species richness and the existence of natural bird communities with forest-core species in the close vicinity of tourist destinations. There are even extreme examples when the negative impacts on bird populations have reduced the attractiveness of the entire tourist destination (Feltwell 1996). Nowadays, tourists use destinations in northern Finland mainly during the winter season, and probably for that reason the effects of tourism on nature are not so harmful. However, winter tourism might increase the abundance of resident avian and non-avian (such as red fox) predators through human subsidies, and correspondingly increase their impact in breeding season as well. In addition, tourist destinations are under pressure to expand their winter season to include the summer. Expansion could change the situation, and good area planning and minimization of habitat modification would probably be the best way to minimize the negative impacts of urban sprawl on bird communities.

ACKNOWLEDGMENTS

Two anonymous reviewers gave good comments for the first version of the manuscript. This study was conducted by the support of the EU LIFE Environment program for the LANDSCAPE LAB project (http://www.arcticcentre.org/?DeptID55672).

LITERATURE CITED

Ahti, T., L. Hämet-Ahti, and J. Jalas. 1968. Vegetation zones and their sections in northwestern Europe. Annales Botanici Fennici 5:169–211.

Allen, C. R. 2006. Sprawl and the resilience of human and nature: an introduciton to the special feature. Ecology and Society 11(1):36. <http:www.ecologyandsociety.org/vol11/iss1/art36/>.

Batten, L. A. 1972. Breeding bird species diversity in relation to increasing urbanization. Bird Study 19:157–166.

Beissinger, S. R., and D. R. Osborne. 1982. Effects of urbanization on avian community organization. Condor 84:75–83.

Bezzel, E. 1985. Birdlife in intensively used rural and urban environments. Ornis Fennica 62:90–95.

Blair, R. 2004. The effects of urban sprawl on birds at multiple levels of biological organization. Ecology and Society 9(2):2. <http:www.ecologyandsociety.org/vol9/iss5/art2/>.

Blair, R. B. 1996. Land use and avian species diversity along an urban gradient. Ecological Applications 6:506–519.

Chace, J. F., and J. J. Walsh. 2004. Urban effects on native avifauna: a review. Landscape and Urban Planning 74:46–69.

Chettri, N., E. Sharma, and D. C. Deb. 2001. Bird community structure along a trekking corridor of Sikkim Himalaya: a conservation perspective. Biological Conservation 102:1–16.

Clergeau, P., S. Crocci, J. Jokimäki, M.-L. Kaisanlahti-Jokimäki, and M. Dinetti. 2006. Avifauna homogenisation by urbanization: analysis at different European latitudes. Biological Conservation 127:336–344.

Clergeau, P., J.-P. L. Savard, G. Mennechez, and G. Falardeau. 1998. Bird abundance and diversity along an urban-rural gradient: a comparative study between two cities on different continents. Condor 100: 413–425.

Diamond, J. M. 1986. Rapid evolution of urban birds. Nature 324:107–108.

Fernández-Juricic, E., and J. Jokimäki. 2001. A habitat island approach to conserving birds in urban landscapes: case studies from southern and northern Europe. Biodiversity and Conservation 10:2023–2043.

Feltwell, J. 1996. Tourisme mort Camargue. Biologist 43:181–183.

Gering, J. C., and R. Blair. 1999. Predation on artificial bird nests along an urban gradient: predatory risk or relaxation in urban environments. Ecography 22:532–541.

Gregory, R. D., and J. H. Marchant. 1996. Population trends of jays, magpies and Carrion Crows in the United Kindom. Bird Study 43:28–37.

Groom, D. W. 1993. Magpie *Pica pica* predation on Blackbird *Turdus merula* nests in urban areas. Bird Study 40:55–62.

Harrison, C. 1975. Field guide to the nests, eggs and nestlings of European birds with North Africa and Middle East. Collins, London, UK.

Haskell, D. G., A. M. Knupp, and M. C. Schnieder. 2001. Nest predator abundance and urbanization. Pp. 243–258. *in* J. M. Marzluff, R. Bowman, and R. Donnelly (editors), Avian ecology and conservation in an urbanizing world. Kluwer Academic Publishers, Boston, MA.

Hosmer, D. W., and S. Lemeshow. 1989. Applied logistic regression. John Wiley and Sons, New York.

Hostetler, M. E., and C. S. Holling. 2000. Detecting the scales at which birds respond to landscape structure in urban landscapes. Urban Ecosystems 4:25–54.

Huhta, E., J. Jokimäki, and P. Helle. 1998. Predation on artificial nests in a forest dominated landcape: the effects of nest type, patch size and edge structure. Ecography 21:464–471.

Huhta, E., T. Mappes, and J. Jokimäki. 1996. Predation on artificial ground nests in relation to forest fragmentation, agricultural land and habitat structure. Ecography 19:85–91.

Järvinen, O., and J. Lokki. 1978. Indices of community structure in bird censuses based on a single visit: effects of variation in species efficiency. Ornis Scandinavica 9:87–93.

Järvinen, O., R. A. Väisänen, and A. Enemar. 1978. Efficiency of the line transect method in mountain birch forests. Ornis Fennica 55:16–23.

Jokimäki, J. 1999. Occurrence of breeding bird species in urban parks: effects of park structure and broad-scale variables. Urban Ecosystems 3:21–34.

Jokimäki, J., and E. Huhta 2000. Artificial nest predation and abundance of birds along an urban gradient. Condor 102:838–847.

Jokimäki, J., M.-L. Kaisanlahti-Jokimäki, A. Sorace, E. Fernández-Juricic, I. Rodriguez-Prieto, and M. D. Jimenez. 2005. Evaluation of the "safe nesting zone" hypothesis across an urban gradient: a multi-scale study. Ecography 28:59–70.

Jokimäki, J., and J. Suhonen 1993. Effects of urbanization on the breeding bird species richness in Finland: a biogeographical comparison. Ornis Fennica 70:71–77.

Jokimäki, J., J. Suhonen, K. Inki, and S. Jokimäki. 1996. Biogeographical comparison of winter bird assemblages in urban environments in Finland. Journal of Biogeography 23:379–386.

Koskimies, P., and R. A. Väisänen. 1988. Monitoring bird populations in Finland: a manual. Helsingin Yliopiston Eläinmuseo, Helsinki.

Laiolo, P., and A. Ronaldo. 2005. Forest bird diversity and ski-runs: a case of negative edge effect. Animal Conservation 8:9–16.

Lehtilä, K., K. Ennola, and K. Syrjänen. 1996. The effect of summer cottages on land bird numbers in a Finnish archipelago. Ornis Fennica 73:49–59.

Major, R. E., G. Gowing, and E. Kendal. 1996. Nest predation in Australian urban environments and the role of the pied currawong *Strepera graculina*. Australian Journal of Ecology 21:399–409.

Mallord, J. W., P. M. Dolman, A. F. Brown, and W. J. Sutherland. 2007. Linking recreational disturbance to population size in a ground-nesting passerine. Journal of Applied Ecology 44:185–195.

Marzluff, J. M. 2001. Worldwide urbanization and its effects on birds. Pp. 19–47 *in* J. M. Marzluff, R. Bowman, and R. Donnelly (editors), Avian ecology and conservation in an urbanizing world. Kluwer Academic Publishers, Boston, MA.

Matthews, A., C. R. Dickman, and R. E. Major. 1999. The influence of fragment size and edge on nest predation in urban bushland. Ecography 22:349–356.

McKinney, M. L. 2002. Urbanization, biodiversity, and conservation. BioScience 10:883–890.

Miller, J. R., and N. T. Hobbs. 2000. Recreational trails, human activity, and nest predation in lowland riparian areas. Landscape and Urban Planning 50:227–236.

Miller, S. G., R. L. Knight, and C. K. Miller. 1998. Influence of recreational trails on breeding bird communities. Ecological Applications 8:162–169.

Nilon, C. H., C. N. Long, and W. C. Zipperer. 1995. Effects of wildland development on forest bird communities. Landscape and Urban Planning 32:81–92.

Regional Council of Lapland. 2003. Lapland tourism strategy 2003–2006. Rovaniemi, Finland.

Rice, W. R. 1989. Analyzing tables of statistical tests. Evolution 43:223–225.

Rolando, A., E. Caprio, E. Rinaldi, and I. Ellena. 2007. The impact of high-altitude ski-runs on alpine grassland bird communities. Journal of Applied Ecology 44:210–219.

Solonen, T. 1985. Suomen linnusto. Lintutieto Oy, Helsinki, Finland.

Sorace, A. 2002. High density of bird and pest species in urban habitats and the role of predator abundance. Ornis Fennica 79:60–71.

Suhonen, J., and J. Jokimäki. 1988. A biogeographical comparision of the breeding bird species assemblages in twenty Finnish urban parks. Ornis Fennica 65:76–83.

Suhonen, J., K. Norrdahl, and E. Korpimäki. 1994. Avian predation risk modifies breeding bird community on a farmland area. Ecology 75:1626–1634.

Sun, D., and D. Walsh 1998: Review of studies on environmental impacts of recreation and tourism in Australia. Journal of Environmental Management 53:323–338.

Thorington, K. K., and R. Bowman 2003. Predation rate on artificial nests increases with human housing density in suburban habitats. Ecography 26: 188–196.

Trexler, J. C., and J. Travis. 1993. Nontraditional regression analyses. Ecology 74:1629–1637.

Vale, T. R., and G. R. Vale. 1976. Suburban bird populations in west-central California. Journal of Biogeography 3:157–165.

Watson, A. 1979. Bird and mammal numbers in relation to human impact at ski lifts on Scottish hills. Journal of Applied Ecology 16:753–764.

Watson, A., and R. Moss. 2004. Impacts of ski-development on ptarmigan (*Lagopus mutus*) at Cairn Gorm, Scotland. Biological Conservation 116:267–275.

Wilcove, D. S. 1985. Nest predation in forest tracts and the decline of migratory birds. Ecology 66:1211–1214.

van der Zande, A. N., J. C. Berkhuizen, H. C. van Latesteijn, W. J. ter Keurs, and A. J. Poppelaars. 1984. Impact of outdoor recreation on the density of a number of breeding bird species in woods adjanced to urban residential areas. Biological Conservation 30:1–39.

Zar, J. H. 1984. Biostatistical analysis. Prentice-Hall, Englewood Cliffs, NJ.

Zeitler, A. 2000. Human disturbance, behaviour and spatial distribution of Black Grouse in skiing areas in the Bavarian Alps. Cahiers d'Ethologie 20:381–402.

APPENDIX 6.1

List of bird species observed in each habitat and their nesting category.

Scientific name	Nesting category[a]	Towns (*n* = 40)	Ski resorts (*n* = 155)	Forests (*n* = 123)
Falco columbarius	T		X	
Lagopus lagopus	G			X
Bonasa bonansia	G		X	X
Tetrao tetrix	G		X	X
Tetrao urogallus	G			X
Grus grus	G		X	X
Charadrius dubius	G	X		
Pluvialis apricaria	G			X
Tringa nebularia	G		X	X
Tringa ochropus	T		X	X
Tringa glareola	G		X	X
Actitis hypoleucos	G		X	
Numenius phaeopus	G		X	
Numenius arquata	G	X		
Gallinago gallinago	G			X
Larus ridibundus	G	X		
Columba palumbus	T		X	X
Columba livia domestica	C	X	X	
Cuculus canorus	T		X	X
Apus apus	C	X		
Dryocopus martinus	C		X	X
Dendrocopos major	C	X	X	X

APPENDIX 6.1 (*continued*)

Scientific name	Nesting category[a]	Towns (n = 40)	Ski resorts (n = 155)	Forests (n = 123)
Picoides tridactylus	C			X
Alauda arvensis	G		X	
Riparia riparia	C	X		X
Hirundo rustica	C		X	
Delichon urbica	C	X	X	
Anthus trivialis	G		X	X
Anthus pratensis	G		X	X
Motacilla alba	G	X	X	X
Motacilla flava	G		X	
Bombycilla garrulus	T		X	X
Prunella modularis	S		X	X
Acrocephalus schoenibanus	S	X	X	
Sylvia borin	S	X		
Phylloscopus trochilus	G	X	X	
Phylloscopus collybita	G			X
Phylloscopus sibilatrix	G			X
Phylloscopus trochiloides	G		X	X
Regulus regulus	T		X	X
Ficedula hypoleuca	C	X	X	X
Muscicapa striata	C	X	X	X
Saxicola rubetra	S	X		X
Oenanthe oenanthe	G		X	
Phoenicurus phoenicurus	C		X	X
Erithacus rubecula	S		X	X
Luscinia svecica	S	X	X	
Turdus pilaris	T	X	X	X
Turdus merula	T	X		
Turdus iliacus	T	X	X	X
Turdus philomelos	T	X	X	X
Turdus viscivorus	T			X
Parus montanus	C	X	X	X
Parus cinctus	C		X	X
Parus caeruleus	C	X		
Parus major	C	X	X	X
Certhia familiaris	C		X	
Perisoreus infaustus	T		X	X

APPENDIX 6.1 (CONTINUED)

Scientific name	Nesting category[a]	Towns ($n = 40$)	Ski resorts ($n = 155$)	Forests ($n = 123$)
Garrulus glandarius	T		X	
Pica pica	T	X	X	X
Corvus monedula	T	X		
Corvus corone cornix	T	X	X	X
Corvus corax	T		X	X
Passer domesticus	C	X	X	
Fringilla coelebs	T	X	X	X
Fringilla montifringilla	T		X	X
Carduelis chloris	T	X	X	
Carduelis spinus	T	X	X	X
Carduelis flammea	S	X	X	X
Loxia leucoptera	T		X	X
Loxia curvirostra	T	X	X	X
Loxia pytyopsittacus	T		X	X
Pyrrhula pyrrhula	T		X	X
Emberiza citrinella	G	X	X	X
Emberiza pusilla	G		X	X
Emberiza schoeniclus	G		X	
Total number of species		32	53	47
Total number of pairs		359	1,798	1,663

[a] G = ground, S = shrubs, T = tree, C = cavities and buildings.

PART TWO

Citizen Science and Demography of Urban Birds

CHAPTER SEVEN

The Use of Citizen Volunteers in Urban Bird Research

Timothy L. Vargo, Owen D. Boyle, Christopher A. Lepczyk, William P. Mueller, and Sara E. Vondrachek

Scientific natural history is one of the few endeavors in which any interested person can make original contributions to science . . . there are just too many kinds of organisms [to study] and too few professional scientists.

E. O. WILSON, 2006

Abstract. Training community volunteers to become citizen scientists for research has become increasingly popular as an effective and cost-saving alternative to, or augmentation of, paid field assistants. Citizen scientists allow for increased data-collection abilities and bring local support to research in their community. However, they also require significant amounts of time and resources in order to be trained effectively. In addition, there are still those in academia who believe citizen science has basic flaws that cannot be overcome. However, as urban areas grow citizen science offers an important and useful approach toward understanding birds in urban ecosystems. To highlight this importance, we present a case study from Milwaukee, Wisconsin (USA), where citizens play a vital role in state-of-the-art urban bird research studying migration stopover habitat use in an urban matrix. Our experiences with citizen volunteers have been overwhelmingly positive despite the added efforts, and have allowed us to tap into their individual strengths and the strengths of the community, as well as providing experiential learning to a community who can claim ownership of local avian research.

Key Words: citizen science, migration, urban parks, volunteers.

Though citizen science has been increasingly formalized over recent decades, the concepts on which the discipline is based are as old as science itself. Much of the earliest astronomical research, for instance, was conducted by clergy in the Catholic Church. While contemporary explanations of citizen science vary slightly, they converge on a definition of *the involvement of citizens from the nonscientific community in academic research* (Trumbull et al. 2000, Lee et al. 2006). By this definition, some of the most renowned scientists and ornithologists could be considered citizen scientists (e.g., Darwin, Mendel, Mayfield, Skutch, etc.); however,

Vargo, T. L., O. D. Boyle, C. A. Lepczyk, W. P. Mueller, and S. E. Vondrachek. 2012. The use of citizen volunteers in urban bird research. Pp. 113–124 *in* C. A. Lepczyk and P. S. Warren (editors). Urban bird ecology and conservation. Studies in Avian Biology (no. 45), University of California Press, Berkeley, CA.

there has been a separation over time between the citizen scientist and the professional scientist. Specifically, the practice of most scientific work is carried out by professionally trained scientists who conduct their research through the auspices of a research institution, government agency, nonprofit organization, or academic institution. In fact, the term "Ivory Tower" came to represent the isolation of academics and professional scientists from the daily life of citizenry (Holton 2002). Inevitably, the idea spread that true science should be kept in the realm of scientists, many of whom undoubtedly scoffed at the idea of a non-specialist participating in academic research without going through the proper channels of academia. Although citizen-based research is slowly gaining acceptance from academic institutions, government agencies, and conservation organizations (Heiman 2007), we have experienced firsthand from within the scientific community the misconception that citizens could not be reliably employed in academic research and therefore should not be used.

Even though citizen scientists have existed since the dawn of science, there has been a groundswell of interest in them in recent years. This interest stems from the confluence of several important situations facing ecology and conservation, namely: (1) the growing recognition of the need to monitor ecosystems and their inhabitants over time and across large areas in a consistent manner (Stevenson et al. 2003); (2) environmental problems, such as habitat loss, global climate change, and species extinction, which are growing ever more serious and outstrip the capacity of traditional academic scientists to address (Wilson 2006); (3) the fact that funding to study and monitor these environmental problems has not kept pace with the need (or has even decreased); and (4) the need to engage the citizenry in the stewardship of their own neighborhoods and to connect scientific research to the community in a more meaningful way. As a result, interest in citizen science monitoring and research has grown dramatically in the past 10 to 15 years, especially with regard to avian ecology and conservation. However, misgivings remain about its utility. Our purpose here is to critically examine the value of citizen scientists and to review the costs and benefits of using citizens in urban bird research. To illustrate our points we present a unique case study from Milwaukee, Wisconsin, USA, where the Milwaukee County Avian Migration Monitoring Partnership (MCAMMP) is carrying out an ambitious, state-of-the-art, stopover migration study that relies heavily on citizen scientists and has a primary goal of training the participants in field ecology techniques.

THE MISCONCEPTIONS OF CITIZEN SCIENCE

The gold standard of scientific research is that it must face the rigor and scrutiny of peer review to ensure objectivity, and that data are collected properly and interpreted correctly. One of the main criticisms of citizen science research stems from the issue of proper data collection, especially in the case of field studies. Specifically, field studies often involve the use of assistants in data collection. This use of assistants is where the misconception surfaces regarding the use of citizens to gather data.

Apart from the inevitable human error, we assert that there should be little difference between data collected by an undergraduate field assistant and data collected by a local resident, given they both have received proper training. Training is the key to success for any study. As long as the data-collection protocol is well thought out, a properly trained field assistant should produce reliable data. A project, therefore, should not be critiqued based solely on the use of citizen scientists, but on *how* they were used. Furthermore, recent studies illustrate that information collected by citizen scientists in such areas as monitoring is comparable to many paid technicians and scientists (Danielsen et al. 2006, Genet et al. 2008).

HOW CITIZEN SCIENTISTS ARE EMPLOYED

Citizen science projects generally fall into one of two categories (Table 7.1). The first category is one we will call the institutional or monitoring-based model, whereby a researcher or major institution initiates a study with a monitoring protocol that can be replicated by people in communities across the country. The institution develops the protocol, creates and monitors a database, and then provides training workshops to leaders who then bring the project to their local communities. Projects such as Cornell University's FeederWatch (Cornell Lab of Ornithology) and the University of Minnesota's Monarch Larvae Monitoring Project

TABLE 7.1
Classification of citizen science projects.

Institutional model	Scientific-training model
Protocol created by centralized institution or researcher (university, government agency) in a way that can be replicated in many different areas	Protocol created by researcher for use at a local site
Volunteer training happens online, or through "train the trainer" workshops	Direct training of volunteers by the researcher
Greater data collection power over larger areas	Data collection power limited in scope and over a smaller area (community)
Requires greater infrastructure to handle a large centralized database and potentially thousands of volunteers	Can be done with smaller infrastructure or fewer resources
Less direct quality control between researcher and volunteer	Greater quality control between researcher and volunteer
Easier for volunteers to participate only through data collection	—
Leans more heavily on self-direction for volunteers to increase scientific literacy	Great potential for in-depth learning of scientific learning through consistent contact with researcher

(University of Minnesota) fall into this category. These projects have their own strengths and weaknesses that have been increasingly critiqued and analyzed as they grow in popularity and produce useful studies (Lepczyk 2005, Genet et al. 2008). However, the primary use of the citizen scientist to the researchers in these cases is strictly for monitoring and data entry.

The second category, and focus of this chapter, is a type of citizen science that we term the scientific training model. Under the scientific training model, researchers bring in volunteers from the community to act as field assistants for their individual projects. The primary differences that set the scientific training model apart from the institutional model is that the citizen scientist is involved in all aspects of research and the scientific method (Trumbull et al. 2000), not just monitoring and data entry. This model adds motivation to the citizen scientist above and beyond helping a local scientist, as they gain experience, develop and fulfill personal interests, and use the research as a stepping-stone to professional proficiency. Additionally, the scientific training model allows the researcher to have more control in the direct training and monitoring of field assistants.

Given the lack of consensus surrounding the definition of the term citizen science, we distinguish the term "citizen volunteer" as a community volunteer who makes themselves available for a research project from a "citizen scientist," who is a community volunteer who gains an understanding and appreciation of the scientific process through their research experience.

CHALLENGES TO EMPLOYING CITIZENS

While no difference should exist among data collected from a student or full-time field assistant and a community volunteer, there are logistical challenges when using citizen volunteers. Specifically, volunteers have full-time careers and/or families, which limit their availability, compared to paid field assistants or student researchers who devote their time to a project as at least a part-time job. Organizing a team of volunteers, each with different skills, schedules, and availabilities, adds a significant amount of time to volunteer coordination and training (see Rotenberg et al., chapter 8, this volume for a similar situation).

Another challenge when dealing with volunteers is retention, especially in ornithological research. Because ornithological studies tend to rely heavily on identification skills for surveying birds (e.g., point counts, transects) and/or technical skills (e.g., mist-netting, banding) that require

TABLE 7.2

MCAMMP volunteer progression and retention from spring 2006 through fall 2007.

Project	Total number of volunteers	Active volunteers through fall 2007	Retention rate	Volunteer hours	$ saved based on $15 per hour to pay field assistants
Banding	93	64	69%	1,420	$21,300
Transects	17	7	41%	221	$3,315
Vegetation	8	8	100%	134	$2,010
Total	124	80	65%	1,775	$26,625

a great deal of training, it is important to retain volunteers over a long period of time. A full-time field assistant may need several weeks or more of intense training before they feel comfortable extracting a bird from a mist net or identifying warblers by song. Volunteer assistants with limited schedules compound this problem manyfold (or may simply be unable to volunteer at the level required in certain circumstances).

In our experience, two types of retention issues emerge when working with citizen scientists. First, the biology student interested in gaining experience has a class schedule to work around, but also has more availability and flexibility than other volunteers. Ideally, the student can work the research into their education in the form of an internship, thesis, or independent study. Often, however, the student graduates or takes on a class load that limits their availability, and eventually he or she drops out of the picture after the researcher has invested a great deal of time and effort in training. While a short-term loss to the project, these students carry the skills they learned to their future endeavors. Thus, such a loss should not be viewed negatively as in the long term we are helping to train future conservation biologists and ornithologists. For example, several of the citizen scientists working on the MCAMMP case study highlighted below have used their skills to advance professionally. Specifically, of the 64 active bird-banding volunteers in the fall 2006 field season (Table 7.2), four are now considering graduate study based on their experiences from the project, while a fifth is using the experience as part of his new position as an environmental educator.

The second type of volunteer who is difficult to retain is the professional whose availability is limited to mornings and weekends. In this case, it might take a year or two to reach the point where the individual can safely extract a bird from a mist net, for example. These volunteers tend to drop out of citizen science projects early because they feel they are not useful, or become bored due to their lack of engagement. Often, however, these volunteers have already developed an environmental ethic through previous experiences and may go through a great deal to adjust their schedules for the good of the project and/or are willing to stick with the project even if used sparingly.

A final challenge for using citizens in research deals with the scale and complexity of methodology, the cost of equipment, and liability issues, which might limit the accessibility of some research to citizenry. This presents a barrier, especially in disciplines such as microbiology or genetics, where it may be impractical to have citizen scientists running an electron microscope or working with hazardous reagents. Ecological research, on the other hand, tends to be much more accessible to employing citizen scientists even though the methodology may be complex.

BENEFITS OF EMPLOYING CITIZENS: OUTREACH

The most powerful case for working with citizens in research brings into play a host of societal and conservation issues. According to Chawla (1998), direct experiences with the natural world are the most important source of environmental sensitivity, and those who are active in conserving the environment most often cite childhood experiences in the natural world as critical antecedents to their later environmental actions. When citizens are directly involved in research, it allows for the interaction with and an appreciation of the value of intact ecosystems and native species in wild settings (Dunn et al. 2006). In other words, "a native bird in the hand is worth thousands on the Discovery Channel" (Dunn et al. 2006). Society as a whole reaps the benefits of getting more people to experience hands-on research, while citizen scientists learn valuable research skills, gain an education, and experience the natural world at a new level.

Another powerful case for working with citizen scientists in research is to connect more meaningfully with the local community. This local connection differs from the more common example of research, where scientists enter an area, carry out fieldwork with their own hired help, share their data in peer-reviewed journals, and leave little, if any, lasting, beneficial connection with the local community. Such ecological colonialism more than likely only leaves behind ill will, or at best indifference. In using local help, researchers benefit by getting the support of a segment of the community which has invested in the project, and who will start to feel pride and "ownership" (Poulsen and Luangleth 2005). Locally based research can help enforce and direct management issues through advisement and legislation (Nerbonne and Nelson 2004), and can lead to changes in the attitudes of locals toward conservation issues (Evans et al. 2005, Danielsen et al. 2006). Such local involvement is important no matter where a research project is carried out, but is particularly important when working in a foreign location (Bassett et al. 2004).

One implicit benefit of employing citizen scientists is that it can be viewed as another form of a college's or university's outreach and extension program. In particular, the training, education, and interaction with the citizen scientists all serve to provide them with more knowledge and skills. Moreover, some citizen science projects are exclusively targeted to measure change in knowledge, behavior, or attitudes of the participants over time.

There are a host of additional benefits in training volunteers to become citizen scientists on a local level. Researchers gain access to free and reliable help, thereby often saving money when compared to utilizing professional assistants (Danielsen et al. 2005), and can greatly increase their sampling areas and expand their data-collection power (McCaffrey and Turner, chapter 9, this volume, Rotenberg et al., chapter 8, this volume). Using local resources, such as nature center volunteers, allows for reliable long-term population studies that might be prohibitively expensive under any other circumstances (Toms and Newson 2006). Furthermore, in the case of urban ecosystems, they provide high densities of potential citizen scientists. To illustrate how a nature center can serve as a local resource for citizen science, we provide a case study dealing with the priority conservation question of how migratory birds use urban parks as refueling depots along the Great Lakes.

CASE STUDY

Urban Ecology Center

In the late 19th century, renowned landscape architect Frederick Law Olmsted was called upon by government leaders in Milwaukee, Wisconsin, to help design a section of their nationally recognized County Parks System. Olmsted saw great potential along a stretch of the Milwaukee River north of downtown, and hence Riverside Park was born. For many decades Riverside Park was the recreational hot spot of the city for activities such as swimming, canoeing, ice skating, picnicking, and curling. Over time, however, a downstream dam began to trap industrial pollutants, eventually causing massive fish die-offs, resulting in a marked decrease in people using the park. Subsequently, crime in the park and neighborhood increased, peaking in the mid 1980s. A grassroots community effort to save the park arose, with the goal of reclaiming Riverside Park for recreational use through environmental education and community support. The outcome of this effort was the creation of the Urban Ecology Center (UEC).

Today the UEC, with an annual budget of $1.9 million and 25 full-time staff, serves 68,000 neighborhood visits in two urban areas and leads tens of thousands of schoolchildren annually through Riverside Park. Concurrently, violent crime in the park has virtually disappeared. A replicable national model in neighborhood-based environmental education programs expanded to include a citizen science program, whereby publishable research is performed by local university students and professors with the support of citizen scientists. This program is run by a full-time research coordinator and draws support from a 30-member-strong Research Advisory Board comprised of university professors, independent researchers, government researchers, and other professionals.

Milwaukee County Avian Migration Monitoring Partnership

Milwaukee County serves as an ideal location for investigating urban bird ecology and conservation within the framework of citizen science. In particular, Milwaukee County has a large park system within an urban matrix that may serve as critical stopover habitat for migratory birds due to its proximity to Lake Michigan, a primary migratory flyway. Although much is understood about the migratory process itself, little is known about how changing stopover habitat conditions may influence avian populations, particularly for those stopovers that lie in urban areas.

Due to the need to gather data on urban migratory stopover sites, the Milwaukee County Avian Migration Monitoring Partnership (MCAMMP) was formed in 2005 by an interdisciplinary group of research scientists, university professors, government agencies, local conservation organizations, and nature centers (i.e., NGOs). MCAMMP's two primary goals were to understand migratory bird use of urban greenspaces and to build support for the nascent citizen science program at UEC. In order to address these goals, MCAMMP divided the research and citizen science training into three major components: (1) mist-netting and bird banding, (2) avian transect counts, and (3) vegetation sampling. This three-pronged approach addresses a variety of structural, physical, and biological characteristics that have been shown to influence bird populations in isolated urban greenspaces (Fernández-Juricic and Jokimäki 2001). Prior to conducting the research, MCAMMP began the process of building a citizen science network by holding an informational meeting for prospective volunteers at the UEC in February 2006.

Research and Training

Mist-netting and Bird Banding

Many studies on bird migration rely on presence/absence data to determine if a given habitat patch is important during migration; however, this approach does not always assess the quality of the habitat in question. That is, a bird in a given patch may have all the food and shelter needed, or it may be in the patch because it was the only place to land and is struggling to find quality food (Mehlman et al. 2005). In order to assess whether or not urban stopover sites provide suitable habitat we used a newly developed protocol that measures circulating metabolites in the plasma (Guglielmo et al. 2005). These metabolites indicate the physiological state of the bird during the 30 minutes prior to sampling, and thus in turn provide an indicator of the quality of the habitat where the bird was found (Guglielmo et al. 2005).

During the spring and fall migrations of 2006 and 2007, MCAMMP researchers and citizen scientists performed constant-effort mist-netting at two urban stopover sites that differ primarily in their degree of vegetative disturbance. In addition, project personnel obtained blood samples from seven species of migrant passerines representing insectivores (American Redstart, *Setophaga ruticilla*; Yellow-rumped Warbler, *Dendroica coronata*; Magnolia Warbler, *Dendroica magnolia*), granivores (Dark-eyed Junco, *Junco hyemalis*; White-throated Sparrow, *Zonotrochia albicollis*), and frugivores (Hermit Thrush, *Catharus guttatus*; Swainson's Thrush, *Catharus ustulatus*). Because metabolite levels change rapidly due to stress associated with capture, blood samples must be taken within 10 minutes of a bird flying into a net. This very short time window proves to be a major logistical challenge, as net runs need to be repeated constantly, and once a target bird is caught, it must be extracted and sampled within the 10-minute time period and then put through the normal banding process. Thus, a prodigious degree of coordination among members of a banding team, each of whom needs to be skilled in at least one of the aforementioned processes, is required.

THE CITIZEN SCIENCE TRAINING PROCESS

MCAMMP trained a dedicated group of citizen scientists to become skilled in net extraction, processing, and blood sampling. This training highlights an important trade-off. Specifically, without these citizen scientists, the project would require significant funding for full-time banding teams. On the other hand, the process leading to the development of skilled citizen scientists took several years and a great deal of patience in training volunteers and student interns, some of whom have already moved on to other projects.

During this one- to two-year process, there was a natural progression in the development of the citizen scientist's skills. Typically, a citizen scientist trainee spent several weeks solely focusing on observing the inner workings of the banding operation. The trainees shadowed members of the banding team, studying each step and asking questions. This shadowing process has proved to be a crucial time in volunteers' development and subsequent attainment of skills. We found that without this direct observational training some volunteers became frustrated with the lack of hands-on work. They felt that their contribution had little value, and as a result they did not stay with the project. Therefore, we have focused a great deal of effort on keeping new citizen scientists engaged by matching them to specific tasks based on skill sets. Of the 93 MCAMMP bird-banding volunteers (Table 7.1), 37 (40%) active volunteers remain in this learning stage (which also includes setting up and taking down mist nets and other equipment) and 29 (31%) are no longer active.

The next step for new volunteers was to learn what has been shown to be perhaps the most important but least technical step in the banding process: scribing. Scribing can be learned quickly but requires serious attention to detail. Because everything goes through the scribe, it is extremely important that this individual does his or her work with diligence and care. While scribing, the citizen scientist witnessed the processing of birds and became familiar with what features to record (e.g., sex, age, molt limits, fat, etc.). Of the 93 MCAMMP volunteers, 27 (29%) reached this stage.

After mastering the art of scribing, trainees learned how to process birds with the help of a trained bander. This accustomed volunteers to the feel and shape of a bird's body and how to properly manipulate a bird, all important skills that came in handy later when learning the extraction process. Fourteen (15%) out of 93 MCAMMP volunteers reached the processing stage, and three volunteers were trained in blood sampling.

Once volunteers mastered their processing skills, they were trained in extraction of birds from the nets. Feeling comfortable with extraction takes a significant amount of practice and one-on-one, hands-on training time. Moreover, extraction has the highest potential of harming birds and therefore must be done with extreme care. Eight (9%) out of 93 MCAMMP volunteers reached the extraction stage.

The time required for an individual to progress from untrained volunteer to skilled citizen scientist varied, depending on the amount of time they were available, the number of volunteers being trained on a given day, how busy the trainers were, the number of birds that were caught in a morning, and their previous knowledge, skills, and enthusiasm. However, we felt it was extremely important to engage trainees in the learning process early, teaching them the skills they needed in order to assist the team.

The utilization of citizen scientists is vital in instances when resources and funding are limited, as is the case with MCAMMP. However, we found three primary issues to address when conducting a citizen science project of this scale. First, it was difficult to find potential citizen scientists who were able to commit the amount of time needed and had the patience to be trained in all aspects of the banding operation. Second, the constraint of refining the net-run process to ensure that target birds were sampled within the 10-minute time period resulted in a steep learning curve for both researchers and trainees. The fact that we were taking blood from birds, a process often unsettling to the public, required us to be proactive in educating both our trainees and the public as to the importance of sampling blood and to the care we go through to minimize the impact on birds. Signage explaining the nature of our research, a rehabilitation box, and supplies, including Pedialyte©, should be part of any banding operation, and was crucial to ours. Third, all citizen scientists were trained using the same protocol at an educational workshop prior to each field season. The nature of this study, however, forced us to continually fine-tune our methodology, thereby requiring us to regularly update the citizen scientists who may have been trained in earlier methodologies. Ongoing training should be expected as a regular component of citizen science and/or monitoring projects (Pilz et al. 2006). Furthermore, because most days produced new combinations of citizen scientists, we fully documented the individuals present, weather conditions, modifications to protocol, and so on. Through our training and collection methods, all data were collected in a consistent manner and had documentation that allowed for flagging potential errors prior to analysis.

Avian Transect Counts

A second component to the MCAMMP study involved censusing the avian community in a number of greenspaces throughout Milwaukee County. We selected eight study sites (including the two sites used for banding) based upon (1) proximity to Lake Michigan (<1 km from the lake or >1 km inland from the lake), (2) upland or riparian plant communities, and (3) degree of vegetative disturbance (as defined by the relative abundance of invasive plant species). We conducted bird censuses (visual and auditory counts) at these eight sites via line-transect surveys, within preselected 250 m × 50 m transects (1.25 ha) during both spring and fall migratory periods of 2006 and 2007. We utilized the transect method as described by Rodewald and Matthews (2005), with modifications that included layout of routes and proximity to watercourses. In the first full year of the study (2006), a total of 111 transect samples (57 in spring and 54 in autumn) were completed within these eight study sites, yielding a total of 91 bird species observed.

THE CITIZEN SCIENCE TRAINING PROCESS

The assistance of trainees was critical in accomplishing the avian censuses. For those who had little or no previous experience with bird censuses, we utilized the online Wisconsin Birder Certification Program website (Cofrin Center for Biodiversity) to begin their training. This website provides a means to study bird identification using both visual and auditory clues, including photographic images and sound files of both songs and calls of Wisconsin birds. Testing for the development of knowledge of these identification clues is built into the website, and provides a method of determining the visual and auditory skills of an observer, with the ultimate goal of certification. Online resources such as these are becoming increasingly important as citizen-based monitoring gains popularity (Stevenson et al. 2003).

Two well-studied trainees were able to run transects on their own, and the rest used their time to study and accompany the MCAMMP transect leader on censuses. But by the end of the second field season, the majority of transect volunteers either had not progressed to the point where they possessed sufficient field skills to perform transect counts alone or had developed the confidence that they could reliably perform transect counts. During 2006, 12 volunteers conducted censuses, of whom three (25%) were retained by the project and participated again during 2007. Six (50%) of the volunteers ultimately did not feel confident enough to do transects on their own, but two of these went on to assist with vegetation sampling and banding, so a total of five were retained by the project as a whole. With the application of additional individual practice utilizing the Wisconsin Birder Certification Program website, our plan is to continue training volunteers to run censuses on their own as the project continues.

Even though many novice birders are enthusiastic and willing to learn, a significant obstacle in engaging them as citizen scientists is that they often become overwhelmed with the perceived degree of difficulty in learning avian songs and calls. Intermediate birders, on the other hand, often feel that they have not progressed far enough in the learning process to go through the certification (which we have deemed critical for data-quality assurances). To assist in providing incentives for both groups, we are seeking local businesses and other entities that will offer incentives (e.g., *Birdwatching*, a nationally recognized birding magazine, offered a free one-year magazine subscription as an incentive) to participants who follow through with certification and then perform a specified number of censuses on our study sites. The unfolding of our progress as the project continues should indicate whether this system of incentives helps to move this aspect of our program forward.

Vegetation Sampling

In order to examine the relationships between body condition of migratory birds and their habitat use with stopover site characteristics, we collected data on plant community composition and structure at the same eight study sites used for bird transects. We are particularly interested in understanding whether relationships exist between exotic plant species abundance at a stopover site and habitat use by migrants. To determine if relationships exist, we sampled plant species composition and percent cover in the herb, shrub, and tree canopy layers in fifty 1-m^2 plots spaced on a regular 20-m grid throughout each site. We also recorded percent cover of other site features such as leaf litter, moss, bare ground, coarse woody debris, and running or standing water in each plot. Finally, we characterized the tree species composition and structure through the use of the random-pairs method (Cottam and Curtis 1949, 1955). Under the random-pairs technique, we randomly selected two trees at each plot, identified them to species, recorded the diameter at breast height, and measured the distance between the pairs. This method provided a quick and reliable way to characterize the tree composition, structure, and distribution of a site (Cottam and Curtis 1949, 1955).

THE CITIZEN SCIENCE TRAINING PROCESS

As with the other areas of MCAMMP, we enlisted citizen scientists to collect vegetation data. They helped with identifying plant species (particularly trees), estimating percent cover of vegetation, measuring tree diameter at breast height and distances between trees, and determining other site characteristics. Volunteer assistance allowed us to double the number of sampling teams and therefore double the amount of vegetation plots we could complete in a field day. Each team consisted of one experienced researcher and one or two citizen scientists. In some cases, volunteers progressed in their ability to identify plants to the point where they could lead a team on their own with only occasional questions of species identification.

Surveying vegetation with citizen scientists presents a slightly different set of challenges from those faced by avian researchers. Where volunteers helping with banding or transects tend to stay on existing trails, vegetation volunteers have to cover an entire site, which may include very steep-sided ravines infested with extremely dense thickets of invasive exotic shrubs like common buckthorn (*Rhamnus cathartica*). Working under these physically demanding conditions for long hours (a typical vegetation sampling field day is 6–8 hours) can be extremely challenging for citizen volunteers not used to working outdoors, especially when summer temperatures push into the 90s (Fahrenheit). However, even if we are only able to collect data for half a day with volunteers, due to strenuous conditions, we are still able to collect the same amount of data as the researchers alone would be able to collect in an entire day, and we have the added benefit of educating volunteers who are engaged in and care for the project. Although vegetation sampling had the smallest number of volunteers (Table 7.1), this portion of the project has had a 100% retention rate.

Another challenge that has occasionally arisen with the vegetation volunteers relates to the concept of random sampling points. At times a citizen scientist has advocated moving a plot "just a little" to capture a new species. This situation was easily turned from a "challenge" to an educational opportunity, and the researcher could reiterate the need for our sampling protocols to remain unbiased. Moments like these where researchers educate citizen volunteers with firsthand examples of the scientific method in action and the power (and limitations) of science are extremely rewarding as we observe a citizen volunteer who has just transcended into a citizen scientist.

Volunteer and Member Strengths of the MCAMMP Project

As with other multifaceted citizen-based research projects, the chances of long-term success and credibility depend on tapping into the strengths of the collaborators (Savan et al. 2003). For MCAMMP, finding and training over 100 active citizen scientists was facilitated by the Urban Ecology Center's strong community presence, large volunteer pool, and the support of two full-time staff members (one a volunteer coordinator and the other a research coordinator). Support from a state agency (Wisconsin DNR) and four collaborating universities (University of Hawai'i at Mānoa, University of Wisconsin-Milwaukee, University of Wisconsin-Madison, and University of Western Ontario) lends considerable credibility, funding support, scientific oversight, and access to high-tech laboratory equipment. The Wisconsin Society for Ornithology (WSO), the Wisconsin Bird Conservation Initiative, Riveredge Nature Center, and the Milwaukee Audubon Society provide local support and highly effective networking opportunities to the birding communities throughout the region. All of the supporting partners are established community organizations with commitment to scientific and environmental education.

We were also extremely fortunate to draw upon a diverse and talented pool of volunteers. For example, who better to train in blood sampling

TABLE 7.3

Benefits and limitations of using citizen scientists in urban bird research.

Benefits	Limitations
Increases data-collection power	Limited availability due to jobs, families, etc.
Enables long-term studies when using local volunteers and organizations are already established in the community	Increased time and effort into training and coordinating volunteers with different schedules
Cost-effective alternative or addition to paid field assistants when funding is limited	Volunteers often drop out if they are too busy or not engaged
Local support and "ownership"	Bird- and plant-related research often requires technical skills and knowledge of bird identification, which may seem daunting
Information collected by citizen scientists is of the same quality as paid technicians	Conditions can be physically demanding for a volunteer not used to spending a long time in the field
Citizen scientists gain an education and learn skills to carry with them	—
Citizen scientists gain an appreciation of the natural world and conservation that can influence legislation and directives	—
Volunteers have unique talents that can help in unexpected ways	—

techniques than a veterinary technologist, or a family doctor? Who better to look for birds in nets and to quickly move a bird in a bag from the net to the blood sampling table then two schoolchildren who could not offer their technical expertise but could offer fresh energy and enthusiasm? And who could have foreseen that our most reliable data scribe would be someone with limited use of her hands, but who deftly records data using a pencil held in her mouth? We drew from the ranks of birders, plant enthusiasts, college students, lawyers, art professors, and high school teachers, each of whom contributed in their own unique way.

CONCLUSION

Urban bird ecology and citizen science have each seen marked growth and interest in recent years. In part, this growth stems from the increasingly urban nature of the world coupled with the need for more data monitoring. However, it can also be seen as an important progression of bringing the public into greater awareness about the environment in which they live. In the case of the MCAMMP project, we sought to conduct an integrated urban bird study directly with citizen scientists using the scientific training model.

Over the first two years of this ongoing project we provided informational workshops to over 65 people, 40 of whom ended up volunteering. The rest of the volunteers were recruited through networking (e.g., Wisbirdn, a Wisconsin ornithology electronic mail listserv), existing volunteers in MCAMMP organizations, and word of mouth. Initial participants ranged from novice to experienced in terms of their knowledge of birds, ecology, and the environment in general. While each participant brought a unique set of skills to the project, they have each been transformed by the experience and training we provided. The degree to which an individual participated in the project undoubtedly affected their transformation. For instance, participants who were involved for less than one field season may only have gained a deeper appreciation for birds and their conservation and not necessarily garnered the skills to truly be considered a citizen doing science. On the other end of the spectrum, we have a number of individuals who have been wholly transformed from non-scientist volunteers to full-fledged citizen scientists, able to readily engage the scientific process. In fact, several of the citizen scientists are now considering returning to school to train as professional scientists based on the MCAMMP experience. Because of the importance of documenting these changes in behavior, we will be adding an evaluation (survey) and longitudinal follow-up component to the 2009 field season, so that we can better evaluate the impact of our project on the citizen scientists.

While our experiences with citizen scientists in urban bird studies have been overwhelmingly positive, we found that the process of training and working with citizen scientists led to a number of trade-offs (Table 7.3). It is critical that any citizen science project be aware of these trade-offs if it is to succeed. For instance, in the case of MCAMMP, not all data collected were of research value, but the process led to improved methodologies, resulting in a better study. Thus, while our research remains ongoing, we have already found great success in taking a citizen science approach to addressing a fundamental question of urban bird ecology. Moreover, the project has been of great value to the community, as witnessed by the great number of people who have observed and commented on the citizen scientists engaged in research at Riverside Park. Such recognition helps to build capacity in the community, as people have a greater understanding of the system they live in and how they affect it. Ultimately, then, our citizen science approach represents a win-win situation as we build capacity in the community and gain new scientific knowledge.

ACKNOWLEDGMENTS

Please note that the authors after Vargo are listed alphabetically. We would like to thank B. Fetterley (UEC), C. Guglielmo (MCAMMP), C. Hagner (*Birdwatching*), and A. Sherkow (MCAMMP) for their comments on a draft version of the manuscript and would like to recognize other MCAMMP partners M. Feider, G. Fredlund, and D. Hartmann. In addition, we greatly appreciate the comments of J. Rotenberg and E. Strauss on a previous version of the manuscript, which helped to focus our work. We gratefully acknowledge the generous financial support of the Wisconsin DNR (Citizen-based Monitoring Grant), the U.S. Fish and Wildlife Service's Neotropical Migratory Bird Conservation Act (NMBCA grant #3503), and the Wisconsin Society for Ornithology. Our chapter is dedicated to the 100+ volunteers who make MCAMMP possible.

LITERATURE CITED

Basset, Y., V. Novotny, S. E. Miller, G. D. Weiblen, O. Missa, and A. J. A. Stewart. 2004. Conservation and biological monitoring of tropical forests: the role of parataxonomists. Journal of Applied Ecology 41:163–174.

Chawla, L. 1998. Significant life experiences revisited: a review of research on sources of environmental sensitivity. Journal of Environmental Education 29:11–21.

Cofrin Center for Biodiversity, University of Wisconsin–Green Bay. Wisconsin Birder Certification Program. <http://www.uwgb.edu/birds/certification/> (26 August 2009).

Cornell Lab of Ornithology. Project Feeder Watch. <www.birds.cornell.edu/pfw> (22 February 2007).

Cottam, G., and J. T. Curtis. 1949. A method for making rapid surveys of woodlands by means of pairs of randomly selected trees. Ecology 30:101–104.

Cottam, G., and J. T. Curtis. 1955. Correction for various exclusion angles in the random pairs method. Ecology 36:767.

Danielson, F., N. D. Burgess, and A. Balmford. 2006. Monitoring matters: examining the potential of locally-based approaches. Biodiversity and Conservation 14:2507–2542.

Dunn, R. R., M. C. Gavin, M. C. Sanchez, and J. N. Solomon. 2006. The pigeon paradox: dependence of global conservation on urban nature. Conservation Biology 20:1814–1816.

Evans, C., E. Abrams, R. Reitsma, K. Roux, L. Salmonsen, and P. P. Marra. 2005. The Neighborhood Nestwatch program: participant outcomes of a citizen-science ecological research project. Conservation Biology 19:589–594.

Fernández-Juricic, E., and J. Jokimäki. 2001. A habitat island approach to conserving birds in urban landscapes: case studies from southern and northern Europe. Biodiversity and Conservation 10:2023–2043.

Genet, K. S., C. A. Lepczyk, R. A. Christoffel, L. G. Sargent, and T. M. Burton. 2008. The use of volunteer monitoring programs for amphibian conservation and management across urban-rural gradients. Pages 565–574 *in* R. E. Jung and J. C. Mitchell, editors. Urban herpetology: ecology, conservation and management of amphibians and reptiles in urban and suburban environments. Society for the Study of Amphibians and Reptiles, Salt Lake City, UT.

Guglielmo, C. G., D. J. Cerasale, and C. Eldermire. 2005. A field validation of plasma metabolite profiling to assess refueling performance of migratory birds. Physiological and Biochemical Zoology 78:116–125.

Heiman, M. K. 2007. Science by the people: grassroots environmental monitoring and the debate over scientific expertise. Journal of Planning Education and Research 16:291–299.

Holton, S. A. 2002. Why an ivory tower? National Teaching & Learning Forum 11(3):10–11.

Lee, T., M. S. Quinn, and D. Duke. 2006. Citizen science, highways, and wildlife: using a web based GIS to engage citizens in collecting wildlife information. Ecology and Society 11(1):11. <http://www.ecologyandsociety.org/vol11/iss1/art11/>.

Lepczyk, C.A. 2005. Integrating published data and citizen science to describe bird diversity across a landscape. Journal of Applied Ecology 42:672–677.

Mehlman, D. W., S. E. Mabey, D. N. Ewert, C. Duncan, B. Abel, D. Cimprich, R. D. Sutler, and M. Woodrey. 2005. Conserving stopover sites for forest-dwelling migratory landbirds. Auk 122:1281–1290.

Nerbonne, J. F., and K. C. Nelson. 2004. Volunteer macroinvertebrate monitoring in the United States: resource mobilization and comparative state structures. Society and Natural Resources 17:817–839.

Pilz, D., H. L. Ballard, and E. T. Jones. 2006. Broadening participation in biological monitoring: handbook for scientists and managers. General Technical Report PNW-GTR-680. U.S. Department of Agriculture, Forest Service, Pacific Northwest Research Station, Portland, OR.

Poulsen, M. K., and K. Luanglath. 2005. Projects come, projects go: lessons from participatory monitoring in southern Laos. Biodiversity and Conservation 14:2591–2610.

Rodewald, P. G., and S. N. Matthews. 2005. Landbird use of riparian and upland forest stopover habitats in an urban landscape. Condor 107:259–268.

Savan, B., A. J. Morgan, and C. Gore. 2003. Volunteer environmental monitoring and the role of the universities: the case of citizens' environment watch. Environmental Management 31:561–568.

Stevenson, R. D., W. A. Haber, and R. A. Morris. 2003. Electronic field guides and user communities in the eco-informatics revolution. Conservation Ecology 7(1):3. <http://www.consecol.org/vol7/iss1/art3>.

Toms, M. P., and S. E. Newson. 2006. Volunteer surveys as a means of inferring trends in garden mammal populations. Mammal Review 36:309–317.

Trumbull, D. J., R. Bonney, D. Bascom, and A. Cabral. 2000. Thinking scientifically during participation in a citizen-science project. Science Education 84:265–275.

University of Minnesota. Monarch larva monitoring project. <www.mlmp.org> (22 February 2007).

Wilson, E. O. 2006. The creation: an appeal to save life on Earth. W. W. Norton & Company, New York.

CHAPTER EIGHT

Painted Bunting Conservation

TRADITIONAL MONITORING MEETS CITIZEN SCIENCE

James A. Rotenberg, Laurel M. Barnhill, J. Michael Meyers, and Dean Demarest

Abstract. Avian citizen science projects can successfully carry out research and/or monitoring programs that are otherwise impossible to carry out by employing field technicians. Reliability of volunteer observations is high, with quality data sets used in many peer-reviewed publications. Using a citizen science approach, we initiated the Painted Bunting Monitoring Project to study the eastern population of Painted Buntings (*Passerina ciris*) in North and South Carolina. Breeding Bird Survey data show that eastern Painted Buntings have declined at least 3.2% annually over a 30-year period, possibly due to increased coastal development and agricultural practices, both of which reduce the shrub-scrub brush vital to breeding Painted Buntings. Since Painted Buntings typically visit backyard bird feeders, citizen scientists can readily participate in a variety of data-generating components that aid us in comparing subpopulations breeding in suburban, rural, and natural habitats. Along with backyard banding of individuals, these data can include quantifying demographic parameters such as population distribution, density, and abundance; productivity and adult survival; and behavioral patterns of site fidelity and habitat use. Here we report the results from our first-year Painted Bunting Observer Team volunteer data sets, and offer recommendations to others who may be considering a citizen science-based data collection program.

Key Words: citizen science, conservation, monitoring, Painted Bunting, *Passerina ciris*.

The citizen science approach, where volunteer observers aid scientists in data collection, has become a useful means of investigating large-scale questions related to a variety of aspects of avian ecology. Programs such as the National Audubon Society's Christmas Bird Count (CBC) and the North American Breeding Bird Survey (BBS) administered by the U.S. Geological Survey and Canadian Wildlife Service, have engaged citizen science volunteers in avian research for years. More recently, programs initiated by the Smithsonian Institution/National Zoo (i.e., the Neighborhood Nestwatch Program) and by the Cornell Laboratory of Ornithology (i.e., Great Backyard Bird Count and Project FeederWatch) have been extremely successful, answering questions covering large temporal and spatial scales that otherwise would be impossible to

Rotenberg, J. A., L. M. Barnhill, J. M. Meyers, and D. Demarest. 2012. Painted Bunting conservation: traditional monitoring meets citizen science. Pp. 125–138 *in* C. A. Lepczyk and P. S. Warren (editors). Urban bird ecology and conservation. Studies in Avian Biology (no. 45), University of California Press, Berkeley, CA.

conduct, as resources to carry out such projects are limited (Wells et al. 1998, Bhattacharjee 2005, Evans et al. 2005). In addition, once engaged in a citizen science-based project, volunteers can increase their awareness of science and the environment (Brossard et al. 2005).

However, citizen science projects have been criticized for a variety of reasons: BBS for not estimating population change for some species (Niven et al. 2004), and for issues regarding roadside sampling and lack of detectability estimation (Link and Sauer 1997); CBC for poor sampling coverage (Dunn 1986); and Project FeederWatch-type projects for their lack of scientific control (Irwin 1995). These programs tend not to include urban areas, survey private lands, and/or limit their scope in these habitats to only a few survey circles and routes (Lepczyk 2005, McCaffrey 2005). Vargo et al. (chapter 7, this volume) provide a comprehensive treatment of the advent of citizen science, its criticisms, and its benefits.

Still, a citizen science monitoring program can be an effective tool for data collection, especially when no viable alternative exists. Specifically, citizen science-based monitoring programs offer project feasibility and economic viability. Many times state and federal wildlife agencies cannot directly carry out monitoring as identified by conservation plans, and therefore some monitoring may be best accomplished by employing citizen science models (Danielson et al. 2006). Examples of such programs are the Mycoplasmal Conjunctivitis Feeder-watch Project for the House Finch (*Carpodacus mexicanus*; Hartup et al. 1998), the Tucson Bird Count (McCaffrey 2005, McCaffrey and Turner, chapter 9, this volume), Road Watch in the Pass (Lee et al. 2006), and the Milwaukee County Avian Migration Monitoring Partnership (Vargo et al., chapter 7, this volume).

Both state comprehensive wildlife conservation strategies (North Carolina Wildlife Resources Commission 2005, Kohlsaat et al. 2006) and national level efforts (North American Bird Conservation Initiative 2007) have emphasized a need for multiscale avian monitoring programs. A key issue for such monitoring programs is that they must be designed and implemented so that the results can assess changes in populations and habitats over time as well as detect species-specific trends of distribution and population size. These data then can be used to guide future management and/or conservation strategies.

One species that may benefit from using the citizen science approach to monitor its population is the Painted Bunting (*Passerina ciris*), because it frequents backyard bird feeders. Past studies using bird feeders and supplemental food have demonstrated the importance of feeders to birds by quantifying a variety of parameters such as: feeding and visitation rates (Brittingham and Temple 1992a, Wilson 2001), possible feeder dependency (Brittingham and Temple 1992b), the utility of mark-recapture methods on birds using feeders (Wilson 2001), survivorship (Brittingham and Temple 1988), and most recently the possibility for feeders as drivers of ecological change (Robb et al. 2008). A set of similar parameters could be collected by using the citizen science model on the eastern population of Painted Bunting, which breeds in a restricted range within the Atlantic Coastal Plain, from North Carolina to Florida (American Ornithologists' Union 1998, Sykes and Holzman 2005). Today, this range can be identified across what is now an urban/suburban-rural/agricultural gradient. Since 1966, BBS data has indicated an annual decline of 3.2% for Painted Buntings in the southeast region (Sauer et al. 1997). The bunting's decline may be caused by a variety of factors, principally increased coastal development (due to a steadily increasing human population over the region) and contemporary agricultural practices, both of which tend to clear the early-successional shrub-scrub habitat that is vital to breeding (Springborn and Meyers 2005). Shrub-scrub vegetation, along with maritime forests, another favored breeding habitat of the Painted Bunting, are among the most threatened habitats along the coastal zone (North Carolina Wildlife Commission 2005). Given the population declines and loss of habitat, Partners in Flight has listed the Painted Bunting as both a highly ranked "Species at Risk" (4.29/5.00; Hunter et al. 1993a, 1993b) and a "Watch List Species" of Continental Importance for the United States and Canada because of restricted distribution or low population size (Rich et al. 2004). Additionally, the U.S. Fish and Wildlife Service listed the Painted Bunting as one of its focal species (USFWS Focal Species Fact Sheet 2005), and states within the eastern range of the species listed Painted Bunting as a species of "highest conservation concern" in their Comprehensive Wildlife Conservation Strategies (North Carolina Wildlife Commission 2005, Kohlsaat et al. 2006).

Because of these threats and declining population trend, the Eastern Painted Bunting Working Group determined that a two-part monitoring program consisting of new protocols specific to Painted Buntings was needed to assess the population's current status and to develop long-term conservation goals. One part is a four-state (North Carolina, South Carolina, Georgia, and Florida) effort, consisting of standardized monitoring (i.e., point counts) conducted mainly by skilled field technicians (i.e., state agency employees, etc.—"traditional monitoring"). The other part is a citizen science volunteer project in North and South Carolina (NC and SC, respectively), established as a way to contribute to and complement the broader four-state monitoring effort, expand our in-the-field workforce, and enable data collection in areas inaccessible with existing surveys. We expect that these data will enable us to gain a greater insight on a state-by-state basis into how Painted Buntings use food resources, how Painted Bunting populations vary over an urban/suburban-rural/agricultural gradient, and how supplemental food at feeders may affect survivorship and productivity across this gradient. If successful, the project may be enlarged to cover Georgia and Florida in future years.

Our main goal is to develop strategies for sustaining eastern Painted Bunting populations, using a citizen science framework where volunteers play a major role in monitoring and collecting data in the field. Based on this goal, our objectives were to: (1) recruit citizen volunteers who would make observations and collect data at backyard bird feeders across the urban/suburban-rural/agricultural gradient within the known breeding range of the species, allowing for monitoring in both private and protected areas; and (2) coordinate mist-netting, mark-recapture and re-sight monitoring of color-banded buntings by agency participants, citizen volunteers, and university student interns at a subset of these private and public protected areas. Given these two objectives, we asked the following research questions:

- Would we be able to motivate volunteers to report data on Painted Buntings at backyard feeders, and if so, would we detect differences in bunting populations?
- How does abundance and distribution of Painted Buntings vary across an urban/suburban-rural/agricultural landscape as determined by surveys of backyard feeders?
- How does variation in human population affect observations in Painted Bunting populations across the urban/suburban-rural/agricultural gradient?
- Are there sex differences in feeder use for Painted Buntings?
- How important are backyard feeders as a food resource for Painted Buntings?
- Have we improved citizen awareness for the conservation of Painted Buntings through our outreach and by involving citizens in monitoring efforts?

METHODS

Study Area

The study area consisted of the entire breeding range of the eastern Painted Bunting in the states of North and South Carolina (Fig. 8.1) based upon the most recent range map (Sykes and Holzman 2005). Across the range is a rural-urban, or more specifically an urban/suburban-rural/agricultural landscape as well as a distinct coastal to inland distribution for the species. For example, in NC Painted Buntings only breed along a thin, constricted area near the coast from SC to approximately Morehead City. In SC, buntings breed along the coast, but also extend well inland, occupying a variety of mixed agricultural and managed young pine stands. Within this matrix are the urbanized areas of Wilmington and Southport, NC, and Columbia and Charleston, SC (Fig. 8.1). For the first year, we did not limit our study to any particular location within this area, as we preferred to see where and at what locations volunteers would respond to our recruitment efforts over the entire range. In addition, we treated the two states as separate data sets because of these habitat differences, and due to the management applications that our findings could have on future migratory conservation plans for each state.

Volunteer Recruitment and Retention

We recruited citizen science volunteers via local birding clubs, press releases, informational flyers targeted to neighborhood retail shops, and word of mouth. Most volunteers indicated that

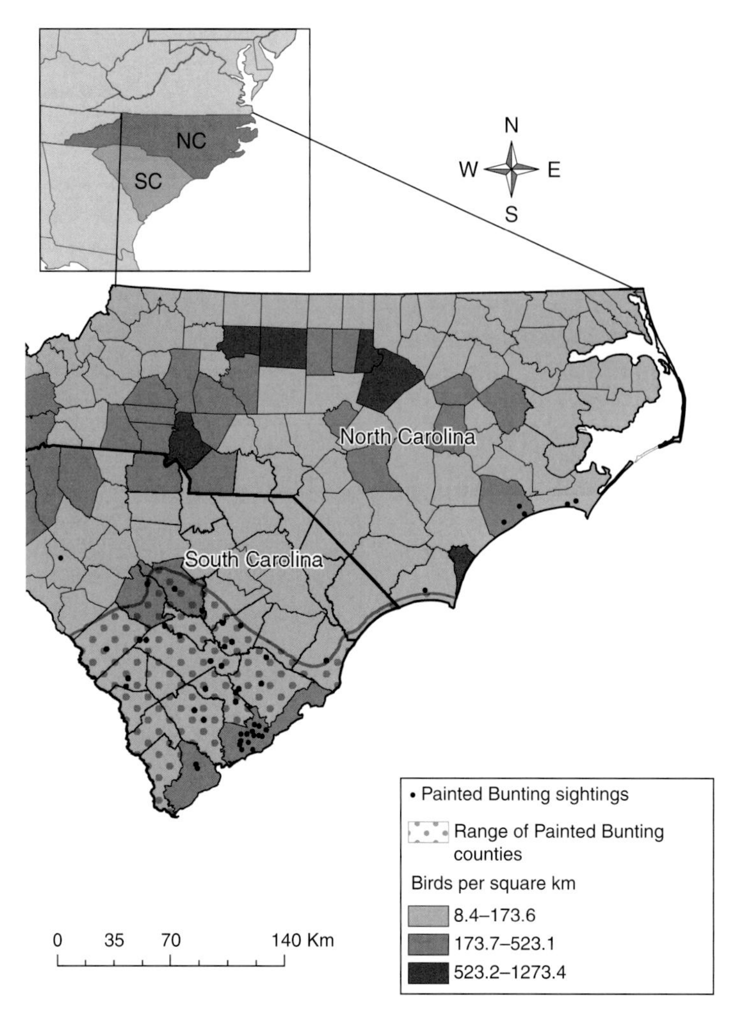

Figure 8.1. Breeding range of the eastern population of Painted Buntings in the North and South Carolina study area. North and South Carolina county populations are based on the 2000 U.S. Census; points represent the 2006 Painted Bunting Observer Team sampling locations (n = 161).

they wanted to participate in our study efforts either by e-mail or telephone, at which time we welcomed them as a member of the Painted Bunting Observer Team (PBOT) and sent them instructions and data observation sheets. To maintain connectedness with PBOT members, monthly updates detailing the program's progress were sent out via e-mail. At the same time, a website was developed by a student team from the Department of Computer Science at the University of North Carolina-Wilmington (UNCW) to inform the public of the project, disseminate data, and facilitate electronic data entry by PBOT volunteers. Due to technical difficulties, this original website could not be used during our monitoring efforts; therefore, we used only e-mail and telephone. However, a new working website, again student-developed, is currently part of our program (http://www.paintedbuntings.org).

In addition to the Internet, we focused on meeting as many of our volunteers in person as possible. We believed that making the citizen science experience personal would enhance volunteer participation and assist with volunteer training. To achieve this, we carried out three activities: (1) visits to PBOT homes to meet volunteers and observe and photograph backyard feeders and birds; (2) banding visits to PBOT volunteer homes (see banding below); and (3) pre-planned instructional presentations at city and county parks and at retail birding supply stores within our region. These presentations were delivered using PowerPoint and geared to audiences of 15–20 individuals. Specifically, the presentations included information about the study and its importance, photos showing the age and sex of buntings, and a demonstration using props, such as a bird feeder and a toy bird, of how to collect data. Home visits also consisted of a brief overview of the program and delivery of our instruction and data recording sheets, and a brief "how-to" demonstration.

Data Collection

Data collection began in mid-April 2006 and continued throughout the breeding season until the end of fall migration for Painted Buntings in October 2006. We focused on three separate metrics for our volunteers to collect: counts, frequency of visits, and visit duration. In order to ensure that our volunteers understood why we asked them to record these three types of data, we translated the metrics into straightforward statements or questions. To control for standardization of data collected, we used a modified version of the feeder counting methods from Project FeederWatch and the Great Backyard Bird Count, developed by the Cornell Laboratory of Ornithology (http://www.birds.cornell.edu). By following the Cornell protocol, we chose an established citizen scientific method that has been tested and accepted in a variety of peer-reviewed publications (see Cornell website for list of publications). Like Cornell's, our data were also quantifiable discrete units, easily tallied by volunteer observers, and we provided easy-to-read, specific instructions on data collection along with direct assistance from project personnel.

We asked volunteers to record data as either a single unique observation (i.e., presence of buntings and how many) or during a block of time (i.e., session), ranging from 15 to 60 minutes, encouraging them to make as many observations per week as possible. For ease of counting, we also asked volunteers to distinguish only between adult male and female/juvenile (referred to hereafter as "green birds") buntings. Male Painted Buntings acquire their multicolored adult plumage only after their second year (ASY), and younger males are green (young males = hatch year (HY) and/or second year (SY) birds), which is the same color as the females. Thus, there is no way to distinguish young male Painted Buntings from females in the field, unless the young male is singing (which is an SY male—females do not sing). We believe that this lack of sexual dimorphism at all ages does not hinder our counts, but acknowledge that it does complicate sex ratio abundance estimation. Because of this complication we used banding data of actual captured individuals to standardize feeder observations (see below). Establishing population ratios of adult males and green birds is significant to bunting ecology and conservation because Painted Bunting declines also may be attributed to harvesting for the international caged bird trade. Specifically, thousands of eastern Painted Buntings are harvested each fall during migration in Cuba, with adult males favored by poachers (Lowther et al. 1999, Iñigo-Elias et al. 2002). We are therefore especially interested in measuring the ratio of adult males to green birds on the breeding grounds in the Carolinas, to determine if there are significant annual declines.

Count data consisted of the largest total number (maximum) of male and green bird Painted Buntings seen at any one time together at the feeder or during any session. For example, if one male bird arrived at the feeder and then left, followed by three male birds feeding all together, the count score is three males. This method avoids double counting individuals, which would result in inflating count numbers (http://www.birds.cornell.edu). We subsequently used these data for our abundance estimates indices. We compared maximum numbers of male and green birds by state using a *t*-test (Sokal and Rohlf 1995), with $P \leq 0.10$ considered significant.

Visit frequency data consisted of a tally of the number of times male and green bird buntings visited feeders per session. This frequency data is different than the count data above, and yielded multiple counts to the feeder by individual birds resulting in a visitation rate (visits/hour). We used the mean number of birds visiting a feeder (i.e., total number visited per session/number of sessions) and converted the number to a mean rate per hour for each feeder site using the time for each session. Unless individual birds are color banded, there is no way to know if a single individual is repeatedly feeding at a particular feeder, or if these data are from multiple individuals. Therefore, the metric yields an index of visit frequency on a per feeder basis, and for the volunteer, translates into the question, "How often do

Painted Buntings use my feeder?" Duration data consisted of the amount of time each bunting stayed at a feeder per visit to the nearest minute, and was averaged across all visits per session to calculate mean duration for each feeder. Again, for the volunteer, this translates into the question, "On average, how much time do Painted Buntings spend feeding at my particular feeder?" Past studies have shown that both visit frequency and duration data can be used as an indicator of nutritional need for supplemental food (Wilson 2001) and as a way to measure specific foraging behaviors that balance feeding efficiency with predation risk avoidance (Lee et al. 2005). Thus the two visit rates translated into, "How important is my feeder to buntings?" We compared the mean visitation rate for Painted Buntings detected at each site by state and tested for differences using a *t*-test (Sokal and Rohlf 1995). Finally, volunteers were asked to record date, time of day, weather, and any other conspecific behavioral data observed.

In addition to volunteer-collected data sets, we banded birds on private lands and at managed public sites (e.g., state parks) within each state ($n = 32$; SC = 19, NC = 13). We used bird feeders as baiting stations to attract buntings to the nets, and used a three-net methodology recommended by Sykes (Sykes 2006), supplemented with a one-way trap to maximize our captures. We banded from 6 July to 11 October 2006. Banding was conducted on days with little wind and no rain, typically during a 5-hour period between sunrise and 12:00 noon, and between 14:00 and sunset EST. Banding was conducted at the same sites as the feeder observations, either using the exact locations of backyard feeders or in the public sites, where we set up feeders at least two weeks prior to banding. In this way, birds were already accustomed to foraging at a site before banding, and our banding could be directly related to the observations from our citizen volunteers. After each banding session, volunteers were asked to record data for the newly banded and unbanded birds. For this first year, our focus was to use banding to examine the accuracy of volunteer indices of abundance with actual numbers of birds banded per site, and use our banding data as a way to estimate females and HY/SY males from our volunteer collected green bird counts. To estimate this relationship, we calculated the mean ratio of females to HY/SY males from the banding data for each state to yield an approximation, on average, of the ratio from the volunteer observer dataset.

Urban-Rural Landscape

Using the backyard observations of our citizen volunteers, we created a distribution map in a GIS (ArcMap ver. 9, ESRI, Redlands, CA) of known Painted Bunting locations by geocoding addresses to latitude/longitude, and plotting these against the breeding range map created by Sykes and Holzman (2005). Our objective was to determine if our volunteer observation sites fell entirely within the breeding range or if buntings occurred elsewhere in the landscape, and if there was a consistent pattern for these sites in relation to human population demographics on a per county basis from U.S. Census data (2000 Tiger data, ESRI). Specifically, we were interested in determining if there were significant correlations ($P < 0.05$) between bunting abundance (count data) and human population, and between bunting abundance and housing density. Past studies on urban bird ecology have shown that these human population parameters can affect bird populations, especially along an urban/suburban-rural/agricultural gradient (Marzluff et al. 2001, Blair 2004, Lepczyk et al. 2008), and that private landowners' activities can influence bird populations (Lepczyk et al. 2004). We also used these housing demographics to determine the location of urban versus rural locations. Since many urban-rural gradients are not smooth but are patchily distributed on the landscape (Pennington and Blair, chapter 2, this volume), we used these data as a better approximation of the continuum that represents this change.

RESULTS

We had 279 PBOT members across North and South Carolina at the end of October 2006, of which 161 (57.7%) responded with usable data (Fig. 8.1). Of the remaining 42.3%, nearly half did not provide data because they either did not live within the bunting breeding area and/or they requested membership so they could receive our monthly updates to learn more about our program. Usable count data ranged from an e-mail or phone call indicating that the volunteer observer had at least one Painted Bunting at their location (essentially presence/absence data), to e-mails with session data including count, visit, and behavioral observations (with at least three or more sessions), from which we calculated duration rates and frequencies. Data that did not meet these criteria were eliminated prior to analysis. Most (79%) of our volunteers in both states reported observing at least one adult male Painted Bunting, while 21% reported having one green bird at their feeders. Only 13% reported having two males and 11% reported having two green birds. All other reports of three or more individuals of either sex accounted for <1% of the observations.

The mean number (from maximum number recorded per site: mean, 90% CI) of male Painted Buntings detected was the same (t = 1.91, df = 94, P = 0.17) in North (1.4, 1.2 to 1.6 males) and South Carolina (1.4, 1.4 to 2.0 males). Similarly, there was no difference in detection of green birds numbers (t = 0.0, df = 60, P = 0.96) between North Carolina (1.9, 1.4 to 2.5 green birds) and South Carolina (1.9, 1.4 to 2.3 green birds). Male and green bird numbers were similar within each state. Total maximum numbers of male and green buntings combined per site was also the same between states (NC: 3.5, 2.5 to 4.4; SC: 4.0, 3.1 to 4.9; t = 0.35, df = 57, P = 0.56).

We banded a total of 359 Painted Buntings in 2006 and uniquely color-marked each individual (three color bands and one metal band from the Bird Banding Laboratory). We captured buntings at a similar rate in both states (NC = 1.75; SC = 1.73 captures/net hour; n = 32 banding locations; SC = 19, NC = 13). Comparing our banding capture data with observer count indices indicated that 53% of observer counts underestimated the number of individuals captured in nets while 41% overestimated and 6% correctly estimated (n = 17 locations with both banding data and count data). The magnitude of those estimates varied from over/underestimating by one individual to underestimating by a total of 9:1. From our banding data, we calculated the mean ratio of females to young males we captured in an attempt to approximate the composition of the green bird population from the volunteer data set. We found our green bird captures in North Carolina consisted of 33% females, 25% green males, and 42% unknowns. In South Carolina we found 30% females, 9% males, and 61% unknowns. The majority (>90%) of unknown

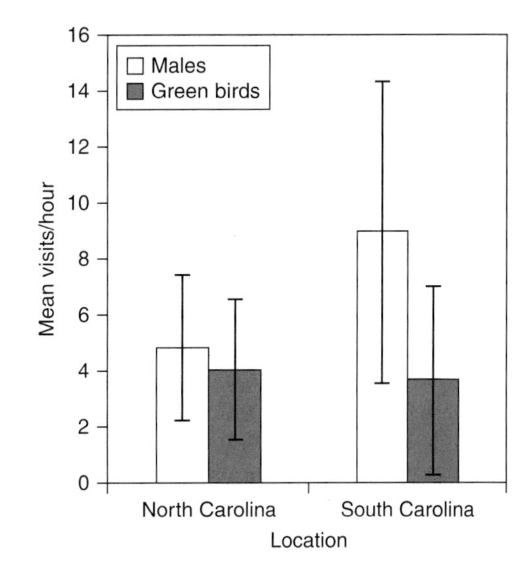

Figure 8.2. Index of Painted Bunting feeder visitation rate (mean visits/hour, 90% CI).

birds were hatch-year birds that are impossible to sex at this age.

We received 20 data sets from volunteers with usable duration and frequency observations, representing 12.4% of our count data. We found no difference in visitation rates for adult males between NC and SC feeders, though adult males tended to make more visits/hr in SC than NC (mean male visits/hr in NC = 4.8, 90% CI 2.3 to 7.4; SC = 9.0, 90% CI 3.6 to 14.3; t = 1.97 df = 18, P = 0.17). We found no differences or trends for green bird visitation rates in the two states (mean green bird visits/hr in NC = 4.0, 90% CI 1.5 to 6.6; SC = 3.7, 90% CI 0.3 to 7.0; t = 0.02 df = 18, P = 0.89; Fig. 8.2). Similar results were found when we compared total numbers of male and green bird visitation rates between states (t = 0.55, df = 18, P = 0.47). Duration at feeder data showed a consistent trend toward longer mean visits in SC (2.78 min) compared to NC (2.04 min); however, a low sample size (n = 5) for SC caused a lack of power for tests. Anecdotal data from volunteers in both states suggests that males and green birds spent longer feeding at feeders in certain sites within specific areas of the landscape. Based on this anecdotal information, we re-analyzed our data by grouping them into four separate areas along a north-south gradient, which yielded a trend for increasing feeder time from south to north in SC, with birds spending only 1 minute on average in more southerly locations compared

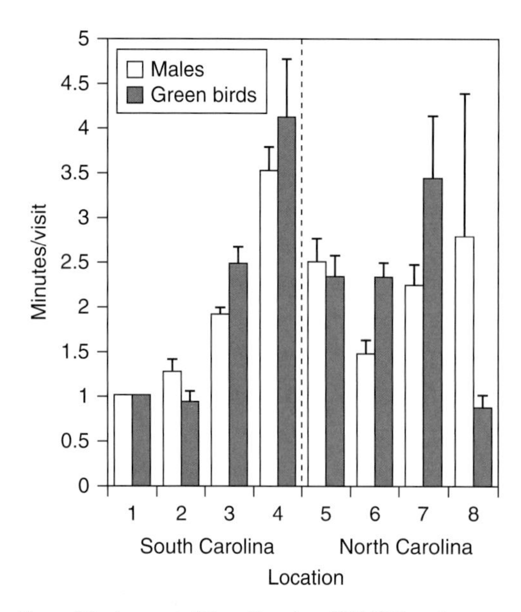

Figure 8.3. Amount of time (duration, 90% CI) buntings stayed at a feeder per visit. Data are grouped into four areas along a north-south gradient for each state.

to up to 4 minutes in more northerly locations. We found no such trend in NC (Fig. 8.3).

Volunteers recorded birds in 22 counties, which ranged in human population size from 11–50 to 401–1,000 persons/mi^2 or 2.59 km^2. These counties were located only along the coast in NC and along the coast and inland in SC (Fig. 8.1). Not surprisingly, most of our volunteers were located in areas with high human population and building densities. The mean number of site locations per county increased with increasing human population density, and mean number of site locations correlated significantly with increased density of housing units by county ($r = 0.618$, $P = 0.002$). No human demographic correlated with bunting abundance overall or with either male or green bird abundances. The breeding range of the eastern Painted Bunting determined by Sykes and Holzman (2005) correctly predicted 89% of volunteer locations in NC and 99% in SC. Of those locations outside of the range, all but two locations were within ≤5 km of the range boundary (one in NC was ca. 29 km and one in SC was ca. 68 km).

DISCUSSION

The first year of the Painted Bunting Observer Team (PBOT) project confirmed that a citizen science approach would generate interest in the species, as measured by the number of PBOT volunteers, and that volunteer data could be used along with traditional techniques to monitor bunting populations. With 57.7% of our PBOT members responding with usable data, we satisfied our first question of whether or not the project would generate enough public interest. For example, our citizen involvement closely compares to the response rate of landowners surveyed on urban birds in southeastern Michigan (Lepczyk 2005). In addition, nearly all of the volunteers were located within the previously determined eastern breeding range for the species, allowing us to confirm volunteer locations as ones with a high probability for Painted Buntings. Next, we were able to determine bird relative abundances for both male and green birds and found no difference, which may indicate that most of the green birds were females on territories with an equal number of males. Because some of the green birds may be first-year males, it was not possible to determine gender, except by capturing and banding. From banding, we were able to determine that, on average, about 30% of green birds recorded by our volunteers in North and South Carolina were females, while about 17% were young males. Therefore, for every ten green birds observed by our volunteers, about three were females and two were SY males. The remainder were hatch-year birds of unknown sex. This breakdown is a rough approximation, as we may have overestimated the amount of HY birds because we extended our banding late into the season. We could assume that because adult males of this species are more colorful, volunteers would tend to focus on males, elevating their abundance. Our data, however, suggest that the volunteers were unbiased observers, correctly identifying and counting these birds. Finally, our banding data suggest that more birds use feeders than reported by our observers, and we anticipate that our data set for 2007 will be more robust as a greater number of individual birds were banded ($n > 900$ for 2007), which should facilitate counts with greater precision. We also need to relate these data to relative abundance or density throughout the range. For example, overall singing male densities for Painted Buntings in a variety of early-successional to old-growth forest habitats in Florida north to North Carolina were found to be 23.3–40.1 birds/km^2 in 2003

(90% CI; J. M. Meyers and K. A. Bettinger, unpubl. data), which is almost identical to home ranges found in these habitats using radiotelemetry (Springborn and Meyers 2005).

The behavioral data for visit frequency and visit duration allowed for some insight as to how important feeders are to Painted Buntings in the Carolinas. For example, using these data together, adult male buntings in NC make about 5 visits/hour to feeders, with each visit lasting an average of approximately 2 minutes. Assuming that the buntings can visit feeders anytime during daylight hours and daylight lasts on average approximately 12 hours (07:00–19:00 EST), then adult males occupy feeders in NC up to 2 hours of every day, or 17% of daylight hours at feeders. Even more noteworthy are feeders and adult males in SC. These birds make 9 visits/hour to feeders, with each lasting an average of approximately 3 minutes. Using the same logic as above, males occupy feeders in SC for 5.4 hours per day, or 45% of the time. We acknowledge that Painted Bunting activity is not constant throughout the day, and that bunting activity typically shows peaks and valleys; however, even if these results are halved, they represent a significant amount of the bird's daily activity budget.

For our landscape and human demography analysis, our volunteer data did not correlate with any of the human population demographics we tested. The number of volunteer sites per county, however, did correlate with increased housing density. These results suggest several possible scenarios for Painted Buntings. First, all else being equal, an area with greater housing density will tend to have more feeders than an area with less housing and thus will attract more buntings because of more available food. For example, some bird populations in the UK increased along with increases in the number of feeders in suburban areas over time, suggesting that birds responded to greater food availability (Chamberlain et al. 2005). Second, greater housing density increases the probability of citizen volunteer participation because human population is greater, and therefore this parameter may be just an artifact of increased human population. This is of potential importance as increased housing development and the subsequent transformation of habitat into non-natural, less desirable habitat for buntings may be one reason for the decline of the species. We can only speculate on these possible effects at this time, but if, on the other hand, housing density in suburban areas in the region does not continue to grow, then a possible "intermediate disturbance" scenario supporting healthy populations of Painted Buntings instead of an "ecosystem-stress" relationship may occur with the supplemental feeding of buntings at feeders (Lepczyk et al. 2008). We hope to gain further insight into these effects as the project progresses.

Recommendations and the Future of the Painted Bunting Observer Team Project

We learned much from our first year of data collection both from the buntings and from our PBOT citizen science volunteers. Several benefits and limitations emerged from our experience. As we conclude our second year (2007), we report a few new insights and offer recommendations to others who may be considering a citizen science-based data collection program.

Private Property Access

Employing the citizen science approach allowed access to survey private lands in areas that were prime Painted Bunting habitat. These lands (e.g., backyards) would not have been accessible to us without volunteers inviting us onto their property. Private property access took place on two levels: indirectly through volunteer observations and directly through our banding efforts. For our project, private property access may be one of the most important aspects of using the citizen science model. A traditional monitoring effort typically avoids private property because of access issues. We are currently dealing with this issue for our four-state traditional monitoring effort. Because we plan on continuing mark-recapture–re-sighting for at least five to seven more years, our data set will be robust enough to determine survivorship and productivity for Painted Buntings over a wide and varied landscape. Having the ability to determine these two key demographic parameters within private property is paramount to our study, especially because issues such as habitat loss due to development and urbanization may be a principal cause of the decline of the species. This alone makes our effort worthwhile.

Website Development

Developing a website that will allow for both the dissemination of information to the public along with an online database for volunteer data entry is imperative. During 2006, we did not have a working website, and instead depended on hundreds of e-mails and phone calls to communicate with our volunteers. In addition, we used e-mail or regular U.S. Postal Service mail to collect volunteer data sets. These methods were cumbersome and may have diminished our ability to collect data from volunteers. Our current website allows for direct data entry by volunteers, including complete instructions on how to enter those data; a frequently asked questions (FAQ) section with information about Painted Buntings, our study, and partner links; and a function that maps and counts bunting sightings on a per location basis using autogeocoding. We plan on enhancing the website both in data entry capabilities and in more interactive maps, tables, and graphs.

Volunteer Recruiting and Retention

Volunteers are the backbone of the project, and without them none of this would be possible. Therefore, it is very important to allocate project resources to recruiting and retaining volunteers. After volunteers are recruited, there needs to be a mechanism in place to retain them. Our experience has shown that personalized service, mainly by way of answering questions via e-mail or by phone, similar to that found in a good customer service department, is essential. Without it, volunteers become disinterested and drop out of the program. Conversely, if it is done well, some volunteers will become more than volunteer observers; several have learned enough with us to become entry-level field technicians. These citizen scientists assisted with scribing and handling birds during banding. Similar to the experience of Vargo et al. (chapter 7, this volume), our training of volunteers has enhanced our program, and has assisted with our efforts to increase public awareness.

Project Coordinator

The project coordinator plans and directs our volunteer home visits, banding dates and locations, and assists with volunteer data management. This position would be important with any monitoring effort, but with volunteers, a research and/or monitoring project changes from a narrowly focused, independent project carried out by, for example, one researcher and a couple of field assistants, to a much broader project that encompasses similar goals, but now includes hundreds of people from the local community and delves into outreach education. The single researcher who is accustomed to having little or no interface with the general public is now faced with hundreds of volunteer questions via e-mail and/or phone, as well as typical duties associated with a project. We found juggling all these duties nearly impossible. We highly recommend a competent project coordinator.

Robustness and Limitations of the Data Set

As mentioned above, critics have labeled citizen science-collected data inadequate for a variety of reasons, but especially because of the problems with observer error. It is thus important to acknowledge the limitations of certain data sets and the utility of others. Taking steps to ensure that volunteers collect uniform data and checking those data entered into the database are vital. For example, conducting group training sessions and developing easy-to-understand methods for making observation allows volunteers to collect discrete data units that they themselves understand. Count, visit frequency, and duration data are examples of these; however, our count data is only an index of abundance and not an absolute measure. It will be interesting to see how these data may compare with estimates generated from either our banding data or our companion four-state monitoring project. In addition, banding data can be used to make more robust estimates from the volunteer-collected data for 2007, and we will have the ability to directly compare nearly all of our banding capture results with feeder observations.

Measuring Project Impact on Public Awareness

Developing a questionnaire to measure the role of citizen science in informing and engaging the public is highly recommended (especially before and after the program is initiated in order to assess change). We have yet to survey our volunteers, and plan to develop our survey in the coming months (based on a control and treatment model). We wish to learn more about our volunteers (e.g., economic demographics), get feedback on the program, and measure the program's impact on educating them about Painted Bunting conservation versus individuals from the general public who were not involved in our program. Though we have yet to quantify our impact, we constantly receive feedback via hundreds of volunteer e-mails, letting us know they believe our conservation work is important to them and that they are contributing to it.

Cost Effectiveness

It can be argued that citizen science projects can be done just as effectively and possibly less expensively by hiring a few technicians rather than employing citizen volunteers. We would respond to this with a resounding No! Several reasons factor into our response. First, on an actual cost basis, we could not hire technicians to equal the contribution of our volunteers. In 2006, our 20 volunteers collected 7,276 hours of behavioral data. This is only the behavioral data and not the contribution from all of our other volunteers who may have only spent 5 minutes or less to record their count data. Similarly, for 2007 we had 46 volunteers who collected 52,036 hours of behavioral data. At $10/hour, we could not afford to pay a seasonal field technician to collect an equivalent data set. Second, intangibles associated with approaching the project as we did are well worth the expense, even if it *were* cheaper-to-use technicians. In other words, "cost-effectiveness" isn't just about dollars and cents and the ability to pay for the work; it is also about the other things that the investments "buy," like ownership, engagement, and institutionalization of commitment toward future citizen science and other monitoring projects (see above).

Citizen Science Monitoring and Research Questions

We anticipate answering a variety of research questions through our data collected by volunteer monitoring and banding. For 2007, we greatly expanded our banding effort, individually color-banding >700 birds (total captures > 900). We also initiated banding site methodology to test more specifically for differences along the urban-rural gradient and between natural areas and private lands, using our PBOT volunteers' backyards for comparison with nearby parks and protected areas. We hope to begin to tease apart questions regarding source and sink relationships between natural and suburban habitat with our banding, mark-recapture–re-sighting effort, and in future years determine trends in bunting demographics across the urban/suburban-rural/agricultural gradient. We also hope to measure variation among habitat classes such as types of urban development and varying disturbance in relation to bunting populations and behaviors. With over 3,700 records recorded by 143 volunteers this year, we look forward to many more PBOTs, and to making further refinements for the 2008 season.

We have demonstrated the merits of the citizen science model for monitoring and for answering future research questions for Painted Buntings. We strongly believe that when carefully planned and implemented, the citizen science model could be applied to a variety of urban bird species studies. Possibly even more important is the model's ability to reconnect science with society, that is, informing and engaging the public, something conventional research efforts do not do. Many times there is a disconnect between public opinion and our ability as scientists to attach value to our goals and objectives as researchers. This phenomenon is perhaps rooted in a society that is shifting toward a "nature-deficit disorder" (Louv 2005). Citizen science—that is, citizens working as and with scientists—bridges that gap.

ACKNOWLEDGMENTS

We thank all of our Painted Bunting Observer Team citizen science volunteers, without whom this research would not be possible. We also thank P. W. Sykes and S. Holzman for the use of their eastern Painted Bunting breeding range map shapefile, L. Glover of the South Carolina Department of Natural Resources, H. Gabriels of the Department of Geography and Geology at the University of North Carolina-Wilmington for assisting with our map figure, the Eastern Painted Bunting Working Group, and the Department of Environmental Studies at the University of North Carolina-Wilmington. This research is funded by a grant from the U.S. Fish and Wildlife Service, Region 4, Atlanta, to J. Rotenberg.

LITERATURE CITED

American Ornithologists' Union. 1998. Check-list of North American birds. 7th ed. AOU, Washington, DC.

Bhattacharjee, Y. 2005. Ornithology: citizen scientists supplement work of Cornell researchers. Science 308:1402–1403.

Blair, R. 2004. The effects of urban sprawl on birds at multiple levels of biological organization. Ecology and Society 9:2. <http://www.ecologyandsociety.org/vol9/iss5/art2> (15 January 2007).

Brittingham, M. C., and S. A. Temple. 1988. Impacts of supplemental feeding on survival rates of Black-capped Chickadees. Ecology 69:581–589.

Brittingham, M. C., and S. A. Temple. 1992a. Use of winter bird feeders by Black-capped Chickadees. Journal of Wildlife Management 56:103–110.

Brittingham, M. C., and S. A. Temple. 1992b. Does winter bird feeding promote dependency? Journal of Field Ornithology 63:190–194.

Brossard, D., B. Lewenstein, and R. Bonney. 2005. Scientific knowledge and attitude change: the impact of a citizen science project. 27:1099–1121.

Chamberlain, D. E., J. A. Vickery, D. E. Glue, R. A. Robinson, G. J. Conway, R. J. W. Woodburn, and A. R. Cannon. 2005. Annual and seasonal trends in the use of garden feeders by birds in winter. Ibis 147:563–575.

Danielson, F., N. D. Burgess, and A. Balmford. 2006. Monitoring matters: examining the potential of locally-based approaches. Biodiversity and Conservation. 14:2507–2542.

Dunn, E. H. 1986. Feeder counts and winter bird population trends. American Birds 40:61–66.

Evans, C., E. Abrams, R. Reitsma, K. Roux, L. Salmonsen, and P. P. Marra. 2005. The Neighborhood Nestwatch Program: participant outcomes of a citizen-science ecological research project. Conservation Biology 19:589–594.

Hartup, B. K., O. M. Hussini, G. V. Kollias, and A. A. Dhondt. 1998. Risk factors associated with mycoplasmal conjunctivitis in house finches: results from a citizen-based study. Journal of Wildlife Diseases 34:281–288.

Hunter, W. C., M. F. Carter, D. N. Pashley, and K. Barker. 1993a. The Partners in Flight species prioritization scheme. Pp. 109–119 *in* D. M. Finch and P. W. Stangel (editors), Status and management of neotropical migratory birds. General Technical Report RM-229. USDA Forest Service, Rocky Mountain Forest and Range Experiment Station, Fort Collins, CO.

Hunter, W. C., D. N. Pashley, R. E. F. Escano, and E. F. Ronald. 1993b. Neotropical migratory landbird species and their habitats of special concern within the Southeast region. Pp. 159–169 *in* D. M. Finch and P. W. Stangel (editors), Status and management of neotropical migratory birds. General Technical Report RM-229. USDA Forest Service, Rocky Mountain Forest and Range Experiment Station, Fort Collins, CO.

Iñigo-Elias, E. E., K. V. Rosenberg, and J. V. Wells. 2002. The danger of beauty. Birdscope (newsletter of the Cornell Laboratory of Ornithology). <http://www.birds.cornell.edu/publications/birdscope/Summer2002/Danger_beauty.html> (10 September 2006).

Irwin, A. 1995. Citizen science: a study of people expertise, and sustainable development. Routledge, London, UK.

Kohlsaat, T., L. Quattro, and J. Rinehart. 2006. South Carolina comprehensive wildlife conservation strategy: 2005–2010. South Carolina Department of Natural Resources. Columbia, SC.

Lee, T., M. S. Quinn, and D. Duke. 2006. Citizen, science, highways, and wildlife: using a web-based GIS to engage citizens collecting wildlife information. Ecology and Society 11:11. <http://www.ecologyandsociety.org/vol11/iss1/art11> (3 February 2007).

Lee, Y. F., Y. M. Kuo, and E. K. Bollinger. 2005. Effects of feeding height and distance from protective cover on the foraging behavior of wintering birds. Canadian Journal of Zoology 83:880–890.

Lepczyk, C. A. 2005. Integrating published data and citizen science to describe bird diversity across a landscape. Journal of Applied Ecology 42:672–677.

Lepczyk, C. A., A. G. Mertig, and J. Liu. 2004. Assessing landowner activities related to birds across rural-to-urban landscapes. Environmental Management 33:110–125.

Lepczyk, C. A., C. H. Flather, V. C. Radeloff, A. M. Pidgeon, R. B. Hammer, and J. Liu. 2008. Human impacts on regional avian diversity and abundance. Conservation Biology 22:405–416.

Link, W. A., and J. R. Sauer. 1997. New approaches to the analysis of population trends in land birds: a comment on statistical methods. Ecology 78:2632–2634.

Louv, R. 2005. Last child in the woods: saving our children from nature-deficit disorder. Algonquin Books, Chapel Hill, NC.

Lowther, P. E., S. M. Lanyon, and C. W. Thompson. 1999. Painted Bunting (*Passerina ciris*). No. 398 *in* A. Poole and F. Gill (editors), The Birds of North America. The Birds of North America, Inc., Philadelphia, PA.

Marzluff, J. M., R. Bowman, and R. Donnelly (editors). 2001. Avian ecology and conservation in an urbanizing world. Kluwer Academic Publishers, Boston, MA.

McCaffrey, R. E. 2005. Using citizen science in urban bird studies. Urban Habitats 3:70–86.
Niven, D., J. Sauer, G. Butcher, and W. Link. 2004. Christmas Bird Count provides insights into population change in land birds that breed in the boreal forest. American Birds 58:10–20.
North Carolina Wildlife Resources Commission. 2005. North Carolina Wildlife Action Plan. Raleigh, NC.
Rich, T., C. Beardmore, H. Berlanga, P. Blancher, M. Bradstreet, G. Butcher, D. Demarest, E. Dunn, W. Hunter, E. E. Iñigo-Elias, J. Kennedy, A. Martell, A. Panjabi, D. Pashley, K. Rosenberg, C. Rustay, J. Wendt, and T. Will. 2004. Partners in Flight North American landbird conservation plan. Cornell Labortory of Ornithology. Ithaca, NY.
Robb, G. N., R. A. McDonald, D. E. Chamberlain, and S. Bearhop. 2008. Food for thought: supplementary feeding as a driver of ecological change in avian populations. Frontiers in Ecology and the Environment 6:476–484.
Sauer, J. R., J. E. Hines, G. Gough, I. Thomas, and B. G. Peterjohn. 1997. The North American Breeding Bird Survey results and analysis, version 96.4. Patuxent Wildlife Research Center, Laurel, MD.
Sokal, R. R., and F. J. Rohlf. 1995. Biometry. W. H. Freeman and Company, New York.
Springborn, E. G., and J. M. Meyers. 2005. Home range and survival of breeding Painted Buntings on Sapelo Island, Georgia. Wildlife Society Bulletin 33:1432–1439.
Sykes, P. W., Jr. 2006. An efficient method of capturing Painted Buntings and other small granivorous passerines. North American Bird Bander 31:110–115.
Sykes, P. W., Jr., and S. Holzman. 2005. Current range of the eastern population of Painted Bunting (*Passerina ciris*), Part 1: Breeding. North American Birds 59:4–17.
U.S. North American Bird Conservation Initiative Monitoring Subcommittee. 2007. Opportunities for improving avian monitoring. U.S. North American bird conservation initiative report. Division of Migratory Bird Management, U.S. Fish and Wildlife Service, Arlington, VA. <http://www.nabci-us.org> (8 January 2008).
Wells, J. V., K. V. Rosenberg, E. H. Dunn, D. L. Tessaglia-Hymes, and A. A. Dhondt. 1998. Feeder counts as indicators of spatial and temporal variation in winter abundance of resident birds. Journal of Field Ornithology 69:577–586.
Wilson, W. H., Jr. 2001. The effects of supplemental feeding on wintering Black-capped Chickadees (*Poecile atricapilla*) in central Maine: population and individual responses. Wilson Bulletin 113:65–72.

CHAPTER NINE

A New Approach to Urban Bird Monitoring

THE TUCSON BIRD COUNT

Rachel E. McCaffrey, Will R. Turner, and Amanda J. Borens

Abstract. With more people living in urban areas, cities are playing an increasingly important role in supporting local biodiversity and maintaining the connection between people and nature. Yet, in order for conservation efforts in urban areas to be effective, information on how to sustain native species in these areas is needed. Few studies have attempted to collect long-term monitoring data on bird populations in urban areas, perhaps due to the high investment in time and manpower required. The Tucson Bird Count (TBC) is a program based on repeated, systematic volunteer bird surveys that represents a novel approach to collecting citywide data and monitoring urban birds. Since the TBC's inception in 2001, volunteers have successfully monitored birds at approximately 1,000 sites throughout the Tucson, Arizona, metropolitan area. Data collected through the TBC have already been used in research studies, planning strategies, and to inform local conservation initiatives. Here we explore how TBC data can be used to assess relationships between species distribution and habitat features, focusing in particular on Abert's Towhee (*Pipilo aberti*). In addition, we compare the two TBC programs, an annual citywide survey and a quarterly survey of specific sites of birding interest. This comparison highlights the importance of isolated habitat islands to the persistence of some species in urban Tucson. We review results from the TBC and the local Christmas Bird Count, finding both considerable overlap between the two counts and unique information contributed by each program to our understanding of the area's avifauna. Finally, we discuss the ability of the TBC to serve as a model for other urban bird monitoring programs and its value in the establishment of a global network of such programs.

Key Words: Abert's Towhee, citizen science, monitoring, *Pipilo aberti*, Tucson, urban birds.

Urbanization stands as one of the world's dramatic trends in the 20th century, with the proportion of the global population living in urban areas increasing from 13% in 1900 to 49% in 2005 (United Nations 2006). By 2030 it is predicted that 4.9 billion people will reside in urban areas. This human population increase has resulted in both an expansion of the spatial extent of urbanized areas and a corresponding decline in agricultural lands and natural ecosystems surrounding cities (Houghton 1994). Thus, urban sprawl has been cited as a prime threat to

McCaffrey, R. E., W. R. Turner, and A. J. Borens. 2012. A new approach to urban bird monitoring: the Tucson bird count. Pp. 139–154 *in* C. A. Lepczyk and P. S. Warren (editors). Urban bird ecology and conservation. Studies in Avian Biology (no. 45), University of California Press, Berkeley, CA.

biodiversity, with urbanization leading to high local extinction rates and the loss of many native species (Czech and Krausman 1997, Marzluff 2001, McKinney 2002). Likewise, many city dwellers, urban planners, and even some conservationists tend to view developed areas as a sort of biological wasteland, with little potential for harboring the rich assemblages of species associated with undisturbed systems.

The city of Tucson, located in Arizona, the fastest growing state in United States (U.S. Census Bureau 2006), in many ways exemplifies how urban sprawl impacts biodiversity (Turner 2003). Between 1990 and 2000 Tucson's population increased 20% while its spatial extent expanded by 24% (Pima Association of Governments 2003). Native vegetation in the Sonoran Desert is being replaced by urban development and nonnative vegetation at a rate of approximately 2,600 ha/yr (Davis 2006). The impacts of urbanization have been well documented for birds in the Tucson area (Emlen 1974, Tweit and Tweit 1986, Mills et al. 1989, Germaine et al. 1998, Boal and Mannan 1999), and are similar to the characteristic responses to urban development demonstrated in other cities (Chace and Walsh 2006). Moving inward along an urbanization gradient, species diversity generally declines as total biomass and avian population density increase. Meanwhile, the original native avifauna is replaced by a smaller group of abundant and often exotic species that take advantage of their ability to flourish in urbanized systems (Beissinger and Osborne 1982, Blair 1996, Melles et al. 2003). In Tucson, this pattern is exemplified by increases in the populations of urban exploiter species such as the Inca Dove (*Columbina inca*), Northern Mockingbird (*Mimus polyglottos*), House Finch (*Carpodacus mexicanus*), and House Sparrow (*Passer domesticus*) and decreases in populations of species adapted to the Sonoran Desert such as the Ash-throated Flycatcher (*Myiarchus cinerascens*), Cactus Wren (*Campylorhynchus brunneicapillus*), and Curve-billed Thrasher (*Toxostoma curvirostre*; Emlen 1974, Tweit and Tweit 1986).

Because urban sprawl affects bird populations in such profound ways, sustaining and promoting native bird diversity in urban areas is of special concern. Conservation efforts of this kind are important not only in the broad scheme of protecting biodiversity, but also because incorporating natural components into urban areas allows for increased human-nature interactions (Turner et al. 2004). While urban landscapes are dissimilar to natural environments, myriad opportunities exist for integrating wildlife into cities (Rosenzweig 2003). Because birds are highly mobile and can respond to long-term changes in habitat structure and composition at relatively fine scales (Knick and Rotenberry 2000, Savard et al. 2000), they are in principle more likely to be able to adjust to urban life than many other types of wildlife. Yet determining ways to prevent or mitigate the typical impacts of urbanization and maintain a diverse assemblage of native birds in urban areas remains a challenge. Meeting that challenge will require a better understanding of how native species are affected by different types and levels of urban development.

As the 21st century began, Tucson continued to expand, with much of the new growth located along the city's edges, and urban ecology information was needed in order to protect natural resources. Despite the existence of 35 years of data from a Christmas Bird Count (CBC) circle centered within Tucson (National Audubon Society 2006), a robust body of local ornithological research (Emlen 1974, Tweit and Tweit 1986, Mills et al. 1989, Germaine et al. 1998, Boal and Mannan 1999) and the presence of an active birding community, Tucson still lacked systematic, citywide data on how populations of birds were changing in response to urban development. Key to developing conservation strategies was the establishment of the Tucson Bird Count (TBC), a volunteer-based program designed to monitor how birds respond to different land use changes and how abundances and distributions of different species vary over both time and space across the city. The TBC is based on the citizen science framework developed by such long-standing programs as the CBC, North American Breeding Bird Survey (BBS), and Britain's Breeding Bird Survey, and extends this model into a program specifically designed to collect information on birds in urban areas. The collection of information on bird abundance and distributions over both broad temporal and spatial scales within an urban area, as provided by the TBC, allows investigation into ways to avoid some of the typical negative impacts of urbanization on birds. Additionally, the urban focus of the TBC provides an opportunity for both participants and other Tucson residents to learn more about the birds near where they live and work, which can result in a greater interest in protecting habitats and reconciling the impacts of urban sprawl. Results from the TBC have already been used to examine how different species respond to various types of land use (Turner 2003) and habitat loss (Turner 2006), and to identify areas within the city that are of particular importance to birds, and therefore are useful targets for conservation efforts.

The goal of this paper is to provide an overview of the design and methodology of the TBC, specific examples of how the TBC data can be used to inform conservation, and an evaluation of the value of the TBC in generating novel data, in order to demonstrate the TBC's potential to serve as a model for other cities interested in developing monitoring programs for urban birds. We illustrate the benefits of the TBC on one species of local conservation concern, Abert's Towhee (*Pipilo aberti*), and how it relates to watercourses in Tucson. Additionally, we compare results from the TBC to those from the Tucson area CBC count circle to assess the contributions of the TBC data to increasing our understanding of how bird populations are changing. Finally, we discuss the advantages and disadvantages of the TBC model, as well as its applicability to other urban areas. With urban sprawl increasing on a global scale, the TBC's urban focus, combined with its citizen science approach, represents a new approach to collecting data that can inform local conservation efforts and help to develop solutions for sustaining populations of native birds in urban areas.

METHODS

The central challenge of the Tucson Bird Count was to design a survey robust enough to produce results rigorous enough to be useful to a wide variety of audiences (e.g., scientists, local citizens, urban planners), yet practical and efficient enough to be sustained over time. To maximize the utility of the data, the TBC encompasses a wide range of sites and land use categories across the Tucson metropolitan area, utilizes a survey methodology designed to detect the maximum number of species, and is repeated at the same sites over time. In order to collect these data over such a broad area, the TBC draws on the skills of experienced birders as volunteer citizen scientists. The TBC initially began in 2001 as an annual citywide spring bird survey, termed the Route Program (RP). Since then the TBC has grown to incorporate a second program, known as the Park Monitoring Program (PMP).

Route Program (RP)

Survey sites for the RP were situated based on a stratified random sampling approach, where one site was randomly located within each cell of a 1-km^2 grid randomly laid over the Tucson area (Turner 2003). Adjacent sites were grouped together into routes that can be reasonably surveyed by an observer in one morning. The number of sites per route varies across the study area (10 to 12 sites/route in the more easily traversed areas, and <5 sites/route in less accessible areas). The initial TBC RP study area (surveyed in spring 2001) encompassed 661 km^2, including industrial and commercial areas, residential neighborhoods, and undisturbed desert. Since then, new survey sites have been added in the developing areas around the city according to the same protocol. By 2006 the TBC RP study area had grown to encompass 981 km^2 (Fig. 9.1). Survey sites were intended to be permanent once established, but a small subset of the TBC sites had to be relocated due to changes in accessibility, excessive noise, or other disruptions. In these cases, a new site was established as close to, and in the same habitat and land use type, as the original site as possible.

Bird surveys were carried out by volunteers qualified as skilled observers, defined as those who can identify the 25 most common Tucson-area species quickly by sight or sound, are familiar with most other birds of the Tucson area, and may need quick reference to a field guide for certain less common or difficult species (Turner 2003). Volunteers who meet this standard use the TBC website (www.tucsconbirds.org) to register for the program, select their routes, and enter their survey results (McCaffrey 2005). Between 15 April and 15 May (the time in the breeding season during which the greatest number of Tucson species are most easily detected), participants survey sites along their route(s), performing a 5-min unlimited radius point count (Blondel et al. 1981, Verner 1985, Ralph et al. 1995). All surveys are conducted between 30 min prior to and 4 hr after local sunrise (Turner 2003).

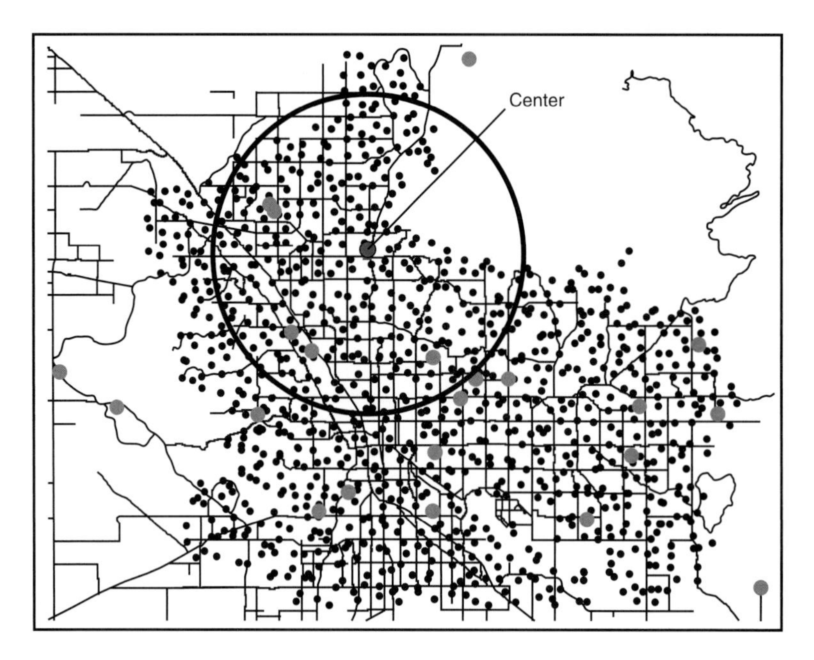

Figure 9.1. Map of the Tucson, Arizona, study area showing 2007 Tucson Bird Count Route Program (●) and Park Monitoring Program (●) survey sites, and the center (●) and extent of the Tucson area Christmas Bird Count circle.

Park Monitoring Program (PMP)

The PMP, designed to complement the annual RP, monitors specific sites of interest in the Tucson area not adequately surveyed by the RP's random sampling design. Additionally, PMP sites are monitored four times a year (winter: 15 January–15 February; spring: 15 April–15 May; summer: 1–31 July; fall: 1–30 September), allowing the detection of species not present during the spring RP survey. The PMP's sites include neighborhood and regional parks, a national park, artificial wetlands, watercourses, neighborhoods, and other locations of scientific, management, conservation, or birding interest (Fig. 9.1).

The PMP also uses skilled volunteer birders to perform surveys. Currently, TBC participants survey 182 sites within 23 parks spread across the Tucson area. In order to allow comparisons among PMP and RP sites, 5-min unlimited radius point counts are performed at the majority of park sites. However, at a few sites with relatively open and uniform environments or at linear sites with watercourses lined with riparian vegetation, participants perform 10-min surveys along 200-m line transects (Bibby et al. 1992). To ensure independence between counts, all points and transects are situated at least 250 m apart from each other (Sutherland et al. 2004).

Web, Database, Geographic Information System (GIS), and Volunteer Tools

Two of the factors that have made the TBC a viable urban bird monitoring program are (1) capitalizing on experienced volunteer skills, and (2) the effective use of technology. In fact, getting the right computer tools in place and recruiting a sufficient number of qualified observers are likely to be the most significant hurdles faced by others initiating citywide bird surveys. However, while programming knowledge is necessary for the initial design and implementation of a web-based program such as the TBC, once established, the program can be run using commercially available software (in the case of the TBC: Microsoft Access, ESRI ArcGIS, Adobe Dreamweaver) that is user friendly and commonly used by researchers.

The TBC website allows prospective volunteers to view available routes, adopt routes, and enter data. The website also communicates the project's background, general information on the local avian community, and TBC results to the public. All dynamic web content interacts with a Microsoft Access relational database via custom server-side JavaScript code. These scripts perform tasks such as registering participants, accepting and performing checks on survey data, and creating interactive web displays of results, including tabular queries by species, site, or year. Maps showing species distributions and abundances are constructed via a script that dynamically places the results for the requested species or time period on top of a standard base map of the study area.

TBC participants are provided with a set of materials that ensure participant survey comfort and data reliability (e.g., count instructions, data forms, site geographic coordinates, descriptions of site locations, and signs indicating that they are participating in a bird survey). Additionally, participants receive custom color route maps generated from GIS coverages of count sites, route boundaries, and relevant features such as roads and watercourses. The GIS layers showing the location of these features were obtained from the Pima County Department of Transportation. In addition to using GIS layers to produce route maps, analyses of the TBC data have been greatly aided by the availability of county-provided GIS data. Coverages that show the locations and qualities of features such as vegetation types, riparian areas, and land use types are valuable tools in assessing the relationships between habitat variables and the bird distribution and abundance patterns generated from the TBC data. Fortunately, such GIS data is becoming more widely available, not just in Tucson, but in many urban areas where such data is valuable not only to researchers, but to city planners and managers.

A script automates much of the TBC's administrative work, normally a potentially time-consuming step. In this case a GIS script (currently Visual Basic in ArcGIS 9.0) zooms to routes, labels sites and roads appropriately, and prints each resulting route. All participant data, site locations and descriptions, and count results are stored in the Access relational database. ArcGIS and GIS scripts are used for survey design, updating of count locations, and spatial analyses. The web-based design of the TBC, combined with the automation of many of the project's organizational tasks, allows the TBC to be run by a single coordinator (historically a graduate student), making the costs of running the program relatively small. The coordinator's main tasks involve maintaining the project's website and database, as well as recruiting and communicating with volunteers.

Data Analysis

To exemplify how data collected through the TBC can be used to assess a species' sensitivity to specific environmental features, we assessed the presence of Abert's Towhees relative to watercourses of varying sizes and riparian vegetation. We used Abert's Towhee data from the 2001–2008 TBC RP results, and defined Abert's Towhee's home range as being 101.5 m in radius (based on Meents et al. 1981). We determined presence of riparian vegetation in the areas from a compilation of regional GAP data sets (North American Warm Desert Wash, North American Warm Desert Lower Montane Riparian Woodland and Shrubland, North American Warm Desert Riparian Woodland and Shrubland, North American Warm Desert Riparian Mesquite Bosque, Open Water, and Invasive Southwest Riparian Woodland and Shrubland; USGS National GAP Analysis Program 2004). The locations and average flow rate classifications of watercourses (CFS 1: <14.2 m^3/sec; CFS 2: 14.2–28.3 m^3/sec; CFS 3: 28.3–56.6 m^3/sec; CFS 4: 56.6–283.2 m^3/sec; CFS 5: 283.2–707.9 m^3/sec; CFS 6: >707.9 m^3/sec) were taken from GIS layers of the Tucson area (Pima County Department of Transportation 2005). We determined whether riparian vegetation and/or each class of watercourse (CFS 1–6) was present within 101.5 m of each survey site using ArcMap (ver. 9.0, ESR Institute, Redlands, CA). A multiple logistic regression was used to model the ability of the seven explanatory variables (riparian vegetation and six classes of watercourse) to predict the presence of Abert's Towhees at a site.

To exemplify the difference between RP and PMP sites, we compared results from the two programs from 2001 to 2006. To compare results between the CBC and TBC, we used CBC results from the Tucson count circle over a six-year period from 2000–2001 to 2005–2006 (National Audubon Society 2002) and TBC results from both the RP and PMPs between spring 2001 and spring 2006. In Tucson, approximately 90% of the CBC count circle overlaps the TBC study area, while about 10% of the CBC circle covers an undeveloped area in the Santa Catalina Mountains that lies above 1,000 m asl (and therefore is not covered by the TBC).

RESULTS AND DISCUSSION

Tucson Bird Count Project

Between 2001 and 2006, 222 volunteers (120 primary observers and 117 assistants) have participated in the TBC project, recording ~219,000 individual birds. A total of 226 bird species have been recorded: 165 species as part of the RP, and 220 species through the PMP surveys. Approximately 35% of participants surveyed more than one route or park. Volunteer retention has remained high, with an average of 79% of volunteers returning year to year between 2001 and 2006.

Species Distributions

Detailed maps showing the Tucson area distribution and abundance of each bird species, based on the data collected through the TBC, are available at the project's website (www.tucsonbirds.org). These maps indicate that many groups of species exhibit similar distribution patterns. For example, Mourning Doves (*Zenaida macroura*), Gila Woodpeckers (*Melanerpes uropygialis*), and House Finches are common throughout the entire study area. Many native desert scrub species, such as Gambel's Quail (*Callipepla gambelii*; Fig. 9.2a), are distributed in a band around the city, with relatively high numbers of individuals observed in less developed areas, and few individuals occupying the city's high-density urban core (Turner 2003, McCaffrey 2005). Conversely, the distribution of the exotic Rock Pigeon (*Columba livia*; Fig. 9.2b) behaves inversely to Gambel's Quail, with high numbers of individuals within the urban core and relatively few birds observed in the less densely developed areas. A similar distribution pattern is evident for the Northern Mockingbird, a native desert species that prefers high-density residential and commercial/industrial areas over low-density residential neighborhoods and natural areas (Turner 2003). The distributions of some species, such as the Phainopepla (*Phainopepla nitens*), are concentrated in areas with relatively dense native woodlands, whereas other species such as the Rufous-winged Sparrow (*Aimophila carpalis*) are only found in small numbers in natural areas remaining at the edge of the city.

Beyond describing how birds are distributed within Tucson, further analyses of the TBC distribution data indicate how birds respond to different habitat features and land uses in urban areas. Turner (2003) used a single year of RP data to quickly assay the sensitivity of more than 70 species to development. Results indicated that a suite of native species are particularly vulnerable to development and merit greater attention in local conservation efforts. One species that was not associated with a specific land use type in Turner's study (2003), but appears instead to be closely linked to the presence of a particular habitat feature, is Abert's Towhee. With one of the smallest distributions of any bird species inhabiting the U.S., Abert's Towhee is restricted to riparian habitats in the southwestern desert (Tweit and Finch 1994). Conservation of Abert's Towhee has been identified as a priority by the National Audubon Society (2002) and the Sonoran Desert Conservation Plan (Pima County 2006) due to the rapid loss of riparian areas. A multiple regression analysis indicates that the probability of an Abert's Towhees being present at a site is significantly related to the presence of riparian vegetation (increases odds of Abert's Towhee presence by 8.4 times after accounting for presence of all watercourses) and the presence of watercourses classified as CFS 4 (increases odds by 10.2 times), CFS 5 (increases odds by 3.3 times), and CFS 6 (increases odds by 2.04 times) within 101.5 m of the site (Table 9.1). There was no evidence that Abert's Towhee presence was associated with watercourses of the smallest 3 classes (CFS 1, 2, and 3).

With a substantial portion of the larger watercourses (CFS 4–6; average flow rates $>56.6\ m^3/sec$) in the study area running through highly developed areas, these results indicate that the Abert's Towhee can persist in urbanized areas if riparian systems are preserved. Unlike many sensitive species that retreat to the edges of the city with development, the case of Abert's Towhee is a hopeful sign that human presence and urbanization need not necessarily spell the demise of native species. A project such as the TBC provides a powerful framework to assist in the conservation of such urban-adaptable species, monitoring them over time as conservation efforts restore key habitats and/or processes to the urban system.

By providing a concise view of how different species are distributed across the Tucson area, the TBC distribution maps have proved useful to both scientists and the public. The generation of real-time maps, immediately updated as participants submit their data online, allows participants to see their results and how they relate to others' observations. Additionally, distribution maps can be viewed by Tucson residents interested in finding out which bird species occur in their neighborhoods or how common particular

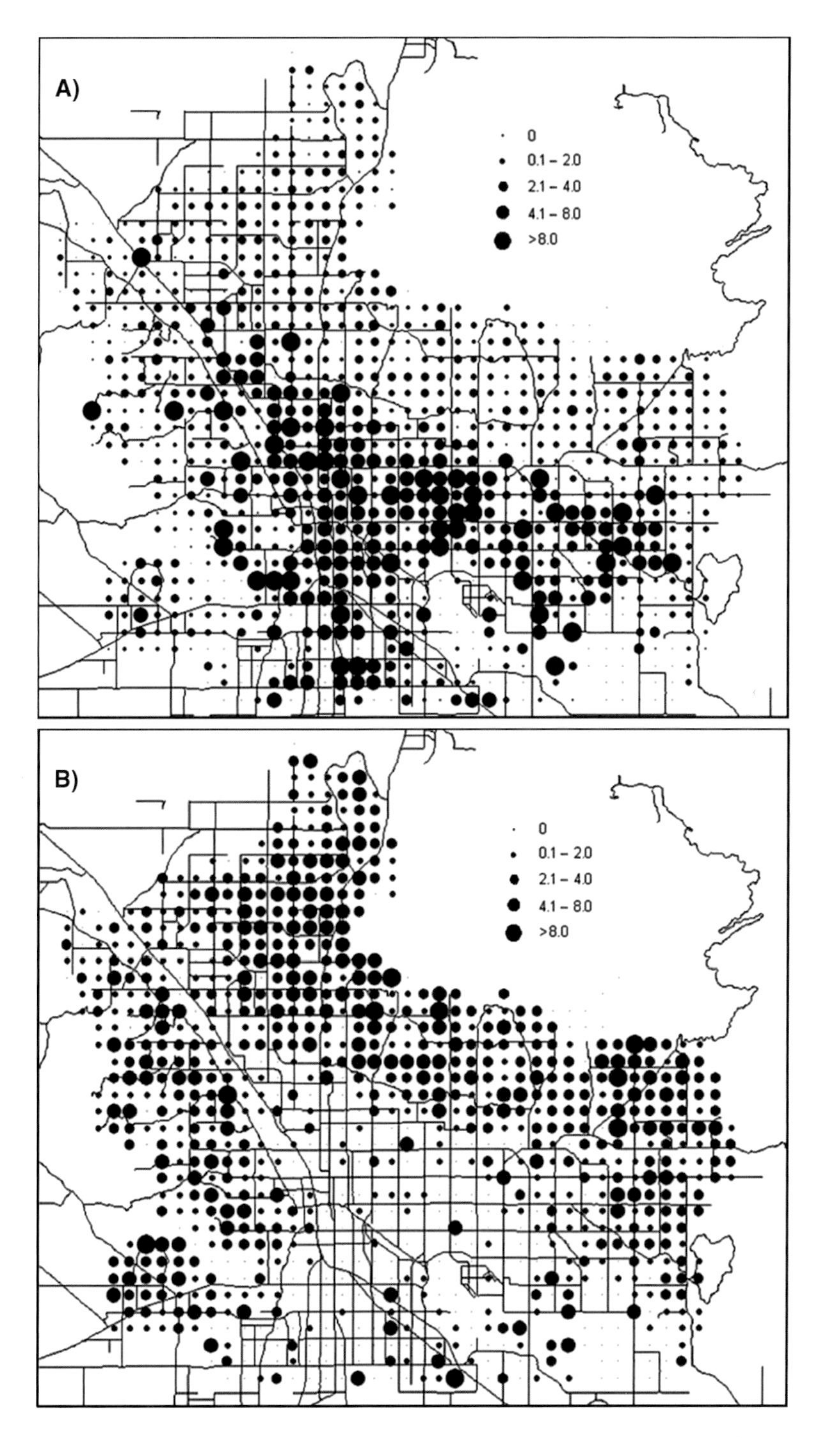

Figure 9.2. Distribution of average abundance (2001–2006) at Route Program sites in the Tucson Bird Count study area for (a) Gambel's Quail and (b) Rock Pigeons. Although each site is randomly positioned within a 1-km² cell, results are shown at cell centers for clarity.

species are. Urban planners interested in targeting sites for habitat preservation have used the TBC data to determine the relative conservation value of different sites (Rosen and Mauz 2001). TBC results have also been used to determine the distributions of vulnerable bird species as part of a regional conservation plan (Pima County 2006). While citywide monitoring results on their own

TABLE 9.1

Logistic regression model relating the probability of Abert's Towhee presence within 101.5 m of a Tucson Bird Count survey site to watercourses of various flow rates and riparian vegetation in Tucson, Arizona.

				95% Confidence intervals	
	Wald's χ^2	$P\leq$	Odds ratio	Lower	Upper
CFS 1	0.04	0.840	1.06	0.57	1.94
CFS 2	0.14	0.710	0.86	0.35	1.87
CFS 3	0.77	0.380	1.44	0.60	3.10
CFS 4	5.76	0.016	2.04	1.12	3.62
CFS 5	4.35	0.037	3.30	1.01	9.71
CFS 6	39.88	0.001	10.20	4.93	21.00
Riparian	35.07	0.001	8.39	4.10	16.80

NOTE: Average flow rate for each class of watercourse: CFS 1: <14.2 m^3/sec; CFS 2: 14.2 m^3/sec to 28.3 m^3/sec; CFS 3: 28.3 m^3/sec to 56.6 m^3/sec; CFS 4: 56.6 m^3/sec to 283.2 m^3/sec; CFS 5: 283.2 m^3/sec to 707.9 m^3/sec; CFS 6: >707.9 m^3/sec.

Riparian vegetation.

can provide sufficient information for a variety of research questions, they are also useful as comparative background data. For example, Hutton (2005) used TBC data to determine the locations of nonnative cavity nesters in the Tucson area, whereas Skagen et al. (2005) used TBC data to study the importance of riparian areas as migration stopover sites in the southwestern U.S.

Park and Route Programs

The annual RP provides citywide data that allow comparisons of species abundances over the entire Tucson area, but data from the PMP enables the monitoring of populations at specific sites year-round. Many park sites are "birding hotspots" that contain habitat features attractive to birds, such as patches of natural desert or water sources. As a result, PMP sites often serve as ecological islands among commercial, industrial, and residential areas, making them particularly important for maintaining populations of birds within the city. In many cases, results from the RP surveys suggest that a species is absent from the study area, but PMP surveys detect the species at sites within these areas.

Of the 220 species recorded during PMP surveys, 51 were not detected in RP surveys. The species unique to the PMP fall into three main groups: (1) water-affiliated species, such as waterfowl and shorebirds; (2) species that are only found in the study area in the winter and fall; and 3) rare species. With naturally occurring, permanent bodies of water nonexistent in the Tucson area, the man-made lakes and ponds at several of the PMP sites serve as stopover and wintering habitats for species such as American Wigeon (*Anas americana*), Green-winged Teal (*Anas crecca*), and Northern Pintail (*Anas acuta*) and shorebirds, including Least Sandpiper (*Calidris minutilla*) and Wilson's Snipe (*Gallinago delicata*). Other species, like the Northern Shoveler (*Anas clypeata*; 2,140 individuals observed at PMP sites and 0 at RP sites), and Sora (*Porzana carolina*; 17 individuals at PMP sites, 0 at RP sites) regularly occur at PMP sites year-round.

While water attracts some species to PMP sites, other species are detected through the PMP because the program's quarterly monitoring schedule permits the inclusion of species only present in the Tucson area in the cooler months, such as Lewis's Woodpecker (*Melanerpes lewis*), Merlin (*Falco columbarius*), and White-breasted Nuthatch (*Sitta carolinensis*). Finally, several of the species unique to the PMP are considered rare in the study area (Corman and Wise-Gervais 2005), and while their detection may be a factor of the specialized habitats available at PMP sites, or the program's year-round monitoring schedule, it may also simply be due to chance observations (several rare species have also been detected through RP surveys, but not PMP surveys). Several of these rare species, such as the American Goldfinch (*Carduelis tristis*), Grasshopper Sparrow (*Ammodramus savannarum*), Mexican Jay (*Apthelocoma ultramarina*), and White-eyed Vireo (*Vireo griseus*), have only been detected at a single site during a single monitoring period. The detection of these species, while too limited to be useful in tracking long-term distribution and abundance trends, is helpful in determining a more complete species list for the Tucson area.

The presence of so many unique species at PMP sites demonstrates the importance of these ecological islands for maintaining bird populations within the Tucson area. Due to the scarcity of natural habitats and permanent bodies of water in developed Tucson, the RP, with its comparatively coarse resolution of ~1 km between sites, misses many of these locations. The finer spatial and temporal resolution of the PMP complements the design of the more extensive RP.

Another benefit of the flexibility in locating PMP sites is that it has enabled the TBC to form partnerships with several other organizations that use the data collected through the project as part of their monitoring programs. In partnership with Tucson Water, for example, the TBC monitors a local watercourse into which the utility recently began periodic discharges of reclaimed waste water. The purpose of the surveys is to determine if the waste water augments bird populations in the area. Similarly, the TBC monitors three sites in conjunction with Pima County Natural Resources, Parks, and Recreation, including an area scheduled to be developed into a regional recreation complex, allowing for the collection of pre- and post-development data. Additional partnerships include the National Park Service, with which the TBC monitors birds at Saguaro National Park, and Tucson Audubon Society, which coordinates its candidate Important Bird Area surveys with the TBC PMP.

The Tucson Bird Count and Christmas Bird Counts

Results of the comparison between the CBC and TBC surveys indicate that in concurrent survey periods (winters of 2000–2001 to 2005–2006 for CBC; spring 2001 to winter 2006 for TBC), both counts showed considerable overlap, yet definitive differences in the species recorded. During this period, 195 species were recorded by the CBC, 165 of them shared with the 226 species recorded by the TBC (see Table 9.2 for species unique to one of the counts). Many of the species recorded by only one of the programs are uncommon in the Tucson area, and were counted only once or twice. Of these, many are shorebirds and waterfowl, such as Baird's Sandpiper (*Calidris bairdii*) and Greater Scaup (*Aythya marila*), that pass through the area, stopping only temporarily at local water sources. Several species, such as Townsend's Solitaire (*Myadestes townsendi*) and Red-breasted Nuthatch (*Sitta canadensis*), are unique to the CBC due to their presence at elevations above 1,000 m, which are not part of the TBC study area but are included in the Santa Catalina Mountains portion of the CBC circle. As expected, many of the species counted as part of the TBC, but not CBC, are typically present in the Tucson area only during the breeding season or pass through during spring and fall migration.

The design of the CBC allows participants to seek out and count rare species and visit a wide variety of habitats within the designated count circle (Dunn et al. 2005) and compare the results to those from other count circles across North America and over time. With the TBC, participants survey the same sites each time, allowing comparisons throughout Tucson and over time. The CBC also provides a general citywide snapshot of Tucson's bird community in the winter, whereas the TBC provides spatially explicit data during the breeding season, and, through the PMP, quarterly throughout the year. The CBC is certainly a useful tool for increasing our information about the avifauna in Tucson, but it was purposefully designed as a broad-scale survey, and the methodology used does not allow for analyses of changes at specific sites within count circles. The TBC, on the other hand, is designed specifically to monitor birds in the Tucson area, and the combination of the RP and PMPs has resulted in a comprehensive approach that allows us to track populations of birds in the area at both broad and fine scales. When combined with information from other local bird studies, which may provide a more in-depth view of particular species (Mannan and Boal 2000), data from

TABLE 9.2

Unique bird species recorded within the Tucson area by the Christmas Bird Count (2000–2001 to 2005–2006) or the Tucson Bird Count (spring 2001–2006).

The 165 species recorded by both programs are not included.

Scientific name (common name)	CBC	TBC	Note
Aythya marila (Greater Scaup)	•		1
Bucephala clangula (Common Goldeneye)	•		1
Cyrtonyx montezumae (Montezuma Quail)		•	1
Aechmophorus occidentalis (Western Grebe)		•	5
Botaurus lentiginosus (American Bittern)	•		1, 3
Ixobrychus exilis (Least Bittern)		•	1
Egretta tricolor (Tricolored Heron)		•	1
Bubulcus ibis (Cattle Egret)		•	1, 3
Plegadis chihi (White-faced Ibis)		•	3
Coragyps atratus (Black Vulture)		•	
Pandion haliaetus (Osprey)		•	3
Asturina nitida (Gray Hawk)		•	1, 2
Buteo swainsoni (Swainson's Hawk)		•	2
Rallus limicola (Virginia Rail)	•		5
Recurvirostra americana (American Avocet)		•	2
Tringa solitaria (Solitary Sandpiper)		•	1, 3
Tringa melanoleuca (Greater Yellowlegs)	•		1, 5
Catoptrophorus semipalmatus (Willet)		•	1, 3
Calidris bairdii (Baird's Sandpiper)		•	1, 3
Calidris melanotos (Pectoral Sandpiper)		•	1
Calidris alpina (Dunlin)	•		3
Phalaropus tricolor (Wilson's Phalarope)		•	1, 3
Larus pipixcan (Franklin's Gull)		•	1, 3
Sterna hirundo (Common Tern)		•	1
Streptopelia risoria (Ringed Turtle-Dove)		•	1
Streptopelia decaocto (Eurasian Collared-Dove)		•	
Columbina passerina (Common Ground-Dove)	•		
Coccyzus americanus (Yellow-billed Cuckoo)		•	1, 2
Glaucidium brasilianum (Ferruginous Pygmy-Owl)		•	1
Micrathene whitneyi (Elf Owl)		•	1, 2
Asio otus (Long-eared Owl)	•		1
Chordeiles acutipennis (Lesser Nighthawk)		•	2
Phalaenoptilus nuttallii (Common Poorwill)		•	2
Chaetura vauxi (Vaux's Swift)		•	3
Amazilia violiceps (Violet-crowned Hummingbird)	•		1

TABLE 9.2 (CONTINUED)

Scientific name (common name)	CBC	TBC	Note
Archilochus colubris (Ruby-throated Hummingbird)	•		1
Archilochus alexandri (Black-chinned Hummingbird)		•	2
Stellula calliope (Calliope Hummingbird)	•		1, 3
Selasphorus platycercus (Broad-tailed Hummingbird)		•	2
Selasphorus rufus (Rufous Hummingbird)		•	3
Sphyrapicus thyroideus (Williamson's Sapsucker)	•		4, 5
Contopus cooperi (Olive-sided Flycatcher)		•	3
Contopus sordidulus (Western Wood-Pewee)		•	2
Empidonax traillii (Willow Flycatcher)		•	1, 3
Empidonax minimus (Least Flycatcher)		•	1
Empidonax hammondii (Hammond's Flycatcher)		•	3
Empidonax oberholseri (Dusky Flycatcher)		•	3
Empidonax difficilis (Pacific-slope Flycatcher)		•	3
Empidonax occidentalis (Cordilleran Flycatcher)		•	2
Myiarchus tuberculifer (Dusky-capped Flycatcher)		•	2
Myiarchus tyrannulus (Brown-crested Flycatcher)		•	2
Tyrannus melancholicus (Tropical Kingbird)		•	2
Tyrannus verticalis (Western Kingbird)		•	2
Vireo bellii (Bell's Vireo)		•	2
Vireo gilvus (Warbling Vireo)		•	2
Cyanocitta stelleri (Steller's Jay)	•		4
Progne subis (Purple Martin)		•	2
Tachycineta bicolor (Tree Swallow)		•	1, 3
Riparia riparia (Bank Swallow)		•	3
Petrochelidon pyrrhonota (Cliff Swallow)		•	2
Sitta canadensis (Red-breasted Nuthatch)	•		1, 4
Sitta carolinensis (White-breasted Nuthatch)	•		4
Certhia americana (Brown Creeper)	•		1, 4
Regulus satrapa (Golden-crowned Kinglet)	•		1, 4
Myadestes townsendi (Townsend's Solitaire)	•		4, 5
Oreothlypsis peregrina (Tennessee Warbler)		•	2
Oreothlypsis ruficapilla (Nashville Warbler)		•	3
Oreothlypsis virginiae (Virginia's Warbler)		•	2
Oreothlypsis luciae (Lucy's Warbler)		•	2
Setophaga americana (Northern Parula)	•		1
Setophaga pensylvanica (Chestnut-sided Warbler)	•		1, 4
Mniotilta varia (Black-and-white Warbler)	•		1, 4

TABLE 9.2 (*continued*)

TABLE 9.2 (CONTINUED)

Scientific name (common name)	CBC	TBC	Note
Setophaga ruticilla (American Redstart)	•		1, 4
Basileuterus rufifrons (Rufous-capped Warbler)	•		1
Icteria virens (Yellow-breasted Chat)		•	2
Piranga ludoviciana (Western Tanager)		•	2
Aimophila cassinii (Cassin's Sparrow)		•	1
Spizella pallida (Clay-colored Sparrow)	•		1
Ammodramus leconteii (Le Conte's Sparrow)	•		1
Passerella iliaca (Fox Sparrow)	•		
Melospiza georgiana (Swamp Sparrow)	•		5
Zonotrichia querula (Harris's Sparrow)	•		1
Pheucticus melanocephalus (Black-headed Grosbeak)		•	2
Passerina caerulea (Blue Grosbeak)		•	2
Passerina amoena (Lazuli Bunting)		•	3
Passerina cyanea (Indigo Bunting)		•	1, 2
Passerina versicolor (Varied Bunting)		•	2
Icterus bullockii (Bullock's Oriole)		•	2
Icterus parisorum (Scott's Oriole)		•	2
Carpodacus cassinii (Cassin's Finch)	•		1, 4
Carduelis tristis (American Goldfinch)	•		5

SOURCE: Comments on species occurrence *sensu* Corman and Wise-Gervais (2005), and Sibley (2003).

1 = only 3 or fewer records.

2 = occurs within the Tucson area only during the breeding season.

3 = migrant.

4 = generally occurs at elevations above 1,000 m in Tucson area.

5 = occurs in the Tucson area only during the winter.

the TBC and CBC provide essential components of an overall picture of Tucson's bird community.

Lessons Learned

The design of the TBC is a trade-off between the need to collect data at many points over broad temporal and spatial scales and the availability of skilled observers. As with many other long-term bird monitoring projects (such as the CBC, the USGS's North American Breeding Bird Survey, and Cornell Lab of Ornithology and Bird Studies Canada's Project FeederWatch) the TBC utilizes volunteer observers to collect data, which both reduces program costs and increases the number of sites that can be surveyed relative to a project that relies solely on a small team of paid researchers (McCaffrey 2005). However, studies based on data collected by citizen science volunteers may be viewed as lacking the necessary scientific rigor to produce valid conclusions (Irwin 1995). In order to avoid this perception, the TBC has several mechanisms in place to ensure the quality of the data gathered by volunteers, beginning with the requirement that participants' birding abilities exceed an established minimum level. Prior to participating in the TBC, potential volunteers are asked to take an online test (which can be viewed at http://www.tucsonbirds.org/current/SelfTest.asp) designed to assess whether or not they are qualified. While there are several additional determinants, a qualified observer will be able to identify most of the species of birds found in the Tucson area both quickly and correctly. Additionally, routes are designed to be as close to linear as possible, rather than clumped; this reduces spatial autocorrelation in observer identity, and facilitates detection of observers whose results deviate from those of similar nearby areas. Moreover, as participants enter their results at the TBC website, count data are subjected to review by TBC staff in the project's Access database, with unusual results immediately flagged for additional investigation.

Yet perhaps the most important way that the TBC manages scientific rigor lies in carefully limiting the ways the data collected are used. The design of the TBC is not appropriate for fine-scale comparisons of population densities among individual species or sites, but rather for examining the broader patterns of how and why species distributions are changing over time and across the study area. This limitation of the data is particularly important for other researchers to consider when designing similar monitoring programs. One approach to at least partially reduce this data limitation is by adding a distance sampling component to the survey protocol, which would allow for corrections to be made based on the differential observability of species (Thomas et al. 2002). Although distance sampling was not included as part of the TBC from the outset, TBC staff have begun to explore its implementation. We recommend that it be considered for other bird monitoring programs from the outset.

A common concern when establishing citizen science programs is whether there will be enough skilled volunteers to adequately carry out the project. Fortunately, one of the benefits of conducting an urban citizen science project is that there is typically a large potential volunteer pool to draw from. The Tucson area has an active birding community, and both professional birding guides and many avid amateurs participate in the TBC (McCaffrey 2005). With a recent study (Cordell and Herbert 2002) indicating that there are more than 2 million birding "enthusiasts" (defined as birders who participate in birding more than 50 days a year) and 3.75 million "active" birders (defined as birders who participate in birding between 6 and 50 days a year) living in counties of more than 1 million people, there are likely sufficient numbers of adequately skilled volunteers in most cities. Additionally, there are options for beginning birders to become involved in programs like the TBC by partnering with more knowledgeable birders or attending training sessions, which will allow them to gain experience and develop their skills so that they may be able to fully participate in the future.

A Global Perspective

The goal of the TBC is to collect information that may suggest ways to maintain and promote populations of native birds in the Tucson area. Yet initial results demonstrate a need for similar efforts to occur elsewhere. Many of the general patterns observed in the TBC data may apply to other cities, but habitat features vary widely across cities, and different species-specific solutions may be appropriate in different areas. Recently, programs such as the Ottawa Breeding Bird Count (http://www.glel.carleton.ca/ottawabirds/) and Fresno Birds (http://www.fresnobirds.org/) have applied the TBC model to other North American cities. While each of these programs is designed to collect information on birds and the impacts of urbanization in their respective cities, there is the potential for an urban bird monitoring network which will allow for comparisons of results among cities.

The impacts of urbanization on birds in tropical areas, in particular, remains understudied (Chace and Walsh 2006), despite the rapid expansion of many tropical cities. Given the richness and diversity of many of the avian communities in these areas, finding ways to limit the impacts of such growth are of particular importance to the conservation of biodiversity. While specific results from the TBC may not provide solutions in these areas, the TBC can serve as a model on which other urban monitoring programs can be based. By resolving many of the time-consuming and effort-intensive issues related to the development of a systematic urban bird survey, such as study design, database development, and count protocols, the TBC provides an example for other cities. With skilled volunteers collecting most of the data, implementing such a monitoring project should also be relatively inexpensive.

The development of other long-term urban monitoring programs, both across the U.S. and around the world, will allow us to develop a better understanding of how we can maintain sustainable populations of native birds in cities. With such variation in environmental conditions, land use practices, and development patterns among cities, we could investigate the impacts of a much wider range of factors, resulting in a more complete picture of which actions and conditions are amenable to supporting birds in an urbanizing world and which are not.

ACKNOWLEDGMENTS

We are grateful to the TBC volunteers, as well as M. Rosenzweig and B. Mannan, for their support and feedback. Thanks to A. J. Borens for her assistance with the A. Towhee's analyses. The Arizona Game and Fish Department, the University of Arizona's School of Natural Resources, and the National Park Service supported this research.

LITERATURE CITED

Beissinger, S. R., and D. R. Osborne. 1982. Effects of urbanization on avian community organization. Condor 84:75–83.

Bibby, C. J., N. D. Burgess, and D. A. Hill. 1992. Bird census techniques. Academic Press, London, UK.

Blair, R. B. 1996. Land use and avian species diversity along an urban gradient. Ecological Applications 6:506–519.

Blondel, J., C. Ferry, and B. Frochet. 1981. Point counts with unlimited distance. Studies in Avian Biology 6:414–420.

Boal, C. W., and R. W. Mannan. 1999. Comparative breeding ecology of Cooper's Hawks in urban and exurban areas of southeastern Arizona. Journal of Wildlife Management 63:77–84.

Chace, J. F., and J. J. Walsh. 2006. Urban effects on native avifauna: a review. Landscape and Urban Planning 74:46–69.

Cordell, K. H., and N. G. Herbert. 2002. The popularity of birding is still growing. Birding 34:54–61.

Corman, T., and C. Wise-Gervais (editors). 2005. The Arizona breeding bird atlas. University of New Mexico Press, Albuquerque, NM.

Czech, B., and P. R. Krausman. 1997. Distribution and causation of species endangerment in the United States. Science 277:1116–1117.

Davis, T. 2006. Feeling crowded? Million reasons why. Arizona Daily Star, 12 November: A9.

Dunn, E. H., C. M. Francis, P. J. Blancher, S. R. Drennan, M. A. Howe, D. LePage, C. S. Robbins, K. V. Rosenberg, J. R. Sauer, and K. G. Smith. 2005. Enhancing the scientific value of the Christmas Bird Count. Auk 122:338–346.

Emlen, J. T. 1974. An urban bird community in Tucson, Arizona: derivation, structure, regulation. Condor 76:184–197.

Germaine, S. S., S. S. Rosenstock, R. E. Schweinsburg, and W. S. Richardson. 1998. Relationships among breeding birds, habitat, and residential development in greater Tucson, Arizona. Ecological Applications 8:680–691.

Houghton, R. A. 1994. The worldwide extent of land-use change. Bioscience 44:305–313.

Hutton, K. A. 2005. Number of bird cavities and wounds in urban saguaro cacti. M.S. thesis, University of Arizona, Tucson, AZ.

Irwin, A. 1995. Citizen science: a study of people, expertise, and sustainable development. Routledge, London, UK.

Knick, S. T., and J. T. Rottenberry. 2000. Ghosts of habitats past: contribution of landscape change to current habitats used by shrubland birds. Ecology 81:220–227.

Mannan, R. W., and C. W. Boal. 2000. Home range characteristics of male Cooper's Hawks in an urban environment. Wilson Bulletin 112:21–27.

Marzluff, J. M. 2001. Worldwide urbanization and its effects on birds. Pp. 19-47 *in* J. M. Marzluff, R. Bowman, and R. Donnelly (editors), Avian ecology in an urbanizing world. Kluwer, Norwell, MA.

McCaffrey, R. E. 2005. Using citizen science in urban bird studies. Urban Habitats 3:70–86.

McKinney, M. L. 2002. Urbanization, biodiversity, and conservation. Bioscience 52:883–890.

Meents, J. K., B. W. Anderson, and R. D. Ohmart. 1981. Vegetation characteristics associated with Abert's Towhee numbers in riparian habitat. Auk 98:818–827.

Melles, S., S. Glenn, and K. Martin. 2003. Urban bird diversity and landscape complexity: species-environment associations along a multiscale habitat gradient. Conservation Ecology 7:5.

Mills, G. S., J. B. Dunning, Jr., and J. M. Bates. 1989. Effects of urbanization on breeding bird community structure in southwestern desert habitats. Condor 91:426–428.

National Audubon Society. 2002. Audubon watch list 2002. <http://www.audubon.org/bird/watchlist/> (4 January 2006).

National Audubon Society. 2006. The Christmas Bird Count historical results. <http://www.audubon.org/bird/cbc> (5 January 2006).

Pima Association of Governments. 2003. Tucson metropolitan community information data summary. 8th ed. <http://www.pagnet.org/TPD/CIDS2003/> (13 January 2006).

Pima County. 2006. Sonoran Desert conservation plan. <http://www.pima.gov/cmo/sdcp/index.html> (23 January 2006).

Pima County Department of Transportation. 2005. Pima County GIS layers. <http://www.dot.pima.gov/gis/maps/mapguide/> (17 November 2005).

Ralph, C. J., S. Droege, and J. R. Sauer. 1995. Managing and monitoring birds using point counts: standards and applications. Pp. 161–168 *in* C. J. Ralph, J. R. Sauer, and S. Droege (editors), Monitoring bird populations by point counts. General Technical Report PSW-GTR-149. USDA Forest Service, Pacific Southwest Research Station, Albany, CA.

Rosen, P. C., and K. Mauz. 2001. Biological values of the West Branch of the Santa Cruz River, with an outline for a potential park or reserve. <http://www.co.pima.az.us/cmo/sdcp/sdcp2/reports/WB/WestB.htm> (12 August 2005).

Rosenzweig, M. L. 2003. Win-win ecology: how earth's species can survive in the midst of human enterprise. Oxford University Press, New York.

Savard, J.-P. L., P. Clergeau, and G. Mennechez. 2000. Biodiversity concepts and urban ecosystems. Landscape and Urban Planning 48:131–142.

Sibley, D. A. 2003. The Sibley field guide to birds of western North America. Knopf, New York, NY.

Skagen, S. K., J.F. Kelly, C. Van Riper, III, R. L. Hutto, D. M. Finch, D. J. Krueper, and C. P. Melcher. 2005. Geography of spring landbird migration through riparian habitats in southwestern North America. Condor 107:212–227.

Sutherland, W. J., I. Newton, and R. E. Green. 2004. Bird ecology and conservation: a handbook of techniques. Oxford University Press, Oxford, UK.

Thomas, L., S. T. Buckland, K. P. Burnham, D. R. Anderson, J. L. Laake, D. L. Borchers, and S. Strindberg. 2002. Distance sampling. Pp. 544–552 *in* A. H. El-Shaarawi and W. W. Piegorsch (editors), Encyclopedia of environmetrics, Vol. 1. John Wiley and Sons, Chichester, UK.

Turner, W. R. 2003. Citywide biological monitoring as a stool for ecology and conservation in urban landscapes: the case of the Tucson Bird Count. Landscape and Urban Planning 65:149–166.

Turner, W. R. 2006. Interactions among scales constrain species distributions in fragmented urban landscapes. Ecology and Society 11:6.

Turner, W. R., T. Nakamura, and M. Dinetti. 2004. Global urbanization and the separation of humans from nature. Bioscience 54:585–590.

Tweit, R. C., and D. M. Finch. 1994. Abert's Towhee (*Pipilo aberti*). No. 111 *in* A. Poole and F. Gill (editors), The birds of North America. Academy of Natural Sciences, Philadelphia, PA, and American Ornithologists' Union, Washington, DC.

Tweit, R. C., and J. C. Tweit. 1986. Urban development effects on the abundance of some common resident birds of the Tucson area of Arizona. American Birds 40:431–436.

United Nations. 2006. Population estimates and projections. <http://un.org/esa/populations/WPP2004/2004Highlights_finalrevised.pdf> (12 January 2006).

U.S. Census Bureau. 2006. Louisiana losses population; Arizona edges Nevada as fastest growing state. <http://www.census.gov/Press-Release/www/releases/archives/population/007910.html> (30 July 2007).

USGS National Gap Analysis Program. 2004. Provisional digital land cover map for the southwestern United States. Version 1.0. RS/GIS Laboratory, College of Natural Resources, Utah State University, Logan, UT.

Verner, J. 1985. Assessment of counting techniques. Pp. 247–302 *in* R. F. Johnston (editor), Current ornithology, Vol. 2. Plenum Press, New York, NY.

CHAPTER TEN

Distribution and Habitat of Greater Roadrunners in Urban and Suburban Arizona

Stephen DeStefano and Charlene M. Webster

Abstract. Greater Roadrunners (*Geococcyx californianus*) are widespread throughout arid regions of the southwestern United States. As a large predatory bird, roadrunners require large home ranges, and their relatively low densities and secretive nature make surveys and observations a challenge. However, roadrunners are among the most well-known and charismatic avian species in the Southwest and draw much attention from residents and visitors. Distribution of roadrunners and habitat features of importance to roadrunners in urban and suburban environments have not been described, and it is not known how development may affect roadrunner populations. Although roadrunners are occasionally seen in southwestern cities, urbanization may eliminate local populations. We used public sighting information to document roadrunner locations in metropolitan Tucson, Arizona. We explored and, when possible, evaluated the types of biases (e.g., species misidentification, erroneous information, nonresponse, and self-selection biases) inherent in public information surveys. We also employed researcher-based, randomly distributed surveys to examine roadrunner distribution and habitat use, to estimate the amount of survey effort required to detect a low-density elusive species like roadrunners, and to establish a sample of sightings based on random sampling. We received 1,449 reports of roadrunner sightings from the public and saw 52 roadrunners on 281 researcher-based surveys. The general pattern of distribution of roadrunner sightings in Tucson was similar in both public and researcher surveys, forming a broad band around the edge of Tucson with few or no sightings in the city center. Tucson residents reported seeing roadrunners most often in areas with housing densities of 2.5–7.5 residences per ha (RHA) and nonnative vegetation, and less often than expected in areas with washes, natural open space, native vegetation, graded vacant land, and housing densities of 7.5–15 RHA. In contrast, during our random surveys we found roadrunners closer to washes and areas with housing densities of 7.5–15 RHA, indicating that roadrunners used these land cover types more frequently than was reported from the public. In general, however, roadrunners used a variety of urban and suburban environments and could be seen almost anywhere in the city except for areas with the highest housing densities (>15 RHA). Public survey information provided a great deal of usable data quickly and inexpensively but reflected where people saw and reported roadrunners. Interpretations of relative abundance or habitat preference need to be made cautiously.

Key Words: Arizona, distribution, *Geococcyx californianus*, Greater Roadrunner, habitat, Tucson, urban.

DeStefano, S., and C. M. Webster. 2012. Distribution and habitat of Greater Roadrunners in urban and suburban Arizona. Pp. 155–166 *in* C. A. Lepczyk and P. S. Warren (editors). Urban bird ecology and conservation. Studies in Avian Biology (no. 45), University of California Press, Berkeley, CA.

Increasing urbanization is characterized by displacement of native vegetation with nonnative vegetation and man-made structures, higher prevalence of exotic species, and increased disturbance and hazards to wildlife because of higher densities of people, roads, and domestic predators such as cats and dogs (DeStefano and DeGraaf 2003). Construction of houses, commercial buildings, parking lots, and roads alters and fragments habitat, and patches of native vegetation become isolated islands within city limits (Shaw et al. 1991, Germaine et al. 1998). As urbanization increases, these islands become farther away from native vegetation surrounding the city. This separation can mean loss of genetic diversity or genetic isolation in species with low mobility (Germaine 1995). These changes in environmental structure and function may lead to changes in species diversity (Knight 1991).

Many species cannot cope with the changes brought on by urbanization and become extirpated locally (Germaine 1995, Harris et al. 1995, Schoenecker and Krausman 2002). Other species, like Cooper's Hawks (*Accipiter cooperii*), are attracted to particular features of urban environments, such as groves of large trees around residential neighborhoods and city parks, and will inhabit developed areas (Boal and Mannan 1998). Still other species remain and use new habitat features as environments are modified. For example, House Finches (*Carpodacus mexicanus*) occupy both undeveloped and developed environments and will readily use native or nonnative vegetation and man-made structures for cover and nesting (Hill 1993). The resulting array of native and nonnative organisms has been called a recombinant community (Soulé 1990)—a completely new community of synanthropic organisms with resulting changes in interspecific relationships and competition (Johnston 2001, Byers 2002).

It is important to understand the changing dynamics in the assemblage of avian species and other wildlife as the landscape undergoes continued human development (Johnson and Klemens 2005). In fact, identifying the species that can inhabit urban and suburban environments and understanding how they coexist with high densities of humans is as important as understanding which species are threatened by human development (DeStefano and Johnson 2005). Greater Roadrunners (*Geococcyx californianus*) are common throughout the southwestern United States and are seen in urban and suburban areas within their geographic range, but the effects of urbanization on roadrunners have not been studied. Roadrunner tolerance for many variables associated with urban areas, such as exotic vegetation and housing density, is unknown. The impact of human disturbances such as urban development has been listed as a priority for future research in roadrunner ecology and conservation (Hughes 1996).

Because of their secretive nature, low densities (<0.1 bird/40 ha), and big home ranges (about 1 km^2), large numbers of roadrunners can be difficult to see (Anderson et al. 1982, Hughes 1996, Webster and DeStefano 2004). However, sightings of roadrunners by the public at large are common and can be amassed into an extensive database (Webster 2000, Webster and DeStefano 2004). Public surveys have been used to acquire information on a variety of wildlife species, particularly those species that are difficult to detect (Quinn 1995, McGuill et al. 1997, Pyrovetsi 1997). For example, in Britain Fussell and Corbet (1992) used a public survey to find bee nests (*Bombus* spp., *Psithyrus* spp.), noting that previous research on bee nests was often based on few records because nests are difficult to find, whereas their public survey generated >400 nest locations. In Washington, Quinn (1995) used 108 public sightings of coyotes (*Canis latrans*) compared to six radiotelemetry-equipped individuals to determine habitat use in a suburban environment. Thus, enlisting the assistance of urban and suburban residents can greatly aid in the collection of data on urban and suburban wildlife.

Although it is true that information from the public can yield large volumes of data, there are many potential problems, including nonresponse, erroneous information, and a variety of other reporting biases. Webster and DeStefano (2004) described and addressed many of these biases related to public information gathered on the presence of roadrunners in Tucson; they found that the public could provide much useful information on roadrunner occurrence, distribution, and nest sites despite some of these biases. In this paper, we assess data on roadrunner distribution and habitat use in urban and suburban Tucson based on public sightings and our own random surveys of the city. Our goal was to answer questions about the distribution and key habitat features for roadrunner occupancy in an

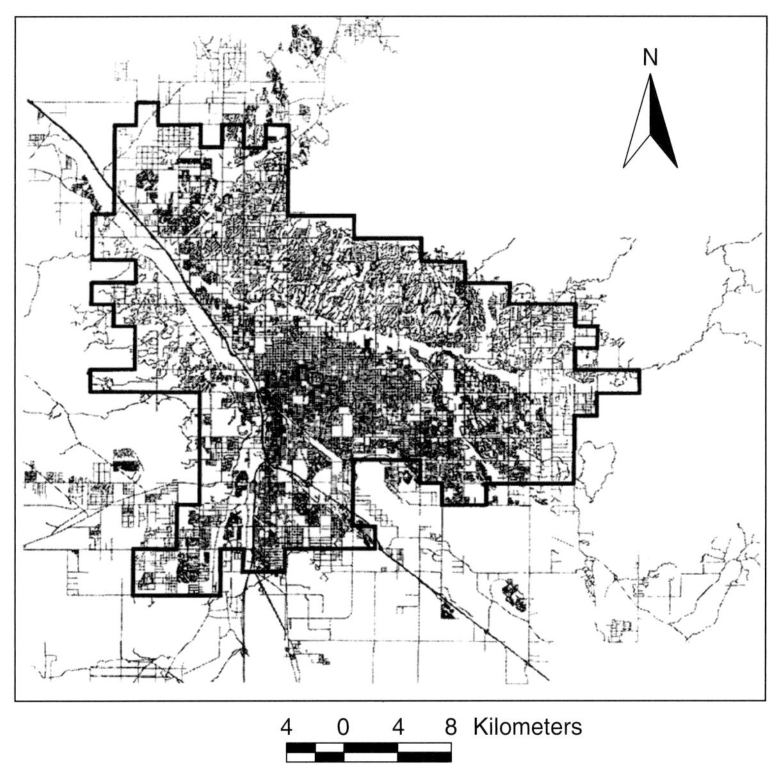

Figure 10.1. Study area (600 km^2) and streets in Tucson metropolitan area, Arizona, used to determine distribution of Greater Roadrunners.

urban and suburban environment while making a critical evaluation of this information as supplied by an interested, but potentially biased, public. Our objectives were to determine the distribution of roadrunners, describe habitat use, and identify habitat variables such as housing densities, washes, and native vegetation that may explain the distribution of roadrunners in metropolitan Tucson.

METHODS

Study Area

Tucson is located in southeastern Arizona, USA (at 32°1′ north latitude and 110°9′ west longitude; Turner and Brown 1982), and is surrounded by mountains to the north, east, and west, and Tucson International Airport and the Tohono O'Odham Nation reservation to the south. In 1990, metropolitan Tucson had a population of almost 667,000 people living in about 24,000 km^2. By 1997, the population had grown to >780,000, and by 2010 the population was projected to be about 890,000 (U.S. Census Bureau 1998). Tucson is part of the Arizona Upland subdivision of the Sonoran Desert. Elevation averages 787 m and annual rainfall is about 280 mm/yr, with 30–60% falling from June to August and 10–40% falling during winter (Turner and Brown 1982).

Our study area covered 600 km^2 in metropolitan Tucson (Fig. 10.1), including areas within the Tucson city limits plus suburbs and adjacent lands, such as incorporated towns and unincorporated Pima County. Industrial, commercial, and residential areas existed within our study area boundaries. Housing densities ranged from ≤2.5 to >15 houses per ha. Open spaces consisted primarily of parks, golf courses, and undeveloped lots. Roadways of various sizes, including major highways, traversed Tucson, as did several major washes and their minor tributaries, all of which were dry most of the year.

Public and Researcher Surveys

Between January 1997 and August 1998, we conducted a public survey to document roadrunner sighting locations in metropolitan Tucson. We solicited information from the public on sightings of roadrunners via fliers, newspaper articles (*Arizona Daily Star*), a 90-second television news segment (KOLD Channel 13), radio broadcasts (KUAT), and word of mouth. We designated a phone number and voice messaging system for callers to leave information about their sightings, including their name, phone number, and address, and the exact location, time, and date of the roadrunner sighting (Quinn 1995). We entered this information into a database and acquired a streets coverage (an address database and digital map of the streets in Pima County) for 1997 from Pima County Government offices.

Between February 1997 and January 1999, we also conducted our own surveys for roadrunners concurrently with the public survey. The objectives for these surveys were to gather information on roadrunner distribution based on a random approach, compare results of that approach to the public survey, and determine the effort required to collect data on randomly distributed observations of roadrunners versus what could be gathered by the public. We also hoped to identify some of the biases inherent in public sighting information and compare the pattern of roadrunner sightings provided by the public to the pattern of sightings from our researcher-based surveys. We divided the study area into 232 blocks of 2.6 km^2 as defined on 7.5-minute topographic maps. Each block contained anthropogenic features such as roads, buildings, and parks over at least 25% of the area. We divided each block into four 0.65-km^2 square sections totaling 928 sections. Days were divided into three parts: AM = sunrise–10:00 MST, mid-day = 10:00–14:00 MST, and PM = 14:00 MST–sunset. Each survey consisted of a randomly selected section, day of the week (1 of 7), and time of day (1 of 3). Because roadrunners were resident birds in the study area, we conducted our surveys throughout the year. Surveys were conducted by skilled bird observers who underwent one day of training. Each survey covered a distance of 3–5 km or 50–60 minutes of walking, whichever came first. If a survey landed on inaccessible property (e.g., gated community), the surveyor selected the next available plot closest to their current location.

Roadrunner Distribution

Using ArcView® (Ver. 3.1, ESRI, Redlands, CA), and ARC/INFO® (Ver. 7.1.2, ESRI, Redlands, CA) geographic information system (GIS) software, we combined the roadrunner sightings database from the public survey and streets coverage with the geocoding feature in ArcView to construct a distribution map of roadrunner sighting locations in metropolitan Tucson. We eliminated sighting locations outside the study area and duplicate sightings. Duplicate sightings were those reported from the same street address on the same day. We used ArcView to produce a relative density map, using a grid cell size of 500 m and search radius of 500 m. We also produced a distribution map of roadrunner sightings resulting from our researcher surveys with the same method used to make the distribution map of public sighting locations. We made a relative density map with a grid cell size of 4,000 m and search radius of 3,400 m and compared the distribution of sighting locations from our researcher surveys to those from the public survey.

Habitat Use

A Wildlife Habitat Inventory Pilot Study (WHIPS) GIS database has been developed for the Tucson area which identifies 30 land use classes and contains housing densities (Shaw et al. 1996). We subsequently reduced the number of land use classes to eliminate those classes that occurred with low frequency (i.e., <5 observations) in association with roadrunner sightings. We also eliminated the roadways classes from our analysis because many of the sighting locations (213 points) that fell within that class (i.e., on roads) may have been an artifact of geocoding. These locations likely fell either on the roads or within land use classes adjacent to the roadways, but we had no way to determine into which class each location fell.

We evaluated roadrunner habitat (land use class) use versus availability with the generalized method described by Neu et al. (1974). We conducted two types of spatial analyses based on Quinn's (1995) study of urban coyotes in Seattle, Washington, which he called "habitat" and "distance zone" analyses. For the first analysis (habitat), we compared the land use classes where each public sighting of roadrunners occurred (used) with the amount of that class based on the WHIPS database (available). We calculated the total area for each land use class and then calculated the expected number of points that would fall in each land use class if use was consistent with availability. We then used chi-square goodness-of-fit tests (Sokal and Rohlf 1995) to compare observed use of land use classes by roadrunners to an expected distribution based on available classes in the study area. For those land use classes that were used by roadrunners more or less than available, we calculated 90% Bonferroni confidence intervals (90% CI) to determine which covers contributed most to that significance (Neu et al. 1974).

For the second analysis (distance zone), we used seven landscape features from the WHIPS database that were significant in the first (habitat) analysis. These seven landscape features included three land use classes and four different housing densities:

- major and minor washes
- nonnative vegetation (areas of open space with a high probability of containing abundant nonnative landscaping, e.g., cemeteries, golf courses, and city parks)
- native vegetation (areas of open space with a high probability of containing abundant native vegetation, e.g., natural open space and state and federal parks and forests)
- ≤2.5 residences per ha (RHA)
- 2.5–7.0 RHA
- 7.0–15 RHA
- >15 RHA

We created buffers (Quinn 1995) around each landscape feature of 0–25 m, 26–50 m, and at 50-m intervals from 51 to 1,000 m for washes and vegetation covers, and 51 to 400 m for housing. We then assigned a value of 1 m for all roadrunner locations or random points that fell within a specific landscape feature, and a mid-point value (e.g., 12.5 m for the 0–25 m category) for each location or point that fell within a distance or buffer. We summed all distances within each of the seven landscape features for the roadrunner locations reported by the public. We repeated this same procedure for 1,200 random points that we generated within the study area, and used *t*-tests to compare distances of roadrunner sightings to random points for each of the seven landscape variables (Sokal and Rohlf 1995).

We conducted habitat and distance zone analyses for both public sightings and researcher sightings. In addition, we compared results from the distance zone analysis for public and researcher sightings.

RESULTS

Public and Researcher Surveys

The public survey generated 1,449 phone calls with roadrunner sighting locations throughout metropolitan Tucson. Of these calls, 1,230 callers responded to three newspaper articles, 140 to fliers, 40 to a KOLD (Channel 13) television news segment, 30 to three KUAT radio broadcasts, and nine to word of mouth. We eliminated 248 locations that fell outside the study area, eight that were duplicate sighting locations, and 178 because information from the public was incomplete (e.g., caller said a roadrunner was seen in their yard but caller did not leave name, address, or phone number). Thus, we obtained 1,015 usable sighting locations within the study area (Fig. 10.2). Most sightings were from callers phoning from home to report a roadrunner in their yard (75%).

Researchers completed 281 random surveys covering 1,250 km^2, about 30% of the study area. We were unable to conduct 10 (3.5%) of our planned surveys on the original plots because they landed on inaccessible property; however, each was relocated to the next nearest accessible plot. Fifty-two roadrunners were seen on 39 different survey blocks, averaging one roadrunner sighting (we counted pairs of roadrunners as one observation) per seven surveys (Fig. 10.3). To obtain the same number of sightings as were acquired from the public would have taken >10,000 hours (Webster and DeStefano 2004).

Roadrunner Distribution

Highest numbers of sightings from the public occurred on the east side of the study area (~50 km^2) and a smaller concentration occurred on the west side (~15 km^2). The fewest sighting locations were in three separate areas: (1) west of a major highway (Interstate 10) that traverses Tucson in the northwest section of the study area (~50 km^2), (2) near the city center (~20 km^2), and (3) in a long, narrow strip of land in the southern portion of the study area (~15 km^2). Highest concentrations of sightings from researcher-based random surveys occurred in four parts of the study area: (1) in the eastern section (~50 km^2), (2) in the western section (~15 km^2), (3) on a golf course in the northern section (~15 km^2), and (4) on a golf course in the southern section (15 km^2). The fewest sightings occurred in two areas: (1) near

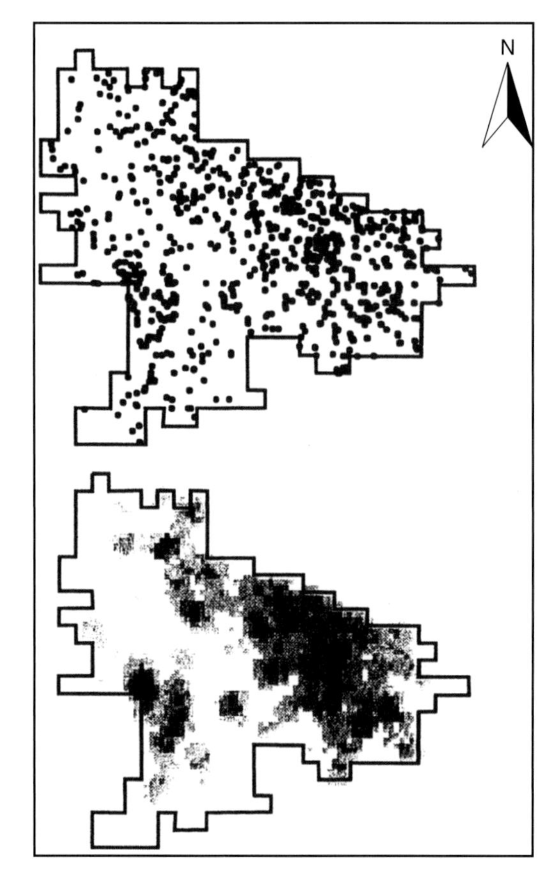

Figure 10.2. Locations (top) and relative distribution from lowest (white = 0) to highest (black = 8) number of sightings of Greater Roadrunners per 0.25 km^2 (bottom) from a public survey in Tucson metropolitan area, Arizona.

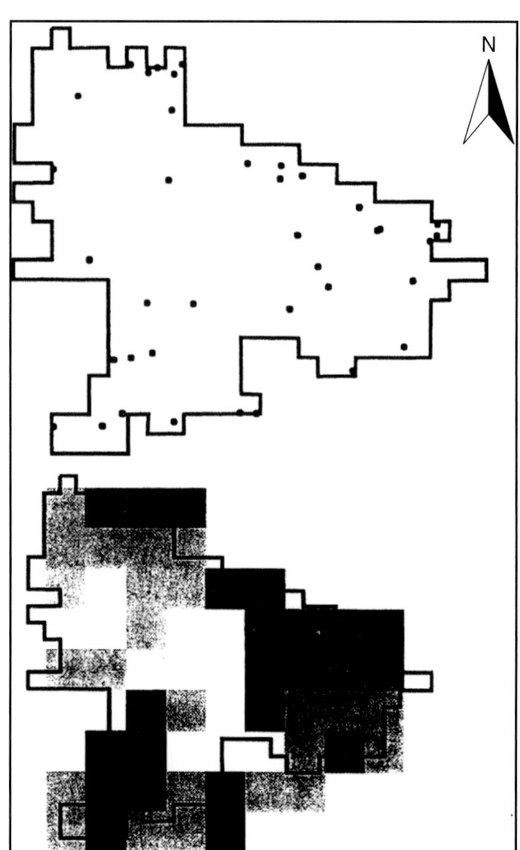

Figure 10.3. Locations (top) and relative distribution from lowest (white = 0) to highest (black = 0.1) number of sightings of Greater Roadrunners per 25 km^2 (bottom) from random surveys covering 30% of the study area in Tucson metropolitan area, Arizona.

the city center (~50 km^2) and (2) west of the major highway throughway near northwest Tucson (~30 km^2). Both surveys (public and researcher) revealed high concentrations of roadrunner sightings, forming a broad band of roughly circular shape around the periphery of the city with a void near the center. The void near the city center corresponded to the highest densities of roads and houses in Tucson (compare Figs. 10.2 and 10.3 to Fig. 10.1).

Habitat Use

Of the 30 land use classes identified in the WHIPS database, we eliminated 13 of these land use covers because of a low frequency of occurrence (0–5 observations) associated with roadrunner sightings (only four points were excluded). The elimination of these four points plus the 213 points that fell on roads left 798 locations available for habitat analysis.

We found a disproportionate distribution of roadrunner sighting locations from the public among the remaining 17 land use classes (χ^2 = 304.18, df = 16, $P < 0.001$, $n = 798$). Roadrunners were reported more frequently than expected in 2.5–7.0 RHA (Exp. = 13.7%, Obs. = 31.1%, 90% CI = 27.4%, 34.8%) and less than expected in natural open space (Exp. = 21.0%, Obs. = 11.4%, 90% CI = 8.9%, 13.9%) and graded vacant land (Exp. = 8.1%, Obs. = 3.5%, 90% CI = 2.1%, 5.0%). Roadrunner sightings from the public were also farther from washes, native vegetation, and 7.0–15 RHA than were random points, but closer to 2.5–7.0 RHA and nonnative vegetation (Table 10.1). We found no evidence of disproportionate distribution of roadrunner sightings from our researcher-based surveys among land use classes

TABLE 10.1

Differences in distance between 1,015 Greater Roadrunner sightings from the public and 1,200 random points from 3 groups of similar land use covers and 4 housing densities in metropolitan Tucson, Arizona.

Positive (negative) values indicate that roadrunners were farther away (closer) than random points.

Land use class	t	SE	$P\leq$	90% CI
Washes	2.98	6.77	0.003	6.90, 33.45
Non-native vegetation	−1.94	14.13	0.053	−55.14, 0.36
Native vegetation	2.75	6.63	0.006	5.23, 31.24
≤2.5 RHA[a]	−0.17	7.71	0.862	−16.49, 13.80
2.5–7.0 RHA[a]	−2.90	7.96	0.004	−38.74, −7.49
7.0–15 RHA[a]	1.98	6.74	0.048	0.14, 26.60
>15 RHA[a]	−0.78	8.83	0.439	−22.84, 9.91

[a] RHA = residences per ha.

(χ^2 = 0.98, df = 4, P = 0.91), nor did we find differences in distances from any landscape feature (P = 0.12). We did not conduct retrospective power analysis but suspect power was low because of low sample size. However, when we compared average distances of roadrunner locations to landscape features from the public survey and the researcher-based surveys, we found a difference in washes and 7.0–15 RHA. Public sightings were farther from these land use covers than sightings from our random, researcher-based surveys (Table 10.2).

DISCUSSION

Sightings of Roadrunners

Our requests to the public to report roadrunner sightings in Tucson were successful and led to a large number of locations of this "low-density and elusive bird" (Hughes 1996). Public surveys allowed us to acquire many more roadrunner sighting locations than was possible with our crew of researchers (Webster and DeStefano 2004). During our study, roadrunners were seen by trained, motivated observers on 39 of 281 randomly surveyed blocks, an observation rate of only 14%, whereas in a short time public participation generated >1,000 observations. Other researchers have espoused the use of public sighting information to generate data on distribution patterns for such diverse groups of organisms as invertebrates (Fussell and Corbet 1992) and large mammalian carnivores (Quinn 1995), especially when those species exist at low densities and are secretive or otherwise difficult to locate. Conventional methods such as capture and marking of animals and logistic problems such as restricted access may be difficult to use or overcome in urban or suburban settings. Including the public can increase not only sample sizes, but stakeholder involvement in and public awareness of wildlife research and conservation as well. The residents of Tucson that we encountered or interviewed were unanimous in their support and interest in our roadrunner research (Webster and DeStefano 2004).

Biases

Despite the large volume of data that can be generated from public sighting information, biases exist. Researchers must be able to identify, assess, and interpret these biases, chief among which are misidentification of species, misinformation, erroneous information, nonresponse, self-selection bias, and limited scope of inference because of non-random sampling (Salant and Dillman 1994, Ramsey and Schafer 1997, Webster and DeStefano 2004). We address these biases as they relate to our study.

Misidentification of wildlife is a common problem among the public at large. It turns out,

TABLE 10.2

Difference in distance between 798 Greater Roadrunner sighting locations from the public survey and 39 sighting locations from our random surveys from 3 groups of similar land use covers and 4 housing densities in metropolitan Tucson, Arizona.

Positive (negative) values indicate that public sightings were farther away (closer) than random survey sightings.

Land use class	t	SE	$P\leq$	90% CI
Washes	2.32	26.23	0.021	9.34, 112.3
Non-native vegetation	0.02	40.74	0.981	−79.25, 81.21
Native vegetation	0.70	24.12	0.482	−30.38, 64.33
≤2.5 RHA[a]	1.07	27.81	0.287	−24.98, 84.30
2.5–7.0 RHA[a]	−0.87	31.66	0.384	−89.75, 34.63
7.0–15 RHA[a]	1.85	34.84	0.065	−3.99, 132.86
>15 RHA[a]	−0.11	34.02	0.909	−70.75, 63.01

[a] RHA = residences per ha.

however, that the roadrunner may be the most recognized avian species in North America. In two field tests of bird identification involving six species and 40 residents of Tucson, only the roadrunner was identified correctly 100% of the time (Webster and DeStefano 2004). The Gambel's Quail (*Callipepla gambelii*) and Rock Dove or Pigeon (*Columba livia*), two very common species in Tucson, were identified correctly only 88% and 68% of the time, respectively. Even residents of Washington, D.C., correctly identified the roadrunner more often than the Bald Eagle (*Haliaeetus leucocephalus*; 65% vs. 35%, respectively; J. Cornett, pers. comm.). In addition, we do not believe there was much misinformation bias (i.e., intentionally misleading information) because there was no political controversy surrounding roadrunners, and the species was generally seen as charismatic and a positive inhabitant of the Tucson community. There was also probably little erroneous information because residents could identify roadrunners correctly and the existence of the city infrastructure (e.g., streets, parks) allowed respondents to ascertain their location accurately based on familiar surroundings and easily transferable details, such as a numbered street address or park names.

Nonresponse bias did exist in our study, likely among the general public and definitely among parts of the Hispanic community in south Tucson. The public seldom reported sightings from the extreme southern section of our study area, which was inconsistent with our researcher-based random surveys. The Hispanic community makes up about 28% of the Tucson metropolitan area (U.S. Census Bureau 1998) and many live in the southern portion of Tucson. All of our requests for information were in English and those who speak only or principally Spanish may not have known about our research or were otherwise reluctant to respond.

Self-selection and the resulting non-random sampling of Tucson were the most important biases in our study. Residents volunteered to respond to our request for information (i.e., self-selection bias), and people did not distribute themselves at random throughout the city during the course of their daily activities (i.e., non-random sampling). Thus, the distribution of roadrunner sightings and associated habitat use resulting from the public survey represents the distribution of people who see and report roadrunners rather than the true distribution of roadrunners in Tucson. Biased reporting is likely the case for all public sighting information in that most, if not all, public information on the distribution of species will represent where people see wildlife rather than the true distribution of the species. In our study, residents of Tucson saw roadrunners throughout the city but were more likely to see and report sightings in their own neighborhoods, where housing densities were 2.5–15 RHA and where cultivation of nonnative ornamental vegetation was common. Tucson residents were unlikely to report roadrunners in washes or undeveloped lots, where native vegetation was often dense, probably because few people ventured into these areas on a regular basis. Roadrunners would also be harder to see in denser vegetation.

We attempted to counter these biases by establishing our own researcher-based, random surveys. However, because roadrunners are infrequently seen on a spatially and temporally constrained basis among a small crew of researchers, our sample of researcher-based roadrunner locations was much lower than the public survey (39 vs. 798), and statistical comparisons were thus restricted. However, when we compared the public sightings to our researcher sightings of roadrunners, we noted that roadrunners were more likely to be seen near washes by researchers than the public at large, supporting our contention that at least this landscape feature was undersampled by the residents of Tucson who participated in our survey.

Resolving Distribution Patterns and Habitat Use

Although we recognize the biases inherent in public survey information and the somewhat limited sample size of our researcher-based surveys, we do believe that both offer important insights into the pattern of distribution of roadrunners in an urban-suburban environment. First, we showed that roadrunners could be found living in close proximity to large numbers of people and pets in fairly developed suburban areas, where housing densities ranged from 2 to 15 RHA. This finding is in contrast to several other researchers who reported that roadrunners were absent from urban and suburban areas (Emlen 1974, Unitt 1984, Tweit and Tweit 1986, Rosenberg et al. 1987). We compared the lack of sightings in Tucson reported by Emlen (1974) to our surveys and found that we had one sighting within 160 m and eight within 1 km of Emlen's study area. We suggest that roadrunners might have been seen during Rosenberg et al.'s survey if more search time had been available; a single observer may spend many days afield before seeing a roadrunner. We agree with Wright (1973), who reported that roadrunners remain in areas of new development, and somewhat with Hughes's (1996) summarization that roadrunners avoid urban areas but can be found occasionally in less densely populated suburban developments.

Secondly, our findings show that roadrunners were distributed throughout most of the study area, except for the most densely settled or developed sections of the city. In fact, roadrunners appeared to be distributed in a broad band around Tucson in established and newly developed suburban areas, and absent or certainly much less common in the city center, which includes the downtown business district, the University of Arizona campus, and where housing densities reach >15 RHA. Both authors lived in this area of Tucson during the course of the study and never saw a roadrunner in their neighborhoods. The fact that the public saw roadrunners less frequently in natural open space, graded lots, native vegetation, and washes is an indication to us that people did not frequent those places, rather than that roadrunners did not use these areas. However, few public sightings in areas with housing densities >15 RHA is a true indication of low or no roadrunner presence; we have no reason to believe that people in these neighborhoods would not report roadrunners. We believe that this pattern of distribution is real and accurate and reflects the relative levels of development throughout the Tucson basin. Furthermore, this "doughnut-shaped" pattern of distribution has been seen for other species of birds in Tucson as well (Turner 2003).

What we cannot say is how relative abundance of roadrunners varied among different levels of development, or which land use classes or landscape features were preferred by roadrunners. Public sighting information alone will not yield this information because the public survey was not a random sample of land use classes in Tucson. Our researcher-based surveys, while random, did not yield nearly the number of roadrunner sightings in the public survey. Although our researcher-based surveys can lend insight into roadrunner ecology, more observations are needed. Larger sample sizes may be difficult to attain for low-density, elusive species in urban settings, but they are essential for determining habitat preferences.

While we cannot make any claims to habitat preferences, it is clear from our data that roadrunners can and do inhabit highly developed suburban areas, avoid (but can occasionally be seen in) the most densely developed urban areas, and use nonnative vegetation (but see Mills et al. 1989). Roadrunners eat a wide range of vertebrate and invertebrate prey as well as some vegetation and will take advantage of the unintentional (e.g., pet food left outside) and intentional (e.g., handouts of meat scraps) food provided by humans (Bryant 1916, Hughes 1996, S. DeStefano and C. M. Webster, pers. obs.). Furthermore, we found or had reported to us by the public several roadrunner nests within the city limits (Webster 2000). Many residents reported that they had roadrunners nesting successfully in carports, ornamental shrubs and trees, and other areas of their yards or neighborhoods for several years. Until demographic parameters such as survival of adults and young are estimated, however, we will not know how urban and suburban development affects roadrunner population viability. Based on our findings, we now know that roadrunners will inhabit areas with some level of suburban development, but we do not know if these areas act as sources or sinks or assist in any way in maintaining the level of local populations.

In general, roadrunners appear to be adaptable to a variety of developed environments, especially suburban areas on the periphery of the city. As development pressure continues to grow in and around the city of Tucson, however, changes in the distribution and abundance of roadrunners and other wildlife can be expected to occur.

Improving Public-generated Data Collection

Maximizing utility and accuracy while minimizing bias and uncertainty in public sighting information and other surveys are predicated upon public education and planning. A better-informed public will yield more reliable information and avoid, or at least ameliorate, such biases as misidentification of species, erroneous information, and nonresponse. A pilot program to introduce a community to upcoming research would likely be extremely helpful in preparing, educating, and training the public and identifying potential problems and biases. Familiarity with local customs and communication with the diversity of cultural groups that inhabit most cities and large towns would also increase public involvement and improve the quantity and quality of information collected.

Within the community of potentially involved residents, it may be helpful to employ some kind of random sampling. Stratifying the city or town by neighborhoods based on economic background (e.g., house prices), housing densities, or other geographic features, and then employing a stratified random sample of participants would be a useful strategy. Public sighting information in conjunction with researcher-based random surveys can lead to important insights into the effects of urbanization on the distribution and ecology of wildlife.

ACKNOWLEDGMENTS

We thank M. A. McLeod, C. K. Kirkpatrick, the Tucson Audubon Society, and students in the Arizona Cooperative Fish and Wildlife Research Unit's minority training program for assistance in the field; C. A. Wissler, A. M. Honaman, and the University of Arizona's Advanced Resource Technology Laboratory for assistance with the geographic information system; C. Bang, D. P. Guertin, W. P. Kuvlesky Jr., W. W. Shaw, and two anonymous reviewers for their comments on the manuscript. We appreciate the assistance we received from the *Arizona Daily Star*, KUAT radio, and Channel 13 (KOLD) News. Special thanks to the residents of Tucson, Arizona, for their cooperation and interest, and P. S. Warren and C. A. Lepczyk for organizing and editing this issue of Studies in Avian Biology. Funding was provided by the University of Arizona's Small Grants Program, Graduate Student Final Project Fund, and Graduate College Unrestricted Fellowship, and the Arizona Game and Fish Department's Heritage Grant Program.

LITERATURE CITED

Anderson, B. W., R. D. Ohmart, and S. D. Fretwell. 1982. Evidence for social regulation in some riparian bird populations. American Naturalist 120:340–352.

Boal, C. W., and R. W. Mannan. 1998. Nest-site selection by Cooper's Hawks in an urban environment. Journal of Wildlife Management 62:864–871.

Bryant, H. C. 1916. Habits and food of the roadrunner in California. University of California Publications in Zoology 17:21–58.

Byers, J. E. 2002. Impact of non-indigenous species on natives enhanced by anthropogenic alteration of selection regimes. Oikos 97:449–458.

DeStefano, S., and R. M. DeGraaf. 2003. Exploring the ecology of suburban wildlife. Frontiers in Ecology and the Environment 1:95–101.

DeStefano, S., and E. A. Johnson. 2005. Species that benefit from sprawl. Pp. 206–235 *in* E. A. Johnson and M. W. Klemens (editors), Nature in fragments, the legacy of sprawl. Columbia University Press, New York.

Emlen, J. T. 1974. An urban bird community in Tucson, Arizona: derivation, structure, regulation. Condor 76:184–197.

Fussell, M., and S. A. Corbet. 1992. The nesting places of some British bumble bees. Journal of Apicultural Research 31:32–41.

Germaine, S. S. 1995. Relationships of birds, lizards, and nocturnal rodents to their habitat in the greater Tucson area, Arizona. Technical Report No. 20. Arizona Game and Fish Department, Phoenix, AZ.

Germaine, S. S., S. S. Rosenstock, R. E. Schweinsburg, and W. S. Richardson. 1998. Relationships among breeding birds, habitat, and residential development in greater Tucson, Arizona. Ecological Applications 8:680–691.

Harris, L. K., P. R. Krausman, and W. W. Shaw. 1995. Human attitudes and mountain sheep in a wilderness setting. Wildlife Society Bulletin 23:66–72.

Hill, G. E. 1993. House Finch. No. 46 *in* A. Poole and F. Gill (editors), The Birds of North America. Academy of Natural Sciences, Philadelphia, PA, and American Ornithologists' Union, Washington, DC.

Hughes, J. M. 1996. Greater Roadrunner (*Geococcyx californianus*). No. 244 *in* A. Poole and F. Gill (editors), The birds of North America. Academy of Natural Sciences, Philadelphia, PA, and American Ornithologists' Union, Washington, DC.

Johnston, R. F. 2001. Synanthropic birds of North America. Pp. 49–67 *in* J. M. Marzluff, R. Bowman, and R. Donnelly (editors), Avian ecology and conservation in an urbanizing world. Kluwer Academic Publishers, Boston, MA.

Johnson, E. A., and M. W. Klemens (editors). 2005. Nature in fragments: the legacy of sprawl. Columbia University Press, New York, NY.

Knight, R. L. 1991. Ecological principles applicable to the management of urban ecosystems. Pp. 24–34 *in* E. A. Webb and S. Q. Foster (editors), Perspectives in urban ecology. Denver Museum of Natural History, Denver, CO.

McGuill, M. W., S. M. Kreindel, A. DeMaria Jr., and C. Rupprecht. 1997. Knowledge and attitudes of residents in two areas of Massachusetts about rabies and an oral vaccination program in wildlife. Journal of the American Veterinary Medical Association 211:305–309.

Mills, G. S., J. B. Dunning Jr., and J. M. Bates. 1989. Effects of urbanization on breeding bird community structure in southwestern desert habitats. Condor 91:416–428.

Neu, C. W., C. R. Byers, and J. M. Peek. 1974. A technique for analysis of utilization-availability data. Journal of Wildlife Management 38:541–545.

Pyrovetsi, M. 1997. Integrated management to create new breeding habitat for Dalmatian Pelicans (*Pelecanus crispus*) in Greece. Environmental Management 21:657–667.

Quinn, T. 1995. Using public sighting information to investigate coyote use of urban habitat. Journal of Wildlife Management 59:238–245.

Ramsey, F. L., and D. W. Schafer. 1997. The statistical sleuth. Duxbury Press, Belmont, CA.

Rosenberg, K. V., S. B. Terrill, and G. H. Rosenberg. 1987. Value of suburban habitats to desert riparian birds. Wilson Bulletin 99:642–654.

Salant, P., and D. A. Dillman. 1994. How to conduct your own survey. John Wiley and Sons, New York, NY.

Schoenecker, K. A., and P. R. Krausman. 2002. Human disturbance in bighorn sheep habitat, Pusch Ridge Wilderness, Arizona. Journal of the Arizona–Nevada Academy of Science 34:65–69.

Shaw, W. W., A. Goldsmith, and J. Schelhas. 1991. Studies of urbanization and the wildlife resources of Saguaro National Monument. Pp. 173–180 *in* C. P. Stone and E. S. Ballantoni (editors), Proceedings of the symposium on research in Saguaro National Monument. National Park Service, Tucson, AZ.

Shaw, W. W., L. K. Harris, M. Livingston, J. P. Carpentier, and C. Wissler. 1996. Pima County habitat inventory: phase II. Final report. Arizona Game and Fish Department, Phoenix, AZ.

Sokal, R. R., and F. J. Rohlf. 1995. Biometry. 3rd ed. W. H. Freeman and Company, New York, NY.

Soulé, M. E. 1990. The onslaught of alien species, and other challenges in the coming decades. Conservation Biology 4:233–239.

Turner, W. R. 2003. Citywide biological monitoring as a tool for ecology and conservation in urban landscapes: the case of the Tucson Bird Count. Landscape and Urban Planning 65:149–166.

Turner, R. M., and D. E. Brown. 1982. Sonoran desertscrub. Pp. 181–221 *in* D. E. Brown (editor), Biotic communities of the American Southwest: United States and Mexico. University of Arizona, Tucson, AZ.

Tweit, R. C., and J. C. Tweit. 1986. Urban development effects on the abundance of some common resident birds of the Tucson area of Arizona. American Birds 40:431–436.

Unitt, P. 1984. The birds of San Diego County. San Diego Society of Natural History, San Diego, CA.

U.S. Census Bureau. 1998. State and metropolitan area data book 1997–1998. Washington, DC.

Webster, C. M. 2000. Distribution, habitat, and nests of Greater Roadrunners in urban and suburban environments. M.S. thesis, University of Arizona, Tucson, AZ.

Webster, C. M., and S. DeStefano. 2004. Using public surveys to determine the distribution of Greater Roadrunners in urban and suburban Tucson, Arizona. Pp. 69–77 *in* W. W. Shaw, L. K. Harris, and L. VanDruff (editors), Urban wildlife conservation. Proceedings of the 4th International Symposium, Tucson, AZ.

Wright, R. E. 1973. Observation on the urban feeding habits of the roadrunner (*Geococcyx californianus*). Condor 75:246.

CHAPTER ELEVEN

Edges, Trails, and Reproductive Performance of Spotted Towhees in Urban Greenspaces

Sarah Bartos Smith, Jenny E. McKay, Jennifer K. Richardson, and Michael T. Murphy

Abstract. Urban habitats present unique ecological challenges for native biota, and understanding effects of urban fragmentation on reproductive success is a question of much concern. We investigated the influence of park size, recreational trails, and habitat edges on the reproductive performance of the Spotted Towhee in urban forest fragments. We studied Spotted Towhees (*Pipilo maculatus*) breeding in two medium (~10 ha) and two large (~20 ha) urban forest fragments, investigating variation in nest success, timing of breeding, clutch size, number of fledglings, or nestling condition between parks of different sizes, or among birds nesting at different distances from recreational trails and habitat edges. We found no evidence for improved reproductive performance in larger parks, and instead found greater fledgling production in medium-sized parks. While the probability of whole-nest success was independent of distance from habitat edge, nests near edges fledged significantly more offspring, were less likely to suffer partial brood losses, and nestlings near neighborhood and road edges were significantly heavier than those in the park interior. The effects may reflect a greater abundance of anthropogenic food resources for Spotted Towhees nesting near edges. Nestling hematocrit was lower near edges, however, which could indicate that these food sources may be of poor quality. Our study emphasizes the importance of both predation and food abundance as factors that influence avian reproductive success in urban areas. We recommend that similar studies investigating multiple components of reproductive performance in other species be undertaken in urban areas.

Key Words: anthropogenic food, edge effects, fragment size, nest predation, recreational trails, Spotted Towhee, urbanization.

The area covered by urban habitats has increased rapidly over the past century, and over half of the United States population now resides in cities (Marzluff et al. 2001a). Despite the expansion of urban landscapes, we remain largely ignorant of the degree to which birds and other organisms can maintain viable populations in these heavily disturbed environments. Urban habitats may present unique ecological challenges for native biota, and understanding specifically how urban fragmentation affects the ability of birds to reproduce successfully

Bartos Smith, S., J. E. McKay, J. K. Richardson, and M. T. Murphy. 2012. Edges, trails, and reproductive performance of Spotted Towhees in urban greenspaces. Pp. 167–182 *in* C. A. Lepczyk and P. S. Warren (editors). Urban bird ecology and conservation. Studies in Avian Biology (no. 45), University of California Press, Berkeley, CA.

is a question of much concern (Marzluff et al. 2001a).

Urban breeding birds ultimately face the same challenges as their counterparts in more natural areas. Adults must acquire sufficient food to produce eggs and feed young, and they must do this without attracting the attention of nest predators. Nest predation accounts for the majority of nest failures in relatively natural areas (Ricklefs 1969, Martin 1993), but whether nest predation increases or decreases with increasing urbanization is still unclear, as studies have yielded conflicting results (Gering and Blair 1999, Matthews et al. 1999, Jokimäki and Huhta 2000, Thorington and Bowman 2003). However, nearly all such studies used artificial nests, and recent evidence suggests that predation on artificial nests differs from predation on real nests in an unpredictable manner (Moore and Robinson 2004). Predator communities in urban environments may reorganize through the loss of large mammalian predators (such as coyotes, *Canis latrans*), resulting in the release of generalist, small- to medium-sized predators such as raccoons (*Procyon lotor*) and other small mammals that frequently depredate nests (the mesopredator release hypothesis; Crooks and Soulé 1999). In addition, burgeoning populations of corvids (frequent nest predators) in urban spaces possibly create additional problems for nesting birds (Marzluff et al. 2001b). Last, domestic pets may generate a new and important cause of mortality for nestling and juvenile birds (Baker et al. 2005).

Fragmentation may lower arthropod food supplies in non-urban (Burke and Nol 2000, Zanette et al. 2000, Doherty and Grubb 2002), and urban forest fragments (Bolger et al. 2000). Despite the potential for an increased food supply in the form of anthropogenic food sources (Partecke et al. 2006), resources may be patchy, and a lack of high quality food sources could lead to food limitation for birds nesting in urban areas (Shawkey et al. 2004, Mennechez and Clergeau 2006). Additionally, because urban bird population densities may be especially high, competition for food in urban areas may be exacerbated (Shochat 2004). Food limitation induces stress on breeding birds (Clinchy et al. 2004), which may negatively influence clutch size, rates of brood losses, nest survival, and annual productivity (Zanette et al. 2003, 2006). Additionally, low food availability and high predator density can act in a negative synergistic manner on nest success (Zanette et al. 2003, 2006; Clinchy et al. 2004), and urban forest fragments may be particularly prone to this phenomenon.

Birds nesting in small forest fragments in both urban and non-urban areas may have a lower reproductive success than those in larger, more intact areas (Andrén 1992, 1995). One potential reason for low reproductive success in small fragments is the oft-described "edge effect" (Yahner 1988). Smaller habitat fragments have proportionally more edge habitat, and consequently less interior core habitat (Saunders et al. 1991). Mammalian and avian nest predator abundances are thought to increase along habitat edges (Donovan et al. 1997, Heske et al. 1999, Bolger 2002), and in urban areas this may be especially important due to the ability of certain predators (e.g., crows, jays, raccoons, domestic cats) to utilize both human-dominated habitats and natural forest habitats for foraging. Most studies of edge effects in urban areas have failed to show an association between nest success and proximity to habitat edges (Matthews et al. 1999, Morrison and Bolger 2002, Patten and Bolger 2003, Thorington and Bowman 2003), but many such studies used artificial nests and more tests need to be conducted using natural nests to assess the strength of edge effects in urban areas.

Urban parks exist not only for wildlife, and arguments for the preservation of greenspaces in urbanizing regions are often based on the recreational benefits to humans. The use of parks by humans is only possible through trail development, but at what point, if any, do trails begin to adversely affect wildlife? Sinclair et al. (2005) found that parks with wider trails and an adjacent landscape matrix containing many buildings supported a higher abundance of mammalian nest predators than parks with narrow trails in a less developed matrix. However, an artificial nest experiment suggested that mammalian predators tended to avoid trails (Miller and Hobbs 2000), while avian predators tend to have increased activity near recreational trails (Miller and Hobbs 2000, Hennings 2001). It is also possible that human use of recreational trails for walking dogs may deter mammalian nest predators because of the latter's aversion to canine scent near trails (Forman 1995). No clear patterns have emerged from studies of the effects of trails on nest success (Miller et al. 1998, Miller and Hobbs 2000,

Langston et al. 2007), and the lack of consensus may exist because not all trails and edges have the same effects on nesting success. If human activity is the primary factor influencing the probability of nest success near trails, lightly traveled, small trails may have little influence on nest success compared to larger, heavily used trails. Similarly, "hard" edges near roads or fields may expose birds to different types of predators than "soft" edges where parks grade into residential neighborhoods of various ages.

A shortcoming of nearly all urban fragmentation/edge studies is that they have focused only on nest predation and nest success (i.e., the percentage of nests fledging at least one offspring), neglecting other aspects of reproductive biology. While nest success is obviously important, documentation of variation in clutch size, fledgling production, and nestling condition can elucidate the factors that most strongly influence reproductive performance. For example, two sites may have identical rates of nest success, but food limitation at one site may lead to delayed breeding, smaller clutch sizes, poorer nestling condition, more frequent nestling starvation, and ultimately lower juvenile recruitment. Ignoring multiple components of reproductive performance may impede researchers from determining the relative effects of predation and food limitation on population productivity. To date such studies are missing for urban breeding birds (but see Morrison and Bolger 2002).

The Spotted Towhee (*Pipilo maculatus*) is an abundant resident species in the Pacific Northwest. It is a socially monogamous and sexually dimorphic passerine that typically nests on the ground (Greenlaw 1996) and can raise up to three broods per season within this portion of its range (S. Bartos Smith, unpubl. data). Spotted Towhees are found in a wide variety of habitats, both forested and unforested, and require dense shrubby vegetation for nesting (Greenlaw 1996). Mammals, birds, and snakes have all been implicated as towhee nest predators (Small 2005), and due to the prevalence of mammalian and avian predators in urban areas, we hypothesized that towhee offspring production would be negatively affected by urbanization. Additionally, southern California populations of towhees have been considered "urbanization sensitive" (Crooks et al. 2004), and Spotted Towhee abundance has been positively related to distance from developed urban edges (Bolger et al. 1997). We therefore hypothesized that Spotted Towhee adults would be more successful in larger greenspaces and when they nest farther from habitat edges and/or recreational trails. We tested our hypotheses in four urban greenspaces in Portland, Oregon, using data on nest success, timing of breeding, clutch size, partial brood losses, nestling condition, and fledging success. We categorized trails and edges to better determine whether certain types of trails (e.g., primary, secondary, tertiary trails, and footpaths) and edges (e.g., those surrounded by residential areas, roads, or open fields) affected towhee reproductive performance.

METHODS

Study Sites and Field Methods

We conducted our study in four greenspaces located in the "West Hills" region within the city limits of Portland, Oregon (Fig. 11.1), between April and August of 2004, 2005, and 2006. Minimum and maximum distances between park centerpoints are 1.2 km and 3.2 km. Maricara and Lesser Parks are both approximately 10 ha in size, while West Portland Park (18 ha) and Springbrook Park (24 ha) are roughly twice the area of the smaller parks. All four greenspaces have recreational trails throughout them. Maricara Park and West Portland Park are surrounded almost entirely by residential development and neighborhood roads, while Lesser and Springbrook Parks are surrounded by residential areas, neighborhood roads, open fields, and major roadways. Although they are located in highly developed landscapes, we emphasize that the parks are undeveloped except for the trail systems, and all four sites are characterized by a closed canopy of mixed coniferous-deciduous forest that is at least 55 years old. Each of the four parks supported a pair of breeding Cooper's Hawks (*Accipiter cooperi*).

Within each greenspace, we captured territorial adult male towhees in mist nets using a mounted specimen and song playback of a conspecific. We captured adult females during the nestling period by positioning nets around the nest. Captured birds were banded with an aluminum U.S. Fish and Wildlife Service band and a unique combination of three colored plastic bands. All adult birds were weighed, and standard morphological

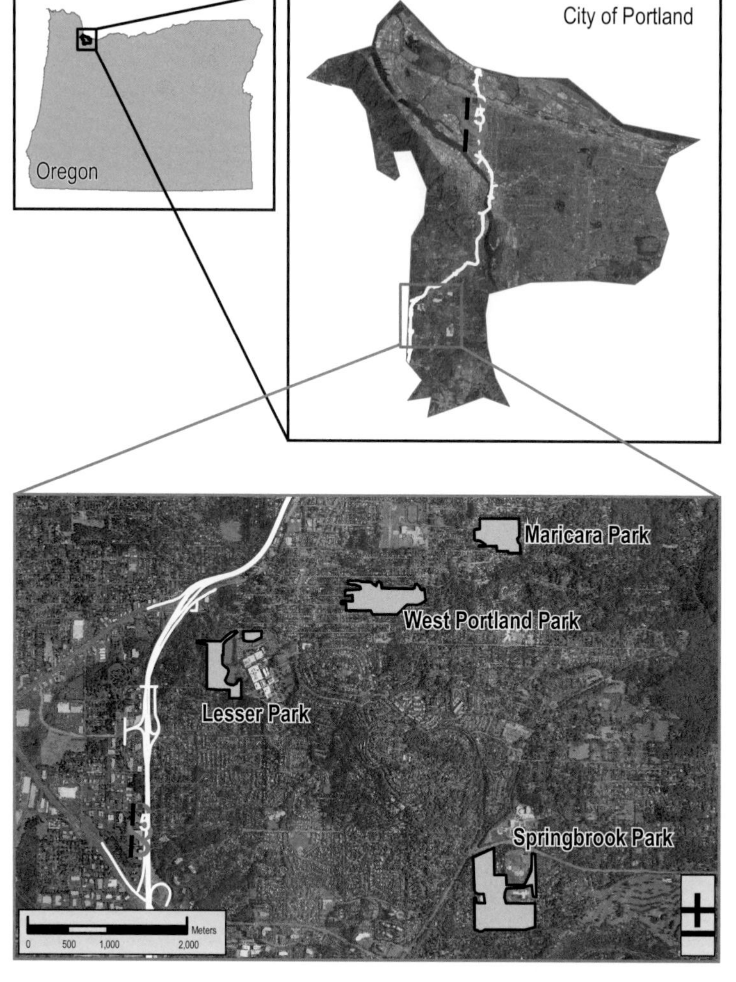

Figure 11.1. Map of the locations of the four study areas within Portland, Oregon.

measurements (tarsus, wing chord, bill length, bill width, and tail length) were taken. We determined the age of each bird (SY or ASY) by the shape of the outer rectrices and the contrast between the greater and lesser wing coverts (Pyle 1997).

We located nests of color-banded pairs in all four greenspaces in 2005 and 2006, and in three of the four (all but West Portland Park) in 2004. Additionally, in 2004 we banded adults and monitored nests only within a subplot of Springbrook Park (the northern half of the park), while in subsequent years we studied the entire park. We checked nests every 2–3 days until the nest failed or fledged young. When nestlings were 7–10 days old, we banded them with a USFW band and three color bands, weighed them, measured their right and left tarsi, and took a small (25–50 μl) blood sample from each nestling. In 2005 and 2006, we spun the blood in a capillary tube in a centrifuge at 11,700 rpm for 5 minutes to measure hematocrit (the proportion of the total blood volume that was made up of packed red blood cells). High hematocrit levels are generally thought to be indicative of good body condition (Merilä and Svensson 1995, Svensson and Merilä 1996).

Nest Sites: Proximity to Edges and Trails

In 2004 we determined distances to the nearest edge and trail manually by using sonic distance measurers to measure the distance between one observer standing at the nest and another standing at the nearest edge or trail. In 2005 and 2006, we collected data on nest location using a Trimble GeoXT GPS receiver equipped with Terrasync software.

In 2005, we also mapped all trail systems within the greenspaces using the same GPS unit. We collected nest location data using at least 30 readings at 3-second intervals, and trail data were collected using 1-second intervals with at least four satellites and a position dilution of precision (PDOP) no greater than 6. A velocity filter was utilized with the trail system in West Portland Park due to the thick coniferous foliage. The standard storage format files were differentially corrected from base stations no further than 70 km from the research sites and exported as shapefiles (Pathfinder Software, ver. 3.0, Sunnyvale, CA). Total estimated accuracies for the differential corrections were 39% within 0.5–1 m, 51% within 1–2 m, and 10% within 2–5 m. All trail and nest shapefiles were re-projected in ArcGIS 9.1 to Lambert Conformal Conic, using the North American Datum 1983 (NAD83), and the High Accuracy Reference Network (HARN) Oregon State Plane North geographic coordinate system. Using the "near" tool, we calculated distances from nests to trail and to edge. The habitat edge data layer was obtained from N. I. Lichti et al. (unpubl. data), who took publicly accessible parks layers and expanded the parks' effective boundaries by including any area contiguous with the greenspace with canopy cover beneath which less than 5% of the ground surface was developed (i.e., buildings, paved roads, etc.).

We view Maricara and Lesser Parks as representative of medium-sized parks for the Portland metropolitan area. West Portland and Springbrook Parks fall into the lower size range of large parks in our system; therefore for statistical purposes we treat the two smaller and two larger parks separately. Within all four parks, we classified the distance of each nest to the habitat edge in two ways. First, we used five distance-to-edge categories (Weldon and Haddad 2005), which assume that edge effects occur within 50 m of the edge (see Batáry and Báldi 2004 for justification): 1 = 0–12.5 m, 2 = 12.6–25 m, 3 = 25.1–37.5 m, 4 = 37.6–50 m, and 5 > 50 m. We refer to this categorical variable as Edge Class 1. Second, we categorized nests into six distance-to-edge categories assuming that edge effects can persist to 100 m from the edge (Edge Class 2): 1 = 0–20 m, 2 = 20.1–40 m, 3 = 40.1–60 m, 4 = 60.1–80 m, 5 = 80.1–100 m, and 6 > 100 m. Measurements for nests in 2004 that were >50 m from the edge were listed as such. Therefore we did not have the necessary data to include these nests in Edge Class 2. In our results, we present data only from Edge Class 1, unless there was a qualitative difference in the results obtained from the two edge classification schemes. We also classified each nest into one of four distance-to-trail categories (Trail Class): 1 = 0–10 m, 2 = 10.1–20 m, 3 = 20.1–30 m, 4 > 30 m. Nests within 30 m of a trail were categorized according to the type of trail they were near (1 = primary, 2 = secondary, 3 = tertiary, 4 = footpath), and those greater than 30 m from a trail were classified as "5" (no trail). Trail categories were determined by trail width (primary ≥ 1 m, secondary = 0.5–0.9 m, tertiary = 0.1–0.4 m, and footpath ≤ 0.1 m). Nests within 50 m of a habitat edge were similarly categorized according to the type of edge they were near (1 = road, 2 = residential, 3 = open space), and those greater than 50 m from an edge were classified as "4" (interior). We did not find any nests that bordered major roadways, so all nests classified as near roads were near edges that were bordered by lightly traveled small neighborhood roads.

Statistical Analysis

We considered a nest successful when at least one offspring fledged. We excluded abandoned nests (n = 12) from all analyses because the cause of abandonment (human or predator induced) could not be determined. Nest success was treated as a binary variable with a value of either 0 (successful) or 1 (unsuccessful). Daily nest mortality rate was calculated using the Mayfield logistic regression method (Hazler 2004), with year, female age, date of clutch completion, park size, distance to trail, distance to edge, and an interaction term (trail × edge) included as independent variables. We performed stepwise removal of the most nonsignificant variable until we arrived at the factor(s) that was (were) the most important predictor(s) of nest success. The Mayfield logistic regression method (Hazler 2004) is similar to Shaffer's logistic regression method (Shaffer 2004) in that the effects of both continuous and categorical variables on nest success can be evaluated while accounting for each nest's number of exposure days.

We included female age as a factor in all analyses because ASY females often have higher reproductive success than do SY females (Perrins and McCleery 1985, Lessels and Krebs 1989). However, to determine if there was any age class difference in habitat use (estimated by nest placement), we compared distance to edge and trail between ASY and SY females using ANOVA. We included year in the model to account for possible annual variation. Each female was represented only once in these analyses, as we used only first nests of the year and randomly chose one year for any female who was present in multiple years.

We pooled all three years for analyses of timing of breeding, clutch size, fledging success, nestling mass, and hematocrit because none varied among years (ANOVA, all $P > 0.05$). Stepwise linear regressions with forward selection were used with female age, date of clutch completion, park size, Trail Class, and either Edge Class 1 or Edge Class 2 as independent variables. We evaluated whether the probability of partial brood losses (i.e., losses of nestlings that did not result in total nest failure) varied with park size, Edge Class, or Trail Class using logistic regression, and included female age, year, and date of clutch completion to remove any effects of these variables. To avoid pseudoreplication, we included only a female's first nesting attempt of the year in all analyses of reproductive performance. As before, we randomly chose one year for females with multiple years of data.

Timing of breeding was determined as the nest's clutch completion date (1 April = day 1). We used clutch completion date rather than the more commonly used clutch initiation date because we observed high rates of partial brood losses (>30% of nests, S. Bartos Smith, unpubl. data); thus we could not be sure of the clutch size for nests found during the nestling period. Additionally, since 95% of the nests for which we do have clutch size data had three and four egg clutches (S. Bartos Smith, unpubl. data), clutches completed on a specific day were unlikely to vary in clutch initiation date by greater than 1 day. For this analysis, we only included first nesting attempts because we were primarily interested in any differences in the start of the breeding season. We also limited our analysis to nests with a clutch completion date prior to 11 May (day 41) to eliminate the possibility of including a large number of replacement nests. Clutch size (i.e., number of eggs laid) was based on nests found no later than the incubation period. We defined fledging success as the number of offspring that fledged from each nest. We conducted the fledging success analyses twice—once including unsuccessful and successful nests, and once including only successful nests. We estimated nestling condition as the average mass and the average hematocrit of nestlings within a nest. For multiple linear regressions of the two indices of nestling condition, we included nestling age as a predictor variable to eliminate the effects of increased mass and hematocrit with nestling age.

To evaluate whether different edge or trail types influenced any of the reproductive success variables described above, we conducted ANOVA with edge type, trail type, year, date of clutch completion, and female age as source variables. Since both nestling mass and hematocrit increased with age, we eliminated age effects by using the residuals from the regression of nestling mass and hematocrit on nestling age. Post hoc comparisons of variables that differed significantly were made using Tukey's test. All analyses were done in Statistix 8.0, and unless stated otherwise, we used $P \leq 0.05$ to establish statistical significance.

RESULTS

Nest Success

We found 69, 137, and 123 nests in 2004, 2005, and 2006, respectively (151 in medium parks and 178 in large parks), and monitored them for 3,500 observation days. Apparent nest success was 46%, and the primary cause of nest failure was predation (90%), followed by nestling starvation (7%) and flooding during periods of heavy rain (3%). We did not observe any parasitism by Brown-headed Cowbirds (*Molothrus ater*). Mayfield logistic regression of all nests yielded a daily survival probability of 0.966 ($\beta = 3.35$, SE = 0.10, $n = 329$). Assuming 26 d in the nesting cycle (based on the average lengths of the laying, incubation, and nestling periods observed for our population), this translates into a percent nest success of 41% (95% confidence interval = 37% to 44%).

Using Mayfield logistic regression with stepwise removal of nonsignificant variables, we determined that the best predictor of nest success was distance to recreational trail, with nests near trails being significantly more likely to fledge young than nests farther from trails ($\beta = -0.19$, SE = 0.08, $n = 328$, $P = 0.02$). With distance to trail in the model, probability of nest success was independent of year ($P = 0.22$), female age ($P = 0.72$), date of clutch completion

TABLE 11.1

Results of stepwise multiple linear regressions of timing of breeding and number of fledglings for Spotted Towhees in Portland, Oregon.

	Variable	Coefficient	SE	$P \leq$	Adj. r^2
Clutch completion date (n = 75)	Trail Class	1.82	0.86	0.04	0.07
	Edge Class 1	1.73	0.72	0.02	
Fledglings (n = 102)	Edge Class 1	−0.20	0.09	0.03	0.06
	Park size	−0.54	0.25	0.03	

(P = 0.15), Edge Class (P = 0.33), park size (P = 0.44), and trail × edge interaction (P = 0.96). We found no difference between SY and ASY females in distance of nest to edge ($F_{1,110}$ = 0.8, P = 0.39), or in park size ($F_{1,110}$ = 1.6, P = 0.21), but ASY females nested significantly closer to trails than SY females ($F_{1,109}$ = 4.5, P = 0.04). We therefore ran separate Mayfield logistic regressions for SY and ASY females and found that the trail effect existed among ASY females (β = −0.28, SE = 0.09, n = 225, P < 0.01), but not SY females (β = −0.05, SE = 0.10, n = 169, P = 0.59). Removal of Trail Class 1 (0–10 m from trail) nests from the ASY analysis resulted in no relationship between distance to trail and nest success (β = −0.15, SE = 0.17, n = 112, P = 0.36), suggesting that improved nest success near trails existed only for ASY females who nested within 10 m of a trail.

Nesting Biology

Stepwise regression analysis of timing of breeding demonstrated that the start of egg laying was significantly earlier for nests near both trails and edges (using Edge Class 1; Table 11.1), and after controlling for distance to edge and trail, breeding date was independent of park size [partial correlation (r_P) = −0.07, P = 0.55] and female age (r_P = −0.12, P = 0.30). When using Edge Class 2, we found that the pattern of earlier breeding near edges became a nonsignificant trend (r = 0.22, n = 66, P = 0.07), and there was no longer any effect of proximity to trails on the timing of breeding (r = 0.17, P = 0.16). Clutch size did not vary with date of clutch completion (r = −0.09, P = 0.50), female age (r = 0.01, P = 0.94), park size (r = −0.11, P = 0.41), distance to trail (r = −0.06, P = 0.65), or Edge Class (r = −0.09, P = 0.49, n = 57 for all).

Stepwise linear regression of the number of fledglings per nest (including both successful and unsuccessful nests) revealed that significantly more fledglings were produced from nests located in medium-sized parks and in nests near habitat edges (using Edge Class 1; Table 11.1), and after removing park size and edge effects, fledgling production did not vary with date of clutch completion (r_P = −0.04, P = 0.66), female age (r_P = −0.12, P = 0.24), or Trail Class (r_P = −0.12, P = 0.22). Repeating this analysis using Edge Class 2 resulted in a stronger relationship between distance to habitat edge and fledgling production (β = −0.25, SE = 0.09, n = 90, P < 0.01, Model r^2_{adj} = 0.07); however, the relationship between park size and fledgling production became a nonsignificant trend (r_P = −0.19, P = 0.07). When we analyzed fledgling production of only successful nests, the number of fledglings in a female's first nest was greater in medium than in large parks (β = −0.43, SE = 0.21, n = 84, P = 0.04, Model r^2_{adj} = 0.04), but fledgling production did not vary with clutch completion date (r_P = −0.02, P = 0.86), female age (r_P = −0.08, P = 0.46), distance to trail (r_P = −0.04, P = 0.70), or Edge Class 1 (r_P = −0.09, P = 0.42). No relationship existed between fledging success and park size with Edge Class 2 in the model (r = −0.19, n = 74, P = 0.10). Despite the absence of an effect of distance to edge on fledging success of successful nests, our analysis of the probability of partial brood losses for successful nests showed that the loss of eggs or nestlings became more frequent as distance from edge increased

TABLE 11.2

Results of multiple logistic regression of the probability of partial brood losses for Spotted Towhees in Portland, Oregon.

Variable	Coefficient	SE	$P\leq$
Edge Class 1	0.60	0.23	0.01
Trail Class	0.14	0.27	0.61
Park size	0.83	0.83	0.12
Date CC[a]	0.01	0.01	0.56
Female age	−0.39	0.52	0.46
Year	0.23	0.35	0.51

[a] Date CC = date of clutch completion for first nests.

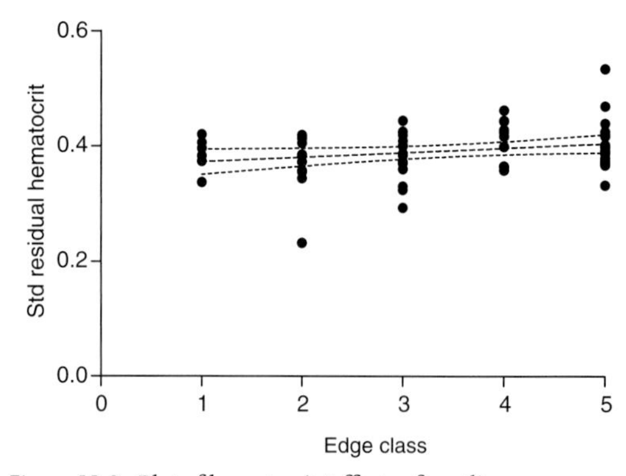

Figure 11.2. Plot of hematocrit (effects of nestling age controlled statistically) against Edge Class 1. Central dashed line shows best fit line; other lines show 95% confidence interval. Hematocrit increased as distance from habitat edge increased ($P = 0.05$).

($n = 80$; Table 11.2). Our results were identical when we used Edge Class 2 ($\beta = 1.01$, SE = 0.60, $n = 70$, $P < 0.01$), and remained significant if the nonsignificant variables were removed from the analysis (Edge Class 1: $\beta = 0.37$, SE = 0.18, $n = 80$, $P = 0.04$).

Average nestling mass was significantly greater for nests located near habitat edges (Table 11.3), but did not vary with park size ($r_P = -0.08$, $P = 0.45$) or with distance to trail ($r_P = -0.15$, $P = 0.17$). Older and earlier breeding females also produced heavier young (Table 11.3). The relationship between nestling mass and proximity to habitat edge was stronger when we used Edge Class 2 in place of Edge Class 1 (Table 11.3); however, the relationship between nestling mass and date of clutch completion become nonsignificant ($r_P = -0.20$, $P = 0.09$). Average nestling hematocrit varied only with distance to edge. Contrary to previous results, hematocrit of nestlings was lowest in edge nests (Table 11.3, Fig. 11.2). There was no relationship between nestling hematocrit and date of clutch completion ($r_P = -0.12$, $P = 0.35$), female age ($r_P = 0.01$, $P = 0.96$), park size ($r_P = -0.06$, $P = 0.64$), or Trail Class ($r_P = 0.17$, $P = 0.21$).

Edge and Trail Types

Clutch size and nestling hematocrit were independent of trail type (clutch size: $F_{4,56} = 1.4$, $P = 0.30$; hematocrit: $F_{4,61} = 1.3$, $P = 0.34$) and edge type (clutch size: $F_{3,56} = 1.0$, $P = 0.40$; hematocrit: $F_{3,61} = 0.4$, $P = 0.75$). Similarly, there was no effect of trail type on timing of breeding, nestling mass, or number of fledglings (Table 11.4), but

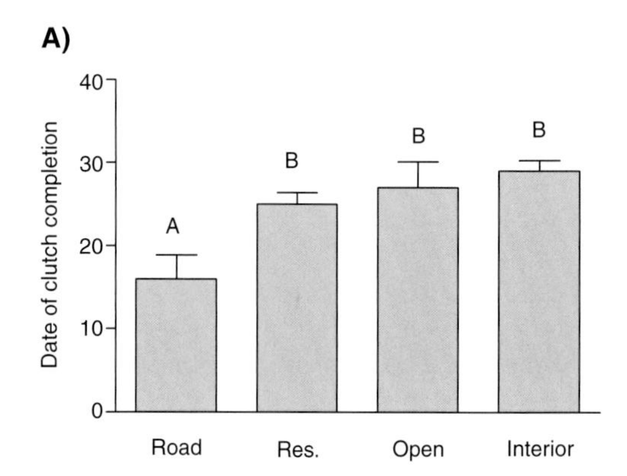

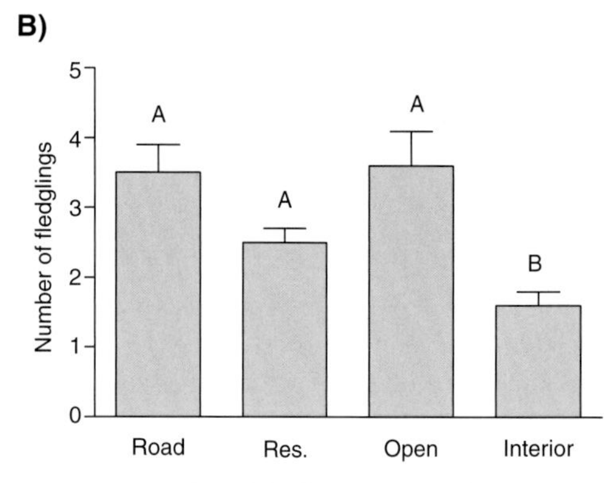

Figure 11.3. Mean date of clutch completion for first nests for each edge type (a) and mean number of fledglings for first nests for each edge type (b). Error bars are ± 1 SE. Edge types are as labeled: "Road" = small residential roads, "Res." = residential, and "Open" = open space. "Interior" indicates nests which were >50 m from edge. Bars with different uppercase letters differ significantly from one another (Tukey post-hoc test).

TABLE 11.3

Results of stepwise multiple linear regressions of nestling mass and hematocrit for Spotted Towhees in Portland, Oregon.

	Variable	Coefficient	SE	$P \leq$	Adj. r^2
Nestling mass[a] ($n = 87$)	Nestling age	1.12	0.41	0.01	0.15
	Female age	1.31	0.55	0.02	
	Date CC	−0.03	0.01	0.05	
	Edge Class 1	−0.45	0.19	0.02	
Nestling mass[b] ($n = 77$)	Nestling age	0.99	0.46	0.03	0.18
	Female age	1.46	0.58	0.01	
	Edge Class 2	−0.69	0.21	0.001	
Nestling hematocrit[a] ($n = 62$)	Nestling age	0.03	0.01	0.01	0.13
	Edge Class 1	0.01	0.00	0.05	

[a] Analysis done with Edge Class 1 (largest distance to edge category >50 m).

[b] Analysis done with Edge Class 2 (largest distance to edge category >100 m).

TABLE 11.4

Results of ANOVA of the effect of trail and edge types on the timing of breeding, nestling mass, and number of fledglings.

	Date of CC[a]			Mass[b]			Fledglings[c]		
Variable	df	F	$P \leq$	df	F	$P \leq$	df	F	$P \leq$
Year	2	2.5	0.09	2	7.7	0.01	2	2.2	0.12
Female age	1	1.3	0.26	1	6.7	0.02	1	0.6	0.43
Date CC	—	—	—	51	1.4	0.19	54	0.7	0.93
Trail type	4	1.2	0.33	4	1.5	0.22	4	0.8	0.56
Edge type	3	5.5	0.01	3	3.0	0.05	3	3.1	0.04
Error	64			25			37		
Total	74			86			101		

[a] Date CC = date of clutch completion for first nests.

[b] Residuals of the regression of nestling mass on nestling age were used in this analysis.

[c] Number of fledglings produced including both successful and unsuccessful.

timing of breeding, nestling mass, and number of fledglings all varied with edge type (Table 11.4). Posthoc pairwise comparisons (Fig. 11.3) revealed that nests within 50 m of neighborhood roads had a significantly earlier start date (mean = 16 April, SE = 2.9 days, $n = 7$) than nests within 50 m of residential homes (mean = 25 April, SE = 1.4 days, $n = 30$), open spaces (mean = 27 April, SE = 3.1 days, $n = 6$), and interior nests (mean = 29 April, SE = 1.3 days, $n = 32$). Nests located in the park interior fledged significantly fewer young (mean = 1.6, SE = 0.2, $n = 45$) than nests in all edge types (road: mean = 3.5, SE = 0.4, $n = 10$; residential: mean = 2.5, SE = 0.2, $n = 38$; and open space: mean = 3.6, SE = 0.5, $n = 9$; Fig. 11.3). The relationship between fledgling production and edge type became nonsignificant when failed nests were omitted from the analysis ($F_{3,83} = 2.4$, $P = 0.09$), indicating that the positive effect of proximity to edge was a result of fewer losses of entire

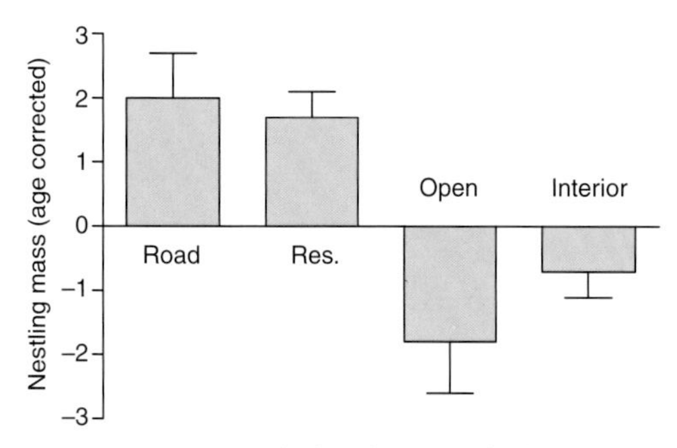

Figure 11.4. Mean residual nestling mass for first nests for each edge type. Error bars are ±1 SE. Edge types are as in Figure 3. Bars with different uppercase letters differ significantly from one another (Tukey post-hoc test).

nests near edges. Last, residual mass (corrected for age) of nestlings in nests within 50 m of neighborhood roads (mean = 2.0, SE = 0.7 g, n = 9; Fig. 11.3) and residential edges (mean = 1.7, SE = 0.4 g, n = 37) was significantly greater than that of young from nests located within 50 m of either open space (mean = −1.8, SE = 0.8 g, n = 7) or in the park's interior (mean = −0.7, SE = 0.4 g, n = 34; Fig. 11.4).

DISCUSSION

Park Size

Timing of breeding and clutch size did not vary with park size, and contrary to our expectations, we found no evidence for higher reproductive success in larger parks. If anything, the trend for first nests of the season was for a greater number of fledglings/nest in medium-sized parks, but whether this translates into season-long differences in productivity between parks of different sizes is unknown. We also acknowledge that high productivity may not be characteristic of smaller habitat fragments (2–3 ha), but our limited data from greenspaces in this size range suggests that towhee productivity does not decline in parks under 10 ha in area (S. Bartos Smith, unpubl. data). Previous work on Spotted Towhee nest success across a fragmentation gradient in urban Southern California found higher nest success in small fragments, and this was attributed to a lack of snakes in small fragments (the primary predator of towhee nests in that system; Patten and Bolger 2003). Daily nest survival in our study was high (0.966 = 41% nest success) compared to work near Sacramento, California (0.928 = 14% nest success assuming a 26-day nesting cycle; Small 2005). Snakes are uncommon in Portland's greenspaces, and their rarity may contribute to the high nest success rates that we recorded. Additionally, Brown-headed Cowbirds do not contribute to nest mortality in our population, whereas they were a major source of nest mortality in California (Small 2005). Alternatively, the high nest success we recorded could be due to a lower abundance of nest predators in urban areas (Shochat 2004). Reduced predation has been observed for urban desert habitats (Shochat 2004), but the generality of this pattern for other regions has yet to be established.

Recreational Trails

Regardless of park size, ASY female Spotted Towhees that nested within 10 m of recreational trails exhibited the highest nest success. We are unsure why this benefit applied only to ASY females, but SY females may be more prone to human disturbance and, as a consequence, exhibit lower nest attendance. Frequent departures from nests might also alert predators to nest locations. Two artificial nest studies also found higher nest success near trails (Boag et al. 1984, Miller and Hobbs 2000), suggesting that this pattern may exist independent of parental behavior. Nest success may be high near trails, particularly for ground-nesting birds, because some mammalian predators may avoid trails (Boag et al. 1984, Forman 1995, Miller and Hobbs 2000). However, improved prospects of nest success near trails may not apply to species whose primary nest predators are birds, as avian nest predators often show increased activity near recreational trails (Miller et al. 1998, Miller and Hobbs 2000, Hennings 2001). Furthermore, aboveground nesting birds may be especially susceptible to nest predation by avian predators (Söderström et al. 1998, Yahner and Scott 1998), and Miller et al. (1998) found nest success to be lower near trails for a number of forest breeding bird species including several neotropical migrants.

We found weak evidence that nesting near trails was associated with early breeding, and this may lead to an extended breeding season for females who initiated first nests near trails. It is possible that earlier nesting occurs near trails because trails provide earlier foraging opportunities for ground-foraging birds. Despite this, any potentially improved food resources afforded by trails did not result in larger clutch sizes or better condition of nestling birds. Additionally, we did not find that successful nests near trails fledged more young than nests far from trails. Therefore, our results suggest that the primary benefit of nesting near trails was decreased probability of nest predation for ASY females.

We did not find an effect of trail type (based on trail width) on any indicator of reproductive success. We assumed that trail width was an indicator of how often trails were used by humans, but because we did not quantify actual trail use, we could not determine whether human activity *per se* had an influence on avian reproduction. It has been suggested that this may be the case, as species nesting in areas frequented by humans have been found to experience lower levels of predation (Tomialojć and Profus 1977, Osborne and Osborne 1980, but see Langston et al. 2007), but further studies with other species are needed.

Habitat Edges

Nest success did not vary with distance from habitat edge, suggesting that predation rate was equally likely at all distances from edge. This is contrary to recent conclusions based on a meta-analysis of nest success studies that demonstrated strong negative effects of nesting within 50 m of habitat edges (Batáry and Báldi 2004). This meta-analysis did not include any urban studies, several of which found no relationship between nest success and distance to edge (Matthews et al. 1999, Morrison and Bolger 2002, Patten and Bolger 2003, Thorington and Bowman 2003). Despite finding no difference in daily nest survival rates between nests near edges and nests in the interior, nests near edges in our study fledged significantly more offspring than nests in the park interior. However, this result only held when we included both successful and unsuccessful nests in the analysis, suggesting that the combination of partial and complete nest loss was lowest near edges. Our finding that partial brood loss was more likely in interior nests is consistent with this interpretation, and probably contributes to the greater fledgling production near edges.

While daily nest survival rates near edges were similar to those in the interior, the earlier breeding, heavier nestlings, and lower incidence of partial brood loss among edge nests suggests an advantage to nesting near edges that is associated with greater food availability. Towhees are omnivorous, and this may allow them to utilize anthropogenic food sources such as seed from bird feeders, as well as the increased quantities of seeds and berries that are often associated with the structurally complex vegetation in edge habitats (Thompson and Willson 1978, Strelke and Dickson 1980). The benefits we found for towhees nesting near habitat edges are all consistent with the idea that food resources in these urban edge habitats are more abundant than those found in the habitat interior. Towhees are commonly found in shrubby habitats (Greenlaw 1996) such as those found near edges; however, we have no evidence that an innate preference for forest edge habitat or settlement by higher-quality individuals near edges is the underlying reason for our results. Towhee pairs were equally abundant near edges and in the park interiors (mean DTE = 45.13, SE = 3.13; 45.6% of pairs nested >50 m from the habitat edge). Furthermore, we found no relationship between adult age and distance to edge for either females (see results) or males (Bartos Smith 2008).

Edges are generally associated with negative effects (Gates and Gysel 1978, Andrén and Angelstam 1998), and it has become somewhat axiomatic to assume that edge effects persist roughly 50 m into the interior (Paton 1994, Batáry and Báldi 2004). However, we did not find any negative edge effects for towhees; instead, we found positive edge effects and that edge effects can persist beyond 50 m from the edge. When we doubled the width of edge categories to evaluate effects up to 100 m from the edge, several variables (e.g., nestling mass) showed stronger associations with distance from edges. Positive edge effects may persist farther from the edge because of the ability of birds to opportunistically visit habitat edges for foraging opportunities regardless of where their nest may be placed. Whatever the reason, a 50-m wide zone may be too narrow to capture all edge effects (Hoover et al. 2006).

Nests near all types of edges fledged more offspring than interior nests, indicating that the reduced probability of nest predation near edges persisted regardless of the edge type. However, the occurrence of heavier nestlings near habitat edges seemed to be most pronounced for edges that bordered residential neighborhoods and their lightly traveled roads. Some landowner activities in residential developments, including bird feeding and natural landscaping, have the potential to positively influence birds (Lepczyk et al. 2004).

The residential developments surrounding the greenspaces in our study were primarily single-family homes, and many of these have bird feeders. Towhees are regular visitors to feeders (S. Bartos Smith, pers. obs.), and reproductive success may have been enhanced by the use of supplemental food sources by adult birds, resulting in greater quantities of natural foods available for feeding young. Many, possibly most, passerine species do not utilize anthropogenic food sources and therefore our results may apply to only a subset of species.

Nevertheless, we found low nestling hematocrit values near edges, and while the utility of hematocrit measurements as indicators of condition has been questioned (Dawson and Bortolotti 1997), high hematocrit is generally thought to indicate good body condition (Merilä and Svensson 1995, Svensson and Merilä 1996). Lower nestling hematocrit values near edges suggest that nesting near edges is not cost free. The exact nature of the costs is unknown, but if edge-nesting towhees supplement their offspring's diet with anthropogenic food sources (seeds), the latter may not be of the same quality as natural foods (arthropods). Anthropogenic foods have been found to reduce avian reproductive success (Annett and Pierotti 1999, Shawkey et al. 2004), often because they are substandard foods for rapidly developing young (Shawkey et al. 2004, Sauter et al. 2006). Adult towhees do sometimes deliver birdseed to nestlings (J. E. McKay, unpubl. data). Thus, nestlings near edges may eat abundant food that allows them to grow rapidly but at a lower overall condition. Similarities between productivity of successful nests at edges and nests in the interior indicate that hematocrit and nesting success were not associated, but the question remains whether the hematocrit differences indicated probability of subsequent recruitment.

We did not compare seasonal productivity of edge-nesting and interior-nesting females, in part because females did not consistently nest near edges or in the interior for the entire breeding season. However, we suspect that, all else being equal, the greater number of fledglings for edge-nesting females, even if only for a single nest during the breeding season, is likely to result in increased seasonal fecundity. Morrison and Bolger (2002) did not find a difference in seasonal productivity between edge- and interior-breeding Rufous-crowned Sparrows (*Aimophila ruficeps*) because snake predation of eggs and young was high at all distances from the edge. Our study suggests that differences in patterns of predation are not the only cause of increased reproductive success near edges for Spotted Towhees: edge effects may also be related to differences in food availability.

An additional consideration is that while overall fledging success was higher near habitat edges, it is possible that fledglings from edge nests suffer a higher post-fledging mortality than fledglings from interior nests. Domestic cat abundances tend to be highest near houses (Crooks and Soulé 1999, Odell and Knight 2001), so it is possible that the risk of predation by domestic cats would be highest near habitat edges, particularly those bordering neighborhoods. While domestic cats may not be major predators at nests (Haskell et al. 2001), they may be significant predators of juvenile birds (Baker et al. 2005). Spotted Towhees leave the nest well before they are able to fly (Baumann 1959, S. Bartos Smith, pers. obs.) and may be easy prey for domestic cats and other mammals in their first few days out of the nest. If this is the case, edges may be ecological traps (Gates and Gysel 1978), such that the perceived increase in fledgling production is eliminated by high levels of post-fledging mortality. Much more work is needed to determine rates of post-fledging mortality in urban areas and to evaluate whether such mortality is influenced by habitat features such as trails and edges.

CONCLUSION

Few studies in urban environments have investigated whether habitat edges and recreational trails affect aspects of reproductive performance other than nest success. While predation may be the main cause of nest failure for many species, we have shown that other factors, such as food availability, may influence the reproductive success of birds. More studies are needed to determine how predation risk and food availability are affected by urbanization, and how these factors cumulatively influence avian reproductive success in urban areas. By determining whether different types of edges and trails affect birds in different ways, we may gain a better mechanistic understanding of the specific features of trails and edges that influence the reproductive success of birds in urban settings and, in so doing, provide better guidelines for management and maintenance of bird populations in urban greenspaces.

ACKNOWLEDGMENTS

A. M. Smith, J. E. Klomp, L. J. Redmond, and many dedicated field assistants provided invaluable help with our field work. JEM would like to thank R. Cody, M. LaFrenz, and J.-D. Geoffery Duh for assistance with GPS and GIS techniques. Portland and Lake Oswego Parks Associations facilitated our work in the greenspaces. We also thank Ç. Şekercioğlu, C. A. Lepczyk, and an anonymous reviewer for helping to improve the original manuscript. This project was funded by a USFW grant to MTM, a Portland State University Forbes-Lea grant to SBS, and an Environmental Protection Agency (EPA) GRO fellowship to SBS. Although this research was funded in part by the EPA, they have not officially endorsed this publication, and the views expressed herein may not reflect the views of the EPA.

LITERATURE CITED

Andrén, H. 1992. Corvid density and nest predation in relation to forest fragmentation: a landscape perspective. Ecology 73:794–804.

Andrén, H. 1995. Effects of landscape composition on predation rates at habitat edges. Pp. 225–255 *in* L. Hansson, L. Fahrig, and G. Merriam (editors), Mosaic lanscapes and ecological processes. Chapman and Hall, London, UK.

Andrén, H., and P. Angelstam. 1988. Elevated predation rates as an edge effect in habitat islands: experimental evidence. Ecology 69:544–547.

Annett, C. A., and R. Pierotti. 1999. Long-term reproductive output in Western Gulls: consequences of alternate tactics in diet choice. Ecology 80:288–297.

Baker, P. J., A. J. Bentley, R. J. Ansell, and S. Harris. 2005. Impact of predation by domestic cats (*Felis catus*) in an urban area. Mammal Review 35:302–312.

Bartos Smith, S. 2008. The reproductive and conservation biology of the Spotted Towhee in urban forest fragments. Ph.D. dissertation, Portland State University, Portland, OR.

Batary, P., and A. Baldi. 2004. Evidence of an edge effect on avian nest success. Conservation Biology 18:389–400.

Baumann, S. A. 1959. The breeding cycle of the Rufous-sided Towhee, *Pipilo erythrophthalmus* (Linnaeus), in central California. Wasmann Journal of Biology 17:161–220.

Boag, D. A., S. G. Reebs, and M. A. Schroeder. 1984. Egg loss among spruce grouse inhabiting lodgepole pine forests. Canadian Journal of Zoology 62:1034–1037.

Bolger, D. T., T. A. Scott, and J. T. Rotenberry. 1997. Breeding bird abundance in an urbanizing landscape in coastal southern California. Conservation Biology 11:406–421.

Bolger, D. T., A. V. Suarez, K. R. Crooks, S. A. Morrison, and T. J. Case. 2000. Arthropods in urban habitat fragments in southern California: area, age, and edge effects. Ecological Applications 10:1230–1248.

Bolger, D. T. 2002. Habitat fragmentation effects on birds in southern California: contrast to the top-down paradigm. Studies in Avian Biology 25: 141–157.

Burke, D. M., and E. Nol. 2000. Landscape and fragment size effects on reproductive success of forest-breeding birds in Ontario. Ecological Applications 10:1749–1761.

Clinchy, M., L. Zanette, R. Boonstra, J. C. Wingfield, and J. N. M. Smith. 2004. Balancing food and predator pressure induces chronic stress in songbirds. Proceedings of the Royal Society of London, Series B 271:2473–2479.

Crooks, K. R., and M. E. Soulé. 1999. Mesopredator release and avifaunal extinctions in a fragmented system. Nature 400:563–566.

Crooks, K. R., A. V. Suarez, and D. T. Bolger. 2004. Avian assemblages along a gradient of urbanization in a highly fragmented landscape. Biological Conservation 115:451–462.

Dawson, R. D., and G. R. Bortolotti. 1997. Are avian hematocrits indicative of condition? American Kestrels as a model. Journal of Wildlife Management 61:1297–1306.

Doherty, P. E., and T. C. Grubb, Jr. 2002. Survivorship of permanent-resident birds in a fragmented forested landscape. Ecology 83:844–857.

Donovan, T. M., P. W. Jones, E. M. Annand, and F. R. Thompson III. 1997. Variation in local-scale edge effects: mechanisms and landscape context. Ecology 78:2064–2075.

Forman, R. T. T. 1995. Land mosaics. Cambridge University Press, Cambridge, UK.

Gates, J. E., and L. W. Gysel. 1978. Avian nest dispersion and fledging success in field-forest ecotones. Ecology 59:871–883.

Gerig, J. C., and R. B. Blair. 1999. Predation on artificial bird nests along an urban gradient: predatory risk or relaxation in urban environments? Ecography 22:532–541.

Greenlaw, J. S. 1996. Spotted Towhee (*Pipilo maculatus*). No. 263 *in* A. Poole and F. Gill (editors), The Birds of North America. Academy of Natural Sciences, Philadelphia, PA, and American Ornithologists' Union, Washington, DC.

Haskell, D. G., A. M. Knupp, and M. C. Schneider. 2001. Nest predator abundance and urbanization.

Pp. 243–258 *in* J. M. Marzluff, R. Bowman, and R. Donnelly (editors), Avian ecology and conservation in an urbanizing world. Kluwer Academic Publishers, Norwell, MA.

Hazler, K. R. 2004. Mayfield logistic regression: a practical approach for analysis of nest survival. Auk 121:707–716.

Hennings, L. A. 2001. Riparian bird communities in Portland, Oregon: habitat, urbanization, and spatial scale patterns. M.S. thesis, Oregon State University, Corvallis, OR.

Heske, E. J., S. K. Robinson, and J. D. Brawn. 1999. Predator activity and predation on songbird nests on forest-field edges in east-central Illinois. Landscape Ecology 14:345–354.

Hoover, J. P., T. H. Tear, and M. E. Baltz. 2006. Edge effects reduce the nesting success of Acadian Flycatchers in a moderately fragmented forest. Journal of Field Ornithology 77:425–436.

Jokimäki, J. and E. Huhta. 2000. Artificial nest predation an abundance of birds along an urban gradient. Condor 102:838–847.

Langston, R. H. W., D. Liley, G. Murison, E. Woodfield, and R. T. Clarke. 2007. What effects do walkers and dogs have on the distribution and productivity of breeding European Nightjar (*Caprimulgus europaeus*)? Ibis 149 (Supplement 1):27–36.

Lepczyk, C. A., A. G. Mertig, and J. Liu. 2004. Assessing landowner activities related to birds across rural-to-urban landscapes. Environmental Management 33:110–125.

Lessells, C. M., and J. R. Krebs. 1989. Age and breeding performance of European Bee-eaters. Auk 106:375–382.

Martin, T. E. 1993. Nest predation among vegetation layers and habitat types: revising the dogmas. American Naturalist 141:897–913.

Marzluff, J. M., R. Bowman, and R. Donnelly. 2001a. A historical perspective on urban bird research: trends, terms, and approaches. Pp. 1–17 *in* J. M. Marzluff, R. Bowman, and R. Donnelly (editors), Avian ecology and conservation in an urbanizing world. Kluwer Academic Publishers, Norwell, MA.

Marzluff, J. M., K. J. McGowan, R. Donnelly, and R. L. Knight. 2001b. Causes and consequences of expanding American Crow populations. Pp. 331–365 *in* J. M. Marzluff, R. Bowman, and R. Donnelly (editors), Avian ecology and conservation in an urbanizing world. Kluwer Academic Publishers, Norwell, MA.

Matthews, A., C. R. Dickman, and R. E. Major. 1999. The influence of fragment size and edge on nest predation in urban bushland. Ecography 22:349–356.

Mennechez, G. and P. Clergeau. 2006. Effect of urbanization on habitat generalists: starlings not so flexible? Acta Oecologica 30:182–191.

Merilä, J., and E. Svensson. 1995. Fat reserves and health state in migrant Goldcrest (*Regulus regulus*). Functional Ecology 9:842–848.

Miller, J. R., and N. T. Hobbs. 2000. Recreational trails, human activity, and nest predation in lowland riparian areas. Landscape and Urban Planning 50:227–236.

Miller, S. G., R. L. Knight, and C. K. Miller. 1998. Influence of recreational trails on breeding bird communities. Ecological Applications 8:162–169.

Moore, R. P., and W. D. Robinson. 2004. Artificial bird nests, external validity, and bias in ecological field studies. Ecology 85:1562–1567.

Morrison, S. A., and D. T. Bolger. 2002. Lack of an urban edge effect on reproduction in a fragmentation-sensitive sparrow. Ecological Applications 12:398–411.

Odell, E. A., and R. L. Knight. 2001. Songbird and medium-sized mammal communities associated with exurban development in Pitkin County, Colorado. Conservation Biology 15:1143–1150.

Osborne, P., and L. Osborne. 1980. The contribution of nest site characteristics to breeding success among blackbirds (*Turdus merula*). Ibis 122:512–517.

Partecke, J., I. Schwabl, and E. Gwinner. 2006. Stress and the city: urbanization and its effects on the stress physiology in European Blackbirds. Ecology 87:1945–1952.

Paton, P. W. C. 1994. The effect of edge on avian nest success: how strong is the evidence? Conservation Biology 8:17–26.

Patten, M. A., and D. T. Bolger. 2003. Variation in top-down control of avian reproductive success across a fragmentation gradient. Oikos 101:479–488.

Perrins, C. M., and R. H. McCleery. 1985. The effect of age and pair bond on the breeding success of Great Tits (*Parus major*). Ibis 127:306–315.

Pyle, P. 1997. Identification guide to North American birds, Part 1: Columbidae to Ploceidae. Slate Creek Press, Bolinas, CA.

Ricklefs, R. E. 1969. An analysis of nesting mortality in birds. Smithsonian Contributions in Zoology 9:1–48.

Saunders, D. A., R. J. Hobbs, and C. R. Margules. 1991. Biological consequences of ecosystem fragmentation: a review. Conservation Biology 5:18–32.

Sauter, A., R. Bowman, S. J. Schoech, and G. Pasinelli. 2006. Does optimal foraging theory explain why suburban Florida Scrub-Jays (*Aphelocoma coerulescens*) feed their young human-provided food? Behavioral Ecology and Sociobiology 60:465–474.

Shaffer, T. L. 2004. A unified approach to analyzing nest success. Auk 121:526–540.

Shawkey, M. D., R. Bowman, and G. E. Woolfenden. 2004. Why is brood reduction in Florida Scrub-Jays

higher in suburban than in wildland habitats? Canadian Journal of Zoology 82:1427–1435.
Shochat, E. 2004. Credit or debit? Resource input changes population dynamics of city-slicker birds. Oikos 106:622–626.
Sinclair, K. E., G. R. Hess, C. E. Moorman, and J. H. Mason. 2005. Mammalian nest predators respond to greenway width, landscape context, and habitat structure. Landscape and Urban Planning 71:277–293.
Small, S. L. 2005. Mortality factors and predators of Spotted Towhee nests in the Sacramento Valley, California. Journal of Field Ornithology 76:252–258.
Söderström, B., T. Pärt, and J. Rydén. 1998. Different nest predator faunas and nest predation on ground and shrub nests at forest ecotones: an experiment and a review. Oecologia 117:108–118.
Strelke, W. K., and J. G. Dickson. 1980. Effect of forest clearcut edge on breeding birds in east Texas. Journal of Wildlife Management 44:559–567.
Svensson, E., and J. Merilä. 1996. Molt and migratory condition in Blue Tits: a serological study. Condor 98:825–831.
Thompson, J. N., and M. F. Willson. 1978. Disturbance and the dispersal of fleshy fruits. Science 200:1161–1163.
Thorington, K. K., and R. Bowman. 2003. Predation rate on artificial nests increases with human housing density in suburban habitats. Ecography 26:188–196.
Tomialojć, L., and P. Profus. 1977. Comparative analysis of breeding bird communities in two parks of Wroclaw and in adjacent Querco-Carpinetum forest. Acta Ornithologica 16:117–177.
Weldon, A. J., and N. M. Haddad. 2005. The effects of patch shape on Indigo Buntings: evidence for an ecological trap. Ecology 86:1422–1431.
Yahner, R. H. 1988. Changes in wildlife communities near edges. Conservation Biology 2: 333–339.
Yahner, R. H., and D. P. Scott. 1988. Effects of forest fragmentation on depredation of artificial nests. Journal of Wildlife Management 52:158–161.
Zanette, L., M. Clinchy, and J. N. M. Smith. 2006. Combined food and predator effects on songbird nest survival and annual reproductive success: results from a bi-factorial experiment. Oecologia 147:632–640.
Zanette, L., P. Doyle, and S. M. Tremont. 2000. Food shortage in small fragments: evidence from an area-sensitive passerine. Ecology 81: 1654–1666.
Zanette, L., J. N. M. Smith, H. van Oort, and M. Clinchy. 2003. Synergistic effects of food and predators on annual reproductive success in Song Sparrows. Proceedings of the Royal Society of London, Series B 270:799–803.

CHAPTER TWELVE

Post-fledging Mobility in an Urban Landscape

Kara Whittaker and John M. Marzluff

Abstract. Habitat loss and fragmentation as a result of urbanization are altering the population dynamics and community composition of songbirds around the world. Successful movement between isolated habitat patches may be necessary for sensitive bird populations to remain viable. The goal of this investigation was to understand how the post-fledging movement process relates to land cover patterns in a heterogeneous urban landscape. We measured the movements of 26 American Robins (*Turdus migratorius*), 15 Swainson's Thrushes (*Catharus ustulatus*), 46 Spotted Towhees (*Pipilo maculatus*), and 35 Song Sparrows (*Melospiza melodia*) during the post-fledging period at various spatial scales across the urban gradient of the Seattle metropolitan area from 2003 to 2005. We used radiotelemetry to track fledglings and test predictions regarding (1) the differences in mobility among species and (2) the relationship between mobility and forest aggregation, percent forest, and percent urban land cover. Juvenile birds varied among species in their mobility, their sensitivity to different land cover types, and the spatial scales at which their movements responded to land cover patterns. The most migratory species, as opposed to the most urban-affiliated species, were more mobile after correcting for body size. More impervious developed areas (heavy–medium urban) had more consistently negative effects on juvenile mobility than less impervious developed areas (light urban), but these effects varied with species and scale. Parts of the urban matrix were permeable to movement and provided food resources, such as fruiting trees and shrubs and bird feeders in residential yards. We suggest urban growth strategies across multiple scales (such as clustered development and matrix management) for maintaining effective bird dispersal in urban ecosystems that focus on maximizing the amount of forest cover and minimizing the amount of impervious urban cover.

Key Words: American Robin, *Catharus ustulatus*, *Melospiza melodia*, natal dispersal, *Pipilo maculatus*, post-fledging movements, radiotelemetry, Song Sparrow, Spotted Towhee, Swainson's Thrush, *Turdus migratorius*.

Dispersal is a critical population process that maintains gene flow within and between populations, allows metapopulation persistence, and supplements population growth (Levins 1970, Merriam 1991, Hanski and Gilpin 1997). Dispersal is also a key parameter in source–sink population models as it contributes to the net gain or loss of individuals through

Whittaker, K., and J. M. Marzluff. 2012. Post-fledging mobility in an urban landscape. Pp. 183–198 *in* C. A. Lepczyk and P. S. Warren (editors). Urban bird ecology and conservation. Studies in Avian Biology (no. 45), University of California Press, Berkeley, CA.

immigration and emigration, respectively, and maintains the existence of sink populations (Pulliam 1988, Pulliam and Danielson 1991). If a species in a given local population is extirpated, the habitat patch may become recolonized by members of the species able to immigrate to the patch (Brown and Kodric-Brown 1977). The viability of forest bird populations after their habitat becomes isolated into separate patches may hinge on the ability of dispersers to move between habitat remnants. Limitation in dispersal can ultimately lead to extinction in species with few, small populations (Macdonald and Johnson 2001).

Dispersal is especially critical in dynamic landscapes. Animal movements through fragmented landscapes are affected by both the connectivity of the remaining habitat and the type of matrix that must be traversed. Studies of a number of forest bird species have demonstrated that gaps between forest fragments may or may not be crossable, depending on the size of the gap, the habitat type that must be crossed, and the behavioral preferences of the organism in question (Dunning et al. 1995, Haas 1995, Machtans et al. 1996, Desrochers and Hannon 1997, Rail et al. 1997, St. Clair et al. 1998, Brooker et al. 1999). The landscape matrix serves as a variable filter to interpatch movement depending on the land use or land cover type that must be crossed and the proportion and configuration of original land cover that remains (Trzcinski et al. 1999, Wiens 2001, Castellón and Sieving 2006). For example, a forest bird species may find a rural agricultural area to be more permeable to interpatch movement than a commercial shopping area. Land cover types within an urban landscape mosaic may differ strongly in their permeabilities due to variation in the risk of predation, hazards of novel environments (such as collision with objects), and abundance of food, which may all have especially strong effects during the post-fledging transition from parental care to independence.

The goal of this investigation was to understand how the post-fledging movement process relates to land cover patterns at various spatial scales in a heterogeneous urban landscape. We measured movements of forest songbird species along different parts of an urban gradient during the post-fledging period. During this period, juvenile birds begin the natal dispersal process, which is characterized by extensive movement (Greenwood and Harvey 1982). The post-fledging period is disproportionately understudied but deserves more attention as it contributes to bird population stability and persistence (McFadzen and Marzluff 1996). As the rapid expansion of urban land cover continues (Brown et al. 2005, Robinson et al. 2005), we will need a better understanding of the dispersal process to maintain viable wildlife populations in urban areas.

Our study was designed to detect differences among species in perceptions of scale of urban landscapes and the effects of different landscape types on movement patterns. Here we tested two predictions:

1. *Post-fledging mobility is influenced by species' life-history strategies.*

A species' relative level of mobility in an urban landscape depends on either its migratory strategy (long- or short-distance migrant, resident; Paradis et al. 1998) or its affinity to urban areas (relative abundance; Marzluff 2001). If migratory strategy is most influential, then the mobility of migratory species should be greater than that of resident species. If urban affinity is most influential, then the mobility of more urban-affiliated species should be greater than that of less urban-affiliated species.

2. *Post-fledging mobility is a function of the extent to which landscape composition and configuration either facilitates or impedes movement.*

A. Birds are more mobile where forest cover is more contiguous. Dispersal is facilitated where there is less resistance to movement (gaps in habitat). Birds can move farther and faster through landscapes with relatively more contiguous forested habitat (Dunning et al. 1995, Haas 1995, Desrochers and Hannon 1997, Rail et al. 1997, St. Clair et al. 1998, Brooker et al. 1999).
B. Birds are more mobile where forest cover is more abundant. In landscapes with a relatively greater percent of forest cover, birds are able to move with greater speed and distance (Bélisle et al. 2001). Forested land cover should be the most suitable land cover type for the dispersal of post-fledging birds of species that breed in forests.
C. Birds are less mobile where urban cover is more abundant. Urban land cover acts as a barrier to bird movement in proportion to its imperviousness (extent paved or built), resulting in relatively lower mobility in highly urban landscapes (Marzluff 2001). The more developed an

area, the more likely a bird is to encounter dangerous objects (cars, windows), novel predators (cats), inhospitable abiotic conditions (water, air, soil, and noise pollution), and disturbance from human activities.

METHODS

Study Sites and Species

We measured the post-fledging movement patterns of juvenile birds using radiotelemetry at sites across the urban gradient of the Seattle, Washington, USA, metropolitan area over a three-year period (2003–2005). We selected 19 forest fragments within a matrix of urban, suburban, exurban, and undeveloped forested landscapes (Marzluff et al. 2001; Fig. 12.1). Sites were selected using a stratified random sample of three landscape metrics calculated for a 1-km^2 area containing each study site: mean patch size, contagion (an estimate of connectivity), and the percent forest and percent urban land cover (see Donnelly and Marzluff 2004, 2006; Blewett and Marzluff 2005 for details). Forest fragments were primarily coniferous, including western hemlock (*Tsuga heterophylla*), Douglas-fir (*Pseudotsuga menziesii*), and western red cedar (*Thuja plicata*), or mixed with deciduous tree species such as red alder (*Alnus rubra*), bigleaf maple (*Acer macrophyllum*), black cottonwood (*Populus trichocarpa*), and Oregon ash (*Fraxinus latifolia*; Franklin and Dyrness 1988). Site elevation varied from sea level to 300 m on the lower slopes of the Cascade Range.

The four study species are all native songbirds that breed in forests: American Robin (*Turdus migratorius*), Swainson's Thrush (*Catharus ustulatus*), Spotted Towhee (*Pipilo maculatus*), and Song Sparrow (*Melospiza melodia*). Swainson's Thrushes are long-distance neotropical migrants, American Robins are short-distance migrants, and Spotted Towhees and Song Sparrows are year-round residents in this region. These four species were chosen for study because they had the highest mass of a set of species we studied previously in this region (Donnelly and Marzluff 2004, 2006) and could thus carry radio transmitters with the longest battery lives (see below).

To assess their sensitivity to urban land cover (Prediction 1), we determined each species' level of urban affinity from relative abundance estimates from point-count survey data at all 19 study sites (for methods see Donnelly and Marzluff 2006). We calculated the relative abundance of a given species at a given site as the number of birds per survey point per visit per year from data spanning 1998–2005 and tested for differences among species with ANOVA and *post-hoc* LSD tests. At forested survey points, species abundance varied ($F_{3,64} = 10.13$, $P < 0.01$), with American Robins equal to Song Sparrows and Spotted Towhees, all of which had greater abundance than Swainson's Thrushes (LSD test, all $P < 0.01$; Table 12.1). At urban matrix survey points, abundance also varied among species ($F_{3,52} = 33.95$, $P < 0.01$), with American Robins more abundant than Song Sparrows, which were equal to Spotted Towhees, and all were more abundant than Swainson's Thrushes (LSD test, all $P < 0.01$; Table 12.1). In paired-t tests of the relative abundance between point types for each species, American Robins were more abundant at matrix than at forest points ($t = -4.0$, $P < 0.01$), and Swainson's Thrushes were more abundant at forest than at matrix points ($t = 2.42$, $P = 0.03$), whereas Song Sparrows and Spotted Towhees were equally abundant in both areas, though slightly more abundant in forests (Table 12.1). We used this information to rank American Robins as the most urban-affiliated species followed by less, but equally urban Song Sparrows and Spotted Towhees. Swainson's Thrushes were the least urban-affiliated species.

Post-fledging Movements and Covariates

We radio-tagged 122 recently fledged birds (~1–20 days after fledging; Table 12.2) and recorded their daily movements until death or transmitter battery expiration. Each transmitter was fixed with an elastic thread leg harness (Rappole and Tipton 1991) that was designed to break and fall off the bird after our observations ended. The battery life of radios was proportional to the maximum weight each species could safely carry (≤3% of body weight, model BD-2, Holohil Systems Ltd., Carp, Ontario, Canada), which was approximately 3 weeks for Song Sparrows, 6 weeks for Swainson's Thrushes, 8 weeks for Spotted Towhees, and 9 weeks for American Robins.

Fledglings were captured either in a hand net shortly after leaving the nest or in mist nets if older, including no more than one young per brood to increase independence of samples. Attempts were made to catch fledglings on their

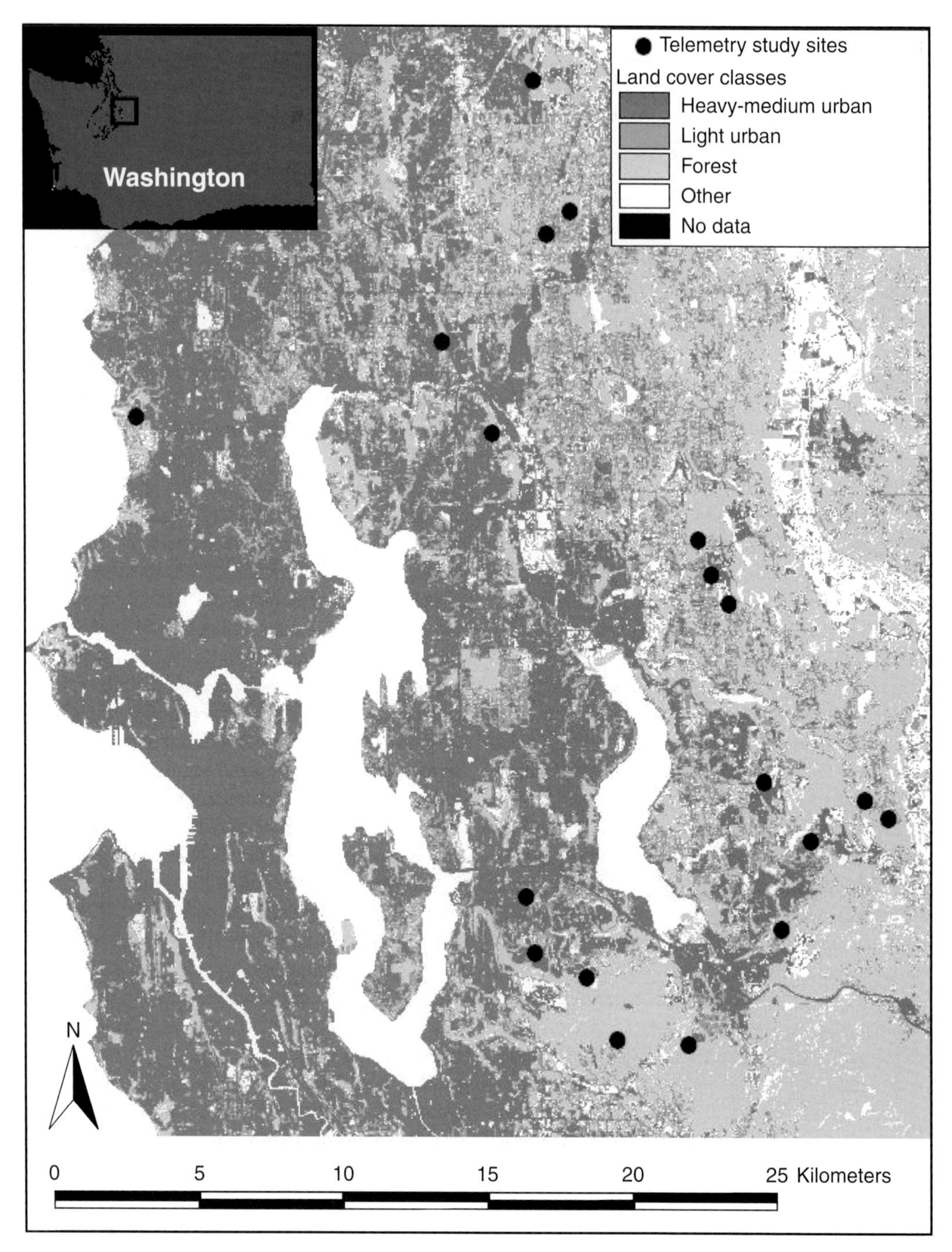

Figure 12.1. The study area includes urban, suburban, exurban, and control study sites (black dots) in the Seattle metropolitan area. Land cover classes (from a 2002 Landsat image) are heavy–medium urban (>50% impervious surface, dark gray), light urban (<50% impervious surface, medium gray), forest (light gray), and other (white).

natal territories (where they were born) while still dependent on their parents, but we also caught older juvenile birds which were behaviorally independent at locations unknown relative to their natal territories. We determined age at capture for each bird based on behavior (begging from adults) and movement patterns (stationary within fragment or actively moving) and classified young birds still dependent on their parents as "fledglings" ($n = 57$) and older behaviorally independent

TABLE 12.1

Relative abundance (RA) and variation (SD) estimates per species at survey points in forest fragments or the urban matrix. The paired t-test was 2-tailed.

Species	Forest RA (SD)	Matrix RA (SD)	t	$P \leq$
American Robin	0.71 (0.20)	0.85 (0.11)	−4.00	0.002
Song Sparrow	0.65 (0.22)	0.56 (0.19)	0.94	0.370
Spotted Towhee	0.72 (0.16)	0.66 (0.15)	1.19	0.260
Swainson's Thrush	0.34 (0.22)	0.23 (0.20)	2.42	0.031

TABLE 12.2

The proportion of individuals (n) of each species assigned to each movement category by the site fidelity test and total sample sizes of radio-tagged birds of each species and age cohort.

Species	Constrained	Dispersed	Random	Fledglings	Juveniles
Song Sparrow	0.49 (17)	0.26 (9)	0.26 (9)	11	24
Spotted Towhee	0.59 (27)	0.07 (3)	0.35 (16)	16	30
Swainson's Thrush	0.15 (2)	0.08 (1)	0.77 (10)	10	5
American Robin	0.12 (3)	0.08 (2)	0.80 (20)	20	6
All species	0.41 (49)	0.13 (15)	0.46 (55)	57	65

NOTE: Test result categories indicate whether a bird's movement pathway was random, more dispersed, or more constrained than 100 random pathways. This test could not be conducted for three birds that had only two observation points each.

birds as "juveniles" ($n = 65$; Table 12.2). Movement pathways of fledglings typically included parts of the periods both before and after parental independence. We analyzed the two age cohorts separately because we viewed them as two somewhat distinct behavioral periods in which birds may respond differently to the same land cover patterns, and combining the two cohorts may confound or obscure any meaningful relationships unique to one cohort or the other.

We attempted to relocate each bird daily with an R-1000 telemetry receiver (Communications Specialists, Inc., Orange, CA) and a hand-held Yagi antenna. We recorded each location (UTM) using a hand-held GPS unit (Garmin 12XL, Olathe, KS) with an accuracy of <10 m estimated position error (GPS EPE). Observers avoided approaching each bird too closely so as to prevent influencing its movement. We recorded only one location per bird per day unless the bird was actively moving while we were tracking it (single movement of >100 m) to allow us to maximize our number of samples rather than follow each bird for an extended period of time (Otis and White 1999). When the bird could not be seen, its position was estimated using triangulation. If the signal did not move from the same location for several days, then the bird was detected visually to ensure it was still alive. In all cases of mortality, the radio was recovered with any evidence of death (e.g., carcass or feathers). If the signal was absent from the same area as the previous day, then the immediate area around the previously known location was searched in an increasingly larger radius using both a hand-held Yagi antenna on foot and a roof-mounted omnidirectional antenna while driving. Searching for an absent signal ceased after the date the expected battery life of the transmitter had passed.

We calculated several metrics of mobility from locations plotted in ArcView GIS (ESR Institute, ver. 3.3, Redlands, CA). Mean daily speed (m/day) and total distance moved (m) were calculated using the Animal Movements 2.0 extension (Hooge et al. 1999). Mean daily speed is the mean number of meters traveled per day (distance/

number of days in data set) and is comparable among individuals and species because it corrects for differences in transmitter battery lives among species and individuals (in cases of mortality). Total distance is a sum of the total length of all daily movements combined and is suitable for comparisons of individuals within the same species with roughly the same transmitter battery lives. We calculated the maximum distance moved as the hypotenuse of the triangle made by the differences between the maximum and minimum X and Y coordinate locations for each individual. Note that this is the maximum distance between the furthest two points recorded per individual (not the maximum distance moved any given day, as calculated by Animal Movements). We also conducted a site fidelity test for each bird to assess whether their movement pathway was random, more dispersed than random, or more constrained than random pathways (Animal Movements 2.0 extension; Hooge et al. 1999). The fidelity test compares the observed movement pattern with 100 random walks using a Monte Carlo simulation and parameters from the original data.

Land Cover Analyses

A 2002 Landsat TM satellite image with 30 m resolution was classified into 12 land cover classes (Hepinstall et al. 2009a), of which three are examined to test our predictions: medium–heavy urban (>50% impervious surface per pixel; includes pavement, buildings, and lawns), light urban (20–50% impervious surface), and forest (deciduous, coniferous, and mixed). These classes represent a gradient of the transition from forest to urban cover, to which species may respond uniquely depending on their habitat needs. We calculated landscape metrics with the program FRAGSTATS 3.3 (McGarigal et al. 2002) for various buffer sizes around the center of each study site to test for differences in landscape extent on movement patterns (105 m, 250 m, 500 m, 1,000 m, 2,000 m, 3,000 m, and 4,000 m; Table 12.3). Buffer sizes were determined from the maximum distance moved by each species and extend in a radius around the mean of the UTM coordinates of the capture locations of all birds at each site. The smallest scale was set at 105 m because it encompasses the land cover area of three full pixels in radius surrounding the central pixel (30-m resolution grid), it was the minimum size at which the aggregation index (see below) was estimated consistently at each site, and it corresponds with an area roughly equal to that of our smallest study sites (3.5 ha). We calculated the percent cover (the percent of pixels in the landscape of the focal land cover class) for each land cover class and the aggregation index (the frequency with which different pairs of patch types appear side by side on the map, taking into account only the like adjacencies involving the focal land cover class) for the forest land cover class. All landscape metrics varied among study sites and spatial scales of analysis (Table 12.3; Fig. 12.2).

Statistical Analyses

We tested for differences among species in post-fledging mobility with one-way ANOVAs and post hoc LSD tests. We log-transformed mean daily speed, total distance, and maximum distance to equalize variances (Zar 1999). Percent forest, percent urban, and forest aggregation were arcsine square root transformed to meet assumptions of normality and equal variances (Zar 1999). Relationships between land cover and movement variables were assessed with multiple linear regression models run separately for each species, age cohort, scale, and mobility metric. Means are reported plus or minus one standard deviation. The significance level (alpha) was 0.05, and P-values were two-tailed. Because each scale was tested separately, our ability to accept or reject each prediction was scale dependent. Sample sizes (n) throughout refer to the number of birds. We completed all statistical analyses using SPSS 13.0 software (ver. 13.0, SPSS, Chicago, IL).

RESULTS

Post-fledging mobility varied among species, which may reflect differences in species' life-history strategies (Prediction 1). Log mean daily speed differed among species for birds of all ages combined ($F_{3,118} = 5.86$, $P < 0.01$; Fig. 12.3a), for fledglings ($F_{3,53} = 3.33$, $P = 0.03$; Fig. 12.3c), and for juveniles ($F_{3,61} = 8.2$, $P < 0.01$; Fig. 12.3e). Fledgling Song Sparrows moved more slowly than fledglings of all other species (LSD tests, all $P < 0.05$; Fig. 12.3a, c, e). Juvenile American Robins and Swainson's Thrushes moved with

TABLE 12.3

Summary statistics of landscape metric values across all spatial scales of analysis.

Scale	Landscape metric	Minimum	Maximum	Mean	SD
105 m	% Heavy–medium urban	0.00	58.62	11.07	19.73
	% Light urban	0.00	41.38	21.23	14.60
	% Forest	3.45	100.00	58.08	28.83
	Forest aggregation	0.00	93.62	74.48	23.50
250 m	% Heavy–medium urban	0.00	51.14	17.47	16.67
	% Light urban	0.91	35.62	19.74	9.85
	% Forest	12.79	96.80	52.58	27.01
	Forest aggregation	59.57	97.72	82.04	11.70
500 m	% Heavy–medium urban	0.00	56.68	21.50	19.24
	% Light urban	9.23	32.14	21.33	7.91
	% Forest	16.30	87.10	47.36	24.52
	Forest aggregation	65.77	94.23	82.63	9.69
1,000 m	% Heavy–medium urban	1.09	64.58	22.89	19.86
	% Light urban	7.98	40.55	22.36	8.17
	% Forest	9.67	87.56	44.94	22.28
	Forest aggregation	60.79	95.61	81.43	9.35
2,000 m	% Heavy–medium urban	2.44	63.74	23.97	17.78
	% Light urban	8.95	39.78	21.45	7.12
	% Forest	12.06	82.93	42.37	19.50
	Forest aggregation	61.08	95.36	81.77	8.56
3,000 m	% Heavy–medium urban	3.70	57.25	23.08	16.38
	% Light urban	12.28	38.02	21.91	7.74
	% Forest	8.59	74.39	41.83	17.78
	Forest aggregation	68.94	93.59	82.17	7.76
4,000 m	% Heavy–medium urban	5.27	48.78	22.24	14.74
	% Light urban	10.41	37.31	22.18	7.19
	% Forest	6.22	65.49	40.81	16.10
	Forest aggregation	66.69	92.50	81.94	7.32
Overall	% Heavy–medium urban	0.00	64.58	20.32	1.94
	% Light urban	7.11	37.83	21.46	8.94
	% Forest	3.45	100.00	46.85	4.77
	Forest aggregation	0.00	97.72	80.92	5.64

NOTE: Scales represent radii around the center of each study site. All metrics on 0–100 scale.

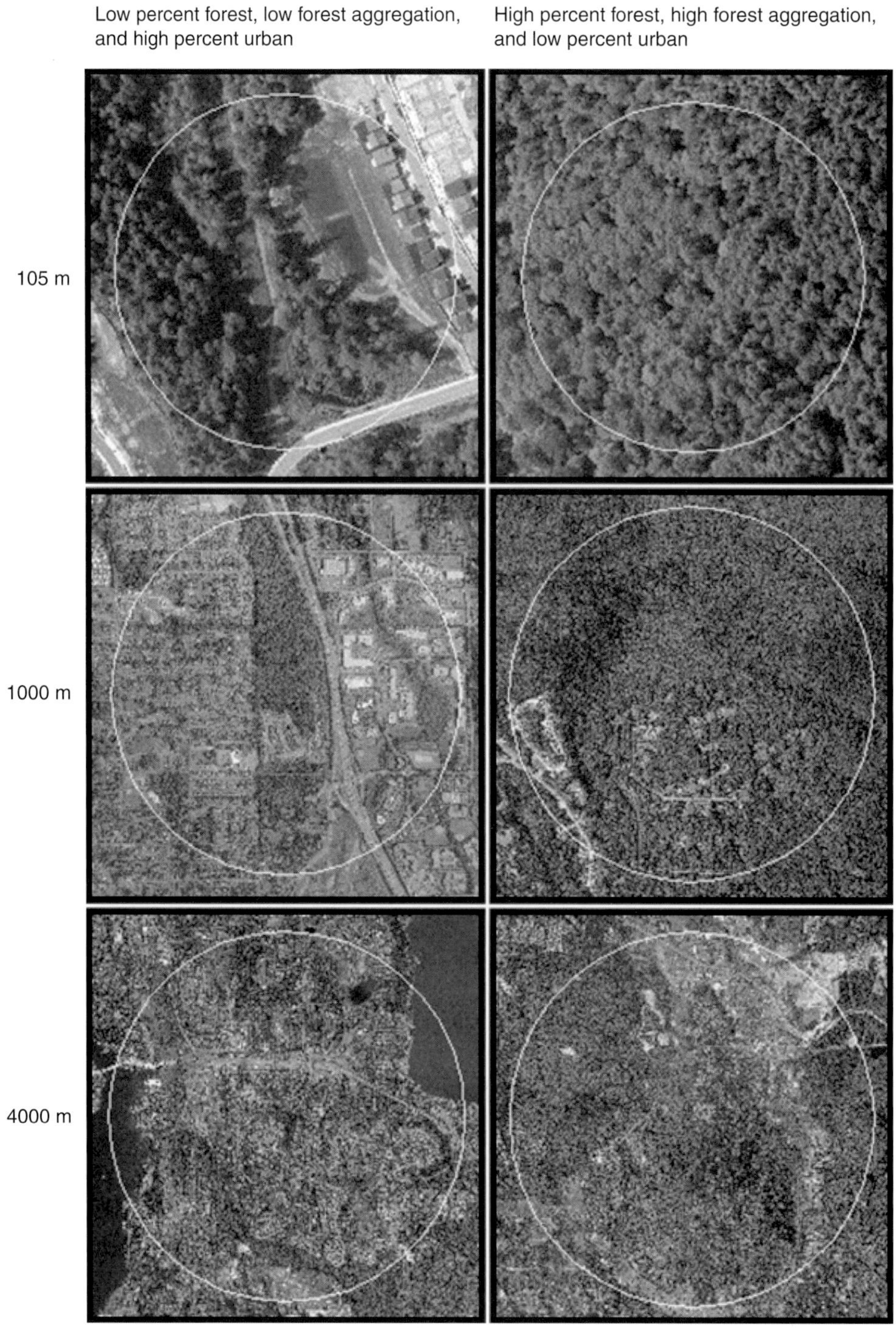

Low percent forest, low forest aggregation, and high percent urban

High percent forest, high forest aggregation, and low percent urban

105 m

1000 m

4000 m

Figure 12.2. Aerial orthophotos of a subset of study sites illustrating variation in land cover patterns with scale (top row: 105 m radius, middle row: 1000 m radius, bottom row: 4000 m radius). Left column shows sites with low percent forest, low forest aggregation, and high percent urban. Right column shows sites with high percent forest, high forest aggregation, and low percent urban. All three landscape metrics are highly correlated at all site scales. For full range of metric values across sites, see Table 12.1.

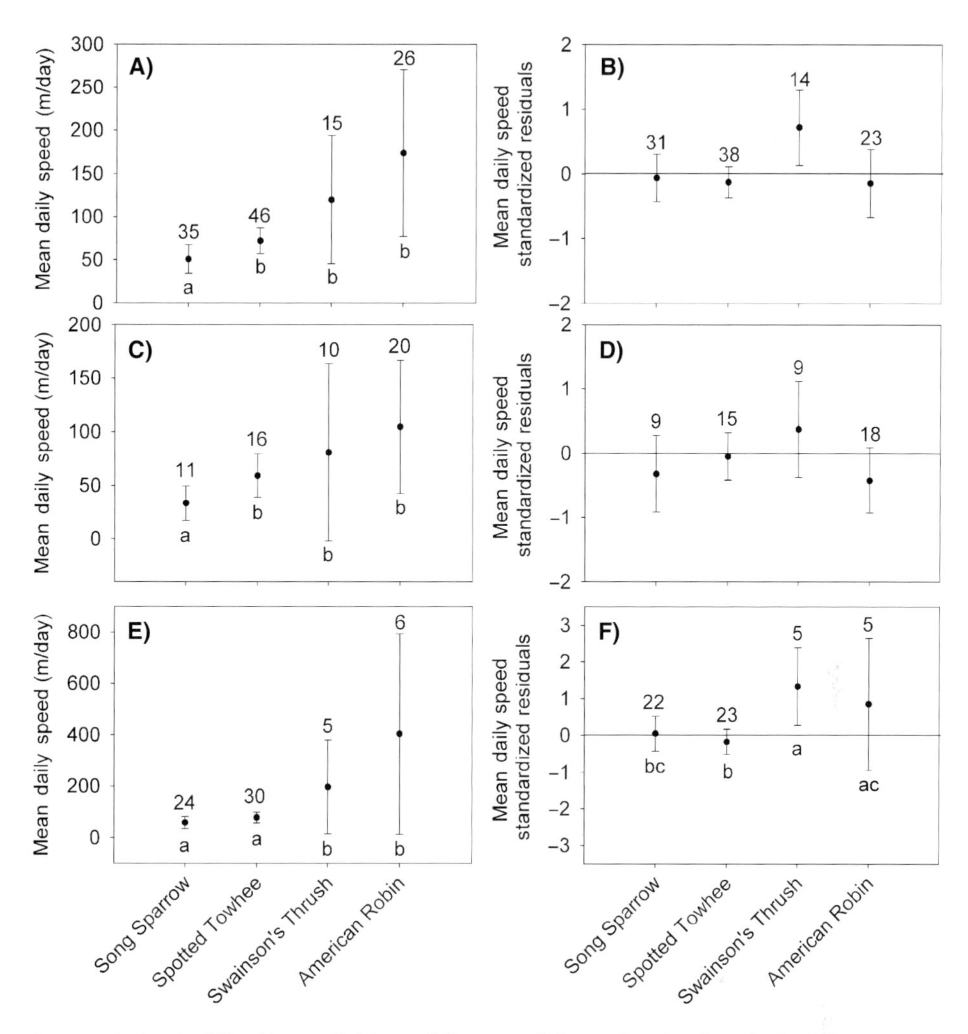

Figure 12.3. Species differed in post-fledging mobility (mean daily speed: m/day) for (a) birds of all ages, (c) fledglings, and (e) juveniles. Species differences were weaker after accounting for variation in body size measured from log-log power functions of all species, shown as standardized residuals of the mean daily speed of (b) birds of all ages, (d) fledglings, and (f) juveniles. Points are means ±95% CI. Numbers above bars indicate sample sizes (number of birds), which are lower on the right side because we did not have measurements of mass for all birds. Letters below bars indicate results of *post-hoc* LSD tests.

greater mean daily speed than juvenile Song Sparrows and Spotted Towhees (LSD tests, all $P < 0.03$; Fig. 12.3a, c, e). Body size was positively correlated with mean daily speed ($r = 0.4$, $n = 106$, $P < 0.01$), as expected from standard allometric relationships (Peters 1983). Therefore, we measured the relative movement of individuals, given their body size, by taking each bird's movement residual from a log-log plot of mean daily speed versus body mass. Using this size-corrected measure of mobility eliminated many of the species differences, with the exception of the juvenile cohort. The mean daily speed of Swainson's Thrush and American Robin juveniles relative to their body size was significantly greater than that of Song Sparrows and Spotted Towhees (Fig. 12.3f). A test of site fidelity revealed predominantly random movement by American Robins (80%) and Swainson's Thrushes (77%) and primarily constrained movement by Song Sparrows (49%) and Spotted Towhees (59%; Table 12.2).

The influence of the amount and pattern of land cover was species- and age-specific (Prediction 2). Song Sparrows were the only species for which measures of mobility covaried with full models of land cover metrics (Table 12.4).

TABLE 12.4

Land cover models and covariates of high relative importance for mobility during the post-fledging period.

Bold indicates partial probability was significant at $P \leq 0.05$ for a 2-tailed test.

Species	Age cohort	Mobility metric	Scale (m)	n (birds)	Adj. r^2	Model P	% Light urban		% Heavy-medium		% Forest urban		Forest aggregation	
							t	$P \leq$	t	$P \leq$	t	$P \leq$	t	$P \leq$
American Robin	Juvenile	MD	250	6	0.99	0.07	3.34	0.19	3.08	0.20	13.44	**0.05**	−12.79	**0.05**
	Juvenile	MD	1000	6	0.99	0.07	5.71	0.11	4.81	0.13	−1.64	0.35	14.18	**0.04**
	Juvenile	MD	2000	6	0.99	0.07	17.79	**0.04**	18.59	**0.03**	19.23	**0.03**	6.65	0.10
	Juvenile	MD	3000	6	0.99	0.07	0.72	0.60	17.77	**0.04**	10.41	0.06	−7.01	0.09
	Juvenile	MD	4000	6	0.99	0.07	13.68	**0.05**	2.43	0.25	−1.97	0.30	6.88	0.09
Song Sparrow	All	MDS	500	35	0.08	0.16	−0.16	0.87	−1.83	0.08	0.77	0.45	−2.57	**0.02**
	All	MD	500	35	0.18	**0.04**	0.18	0.86	−2.40	**0.02**	0.56	0.58	−2.32	**0.03**
	All	TD	500	35	0.08	0.17	0.07	0.94	−2.16	**0.04**	0.20	0.84	−2.17	**0.04**
	Fledgling	TD	105	11	0.61	**0.04**	−2.24	0.07	−3.67	**0.01**	−1.00	0.35	−3.97	**0.01**
	Fledgling	MDS	105	11	0.44	0.11	−1.66	0.15	−3.17	**0.02**	−1.25	0.26	−3.20	**0.02**
	Fledgling	MD	105	11	0.52	0.07	−1.82	0.12	−2.81	**0.03**	−0.79	0.46	−2.86	**0.03**
	Fledgling	MDS	2000	11	0.37	0.16	−1.79	0.12	−1.41	0.21	−0.74	0.49	−2.73	**0.03**
	Fledgling	TD	3000	11	0.38	0.15	1.14	0.30	2.00	0.09	2.78	**0.03**	−1.68	0.14
	Fledgling	MDS	4000	11	0.57	0.06	−2.82	**0.03**	0.72	0.50	1.65	0.15	−2.24	0.07
	Juvenile	TD	500	24	0.08	0.25	0.19	0.85	−2.18	**0.04**	−0.77	0.45	−1.32	0.20
	Juvenile	MDS	500	24	0.10	0.20	0.55	0.59	−2.19	**0.04**	−0.68	0.51	−1.16	0.26
	Juvenile	MD	500	24	0.13	0.16	0.86	0.40	−2.10	**0.05**	−0.66	0.52	−0.96	0.35
	Juvenile	MD	4000	24	0.27	**0.04**	1.05	0.31	−0.94	0.36	−0.50	0.62	0.17	0.87
Spotted Towhee	Juvenile	TD	250	30	0.18	0.06	1.30	0.21	2.70	**0.01**	3.17	**0.00**	−1.38	0.18
	Juvenile	TD	500	30	0.06	0.23	1.33	0.19	1.63	0.11	2.44	**0.02**	−0.79	0.43
	Juvenile	MD	2000	30	0.07	0.21	−1.87	0.07	−1.10	0.28	1.57	0.13	−2.48	**0.02**
	Juvenile	TD	2000	30	0.05	0.28	−1.17	0.25	−0.42	0.68	1.95	0.06	−2.22	**0.04**

MDS = mean daily speed, TD = total distance, MD = maximum distance.

Across age cohorts and scales from 105 to 2,000 m, Song Sparrows' mobility consistently increased with decreasing forest aggregation and percent heavy–medium urban land cover. Additional important land cover covariates for this species included the percent forest (positively related to total distance at the 3,000-m scale) and percent light urban (negatively related to mean daily speed at the 4,000-m scale; Table 12.4). For American Robins and Spotted Towhees, none of the full multiple regression models were statistically significant, but individual covariates were related to different mobility metrics. For example, the maximum distance moved by American Robin juveniles was inversely related to forest aggregation at the 250-m scale, but increased with forest aggregation at the 1,000-m scale (Table 12.4). Maximum distance also increased with the percent forest, heavy–medium urban, and light urban across multiple scales (Table 12.4). Because the juvenile robin sample size is so small ($n = 6$), we interpret the results of their models only to indicate the relative influence of the various metrics, not their absolute effect. The total distance moved by Spotted Towhee juveniles increased with the percent forest and heavy–medium urban at the 250–500-m scales, whereas their total and maximum distances increased with decreasing forest aggregation at the 2,000-m scale (Table 12.4). The mobility of Swainson's Thrushes was unrelated to any land cover variables.

DISCUSSION

Post-fledging mobility appeared to be influenced by species' migratory strategies more than their level of urban affinity (Prediction 1). If migratory strategy is most influential, then the mobility of migratory species should be greater than for resident species (Swainson's Thrush > American Robin > Spotted Towhee = Song Sparrow). If urban affinity is most influential, then the mobility of more urban-affiliated species should be greater than for less urban-affiliated species (American Robin > Spotted Towhee = Song Sparrow > Swainson's Thrush). For birds caught during the most mobile juvenile stage, the observed relative mean daily speed among species was similar before and after mobility was corrected for body size (Swainson's Thrush = American Robin ≥ Spotted Towhee = Song Sparrow; Fig. 12.3e, f). This result suggests that migratory ability influenced juvenile mobility more than urban affinity, because the long-distance migrant (Swainson's Thrush) and short-distance migrant (American Robin) were more mobile than both resident species (Song Sparrows and Spotted Towhees). Phylogenetic inertia also may have contributed to this pattern of mobility, but a better test of this hypothesis would involve more closely related species or subspecies that vary in their migratory strategies. We may have observed the beginning of the migratory period for a few Swainson's Thrushes who made relatively large southward movements before disappearing from the study area, which may have slightly inflated our estimates of mobility. However, the importance of migratory behavior to dispersal was similar in 75 terrestrial bird species of Great Britain, where migratory species dispersed farther than resident species regardless of body size (Paradis et al. 1998).

In addition to the effect of migratory behavior, we suspect species differences in post-fledging movements may also be driven by variation in foraging requirements. Song Sparrows and Spotted Towhees commonly used bird feeders in residential yards, whereas American Robins and Swainson's Thrushes often moved more widely to ephemeral food sources (fruiting trees and shrubs, invertebrate concentrations; K. Whittaker, pers. obs.). Consistent with these observations, American Robins and Swainson's Thrushes exhibited predominantly random movement pathways, and Song Sparrows and Spotted Towhees exhibited primarily constrained movement pathways (Table 12.4). We observed a mixture of fast, linear movements and slow, pulsed movements by all species, which we suspect may have been driven by differences in food resource availability across the landscape. Sallabanks and James (1999) reported that in autumn and winter American Robin juveniles and adults track sources of berries in large flocks. White et al. (2005) found that habitat use by juvenile Swainson's Thrushes in central coastal California was best predicted by fruit abundance variables. Habitat use by juvenile Wood Thrushes (*Hylocichla mustelina*) in the Virginia Piedmont was best described by the optimal foraging hypothesis (Vega Rivera et al. 1998). The movement patterns of juvenile Eastern Meadowlarks (*Sturnella magna*) also seemed to be driven by immediate needs for food resources (Kershner et al. 2004). We suggest that life-history differences between frugivores and granivores influence the extent of movement during the post-fledging period, relative to the spatial pattern of food resources in the urban matrix. Grains provided at feeders were an important driver of spatial food patterns in our study area, suggesting one factor important to birds that is unique to human-dominated landscapes. Our choice of species did not allow us to separate the effects of migratory guild and foraging guild on juvenile mobility, but future research can test for this difference in a different set of study species representing each guild combination.

Post-fledging mobility varied with all land cover patterns examined (Prediction 2), but species and age cohorts showed considerable differences in the landscape metrics and scales at which their movements responded. Movements of the juvenile age cohort were more sensitive to land cover patterns than the fledgling cohort, with nearly twice as many significant effects, but this is to be expected due to the criteria for group definition (pre- versus post-parental independence at the time of capture). The movement patterns of Song Sparrows were more tuned to land cover than were those of other species. We found effects of land cover patterns on mobility at all scales considered, but the most consistently important scales were at 105–500-m radii around study sites. The extent of forest aggregation and heavy–medium urban land cover appear to be the most important variables influencing post-fledging mobility across scales, species, and ages. Swainson's Thrushes were an exception among species in that they showed the highest mobility, but their movements were independent of land cover patterns.

Post-fledging mobility was related to both forest metrics examined (Predictions 2a, b), but not consistently in the predicted direction. Mobility generally increased with the percent forest cover but decreased with increasing forest aggregation to a greater extent across scales and species. Thus, the amount of forest present in the landscape is important for juvenile dispersal, but juveniles of the species we studied do not appear to need highly contiguous forest cover for dispersal. This pattern was especially prevalent in Song Sparrows, which are sometimes classified as an "edge species" for which an interspersed pattern of forest cover would not be expected to hinder movement, except where it is interspersed with heavy–medium urban cover. Alternatively, higher levels of mobility in more fragmented forest landscapes could reflect faster and farther movements through unsuitable habitats such as the urban matrix. Contrasting results were found in a study of bird communities conducted at the same sites as this study, in that habitat quantity and structure (percent land cover) were stronger predictors of species presence or absence than were habitat pattern variables (patch size and aggregation; Alberti and Marzluff 2004, Donnelly and Marzluff 2006). Surveys of forest-breeding bird presence or absence in southwest Ontario and Québec, Canada (Traczinski et al. 1999), and Oregon (McGarigal and McComb 1995) also showed a stronger effect of forest abundance than forest fragmentation on species richness. Other studies have demonstrated reduced dispersal ability of birds in response to habitat gaps or in landscapes with low amounts of habitat remaining, such as agricultural areas (Haas 1995, Brooker et al. 1999, Martin et al. 2006) or forest clearcuts (Dunning et al. 1995, Machtans et al. 1996, Desrochers and Hannon 1997, Rail et al. 1997, St. Clair et al. 1998, Castellón and Sieving 2006), but this is the first study to our knowledge to measure the effects of urban landscape patterns on bird dispersal.

It is important to note that the landscape metrics we included in this study were highly correlated at the scales considered, so the individual effects of different landscape metrics cannot be clearly separated. We interpret these results cautiously because the percent and aggregation of forest may interact to affect movement (Andrén 1994). For example, our study sites average 47% ±5% forest cover with 81% ± 6% aggregation (Table 12.3), and aggregation may not be a limiting factor until the amount of forest drops below a critical threshold level. Results of a linked spatially explicit land use and land cover change–ecological process model predict that the amount of forest in this region will drop from 60% in 2003 to 38% in 2027 (Hepinstall et al. 2009b), largely due to increasing urban cover. Our sites are most representative of suburban and exurban landscapes, so our results may have differed had we measured patterns of bird movement at the more urban end of the gradient. Future studies can be designed to sample populations from more distinctive landscapes to better tease apart the independent effects of various landscape patterns on bird movements.

Both urban land cover classes were related to measures of post-fledging mobility, but the heavy–medium urban class had more consistently negative effects than the light urban class. Negative effects offer support for Prediction 2c, that urban land cover acts as a barrier to bird movement in proportion to its imperviousness, but only for Song Sparrows. It was not surprising that the mobility of American Robins increased with the amount of heavy–medium and light urban cover, considering they were the most urban-affiliated species we studied and commonly utilized lawns as foraging sites. Spotted Towhees' mobility also increased with the amount of heavy–medium urban cover, which could reflect their movements between feeders in urban yards or higher mobility through unsuitable parts of the urban matrix. The light urban land cover class (<50% impervious surface) characterizes the ecotone between forest fragments and residential yards (where supplemental food is often found). We suspect that abundant light urban land cover may lead to low mobility if bird feeders or fruiting trees and shrubs are common, or high mobility if light urban land cover is safer from predation than forest cover. Predation risk is consistent with our observations of feeder use by juveniles in residential yards and observations of the locations of depredation. Only 8% (n = 2/24) of depredation events occurred in residential yards, and the remaining 92% (n = 22/24) of depredations occurred in forest fragments (Whittaker and Marzluff 2009).

A number of parameters contribute to population viability, including reproduction, survival, and dispersal, which often vary in the strength of their effects on population persistence. Dispersal can be an important limiting factor of population persistence (Brooker and Brooker 2002), but in our urban landscape the birds we studied during the post-fledging period are dispersing and surviving well, with the possible exception of Song Sparrows. Juvenile movements were limited in highly urban areas for some species, but low mobility in some species may be explained by the abundance of patchy food resources. We observed only 20% mortality for radio-tracked birds during the post-fledging period (Whittaker and Marzluff 2009). Our study species and other native forest songbird species have average levels of reproductive success (52% of nests and 49% of territories successful; Marzluff et al. 2007). We suspect that adult survival may be a more important factor limiting bird population persistence in this urbanizing region. Of the species in this study, Swainson's Thrushes had the lowest relative abundance, were the least urban-affiliated species, and did not breed at some of our study sites. Yet their young are highly mobile and survive the post-fledging period well (7%, or 1/15 mortalities; Whittaker and Marzluff 2009). Dispersal does not appear to limit their populations. The species we chose to measure in this study may be less sensitive to urban land cover than other less abundant or specialized species, so it would be inappropriate to extrapolate our results to all species. Our findings represent a new direction in urban bird studies in that they fill an important gap in existing urban bird literature in a habitat where songbird post-fledging ecology has not previously been investigated.

MANAGEMENT AND PLANNING IMPLICATIONS

We suggest several strategies for maintaining effective bird dispersal in an urban ecosystem that focus on maximizing the amount of forest and minimizing the amount of impervious urban cover. Habitat improvements can be achieved through restoration of degraded, fragmented habitat in developed areas or maintenance of existing forest cover during future urban development. The latter alternative allows for the most flexibility in design and the least potential conflict with existing land uses. In either scenario, the goal is to maximize the retention of native vegetation and to encourage native landscape plantings to make the matrix conditions more like those of the native habitat fragments (Marzluff and Ewing 2001). Improvement may take the form of stepping-stones (intact forest patches), corridors (linear strips of forest connecting patches), or maintaining forest cover in the urban matrix, whether it be residential, commercial, or industrial (Beier and Noss 1998, Bennett 1999). Our priorities should be focused on species whose mobility is most sensitive to urban land cover (Song Sparrows) as opposed to those which seem to have adapted to moving through the urban matrix (American Robins and Spotted Towhees), because urban cover is expanding while forest cover is shrinking in many of our growing cities.

The most consistently important landscape scales in our study were at 105–500-m radii around study sites. This area corresponds to roughly the area of a small forest fragment or subdivision (3–80 ha), and our findings suggest that maintaining forest cover and contiguity within such an area will benefit the movement of dispersing birds at this critical life-history stage. Landscape patterns at the 500-m radius (or 1 km^2) scale were also related to bird species richness in the same study area (Donnelly and Marzluff 2004, 2006; Blewett and Marzluff 2005). A multiscaled approach is most appropriate because no single strategy will meet the requirements of all species in heterogeneous urban landscapes with differences in the scale, habitat specificity, and mobility among species (Lindenmayer and Franklin 2002, Opdam and Wiens 2002). Providing landscape heterogeneity through a variety of different settlement and open space configurations is needed to conserve the full diversity of native bird species (Donnelly and Marzluff 2004, 2006; Blewett and Marzluff 2005). By managing our forests across multiple scales utilizing multiple strategies by diverse stakeholders, we can facilitate the effective dispersal of birds and maintain viable bird populations in urban ecosystems.

ACKNOWLEDGMENTS

We would like to thank the following people for their help in data collection and netting efforts: J. DeLap, D. Oleyar, S. Rullman, T. Unfried, E. Tompkins, I. Webb, S. Hudson, and R. Silverstein. Many private landowners graciously gave us their permission to conduct research on their property. M. Alberti and J. Hepinstall provided the classified Landsat TM satellite image and technical support. T. Unfried, B. Webb, S. Rullman, and D. Oleyar made helpful comments to this manuscript. Funding was provided by NSF awards BCS0120024 and IGERT0114351 and an EPA STAR Fellowship FP916383.

LITERATURE CITED

Alberti, M., and J. M. Marzluff. 2004. Ecological resilience in urban ecosystems: linking urban patterns to human and ecological functions. Urban Ecosystems 7:241–265.

Andrén, H. 1994. Effects of habitat fragmentation on birds and mammals in landscapes with different proportions of suitable habitat: a review. Oikos 71:355–366.

Beier, P., and R. F. Noss. 1998. Do habitat corridors provide connectivity? Conservation Biology 12:1241–1252.

Bennett, A. F. 1999. Linkages in the landscape: the role of corridors and connectivity in wildlife conservation. IUCN, Gland, Switzerland.

Bélisle, M., A. Desrochers, and M. J. Fortin. 2001. Influence of forest cover on the movements of forest birds: a homing experiment. Ecology 82:1893–1904.

Blewett, C. M., and J. M. Marzluff. 2005. Effects of urban sprawl on snags and the abundance and productivity of cavity-nesting birds. Condor 107: 678–693.

Brooker, L., and M. Brooker. 2002. Dispersal and population dynamics of the Blue-breasted Fairy-Wren, *Malurus pulcherrimus*, in a fragmented habitat in the Western Australian wheatbelt. Wildlife Research 29:225–233.

Brooker, L., M. Brooker, and P. Cale. 1999. Animal dispersal in fragmented habitat: measuring habitat connectivity, corridor use, and dispersal mortality. Conservation Ecology 3:article 4.

Brown, D. G., K. M. Johnson, T. R. Loveland, and D. M. Theobald. 2005. Rural land-use trends in the conterminous United States, 1950–2000. Ecological Applications 15:1851–1863.

Brown, J. H., and A. Kodric-Brown. 1977. Turnover rates in insular biogeography: effect of immigration on extinction. Ecology 58:445–449.

Castellón, T. D., and K. E. Sieving. 2006. An experimental test of matrix permeability and corridor use by an endemic understory bird. Conservation Biology 20:135–145.

Desrochers, A., and S. J. Hannon. 1997. Gap crossing decisions by forest songbirds during the post-fledging period. Conservation Biology 11: 1204–1210.

Donnelly, R., and J. M. Marzluff. 2004. Importance of reserve size and landscape context to urban bird conservation. Conservation Biology 18:733–745.

Donnelly, R., and J. M. Marzluff. 2006. Relative importance of habitat quantity, structure, and spatial pattern to birds in urbanizing environments. Urban Ecosystems 9:99–117.

Dunning, J. B., Jr., R. Borgella, Jr., K. Clements, and G. K. Meffe. 1995. Patch isolation, corridor effects, and colonization by a resident sparrow in a managed pine woodland. Conservation Biology 9:542–550.

Franklin, J. F., and C. T. Dyrness. 1988. Natural vegetation of Washington and Oregon. Oregon State University Press, Corvallis, OR.

Greenwood, P. J., and P. H. Harvey. 1982. The natal and breeding dispersal of birds. Annual Review of Ecology and Systematics 13:1–21.

Haas, C. A. 1995. Dispersal and use of corridors by birds in wooded patches on an agricultural landscape. Conservation Biology 9:845–854.

Hanski, I., and M. E. Gilpin. 1997. Metapopulation biology: ecology, genetics, and evolution. Academic Press, San Diego, CA.

Hepinstall-Cymerman, J., M. Alberti, S. Coe. 2009a. Using urban landscape trajectories to develop a multi-temporal land cover database to support ecological modeling. Remote Sensing 1:1353–1379.

Hepinstall, J., J. M. Marzluff, and M. Alberti. 2009b. Modeling the responses of birds to predicted changes in land cover in an urbanizing region. Pp. 625–659 *in* J. Millspaugh and F. R. Thompson III (editors), Models for planning wildlife conservation in large landscapes. Elsevier Science, Amsterdam, NL.

Hooge, P. N., W. Eichenlaub, and E. Solomon. 1999. The animal movement program. USGS, Alaska Biological Science Center, Anchorage, AK.

Kershner, E. L., J. W. Walk, and R. E. Warner. 2004. Postfledging movements and survival of juveniles Eastern meadowlarks (*Sturnella magna*) in Illinois. Auk 121:1146–1154.

Levins, R. 1970. Extinction. Pp. 77–107 *in* M. Gerstenhaber (editor), Some mathematical questions in biology: lectures on mathematics on the life sciences, Vol. 2. American Mathematical Society, Providence, RI.

Lindenmayer, D. B., and J. F. Franklin. 2002. Conserving forest biodiversity: a comprehensive multiscaled approach. Island Press, Washington, DC.

Macdonald, D. W., and D. D. P. Johnson. 2001. Dipsersal in theory and practice: consequences for conservation biology. Pp. 358–372 *in* J. Clobert, E. Danchin, A. A. Dhondt, and J. D. Nichols (editors), Dispersal. Oxford University Press, New York, NY.

Machtans, C. S., M. A. Villard, and S. J. Hannon. 1996. Use of riparian buffer strips as movement corridors by forest birds. Conservation Biology 10:1366–1379.

Martin, J., J. D. Nichols, W. M. Kitchens, and J. E. Hines. 2006. Multiscale patterns of movement in fragmented landscapes and consequences on demography of the Snail Kite in Florida. Journal of Animal Ecology 75:527–539.

Marzluff, J. M. 2001. Worldwide urbanization and its effects on birds. Pp. 19–47 *in* J. M. Marzluff, R. Bowman, and R. Donnelly (editors), Avian ecology and conservation in an urbanizing world. Kluwer Academic, Norwell, MA.

Marzluff, J. M., R. Bowman, and R. E. Donnelly. 2001. A historical perspective on urban bird research: trends, terms, and approaches, Pp. 1–17 *in* J. M. Marzluff, R. Bowman, and R. Donnelly (editors), Avian ecology and conservation in an urbanizing world. Kluwer Academic Publishers, Norwell, MA.

Marzluff, J. M., and K. Ewing. 2001. Restoration of fragmented landscapes for the conservation of birds: a general framework and specific recommendations for urbanizing landscapes. Restoration Ecology 9:280–292.

Marzluff, J. M., J. C. Withey, K. A. Whittaker, M. D. Oleyar, T. M. Unfried, S. Rullman, and J. DeLap. 2007. Consequences of habitat utilization by nest predators and breeding songbirds across multiple scales in an urbanizing landscape. Condor 109:516–534.

McFadzen, M. E., and J. M. Marzluff. 1996. Mortality of Prairie Falcons during the fledging-dependence period. Condor 98:791–800.

McGarigal, K., S. A. Cushman, M. C. Neel, and E. Ene. 2002. FRAGSTATS: spatial pattern analysis program for categorical maps. University of Massachusetts, Amherst, MA. <www.umass.edu/landeco/research/fragstats/fragstats.html>.

McGarigal, K., and W. C. McComb. 1995. Relationships between landscape structure and breeding birds in the Oregon coast range. Ecological Monographs 65:235–260.

Merriam, G. 1991. Corridors and connectivity: animal populations in heterogeneous environments. Pp. 133–142 *in* D. A. Saunders and R. J. Hobbs (editors), Nature conservation 2: The role of corridors. Surrey Beatty & Sons, Chipping Norton, New South Wales, Australia.

Opdam, P., and J. A. Wiens. 2002. Fragmentation, habitat loss and landscape management. Pp. 202–223 *in* K. Norris and D. Pain (editors), Conserving bird biodiversity. Cambridge University Press, Cambridge, UK.

Otis, D. L., and G. C. White. 1999. Autocorrelation of location estimates and the analysis of radiotracking data. Journal of Wildlife Management 63: 1039–1045.

Paradis, E., S. R. Baillie, W. J. Sutherland, and R. D. Gregory. 1998. Patterns of natal and breeding dispersal in birds. Journal of Animal Ecology 67:518–536.

Peters, R. H. 1983. The ecological implications of body size. Cambridge University Press, Cambridge, UK.

Pulliam, H. R. 1988. Sources, sinks, and population regulation. American Naturalist 132:652–661.

Pulliam, H. R., and B. J. Danielson. 1991. Sources, sinks, and habitat selection: a landscape perspective on population dynamics. American Naturalist 137: S50–S66.

Rail, J. F., M. Darveau, A. Desrochers, and J. Huot. 1997. Territorial responses of boreal forest birds to habitat gaps. Condor 99:976–980.

Rappole, J. H., and A. R. Tipton. 1991. New harness design for attachment of radio transmitters to small passerines. Journal of Field Ornithology 62:335–337.

Robinson, L., J. P. Newell, and J. M. Marzluff. 2005. Twenty-five years of sprawl in the Seattle region: growth management responses and implications for conservation. Landscape and Urban Planning 71:51–72.

Sallabanks, R., and F. C. James. 1999. American Robin (*Turdus migratorius*). No. 462 *in* A. Poole and F. Gill (editors), The Birds of North America. The Birds of North America, Inc., Philadelphia, PA.

St. Clair, C. C., M. Belisle, A. Desrochers, and S. Hannon. 1998. Winter responses of forest birds to habitat corridors and gaps. Conservation Ecology 2:article 13.

Trzcinski, M. K., L. Fahrig, and G. Merriam. 1999. Independent effects of forest cover and fragmentation on the distribution of forest breeding birds. Ecological Applications 9:586–593.

Vega Rivera, J. H., J. H. Rappole, W. J. McShea, and C. A. Haas. 1998. Wood Thrush postfledging movements and habitat use in northern Virginia. Condor 100:69–78.

Whittaker, K. A., and J. M. Marzluff. 2009. Species-specific survival and relative habitat use in an urban landscape during the postfledging period. Auk 126:1257–1276.

Wiens, J. A. 2001. The landscape context of dispersal. Pp. 96–109 *in* J. Clobert, E. Danchin, A. A. Dhondt, and J. D. Nichols (editors), Dispersal. Oxford University Press Inc., New York, NY.

White, J. D., T. Gardali, F. R. Thompson III, J. Faaborg. 2005. Resource selection by juvenile Swainson's Thrushes during the postfledging period. Condor 107:388–401.

Zar, J. H. 1996. Biostatistical analysis. 4th ed. Prentice Hall, Upper Saddle River, NJ.

PART THREE

Human-Avian Interactions and Planning

CHAPTER THIRTEEN

Avian Conservation in Urban Environments

WHAT DO ECOLOGISTS BRING TO THE TABLE?

James R. Miller

Abstract. Many of the papers describing urban bird studies are published in journals that require authors to emphasize the practical implications of their work for practitioners. I surveyed papers published from 1996 to 2006 that focus on avian ecology in urban areas in North America ($n = 51$) to assess (1) what aspects of avian ecology and features of urban environments were studied, (2) which factors in the design of these studies could provide links to practice, and (3) the utility of guidelines and recommendations for the target audiences. Most of these studies (85%) focused on habitat use, and 35% addressed topics related to reproduction. More papers (59%) included study sites on undeveloped land than any other type of land use or land cover. Eighty percent of the articles examined the effects of both local habitat features and the surrounding landscape on bird communities, while the remainder focused on one of these two scales. The majority of the studies (67%) seemed to be designed to meet the needs of practitioners, but in most cases these needs appeared to be assumed by the researchers. Less than 20% of the papers cited planning documents, and only 6% indicated that practitioners had been consulted during the study. Of those papers offering guidelines, the majority targeted land managers (69%) and/or planners (66%). Eighty-eight percent of the papers offering recommendations included at least some advice on features not examined in the study, and there was a mismatch in several papers between the spatial scale of the recommendations and either the scale of the study or the scale within which the target audience tends to operate. About a third of the papers included guidelines that were quite specific numerically, and the majority attached caveats to their advice, based on the limitations of their data. Overall, these results indicate that although some authors take great care in designing their studies to meet the needs of practitioners and devising meaningful recommendations to meet these needs, there is much room for improvement.

Key Words: conservation implications, land use planning, literature survey, management recommendations, urban ecology.

Miller, J. R. 2012. Avian conservation in urban environments: what do ecologists bring to the table? Pp. 201–214 *in* C. A. Lepczyk and P. S. Warren (editors). Urban bird ecology and conservation. Studies in Avian Biology (no. 45), University of California Press, Berkeley, CA.

Ecological research on birds in urban environments dates at least to Pitelka's (1942) comparison of the avifauna in a Southern California resort town with the bird community in the surrounding hills. There was subsequently little interest in this area of research until the 1970s, when a sharp rise in published studies of urban impacts on birds was followed by another strong surge in the following decade (Marzluff et al. 2001a). In the 1980s, coincident with the ascendance of conservation biology (Soulé and Wilcox 1980, Soulé 1985), some authors departed from the template established by Pitelka of simply describing avian response to patterns of development and began to include recommendations for maintaining or enhancing populations of native birds in cities and suburbs (Askins and Philbrick 1987, DeGraaf 1987, Johnson 1988, Adams and Dove 1989). This trend has accelerated in recent years, as researchers have increasingly sought not just to understand the ecology of birds in urban environments, but also to devise strategies for their conservation (Marzluff et al. 2001b).

Recommendations for conservation action typically target practitioners such as policymakers, planners, and natural resource managers. A substantial portion of the literature on birds and urbanization is found in journals which are trying to cultivate a readership that includes these practitioners. Authors submitting manuscripts to these journals are therefore strongly encouraged (if not required) to emphasize the practical implications of their work for environmental management and policy.

If avian ecologists and the journals that publish their work value conservation in areas of human settlement, it is worth considering ways to improve the quality and relevance of conservation guidelines being proposed so they are of greater utility to practitioners. A logical starting point in this endeavor is to survey the literature, examine the topics being addressed, and pinpoint the strengths and weaknesses of recommendations made thus far. The goal of this paper is to do just that, with the hope that such an examination will inform future efforts. The specific objectives of this study were to: (1) characterize the aspects of avian ecology and features of urban environments being studied in papers published in journals that strongly encourage or require an emphasis on links to practice; (2) identify factors in the design or conceptualization of these studies that link to practice; and (3) determine the number of papers that actually include recommendations or guidelines for practitioners and to assess the utility of this advice for the intended audiences.

METHODS

To identify papers for review, I first examined the mission statements and guidelines for authors provided by major journals that publish avian studies in North America. From these, I selected journals that placed a strong emphasis on the links between research and practice, and the application of research findings to environmental management or policy. This set of selected journals included *Biological Conservation, Conservation Biology, Ecology and Society* (formerly *Conservation Ecology*), *Ecological Applications, The Journal of Applied Ecology, The Journal of Wildlife Management, The Wildlife Society Bulletin, Landscape and Urban Planning, Restoration Ecology*, and *Urban Ecosystems*. I complemented this first set with a second set of journals whose guidelines encouraged the submission of manuscripts that emphasized the application of research findings, but also invited manuscripts with a more theoretical or purely ecological focus. This second set included *The Condor, Ecography, Environmental Management*, and *Landscape Ecology*. I examined volumes of the first set of journals from 1996 to 2006, selecting research papers focused on birds in urban or suburban environments (Set 1). I repeated this process for the second set of journals, but only chose papers that contained recommendations or implications for practice (Set 2).

For both sets, I first characterized each study in terms of the aspects of avian ecology which were addressed (e.g., nest predation, habitat selection) and the land use type(s) where the research was conducted. Examples of the latter might include nature reserves, formal parks, open space and greenways, and residential or commercial areas. I then noted the specific features of these environments that were investigated, such as local habitat structure and composition, landscape structure, or predator assemblages.

To identify factors in the design or conceptualization of these studies that may link to practice, I first tried to determine whether a project was developed to address specific management or policy needs, or whether it was intended to test or advance ecological theory. Whereas the needs of practitioners may be better served by the former, some journals (and funding agencies) tend to favor the latter. I also noted any language suggesting that practitioners had been consulted during the course of a given study and whether any non-academicians, or academicians from other disciplines, had served as co-authors on the paper. References to planning documents (e.g., comprehensive plans, open space management plans, state wildlife action plans, and endangered species recovery plans) were also recorded.

Next, I examined each of the papers in Set 1 to determine whether they included recommendations and/or implications for practice (those in Set 2 did by definition). To assess the utility of recommendations offered in papers from both sets, I first asked whether the guidelines were clearly stated and free of jargon so that a non-ecologist would be likely to understand them. I also estimated the percentage of the text (excluding literature cited and appendices) devoted to guidelines in each paper, as it is one measure of the effort invested by the author in devising guidelines, although admittedly the relationship is not necessarily linear. A few sentences may convey very useful information in some cases, whereas a longer interpretation of results from a poorly designed study might contain little of practical use. Here, I was most interested in the tail of the distribution, or instances in which effort was truly minimal.

I then asked several questions to assess the degree of correspondence between the results of the study and the recommendations or guidelines being offered. Was inference extended beyond the spatial extent of the study area, and if so, was it reasonable to do so? Did the recommended guidelines refer to features of urban areas not examined in the study, and again if this was the case, could it be reasonably justified? Finally, did the spatial scale represented in the guidelines reflect the scale at which the study was conducted?

Each paper was examined to see if there was a clear indication of the intended audience for the recommendations. Next, I tried to determine if the spatial scale of the recommendations was reasonably well matched to the scope within which that audience operates.

Ecologists have a reputation for being somewhat vague regarding the particulars of their recommendations—the "it depends" syndrome (Perlman and Milder 2005). Although conservation scientists are understandably wary about providing rules of thumb, given the dynamic and unpredictable nature of ecological systems (Soulé and Orians 2001), this reluctance is a source of frustration for

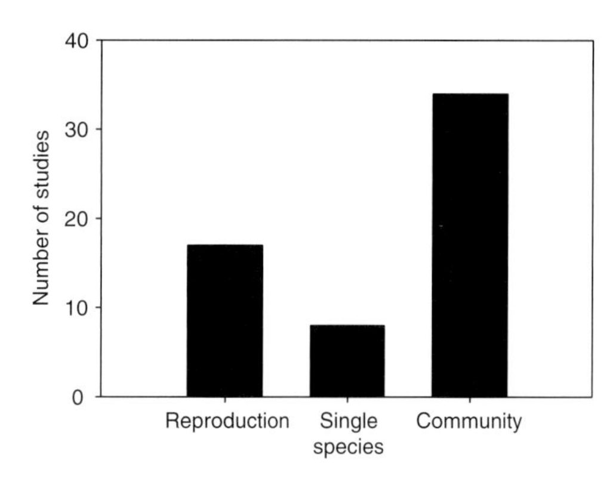

Figure 13.1. Topics in avian ecology addressed in studies conducted in urban and suburban environments in North America, published from 1996 to 2006 ($n = 49$). Some papers addressed >1 topic.

practitioners (Musacchio 2008). I therefore recorded instances in which guidelines were specific, particularly regarding the inclusion of numerical values for patch sizes, buffer widths, and the like. Finally, because researchers are in the best position to recognize the limitations of their work, I noted whether any caveats were offered to help non-ecologists interpret the results on which recommendations were based.

RESULTS

Topics and Study Locations

A total of 51 articles met the selection criteria, with Set 1 comprising 45 articles and Set 2 comprising six articles (see Appendix 13.1 for complete citations). Two of the 51 articles were essentially reviews of other studies encompassing a variety of topics in urban and suburban areas (Marzluff and Ewing 2001, Chace and Walsh 2006). Of the remaining 49 articles, the topic most often addressed in terms of avian ecology was habitat use, with 69% of the articles focused at the community level and 16% examining a single species or, in one case, a single taxonomic group—corvids (Marzluff and Neatherlin 2006; Fig. 13.1). Twenty-one of the community-level articles included analyses for individual species; some of these and the other 13 community-based articles also conducted guild-based analyses or emphasized species diversity. Other topics addressed all fell under the general theme of avian reproduction (Fig. 13.1). Specifically, eight articles focused on nest success (of which two also examined annual productivity and one tracked fledgling survival) and nine articles described studies of nest predation (of which

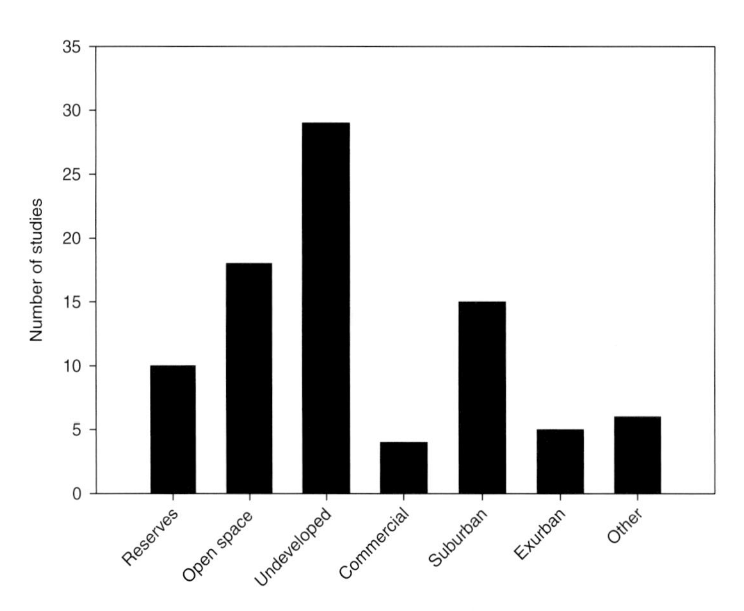

Figure 13.2. Study site locations in urban and suburban environments for papers focused on avian ecology in North America and published from 1996 to 2006 ($n = 49$). Some studies were conducted in >1 location.

two also examined nest parasitism, as did two additional investigations).

More articles included study sites on undeveloped land, both public and private, than any other type of land use or land cover (Fig. 13.2). The next two leading land use categories were open space (which included parks and greenways) and suburban areas, both investigated by approximately one-third of the 49 studies. Twenty percent of the articles reported study sites in nature reserves, and about 10% conducted work in commercial development or exurban areas. Other articles included golf courses ($n = 4$), or simply characterized their sites as encompassing all urban areas ($n = 3$) or the urban-rural gradient ($n = 2$).

Regarding features of the urban environment that were examined, the effects of local habitat characteristics (vegetation features, aspects of the built environment, etc.) and landscape structure (patch size, distance to edge, isolation, etc.) on birds were each quantified in nearly two-thirds of the articles (Fig. 13.3). Most of these articles examined features at both scales, but ten articles focused solely on one or the other. In 11 articles, study sites were characterized on the basis of land use or cover, but specific features of these sites or the surrounding landscape matrix were not quantified. Predator assemblages, food resources, and the effects of human activity were also examined, but to a lesser degree (Fig. 13.3).

Links to Practice in Study Design

Of the 51 articles, seven included explicit statements indicating that the study was in response to the stated needs of managers, policymakers, or planners. In some cases, this involved quantifying the association between avian abundance and different habitats or land uses (Germaine et al. 1998, Haire et al. 2000), and using such data to develop predictive models for avian response to development based on habitat use (Lussier et al. 2006) or reproductive success (Millsap and Bear 2000). Other articles were more narrowly focused, comparing features of specific development types (Lenth et al. 2006) or striving to inform policy (Chace et al. 2003, Blewett and Marzluff 2005). Twenty-seven additional articles could be construed as responding to the needs of practitioners, but this seemed to be an assumption on the part of the authors. Many of the topics addressed in these articles were similar to those described above. Others explored themes that could potentially be of interest in a management or planning context, ranging from the use of surrogate or indicator species (Blair 1999, Nelson and Nelson 2001) to establishing a disturbance zone for individual houses (Odell and Knight 2001) to the effects of recreational trails (Russo and Young 1997, Miller and Hobbs 2000). Whether or not practitioners agreed that these topics were research priorities was unclear.

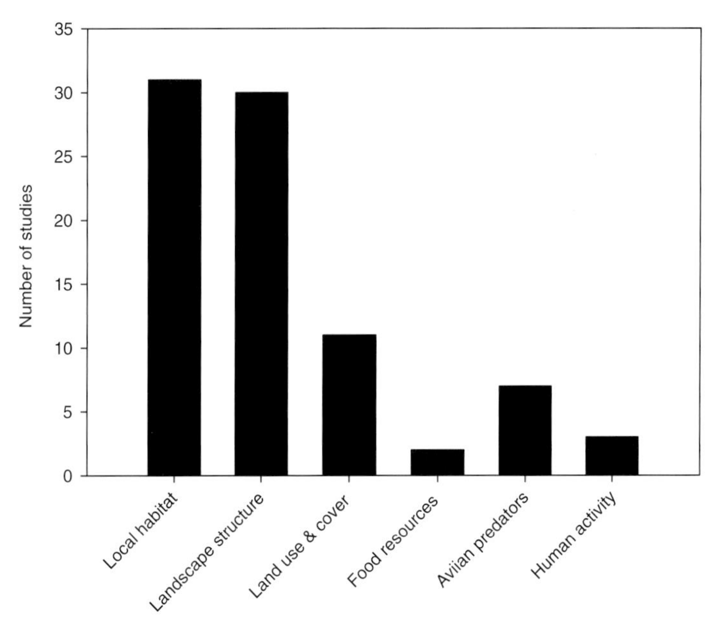

Figure 13.3. Features of urban and suburban environments examined in papers focused on avian ecology in North America and published from 1996 to 2006 ($n = 49$). In some studies, >1 feature was examined.

In six articles, the authors were explicit in stating that their studies were intended to test or extend specific theories, and for the most part made little reference to the needs of practitioners in framing their objectives. These articles focused on island biogeography (Crooks et al. 2001, Marzluff 2005), source-sink and ecological trap dynamics (Vierling 2000, Leston and Rodewald 2006), and the relationship between spatial scale and habitat selection or habitat use (Cam et al. 2000, Hostetler and Holling 2000). The author of one article (Marzluff 2005) was explicit in his intention to explore island biogeography theory, while at the same time addressing themes that were clearly oriented to the needs of practitioners. Nine more articles seemed to have a more theoretical orientation, again with little reference to management or design, but the specific theory being tested was unclear.

Nine articles cited references of the sort that commonly describe the goals or research needs of managers, planners, and policymakers. These citations included municipal or county open space planning documents and conservation plans from state (e.g., Department of Natural Resources) or federal (e.g., U.S. Forest Service) agencies, or nongovernmental organizations (NGOs; e.g., Partners In Flight, The Nature Conservancy). Three articles included authors from the faculties of university planning departments. In addition, non-academics were listed as authors on an additional 15 articles. The majority of these non-academic authors appeared to have research appointments with federal agencies ($n = 12$); also included were an employee of a state Fish and Game Department, an NGO employee, a consultant, and a planner. The authors of four articles stated that they had consulted with NGOs, state conservation agencies, or planners during the course of their study.

Utility of Recommendations

Of the 45 articles in Set 1, 13 offered no recommendations to practitioners, nor was there a discussion of the conservation implications of the work. Of the remaining 38 articles (32 in Set 1 and 6 in Set 2), four dedicated ≤1% of the article length to recommended actions, which was equivalent to one or two sentences of recommendation, and notably lacked substance and seemed tacked on as an afterthought. The largest portion of an article dedicated to recommendations or guidelines was 33% in the review by Marzluff and Ewing (2001). In contrast, the other review (Chace and Walsh 2006) offered no specific recommendations to practitioners, but rather summarized the findings of a variety of avian studies conducted in urban areas. Aside from Marzluff and Ewing (2001), the average portion dedicated to recommendations or guidelines in articles that provided them was approximately 4%. In a ten-page article (excluding references and appendices), this is about two paragraphs.

Twenty-eight of the articles that offered recommendations extended inference from their results beyond the spatial extent of their study. These inferences seemed reasonable in the majority of cases, as the recommendations were extrapolated to other urban areas in the same ecoregion or were general enough that they would likely apply in a wide range of situations. There were, however, a number of articles that did not restrict their guidelines to the type of land cover represented in their study, and this may lead to unintended consequences. For example, guidelines derived from research conducted in a forested matrix may be inappropriate in a region dominated by grasslands.

Authors of 20 articles made recommendations for practices or features in metropolitan environments that were not examined in their particular study. These were reasonable in most cases, such as advocating regional-scale conservation planning (Fraterrigo and Wiens 2005) or greater levels of collaboration with stakeholders (Marzluff 2005). The authors of six articles recommended a greater emphasis on outreach or education; although this advice did not have direct empirical support in these studies, it would be difficult to argue against it.

Another unstudied feature that was emphasized in the guidelines of several articles was native vegetation. Again, few would argue against retaining native plants, but recommendations to plant more native vegetation is of limited value in the absence of details on the particular species and locations for planting (Germaine et al. 1998, Clergeau et al. 2001, Melles et al. 2003, Turner 2003). Restoring additional habitat features or ecological processes to suburban or exurban areas was likewise advised, although no data had been collected or analyzed on these factors, and again, specifics to guide the restoration process were lacking (Clergeau et al. 2001, Melles et al. 2003, Lenth et al. 2006). Some studies conducted in reserves included suggestions pertaining to development patterns or landscaping in the matrix (Donnelly and Marzluff 2004, Smith and Wachob 2006), whereas others conducted in developed areas offered advice on habitat features on reserved land (Germaine et al. 1998). Some authors cited additional studies in support of such recommendations (Germaine et al. 1998, Clergeau et al. 2001, Lenth et al. 2006). Yet some did not (Melles et al. 2003, Turner 2003), suggesting that practitioners should accept the advice on faith.

More than half of the articles that offered recommendations directed their remarks to land planners and nearly as many targeted land managers (Fig. 13.4). Other target audiences included homeowners, developers, policymakers, educators, graduate students, other ecologists, and the conservation community in general. It was unclear who the intended audience was in six of the articles.

The content of the recommendations seemed reasonably well matched to the audiences in most cases. There were, however, several exceptions to this pattern. For example, some authors suggested that natural resource managers can determine the type or pattern of residential development in an area (Lussier et al. 2006). Land managers and urban planners were advised to restrict bird feeders and water usage for landscaping in residential areas (Chace et al. 2003) and to increase canopy cover on private land (Hennings and Edge 2003). While such guidelines may have merit, it seems unlikely that they would fall under the purview of these practitioners. Nor do city planners typically seem well suited to the task of identifying key habitat features (Forrest and St. Clair 2006). The authors of several articles addressed their recommendations to regional planners (Blair 1996, Fraterrigo and Wiens 2005, Lenth et al. 2006), but the level of government at which these planners operate was unclear, with the exception of Lenth et al. (2006), who noted their involvement with planners at the county level. Because a regional approach to land use planning is increasingly rare (Duerkson and Snyder 2005, Lawrence 2005), in the absence of further details I wonder if such recommendations really reflect wishful thinking.

Of the 38 articles that offered recommendations, 15 examined factors at both local and landscape scales, and suggested guidelines at both scales. Authors of an additional nine articles that were conducted at both scales elected to provide guidelines at only one of the two scales. Eight articles examined only local factors and one only landscape features, and the associated recommendations were matched accordingly. Four articles focused at local scales but included guidelines that addressed landscape or regional goals (Blair 1996, Nelson and Nelson 2001, Odell and Knight 2001, Lenth et al. 2006), and one article conducted at landscape scales offered advice emphasizing local habitat features (Turner 2003).

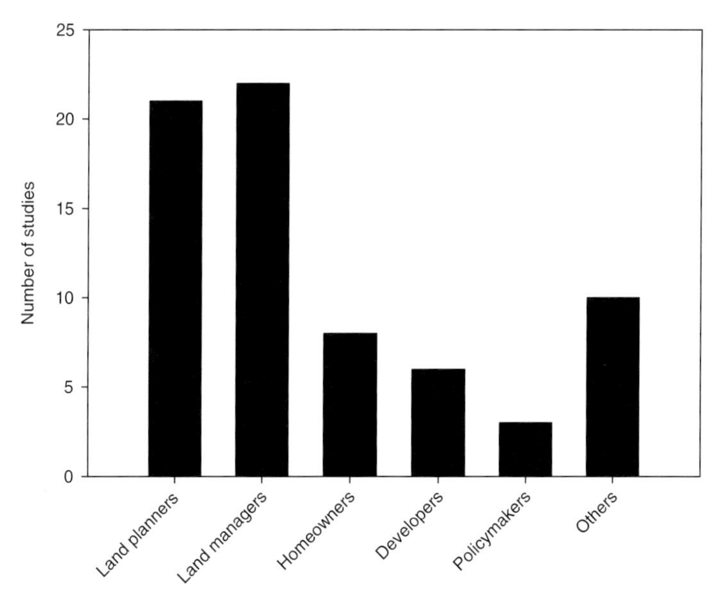

Figure 13.4. The audience targeted by recommendations or implications for practice in papers focused on avian ecology in urban and suburban environments in North America, published from 1996 to 2006 ($n = 51$). Numerous papers provided recommendations to >1 group.

Thirteen articles offered very specific numeric guidelines for local habitat features (Blewett and Marzluff 2005), buffer widths (Millsap and Bear 2000, Odell and Knight 2001, Chace et al. 2003, Hennings and Edge 2003), and habitat patch sizes (Germaine et al. 1998, Donnelly and Marzluff 2004, Blewett and Marzluff 2005). Several articles specified thresholds in spatial scale or the extent of development below which particular conservation goals could be achieved or ecological relationships could be expected to hold (Berry et al. 1998, Stratford and Robinson 2005). Some authors offered numeric guidelines that were less specific. For example, recommendations from two articles each suggested a range of buffer sizes that spanned an order of magnitude (Hostetler and Knowles-Yanez 2003, Dunford and Freemark 2004). For the most part, guidelines were supported by the results of these studies, but in some cases they seem to be overstated. For example, Mancke and Gavin (2000) recommended preserving intact woodlands >5,000 ha because they detected declines in the abundance of a forest species in smaller woodlots with increasing distance from these big woods. It would have been informative to emphasize that the next largest woodlot examined in this study was <500 ha, so the evidence was inconclusive as to whether contiguous wooded areas of, say, 3,000 ha or 4,000 ha may have served the same function.

The authors of 20 articles did attach caveats to the results of their work and the suggestions being made. Most often these caveats highlighted the limitations of abundance data (e.g., abundance may not infer habitat quality or the possible inclusion of unpaired males or immature birds); difficulties in interpretation imposed by correlations among variables; the pitfalls of extrapolating results to other locations, across spatial scales, or to other species; the generality of results from studies of limited duration or time of the year; and the possible influence of unstudied factors.

DISCUSSION

Clearly, not every article focused on avian ecology in urban environments should be required to offer guidance to land managers, planners, and the like. There are plenty of interesting and important questions that need to be addressed but may not directly inform conservation action. What is unclear, however, is why such a large percentage (nearly a third) of the articles published in journals that stress the importance of linking research results to practice apparently did not attempt to do so. To that, add another 10% in which a token effort conveyed the impression that this part of the article was viewed simply as minor hurdle to be cleared on the path to publication.

I initially thought that studies designed to test theories would be less likely to provide recommendations for practice, but this was not the case. In fact, a number of these articles provided suggestions to practitioners, just as a number of more management-oriented articles did not. Particularly noteworthy in this regard was one article (Marzluff 2005) in which the study was designed with both theoretical considerations and the needs of practitioners in mind. Marzluff's (2005) work serves to remind us that it is possible to make headway on both fronts simultaneously.

In reviewing the topics in avian ecology addressed in this collection of articles, there were no obvious patterns suggesting that certain subjects were more conducive to developing guidelines. The predominant emphasis on habitat occupancy in these articles is understandable, as such data are more easily acquired than those necessary to quantify reproductive success or survival. It is also easier to gather information on occurrence or abundance over broad spatial scales, and this is clearly a popular approach in urban studies of birds, as it is in avian ecology generally. Still, more than two decades after the often-cited admonition of Van Horne (1983) that habitat use is not necessarily indicative of habitat quality, Bock and Jones (2004) presented evidence that count data are particularly unreliable for evaluating habitats in areas of human disturbance. And what type of land use or land cover is subjected to human disturbance more than cities? A number of authors said as much in caveats attached to recommendations based on survey data.

It was somewhat surprising that there were not more single-species studies, or studies of smaller groups of ecologically related species. A single-species approach would be more likely to lead to the type of guidelines that practitioners need because a tighter focus should translate to greater detail. The survey data also suggest that there is a need for a greater emphasis on resource use, as this topic was explored in only a few studies, unless one considers vegetation to be synonymous with resources—a tenuous assumption (Hall et al. 1997, Mitchell and Powell 2003, Miller and Hobbs 2007). Again, this sort of information would potentially be of great value to those crafting the specifics of conservation plans, particularly given the novel resources that often exist in human-dominated areas.

A greater balance among the land uses where research is conducted seems warranted. It is likely that the greatest potential for conservation gains is in undeveloped as opposed to developed areas in the metropolitan landscape, at least for most native species. Nevertheless, it is worth exploring whether these prospects can be enhanced by softening the matrix around undeveloped areas, as a number of authors suggested, through more widespread use of native plantings and fewer impervious surfaces or closely cropped lawns. There are design professionals interested in trying to achieve this goal of a softer matrix, but again they need detailed recommendations on exactly what this sort of matrix would look like.

One obstacle that may limit the capacity for ecological research to inform the efforts of practitioners has to do with the increased emphasis on broad-scale investigations over the last decade. A broad-scale focus has been enabled by technological advances, particularly in the areas of remote sensing and geographic information systems, and fostered by conceptual frameworks developed in the field of landscape ecology. Much has been gained by broadening our focus spatially, such as a deeper appreciation for the importance of connectivity and context. These factors are highly relevant in urban and suburban landscapes, given the fine-grained patterns that characterize them. But landscape ecology has also fostered a worldview defined by corridors and internally homogeneous patches (Forman 1995, Dramstad et al. 1996), and this vision is evident in many of the studies reviewed here.

While a patch-matrix-corridor worldview is useful in some ways, it may blind us to the importance of the arrangement of elements, including resources, within patches (Mitchell and Powell 2003). This sentiment is echoed by planners in the state of Washington, the majority of whom worked for city governments, observing that they were mostly involved with site-scale projects, whereas guidelines from the Washington Department of Fish and Wildlife were better suited for landscape-scale efforts (Azerrad and Nilon 2006). Forman (2002) acknowledged that a landscape perspective must be balanced with careful attention to site ecology, but our understanding of the latter is still rudimentary for the majority of species.

One site-scale element that receives much attention from planners and designers is human activity, or the ways that people use or move through space. Few authors explicitly addressed human activity in their study designs or recommendations, which was unexpected. Perhaps a lack of attention stems from our empirical roots, as humans were rarely a consideration in field ecology historically, but this sense of unfamiliarity may indicate a reason to collaborate with social scientists. In any event, it is hardly a revelation to note that people are a defining feature of commercial and residential areas, and it is also true that undeveloped sites where avian conservation may be a goal are also often dedicated to recreational activities. There is much opportunity here, as the notion of completely excluding people from bird habitats has its limitations as a workable strategy in areas of human settlement. Excluding people from bird habitats may be appropriate for species that are threatened or endangered, but in many cases avian conservation may be better served in a broad sense by trying to create opportunities for people to experience native birds close to home and on a regular basis (Pyle 1993, Miller and Hobbs 2002, Turner et al. 2004, Miller 2005, Stokes 2006). Some have gone so far as to suggest that the future of many native species may depend on such interactions (Dunn et al. 2006).

Recommendations

Based on the results of this literature survey, it is clear that there is much room for improvement in developing stronger links between research and practice. Toward this end, I offer some suggestions for improving the advice that we offer to practitioners.

- *Keep the target audience in mind.* Identify the practitioners you wish to address early on. A nice example is provided by Lenth et al. (2006), who identified their audience (county planners) during the early phases of the project and sought input from them when designing the study. A high level of involvement could go a long way toward ensuring that the products of a study are relevant to the needs of the target audience.
- *Know something about their world.* The recommendations reviewed here sometimes conveyed the impression that ecologists cannot be expected to delve deeply into the nature of their target audience's profession. To negate this impression, we must begin with the training of our students. As Marzluff (2005) noted, there are a growing number of programs whose aim is to foster a high level of interaction among students in ecology, planning, design, policy, and so on. The next best option would be individual courses where these students could share their perspectives and interact in a collaborative fashion, and this approach could also enrich undergraduate programs in applied ecology. Those of us who were not fortunate enough to have had these opportunities in our post-secondary education must get up to speed on our own. To help achieve this goal, there are a growing number of ecologist-friendly publications that are useful primers on land use planning and policy (Duerkson et al. 1997, Nolan 2003, McElfish 2004).
- *Provide information that's relevant to that world.* Identifying land managers as the target audience and then recommending ways to landscape residential properties, or targeting planners and offering guidelines for managing reserves tends to convey a sense of carelessness and seems unlikely to inspire action on the part of practitioners. It may also serve to erode the perceived value of the more relevant recommendations being offered. Be sure that recommended actions can be implemented within the sphere in which the target audience operates. Along these same lines, ecologists should be aware of any stated research needs in regional or local planning documents. A particularly useful source of information may be state wildlife action plans, which have been developed for all 50 states and list research needs by habitat type as well as species distributions and ways to mitigate threats to biodiversity posed by development.
- *Make recommendations as specific as possible.* Ecologists work in complex and dynamic systems, perhaps none more so than urban environments. We are conditioned to be more comfortable dealing with probabilities than with certainty (Perlman and Milder 2005). Nonetheless, there is an inverse relationship between the usefulness of the guidelines we propose and their ambiguity. Practitioners need specifics, and to the best of our ability, it is up to those conducting the research to develop recommendations and then offer whatever caveats seem warranted.
- *Consider the limitations of research methods and attach caveats to the guidelines as appropriate.* Several of the investigations reviewed here were remarkable in the comprehensive nature

of their design, involving data collection for both avian abundance and demography, habitat assessments at multiple scales, and other key factors. Yet no single study can cover all the bases, and no matter how good the data may be, recommended actions may fail to produce the desired effect in another location or under different conditions. It is therefore crucial to provide an honest assessment of the limitations of one's study and the conditions under which patterns in the results may not hold. Caveats reinforce the notion that while recommendations are based on a careful analysis and interpretation of the data and one's best professional judgment, monitoring of outcomes after implementation is still necessary.

- *Stick to the data at hand.* Or at the very least, provide supporting references for recommendations that are relevant to the study but go beyond the data that were collected. Be mindful that suggested actions considered *de rigueur*, such as "restore native vegetation," may not be what is most needed in a given situation and may divert attention and resources from more pressing concerns.
- *Avoid laundry lists.* There was a tendency on the part of some authors to end their papers with advice that covered any and all actions that could potentially improve bird conservation in cities and suburbs. Yet guidelines that are all over the map in terms of their scope and the amount of effort required to implement them may be more discouraging than empowering to practitioners. This is not to say that there is a negative correlation between the number of recommendations and their utility, but rather reinforces the notion that we need to consider carefully our target audience and their sphere of influence. It may be helpful to prioritize recommended actions on longer lists and if one activity needs to be predicated on another, to make that clear (Miller and Hobbs 2007).
- And finally, journal editors and reviewers—*hold our (authors') feet to the fire!* What if the same standards applied to methods and data analyses, which tend to be reviewed in great (and sometimes excruciating) detail, were also applied to the sections on "Implications for Practice" or "Conservation Implications"? Greater attention could go a long way toward eliminating the inconsistencies observed in this survey.

I offer these suggestions in full recognition that although some of my own papers were included in this review, they were not shining examples lighting the way forward. My hope is that this assessment will serve to stimulate further dialogue about ways to elevate the relevance of our work for practitioners, and that efforts to conserve avian diversity in metropolitan areas will be enhanced as a result.

ACKNOWLEDGMENTS

D. N. Pennington's assistance in the selection and initial evaluation of some of the papers is very much appreciated. Comments by C. A. Lepczyk and two anonymous reviewers substantially improved an earlier draft of the manuscript. Many thanks to C. A. Lepczyk and to P. S. Warren for organizing the session in Veracruz and for inviting me to participate.

LITERATURE CITED

Adams, L. W., and L. E. Dove. 1989. Wildlife reserves and corridors in the urban environment. National Institute for Urban Wildlife, Columbia, MD.

Askins, R. A., and M. J. Philbrick. 1987. Effect of changes in regional forest abundance on the decline and recovery of a forest bird community. Wilson Bulletin 99:7–21.

Azerrad, J. M., and C. H. Nilon. 2006. An evaluation of agency conservation guidelines to better address planning efforts by local governments. Landscape and Urban Planning 77:255–262.

Berry, M. E., C. E. Bock, and S. L. Haire. 1998. Abundance of diurnal raptors on open space grasslands in an urbanized landscape. Condor 100:601–608.

Blair, R. B. 1996. Land use and avian species diversity along an urban gradient. Ecological Applications 6:506–519.

Blair, R. B. 1999. Birds and butterflies along an urban gradient: surrogate taxa for assessing biodiversity? Ecological Applications 9:164–170.

Blewett, C. M., and J. M. Marzluff. 2005. Effects of urban sprawl on snags and the abundance and productivity of cavity-nesting birds. Condor 107:678–693.

Bock, C. E., and Z. F. Jones. 2004. Avian habitat evaluation: should counting birds count? Frontiers in Ecology and the Environment 2:403–410.

Cam, E., J. D. Nichols, J. R. Sauer, J. E. Hines, and C. H. Flather. 2000. Relative species richness and community completeness: birds and urbanization in the mid-Atlantic states. Ecological Applications 10:1196–1210.

Chace, J. F., and J. J. Walsh. 2006. Urban effects on native avifauna: a review. Landscape and Urban Planning 74:46–69.
Chace, J. F., J. J. Walsh, A. Cruz, J. W. Prather, and H. M. Swanson. 2003. Spatial and temporal activity patterns of the brood parasitic Brown-headed Cowbird at an urban/wildland interface. Landscape and Urban Planning 64:179–190.
Clergeau, P., J. Jokimäki, and J.-P. L. Savard. 2001. Are urban bird communities influenced by the bird diversity of adjacent landscapes? Journal of Applied Ecology 38:1122–1134.
Crooks, K. R., A. V. Suarez, D. T. Bolger, and M. E. Soulé. 2001. Extinction and colonization of birds on habitat islands. Conservation Biology 15:159–172.
DeGraaf, R. M. 1987. Urban wildlife habitat research: application to landscape design. Pp. 107–111 *in* L. W. Adams and D. L. Leedy (editors), Integrating man and nature in the metropolitan environment. National Institute for Urban Wildlife, Columbia, MD.
Donnelly, R., and J. M. Marzluff. 2004. Importance of reserve size and landscape context to urban bird conservation. Conservation Biology 18:733–745.
Dramstad, W. E., J. D. Olson, and R. T. Forman. 1996. Landscape ecology principles in landscape architecture and land-use planning. Island Press, Washington, DC.
Duerksen, C. J., D. L. Elliot, N. T. Hobbs, E. Johnson, and J. R. Miller. 1997. Habitat protection planning: where the wild things are. Planning Advisory Service Report No. 470–471. American Planning Association, Chicago, IL.
Duerksen, C., and C. Snyder. 2005. Nature-friendly communities: habitat protection and land use planning. Island Press, Washington, DC.
Dunford, W., and K. Freemark. 2004. Matrix matters: effects of surrounding land uses on forest birds near Ottawa, CA. Landscape Ecology 20: 497–511.
Dunn, R. R., M. C. Gavin, M. C. Sanchez, and J. N. Solomon. 2006. The pigeon paradox: dependence of global conservation on urban nature. Conservation Biology 20:1814–1816.
Forman, R. T. 1995. Land mosaics. Cambridge University Press, Cambridge, UK.
Forman, R. T. 2002. The missing catalyst: design and planning with ecology roots. Pp. 85–110 *in* B. R. Johnson and K. Hill (editors), Ecology and design: frameworks for learning. Island Press, Washington, DC.
Forrest, A., and C. C. St. Clair. 2006. Effects of dog leash laws and habitat type on avian and small mammal communities in urban parks. Urban Ecosystems 9:51–66.
Fraterrigo, J. M., and J. A. Wiens. 2005. Bird communities of the Colorado Rocky Mountains along a gradient of exurban development. Landsacpe and Urban Planning 71:263–275.
Germaine, S. S., S. S. Rosenstock, R. E. Schweinsburg, and W. S. Richardson. 1998. Relationships among breeding birds, habitat, and residential development in greater Tucson, Arizona. Ecological Applications 8:680–691.
Haire, S. L., C. E. Bock, B. S. Cade, and B. C. Bennett. 2000. The role of landscape and habitat characteristics in limiting abundance of grassland nesting songbirds in urban open space. Landscape and Urban Planning 48:65–82.
Hall, L. S., P. R. Krausman, and M. L. Morrison. 1997. The habitat concept and a plea for standard terminology. Wildlife Society Bulletin 25:173–182.
Hennings, L. A., and D. W. Edge. 2003. Riparian bird community structure in Portland, Oregon: habitat, urbanization, and spatial scale patterns. Condor 105:288–302.
Hostetler, M., and C. S. Holling. 2000. Detecting the scales at which birds respond to structure in urban landscapes. Urban Ecosystems 4:25–54.
Hostetler, M., and K. Knowles-Yanez. 2003. Land use, scale, and bird distributions in the Phoenix metropolitan area. Landscape and Urban Planning 62:55–68.
Johnson, C. W. 1988. Planning for avian wildlife in urbanizing areas in American desert/mountain valley environments. Landscape and Urban Planning 16:245–252.
Lawrence, B. 2005. The context and causes of sprawl. Pp. 3–17 *in* E. A. Johnson and M. W. Klemens (editors), Nature in fragments: the legacy of sprawl. Columbia University Press, New York, NY.
Lenth, B. A., R. L. Knight, and W. C. Gilgert. 2006. Conservation value of clustered housing developments. Conservation Biology 20:1445–1456.
Leston, L. F. V., and A. D. Rodewald. 2006. Are urban forests ecological traps for understory birds? An examination using Northern Cardinals. Biological Conservation 131:566–574.
Lussier, S. M., R. W. Enser, S. N. Dasilva, and M. Charpentier. 2006. Effects of habitat disturbance from residential development on breeding bird communities in riparian corridors. Environmental Management 38:504–521.
Mancke, R. G., and T. A. Gavin. 2000. Breeding bird density in woodlots: effects of depth and buildings at the edges. Ecological Applications 10:598–611.
Marzluff, J. M. 2005. Island biogeography for an urbanizing world: how extinction and colonization may determine biological diversity in human-dominated landscapes. Urban Ecosystems 8:157–177.

Marzluff, J. M., R. Bowman, and R. Donnelly. 2001a. A historical perspective on urban bird research: trends, terms, and approaches. Pp. 1–18 *in* J. M. Marzluff, R. Bowman, and R. Donnelly (editors), Avian ecology and conservation in an urbanizing world. Kluwer Academic Publishers, Norwell, MA.

Marzluff, J. M., R. Bowman, and R. Donnelly (editors). 2001b. Avian ecology and conservation in an urbanizing world. Kluwer Academic Publishers, Boston, MA.

Marzluff, J. M., and K. Ewing. 2001. Restoration of fragmented landscapes for the conservation of birds: a general framework and specific recommendations for urbanizing landscapes. Restoration Ecology 9:280–292.

Marzluff, J. M., and E. Neatherlin. 2006. Corvid responses to human settlements and campgrounds: causes, consequences, and challenges for conservation. Biological Conservation 130:301–314.

McElfish, J. M., Jr. 2004. Nature-friendly ordinances. Environmental Law Institute, Washington, DC.

Melles, S., S. Glenn, and K. Martin. 2003. Urban bird diversity and landscape complexity: species–environment associations along a multiscale habitat gradient. Conservation Ecology 7:5.

Miller, J. R. 2005. Biodiversity conservation and the extinction of experience. Trends in Ecology & Evolution 20:430–434.

Miller, J. R. 2008. Conserving biodiversity in metropolitan landscapes: a matter of scale (but which scale?). Landscape Review 9:167–170.

Miller, J. R., and N. T. Hobbs. 2000. Recreational trails, human activity, and nest predation in lowland riparian areas. Landscape and Urban Planning 50: 227–236.

Miller, J. R., and R. J. Hobbs. 2002. Conservation where people live and work. Conservation Biology 16:330–337.

Miller, J. R., and R. J. Hobbs. 2007. Habitat restoration: do we know what we're doing? Restoration Ecology 15:382–390.

Millsap, B. A., and C. Bear. 2000. Density and reproduction of burrowing owls along an urban development gradient. Journal of Wildlife Management 64:33–41.

Mitchell, M. S., and R. A. Powell. 2003. Linking fitness landscapes with the behavior and distribution of animals. Pp. 93–124 *in* J. A. Bissonette and I. Storch (editors), Landscape ecology and resource management: linking theory with practice. Island Press, Washington, DC.

Musacchio, L. R. 2008. Metropolitan landscape ecology: using translational research to increase sustainability, resilience, and regeneration. Landscape Journal 27:1–8.

Nelson, G. S., and S. M. Nelson. 2001. Bird and butterfly communities associated with two types of urban riparian areas. Urban Ecosystems 5:95–108.

Nolan, J. 2003. Open ground: effective local strategies for protecting natural resources. Environmental Law Institute, Washington, DC.

Odell, E. A., and R. L. Knight. 2001. Songbird and medium-sized mammal communities associated with exurban development in Pitkin County, Colorado. Conservation Biology 15:1143–1150.

Perlman, D. L., and J. C. Milder. 2005. Practical ecology for planners, developers, and citizens. Island Press, Washington, DC.

Pitelka, F. A. 1942. High populations of breeding birds within an artificial habitat. Condor 44:172–174.

Pyle, R. M. 1993. The thunder tree: lessons from an urban wildland. Houghton Mifflin, Boston, MA.

Russo, C., and T. Young. 1997. Egg and seed removal at urban and suburban forest edges. Urban Ecosystems 1:171–178.

Smith, C. M., and D. G. Wachob. 2006. Trends associated with residential development in riparian breeding bird habitat along the Snake River in Jackson Hole, WY, USA: implications for conservation planning. Biological Conservation 128:431–446.

Soulé, M. E. 1985. What is conservation biology? Bioscience 35:727–734.

Soulé, M. E., and G. H. Orians. 2001. Conservation biology research: its challenges and contexts. Pp. 271–285 *in* M. E. Soulé and G. H. Orians (editors), Conservation biology: research priorities for the next decade. Island Press, Washington, DC.

Soulé, M. E., and B. A. Wilcox (editors). 1980. Conservation biology: an evolutionary-ecological perspective. Sinauer Associates, Inc., Sunderland, MA.

Stokes, D. L. 2006. Conservators of experience. Bioscience 56:6–7.

Stratford, J. A., and W. D. Robinson. 2005. Distribution of neotropical migratory bird species across and urbanizing landscape. Urban Ecosystems 8:59–77.

Turner, W. R. 2003. Citywide biological monitoring as a tool for ecology and conservation in urban landscapes: the case of the Tucson Bird Count. Landscape and Urban Planning 65:149–166.

Turner, W. R., T. Nakamura, and M. Dinetti. 2004. Global urbanization and the separation of humans from nature. Bioscience 54:585–590.

Van Horne, B. 1983. Density as a misleading indicator of habitat quality. Journal of Wildlife Management 47:893–901.

Vierling, K. T. 2000. Source and sink habitats of Red-winged Blackbirds in a rural/suburban landscape. Ecological Applications 10:1211–1218.

APPENDIX 13.1

Papers reviewed in this study.

Author(s)	Year	Journal
Berry, M. E., C. E. Bock, and S. L. Haire[a]	1998	Condor 100:601–608
Blair, R. B.	1996	Ecological Applications 6:506–519
Blair, R. B.	1999	Ecological Applications 9:164–170
Blair, R. B.	2004	Ecology and Society 9:2 [online]
Blewett, C. M., and J. M. Marzluff[a]	2005	Condor 107:678–693
Boal, C. W., and R. W. Mannan	1999	Journal of Wildlife Management 63:77–84
Bolger, D. T., T. A. Scott, and J. T. Rotenberry	1997	Conservation Biology 11:406–421
Borgmann, K. L., and A. D. Rodewald	2004	Ecological Applications 14:1757–1765
Chace, J. F., J. J. Walsh, A. Cruz, J. W. Prather, and H. M. Swanson	2003	Landscape and Urban Planning 64:179–190
Chace, J. F., and J. J. Walsh	2006	Landscape and Urban Planning 74:46–69
Clergeau, P., J. Jokimäki, and J.-P. L. Savard	2001	Journal of Applied Ecology 38:1122–1134
Cooper, D. S.	2002	Biological Conservation 104:205–210
Crooks, K. R., A. V. Suarez, D. T. Bolger, and M. E. Soulé	2001	Conservation Biology 15:159–172
Danielson, W. R., R. M. DeGraaf, and T. K. Fuller	1997	Landscape and Urban Planning 38:25–36
Donnelly, R., and J. M. Marzluff	2004	Conservation Biology 18:733–745
Dunford, W., and K. Freemark[a]	2004	Landscape Ecology 20:497–511
Forrest, A., and C. C. St. Clair	2006	Urban Ecosystems 9:51–66
Fraterrigo, J. M., and J. A. Wiens	2005	Landscape and Urban Planning 71:263–275
Gale, G. A., L. A. Hanners, and S. R. Patton	1997	Conservation Biology 11:246–250
Germaine, S. S., S. S. Rosenstock, R. E. Schweinsburg, and W. S. Richardson	1998	Ecological Applications 8:680–691
Green, D. M., and M. G. Baker	2003	Landscape and Urban Planning 63:225–239

APPENDIX 13.1 *(continued)*

APPENDIX 13.1 (CONTINUED)

Author(s)	Year	Journal
Haire, S. L., C. E. Bock, B. S. Cade, and B. C. Bennett	2000	Landscape and Urban Planning 48:65–82
Hennings, L. A., and D. W. Edge[a]	2003	Condor 105:288–302
Hostetler, M.	1999	Landscape and Urban Planning 45:15–19
Hostetler, M., and C. S. Holling	2000	Urban Ecosystems 4:25–54
Hostetler, M., and K. Knowles-Yanez	2003	Landscape and Urban Planning 62:55–68
Kristan, W. B., A. J. Lynam, M. V. Price, and J. T. Rotenberry[a]	2003	Ecography 26:29–44
Lenth, B. A., R. L. Knight, and W. C. Gilgert	2006	Conservation Biology 20:1445–1456
Leston, L. F. V., and A. D. Rodewald	2006	Biological Conservation 131:566–574
Lussier, S. M., R. W. Enser, S. N. Dasilva, and M. Charpentier[a]	2006	Environmental Management 38:504–521
Maestas, J. D., R. L. Knight, and W. C. Gilgert	2003	Conservation Biology 17:1425–1434
Mancke, R. G., and T. A. Gavin	2000	Ecological Applications 10:598–611
Marzluff, J. M.	2005	Urban Ecosystems 8:157–177
Marzluff, J. M., and K. Ewing	2001	Restoration Ecology 9:280–292
Marzluff, J. M., and E. Neatherlin	2006	Biological Conservation 130:301–314
Melles, S., S. Glenn, and K. Martin	2003	Conservation Ecology 7:5 [online]
Miller, J. R., and N. T. Hobbs	2000	Landscape and Urban Planning 50:227–236
Miller, J. R., J. A. Wiens, N. T. Hobbs, and D. M. Theobald	2003	Ecological Applications 13:1041–1059
Millsap, B. A., and C. Bear	2000	Journal of Wildlife Management 64:33–41
Morrison, S. A., and D. T. Bolger	2002	Ecological Applications 12:398–411
Nelson, G. S., and S. M. Nelson	2001	Urban Ecosystems 5:95–108
Odell, E. A., and R. L. Knight	2001	Conservation Biology 15:1143–1150
Rodewald, A. D., and M. H. Bakermans	2006	Biological Conservation 128:193–200
Rottenborn, S. C.	1999	Biological Conservation 88:289–299
Russo, C., and T. Young	1997	Urban Ecosystems 1:171–178
Smith, C. M., and D. G. Wachob	2006	Biological Conservation 128:431–446
Stratford, J. A., and W. D. Robinson	2005	Urban Ecosystems 8:59–77
Traut, A. H., and M. E. Hostetler	2004	Landscape and Urban Planning 69:69–85
Turner, W. R.	2003	Landscape and Urban Planning 65:149–166
Vierling, K. T.	2000	Ecological Applications 10:1211–1218

[a] Papers included in Set 2.

CHAPTER FOURTEEN

How Biologists Can Involve Developers, Planners, and Policymakers in Urban Avian Conservation

Mark Hostetler

Abstract. Conserving avian biodiversity in urban environments is contingent on the collective decisions made by homeowners, developers, and planners/policymakers. A connection exists between habitat selection by birds and the hierarchy of decisions made by different sectors within a society. Often, decisions made by developers and planners/policymakers have the broadest impact on the land and set the stage for how municipalities conserve avian habitat for decades into the future. In this paper, I outline strategies for biologists to engage developers, planners, and policymakers in adopting conservation methodologies that would promote avian diversity. I give background on a University of Florida academic group, the Program for Resource Efficient Communities (PREC), and how they have engaged the design-build community to create more resource-efficient, bird-friendly, residential neighborhoods. The lessons learned from PREC activities could help biologists to engage any community to adopt practices that will conserve urban avian diversity.

Key Words: conservation planning, land use planning, management recommendations, urban ecology.

Conserving avian habitat in urban environments is directly related to the myriad decisions made by people. The decisions made by homeowners, policymakers, and design-build professionals (a term that encompasses firms and professionals that engage in both planning and construction, e.g., developers and landscape architects) interact in dynamic ways to either enhance or inhibit biodiversity (Grimm et al. 2000, Pickett et al. 2001, Hostetler and Knowles-Yanez 2003). For example, developers would have to preserve habitat of good quality and residents that live around these areas would have to appropriately manage their own properties so as not to have a negative impact on these conserved habitats (Fuller et al., chapter 16, Lepczyk et al., chapter 17, this volume). Although many urban studies on birds have been conducted (Marzluff et al. 2000), generating various best design and management strategies, often these strategies are not adopted by policymakers, developers, and homeowners (Hostetler et al., 2008). Even if we know design and management strategies that would increase urban biodiversity or perhaps enhance the success of certain species, how do we help urbanites to adopt these recommendations?

Hostetler, M. 2012. How biologists can involve developers, planners, and policymakers in urban avian conservation. Pp. 215–222 *in* C. A. Lepczyk and P. S. Warren (editors). Urban bird ecology and conservation. Studies in Avian Biology (no. 45), University of California Press, Berkeley, CA.

In particular, how do we assist policymakers to adopt bird-friendly policies, developers to build bird-friendly residential and commercial properties, and homeowners to appropriately manage their yards and neighborhoods?

My goal in this paper is to present some ideas about how scientists could reach policymakers and developers to help a city move toward adopting conservation practices that would benefit birds. To illustrate how scientists could get involved, I give background information about a unique academic group, called the Program for Resource Efficient Communities (PREC) at the University of Florida (UF); this group of scientists is partnering with municipalities and developers to create "green communities." PREC has learned many valuable lessons about reaching developers and policymakers, and I will focus my discussion on how scientists could foster meaningful partnerships to create communities that conserve avian habitat. But before I discuss PREC and how scientists can establish a dialogue with policymakers and developers, I think it is important to discuss the interaction among policymaker, developer, and even homeowner decisions because each decision level can affect the distribution of different types of birds. Effects depend on how broad an area is affected by a decision and how much it limits the opportunities (of other people) to conserve or restore avian habitat within an urban community.

DECISION HIERARCHIES OF PEOPLE AND BIRDS

The physical design of cities is the result of decisions made by various people within a community. I divide the major decision makers into three nested, hierarchical levels where decisions made at one level are constrained by decisions made at other levels. The design and management of a small lot is primarily the decision of a homeowner, but the lot is set in a neighborhood that was formed by the decisions of a developer. In turn, the development of that neighborhood and the surrounding landscape matrix is governed by zoning and regulations set forth by planners and politicians. The decisions made by homeowners are constrained by the decisions of the developer, which are constrained by the decisions of planners and politicians. The collective influences of these decisions impact the landscape from fine (homeowner) to broad (planner/policymaker) scales, and these decisions are modified by socioeconomic components of a city (Grimm et al. 2000, Pickett et al. 2001, Hostetler and Knowles-Yanez 2003).

When selecting habitat, birds also go through a hierarchy of "decisions" where environmental cues at different scales influence the distribution of species across a region (Johnson 1980, Kotliar and Wiens 1990, Holling 1992, Hostetler 2000, Hostetler and Holling 2000, Hostetler and Knowles-Yanez 2003, Pennington and Blair, chapter 2, this volume, González-Oreja et al., online material). When searching for food within an urban area, consider the nested set of hierarchical decisions of a Carolina Wren (*Thyothorus ludovicianus*) and a Red-tailed Hawk (*Buteo jamaicensis*; Hostetler 2000, Hostetler and Holling 2001). Each bird probably responds to landscape attributes at different scales as a function, in part, of its body size, and the decisions made by homeowners, design–build professionals, and planners/policymakers can have different impacts on certain groups of birds (Fig. 14.1). A homeowner's decision to landscape his or her yard could attract small birds such as wrens and hummingbirds; a developer's decision could primarily affect the distribution of medium-sized birds such as thrushes; and policymaker/planner decisions could primarily affect the distribution of large birds such as hawks (Hostetler 2000; Fig. 14.1). Of course, all decision levels have the potential to affect the distribution of any bird, but, as mentioned previously, higher-level decisions made by developers and policymakers can have lasting effects in both space and time. For example, a subdivision that has all lots with 90% lawn, no open space conservation, and deed restrictions preventing people from converting lawns to native vegetation would hinder the creation of medium-sized habitat patches within the neighborhood. In order to promote avian diversity in urban environments, design-build professionals and policymakers/planners are critical sectors of society to reach.

PROGRAM FOR RESOURCE EFFICIENT COMMUNITIES (PREC)

With an eye on how expanding metropolitan areas impact the environment and the livelihoods of people, municipalities are trying to implement

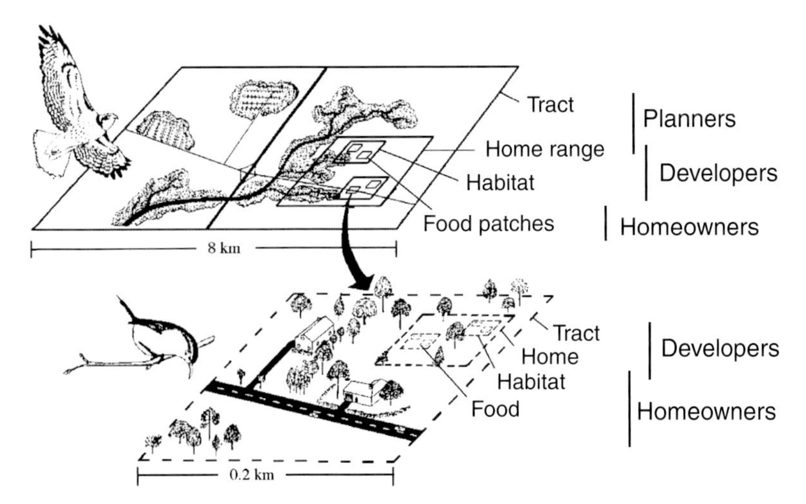

Figure 14.1. Hierarchical responses to landscape structures by a Carolina Wren and a Red-tailed Hawk. This illustration is based on estimated home range sizes of both birds. Responses that occur at broader levels generally have more influence on the distribution of a species. For example, the distribution of landscape features at the tract level for the hawk helps determine where a hawk establishes its home range. Whether a bird appears in a backyard is much more contingent on the features that surround the yard than on what is contained within the yard. Notice that the only overlap in the types of objects sampled by the two birds is at the food patch level for the hawk and at the tract level for the wren. Because planner/policymaker decisions operate at broader levels than those of developers and homeowners, the decisions made by planners and policymakers have the greatest impact on the distribution of Red-tailed Hawks, much more than on the distribution of Carolina Wrens (adapted from Hostetler and Holling 2000).

sustainable practices in their communities (Arendt 1999, Calthorpe and Fulton 2001). With regard to residential developments, best management and design recommendations must be practical and understood by design-build professionals, planners, policymakers, and homeowners in order to have successful implementation (Youngentob and Hostetler 2005). In Florida over the last decade, ~100,000 new single-family, detached homes have been built annually, with over 200,000 homes being built in 2005 (U.S. Census Bureau 2006). Research-based information that could reduce adverse environmental effects of residential developments on birds is contingent on the adoption of best management practices and designs by planners, developers, and even homeowners.

Realizing that much of the sustainability recommendations (based on research) coming out of the University of Florida have had only limited uptake by the public (Hostetler et al. 2008), a group of UF scientists (e.g., energy, wildlife, water, and horticulture specialists) saw the need to form the Program for Resource Efficient Communities (PREC; www.buildgreen.ufl.edu) as a way to interface with the public, particularly design-build professionals and practicing planners and policymakers. PREC's mission is to integrate and apply UF's educational and analytical assets to promote the adoption of best design, construction, and management practices in new residential community developments that measurably reduce energy and water consumption and conserve biodiversity. The program's focus extends from the individual home and lot level through site development to surrounding lands and ecological systems. PREC was established in 2004 to identify and coordinate educational and analytical resources available at UF to support the design, construction, and management of more resource efficient residential developments. Many of our activities include organizing workshops, consultations, and offering continuing education courses. The purpose of these gatherings is not only to convey information and promote networking; we also gather input from practitioners about the barriers to sustainable development and about which practices work when considering economic, political, and cultural factors.

STEPS FOR PARTNERSHIPS TO CONSERVE AND RESTORE HABITAT

Consider the goal of conserving avian habitat in a growing city. Habitat protection and management could enhance overall avian diversity or could be targeted to a particular species. How would one encourage developers and municipalities to conserve such habitat or adopt particular management practices? I discuss below some steps that academics (and others) could take in order to engage various sectors within a growing community. These strategies come from experiences of working with growth management issues in the state of Florida through the PREC program.

Partner with a Developer to Create a Model Community

One of the first steps to reach the development community is to find that first developer who is willing to listen and to try new things. This is an important but difficult step. Having a model community in the area will provide opportunities for other developers to see how conserving bird habitat could be done in today's marketplace. How to find this developer? One way is to interview a municipal regulatory agency or planning board. These regularly work with developers and get to know which ones are sympathetic to environmental issues. Another strategy is to communicate with environmental consultants. Consulting firms know local developers and which projects have the best chance to perhaps save habitat.

Having local examples of developments that incorporated bird-friendly practices goes a long way to promote the conservation of avian habitat. A developer who has tried a certain practice paves the way for other developers to replicate it, especially if the development received favorable press, had an easier time through the permitting process, or saved money when implementing a strategy to conserve habitat. In one example, we collaborated with a developer (Madera) on a community in Gainesville, Florida. Because of this partnership, the developer designed the community around large trees, preserved several forested patches, and kept many snags for the benefit of woodpeckers and other birds. The lots also had minimal lawns, native plants, and preserved much of the existing natural vegetation. As a result of these practices, the developer received local support from nearby neighborhood activists, an easier time in the permitting process, and saved several thousand dollars per lot in terms of landscaping costs (G. Acomb, pers. comm.).

As another example, we collaborated with the Town of Harmony (www.harmonyfl.com) to implement a variety of design and management practices to conserve avian habitat. The developer, for example, created littoral shelf zones in most of the retention ponds across the development, providing foraging access to wading birds. Even the Florida Sandhill Cranes (*Grus canadensis pratensis*), a sensitive species, are nesting in these areas (G. Golgowski, pers. comm.). In particular, the developer left the shoreline around a 500-acre lake undeveloped to enhance wildlife habitat. Because of these environmental measures (and others), the developer had a much easier time going through the development review process from both local and regional agencies (G. Golgowski, pers. comm.). Further, their marketing has a strong environmental theme, highlighting local bird species, and the developers have said this is an important reason why the homes have sold so well.

We currently use the Harmony and Madera communities as examples of how developers can conserve avian habitat in residential communities and have used these examples in several consultations with developers across the state. We have found that it is much easier to make the case for conserving avian habitat when we can show financially successful, local examples of conserved or restored habitat. Sometimes developers and their staff tour the Madera and Harmony communities or they listen to one of our presentations about the practices adopted in these unique communities. As a result, these developers are incorporating bird-friendly practices into plans for their own projects.

Caution should be exercised when engaging certain developers, because some will use wildlife experts to help them through permitting difficulties. For example, a development order (permission to build) may only be issued if the developer comes up with acceptable plans to minimize their impact on the environment. Conserving avian habitat may be one practice that helps the development order to be issued. The developer may or may not be serious about conserving avian habitat and things may go astray once construction begins. How does one ascertain how committed a developer is? First, have a conversation with local authorities to see if there are repercussions in place (e.g., a construction permit is revoked) if a developer does not implement (agreed upon) design and management practices concerning avian habitat. Second, look at the history of this developer and relations with local planning staff. You can learn a lot by how much other people trust them. Third, within the architectural and management plans submitted to a planning board, see if the developer will include specific designs and management practices that would benefit birds.

The above actions will, at the least, provide some resources to work with and a token of commitment shown by the developer. With the Madera development, I had extensive conversations with the developer about on-site avian habitat conservation and the developer did incorporate many of my suggestions in the site layout when plans were submitted to the local planning board. As an example, the developer left a 30-ft vegetative buffer on the backside of lots that bordered a field where Greater Sandhill Cranes (*Grus canadensis*) foraged. This buffer served as a visual barrier to prevent the disturbance of adjacent Sandhill Cranes. If the buffer was not included during construction, the construction permit could be revoked. In the end, there is no formula to guarantee a meaningful relationship with a developer, but from our experience, the more face-to-face time you have, the better able you are to establish a trusting relationship.

All Individuals Must Have Understanding and Buy-in

All individuals associated with the development must understand and have buy-in to conserve avian habitat. Consider the three phases of a development: design, construction, and post-construction. Each phase has individuals who make decisions that determine the success of conserving avian habitat. Typically, the design phase involves architects, landscape architects, civil engineers, and developers. The construction phase involves a host of contractors and subcontractors, and the post-construction phase is dominated by homeowners and the realtors who sell the homes. To conserve avian habitat, a good design needs to be implemented on paper during the planning phase. For the construction phase, contractors need to subscribe to specific practices that will minimize their impact on avian habitat and perhaps resident populations (e.g., proper placement of barriers around trees, minimize activity near nesting or roosting sites). During post-construction, homeowners should be made aware of local birds (Fuller et al., chapter 16, this volume) and the importance of the protected areas so that residents can manage them appropriately (e.g., no ATVs running through preserved habitat). In addition, realtors should be informed about the designs and intent of the neighborhood and convey this to potential homebuyers. If people involved in each of the three stages of building a community have not signed on, the long-term persistence of bird populations and their habitat is doubtful. I stress these phases because in reviewing a number of different development projects in Florida, I have found that few of the site plans include contingencies for proper management during the construction and post-construction phases. Most only stress the design phase.

How to engage people involved with the construction and post-construction phases of a development? First, all built-environment professionals involved with the construction of the development should have meetings at the beginning of the project and at various interludes during construction to review progress and to address problems that arise. These meetings could be facilitated by environmental consultants and/or academics. Second, contractors and other environmental building (sometimes called "green building") professionals should take continuing education classes that help them understand the design features and management practices concerning avian habitat conservation. As academics, we could design and offer such courses to train professionals associated with the development. PREC offers over 25 continuing education courses involving sustainable development issues and practices. Several intersect with avian habitat concerns such as keeping trees alive during construction, preserving wildlife habitat in residential communities, and even outdoor water conservation practices that would prevent runoff in nearby wetlands or consumption of local water sources that could impact water levels in wetlands (see www.buildgreen.ufl.edu/continuing_education.htm). We charge for all classes, and funds generated from such activities are funneled back into a program to fund graduate students, conduct research, to further develop continuing education courses and other outreach activities, such as construction of demonstration sites. To encourage design-build professionals to take such courses, policymakers should create incentives for practitioners to receive training (see further discussion below), and all courses need to be approved for continuing education credits from relevant professional societies. Many professional organizations require that their members take continuing education courses at regular intervals to maintain their accreditation. As an example, all PREC courses have been reviewed and approved by professional society boards, including the Construction Industry Licensing Board (CILB), Board of Architecture and Interior Design (ARID), Building Construction Administrators and Inspectors (BCAI), Electrical Contractors Licensing Board (ECLB), and Board of Landscape Architects (BOLA).

To engage homeowners, the developer must implement a robust education program. A plan should consist of educational signs, a website, and a brochure. Part of the program would address environmental issues concerning bird conservation within a community. In any green development, there is the potential for residents to resume conventional, non-environmentally oriented behaviors (DeLorme et al. 2003, Youngentob and Hostetler 2005). A study in Gainesville, Florida, found that residents of a "green" community lacking an educational program did not differ or scored lower on several questions about environmental knowledge, attitudes, and behaviors than residents of conventional community types (Youngentob and Hostetler 2005). In another study, new homebuyers of green and conventional communities had little difference in terms of environmental knowledge, attitudes, and behavior, and overall measures were quite low for all respondents, regardless of community type (Hostetler and Noiseux 2010). Thus, green communities may not be attracting or encouraging environmentally sensitive residents, and in the absence of an environmental education program, a community probably does not retain the long-term ability to manage avian habitat of good quality. In a recent study that compared residents of a community (Town of Harmony) that had an educational program (website, kiosks, and brochure; www.wec.ufl.edu/extension/gc/Harmony/index.htm) to residents of a community lacking such a program, the educational program did have a positive impact on residents' knowledge. Harmony residents displayed improved awareness about bird conservation and other related natural resource conservation practices (Hostetler et al., 2008). Development of such educational programs is a key role for academics.

Reach Out to Planners and Policymakers

Because planners and policymakers constrain or encourage what developers can do, it is important that policy barriers to unique practices be removed and bird-friendly policies be adopted. With barriers, local ordinances and regulations may prohibit certain sustainable practices, limiting opportunities to conserve avian habitat. For example, Florida law requires that stormwater must be retained on site for a development. Concrete curbs, culverts, and retention ponds are typical practices used to keep water on site. In the Madera development, we promoted alternative practices such as swales, narrow roads, and use of a forested retention area to meet stormwater regulations. All of the above actions not only served the function of keeping water from going off site, but helped provide additional bird habitat through conservation of open space and the use of native vegetation in the swales. These proposals were met with a lot of opposition from the local regulatory agency because they were more familiar with the concrete curbs and retention ponds. The best way to avoid such confrontations is to get to know the local regulatory agency and see how certain sustainable practices fit in with local regulations. There may be unforeseen problems associated with conserving avian habitat on a property. From our experience, most developers do not want to jump through extra hoops to get a project approved.

In addition to removing policy barriers, tone can encourage municipalities to adopt bird-friendly policies. Through PREC, I have presented the importance of conserving natural habitat at several different county workshops across Florida. These counties are now looking into policies that would promote habitat conservation in residential development. Developers and landowners, though, are wary of any new regulations that restrict the way they could develop their property. In my experience, voluntary or incentive-based policies are a good way to introduce new development practices. However, from a review of incentive-based ordinances tried in Florida, most had a limited effect (Romero and Hostetler 2007). Successful ones tended to have good stakeholder input, a "meaningful" economic incentive, and a substantial marketing and education campaign. From our conversations with developers, meaningful incentives mean economic incentives, such as expediting permits, permit fee reductions, and density bonuses (i.e., more units per acre). In most cases, municipalities should first try an incentive-based ordinance, giving an opportunity for developers (and the local government) to try out the ordinance and help set up a culture of acceptance for a new design-build practice.

Form a Multidisciplinary Team

We have found that every development site is different. Sometimes there are major wildlife issues, sometimes water, and sometimes energy/transportation issues. A developer or municipality may come to you with one major concern, but this concern can lead to meaningful conversations about other conservation issues. Many connections exist among the various natural resource issues. For example, in Madera, concern about home energy conservation led to saving trees to shade the home, which fostered more discussions about conserving avian habitat across the site. In forming the team, I recommend having several state extension specialists and county agents who work with the public on a day-to-day basis. The extension service is part of the mission of land-grant universities and has a network of faculty and resources positioned in most counties in a state. In our experience, the design-build community operates on a much faster time frame than academia. Having extension faculty either leading the team or directly involved helps the academic group respond to questions and concerns in a timely manner.

Organize a Summit on Sustainable Development

Invite design/build professionals, politicians, planners, developers, environmental agencies, concerned citizens, and landowners to a sustainable development summit. A summit could be local, regional, or statewide. The sustainable development summit would cover several topic areas, one of which would be general wildlife habitat conservation. PREC partnered with Audubon International to hold a statewide summit on sustainable development in 2005. About 90 people attended this summit, and we found that it created many networking opportunities to work with people across the state. Many of our subsequent collaborations with municipalities and developers stemmed from this initial interaction.

Become Familiar with Certification Programs

It is important to become familiar with and incorporate local, state, or national certification programs: From our experience, developers are looking for certification programs in order to market their communities as "green." Having plans to conserve avian habitat within a community can meet some of the requirements for a certification program. Examples of certification groups include Audubon International (www.auduboninternational.org) and the U.S. Green Building Coalition and its LEED standards (www.usgbc.org/). In some cases, collaborating with these certification groups can lead to modification of standards for certification. Such was the case when PREC became involved with the Florida Green Building Coalition (floridagreenbuilding.org). Many of our suggestions were incorporated in their guidelines for conserving open space and wildlife habitat. Each certification has a distinct set of standards, and each program has both strength and weaknesses. For example, we found that several Florida communities, certified by Florida Green Building Coalition, made only minimal efforts to conserve open space and provide habitat for wildlife.

CONCLUSIONS

Creating opportunities for municipalities or developers to change their practices is not easy. Many growth-management policies and developer practices have become entrenched. It is a messy problem, and shifting current conventional policies and practices requires efforts from a number of different groups. Urban avian conservation and most other environmental problems are not precisely defined, and each locality will probably have different issues that need to be resolved. I hope the steps outlined in this chapter will help scientists to partner with planners, policymakers, and the design-build community to create new ways for conserving and managing urban avian habitat. If done well, the potential impacts can be high, as the decisions made by policymakers, planners, and design–build professionals can affect broad spatial and temporal scales.

ACKNOWLEDGMENTS

I would like to thank my colleagues at the Program for Resource Efficient Communities, especially P. Jones, H. Knowles, G. Acomb, M. Dukes, M. Clark, and K. Ruppert. My interactions with them helped form many of the ideas contained within this chapter. In addition, their support helped this biologist break out of the conventional mode and into the realm of collaborations with planners, policymakers, and developers.

LITERATURE CITED

Arendt, R. 1999. Growing greener: putting conservation into local plans and ordinances. Island Press, Washington, DC.

Calthorpe, P., and W. Fulton. 2001. The regional city: planning for the end of sprawl. Island Press, Washington, DC.

DeLorme, D., S. Hagen, and I. Stout. 2003. Consumers' perspctives on water issues: directions for educational campaigns. Journal of Environmental Education 34:28–35.

Grimm, N. B., J. M. Grove, S. T. A. Pickett, and C. L. Redman. 2000. Integrated approaches to long-term studies of urban ecological systems. Bioscience 50:571–584.

Holling, C. S. 1992. Cross-scale morphology, geometry and dynamics of ecosystems. Ecological Monographs 62:447–502.

Hostetler, M. E. 2000. The importance of multi-scale analyses in avian habitat selection studies in urban environments. Pp. 139–154. *in* J. M. Marzluff, R. Bowman, and R. Donnelly (editors), Avian ecology and conservation in an urbanizing world. Kluwer Academic Publications, Boston, MA.

Hostetler, M. E., and C. S. Holling. 2000. Detecting the scales at which birds respond to structure in urban landscapes. Urban Ecosystems 4:25–54.

Hostetler, M. E., P. Jones, M. Dukes, H. Knowles, G. Acomb, and M. Clark. 2008. With one stroke of the pen: how can extension professionals involve developers and policymakers in creating sustainable communities? Journal of Extension 46(1):online.

Hostetler, M. E., and K. Knowles-Yanez. 2003. Land use, scale and bird distributions in the Phoenix metropolitan area. Landscape and Urban Planning 62:55–68.

Hostetler, M., E. Swiman, A. Prizzia, and K. Noiseux. 2008. Reaching residents of green communities: evaluation of a unique environmental education program. Applied Environmental Education & Communication 7:11–124.

Hostetler, M. E., and K. Noiseux. 2010. Are green residential developments attracting environmentally savvy homeowners? Landscape and Urban Planning 94:234–243.

Johnson, D. H. 1980. The comparison of usage and availability measurements for evaluating resource preference. Ecology 61:65–71.

Kotliar, N. B., and J. A. Wiens. 1990. Multiple scales of patchiness and patch structure: a hierarchical framework for the study of heterogeneity. Oikos 59:253–260.

Lawrence, F. D., and P. Engelke. 2005. Multiple impacts of the built environment on public health: walkable places and the exposure to air pollution. International Regional Science Review 28:193–216.

Marzluff, J. M., R. Bowman, and R. Donnelly (editors). 2000. Avian ecology and conservation in an urbanizing world. Kluwer Academic Publications, Boston, MA.

Pickett, S. T. A., M. L. Cadenasso, J. M. Grove, C. H. Nilon, R. V. Pouyat, W. C. Zipperer, and R. Costanza. 2001. Urban ecological systems: linking terrestrial ecological, physical, and socioeconomic components of metropolitan areas. Annual Review of Ecology and Systematics 32:127–157.

Romero, M., and M. E. Hostetler. 2007. Policies that address sustainable landscaping practices. <http://edis.ifas.ufl.edu/UW253>.

U.S. Census Bureau. 2006. <http://www.census.gov>.

Wiens, J. A. 1989. The ecology of bird communities, processes and variations. Cambridge University Press, New York, NY.

Youngentob, K., and M. Hostetler. 2005. Is a new urban development model building greener communities? Environment and Behavior 37: 731–759.

CHAPTER FIFTEEN

Predicting Avian Community Responses to Increasing Urbanization

Jeffrey Hepinstall-Cymerman, John M. Marzluff, and Marina Alberti

Abstract. We predicted changes in avian communities that are likely to occur given predicted land cover and land use for 2003, 2015, and 2027 in the lowlands of the central Puget Sound of western Washington State, USA. The 8,800-km^2 study area is undergoing significant urban development and landscape change. We combined a microeconomic development model of human behavior (UrbanSim) with a land cover change model (LCCM) to predict land use and land cover 28 years into the future. We developed regression models of avian species richness and relative abundance derived from seven years of field studies with over 6,000 point count surveys spread across the urban to exurban gradient. We applied the avian models to the predicted land use and land cover to estimate future avian community measures. We found that observed avian diversity was sensitive to both the amount and the pattern of land use and land cover. As the transition zone between landscapes dominated by human development and exurban areas is transformed into dense development, we predict that bird communities will become spatially partitioned and dominated by either adaptable, synanthropic species (in dense developments) and early successional species that can tolerate some human development or resilient native forest birds (in the exurban zone). Native forest birds will likely become increasingly reliant on higher-elevation forests because most low-elevation forests will be converted to development too dense to sustain populations. While these findings are not unexpected, the high spatial resolution of our land use and land cover predictions (landscape pattern metrics can be calculated for 1-km^2 "windows") allowed us to explore how the amount and pattern of future development may influence avian communities. These spatially explicit predictions of future land use, land cover, and avian communities provide an integrated picture of how the landscape might look in ~25 years. Additionally, the potential impacts of a proposed policy regulating development location, intensity, or configuration can be modeled to show how it may change land use, how that will likely influence land cover, and how both of these will influence avian communities.

Key Words: avian abundance, avian species richness, land cover change, land use change, modeling, predictive models, urban development, urban ecology.

Hepinstall-Cymerman, J., J. M. Marzluff, and M. Alberti. 2012. Predicting avian community responses to increasing urbanization. Pp. 223–248 *in* C. A. Lepczyk and P. S. Warren (editors). Urban bird ecology and conservation. Studies in Avian Biology (no. 45), University of California Press, Berkeley, CA.

As human populations grow, the urbanized area of earth also increases (Meyer and Turner 1992, Houghton 1994). Depending on economics, social preferences, and land use policies, the growth of urban populations causes cities, and even more profoundly their suburbs, to spread across large expanses of former agricultural and natural lands (Ewing 1994). The worldwide extent of sprawling settlement is visible in the nighttime images of earth from space (Elvidge et al. 1997). These images reveal that substantial portions of the north temperate zone are heavily settled, most ice-free coastlines are settled, our most fertile lands are quickly being developed, and overall about 3% of earth's land area is urban (Lawrence et al. 2002, Imhoff et al. 2004).

The changes in land cover associated with urban development (defined as a change from non-built land uses, e.g., forest, to land uses with supporting infrastructure and buildings, e.g., commercial, industrial, residential) threaten biological diversity now and increasingly in the future (Sala et al. 2000). As a result of increased urban population growth and immigration, agricultural lands and undeveloped lands that hold great stores of biodiversity are being converted into developed land uses (Foley et al. 2005, Mörtberg et al. 2007). Urban development changes species extinction and colonization rates (Marzluff 2001), likely leading to the loss of native species and replacement by nonnative species (Kowarik 1995; Blair 1996, 2001a, 2001b), thereby changing local standing diversity (Marzluff 2005).

While we are beginning to understand how local urbanization processes influence biodiversity, we know much less about how these altered rates of extinction and colonization will play out through time. Understanding and predicting how biodiversity responds to land cover change requires large-scale modeling (Sala et al. 2000, Schumaker et al. 2004, Hepinstall et al. 2008). While there is no generally accepted single method for predicting how biodiversity will change with landscape change (Doak and Mills 1994, Ruckelshaus et al. 1997), there is a rich literature on how species respond to loss and fragmentation of their habitat (Fahrig and Merriam 1994, McGarigal and McComb 1995, Villard et al. 1999, Rodewald, chapter 5, this volume). In addition, several methods exist for developing spatially explicit habitat models, either at coarse scales with coarse input such as habitat-association models (Scott et al. 1993, White et al. 1997, Schumaker et al. 2004) or individually based finer-resolution models (Dunning et al. 1995, Schumaker et al. 2004). Many studies have attempted to model future land use change and urban development (Waddell 2002, Waddell et al. 2003,Verburg et al. 2004). What is missing from previous modeling efforts, generally, are links between models that predict how species respond to change and models that predict the future extent, configuration, and intensity of urban development (Pickett et al. 2001, Schumaker et al. 2004, Van Sickle et al. 2004).

Our objectives in this article are to: (1) provide a case study linking output from urban development and land cover change models to models of avian community dynamics and individual species relative abundance; (2) explore how different species assemblages within habitat guilds affect future projections of species richness and individual species' relative abundance; and (3) compare guild predictions of species richness with individual species predictions of relative abundance. Linking ecological models with urban development and land cover change models is important for understanding how landscape change will affect ecosystem services in urbanizing environments. In this article, we model the influence of urbanization on biodiversity, building upon modeling approaches from several different fields, including urban development modeling (Wear and Bolstad 1998, Waddell 2002), land cover change modeling (Mladenoff 2004), and wildlife habitat modeling (Schumaker et al. 2004). We combine output from spatially explicit models that simulate (1) urban development and (2) land cover change with (3) models of avian species richness and relative abundance. We demonstrate how urban development in the rapidly urbanizing, Seattle, Washington, region will likely change land cover and how these changes in land use and land cover may change the abundance and diversity of birds. By modeling the potential responses of birds to changing landscapes, we demonstrate how one component of biological diversity might respond to urbanization.

Our study area, the central Puget Sound, containing Seattle and several other large cities, has experienced dramatic urban growth, especially during the last 30 years (Hansen et al. 2005, Robinson et al. 2005) and in 2002 was projected to grow from ~3 million (year 2000) to 4 million

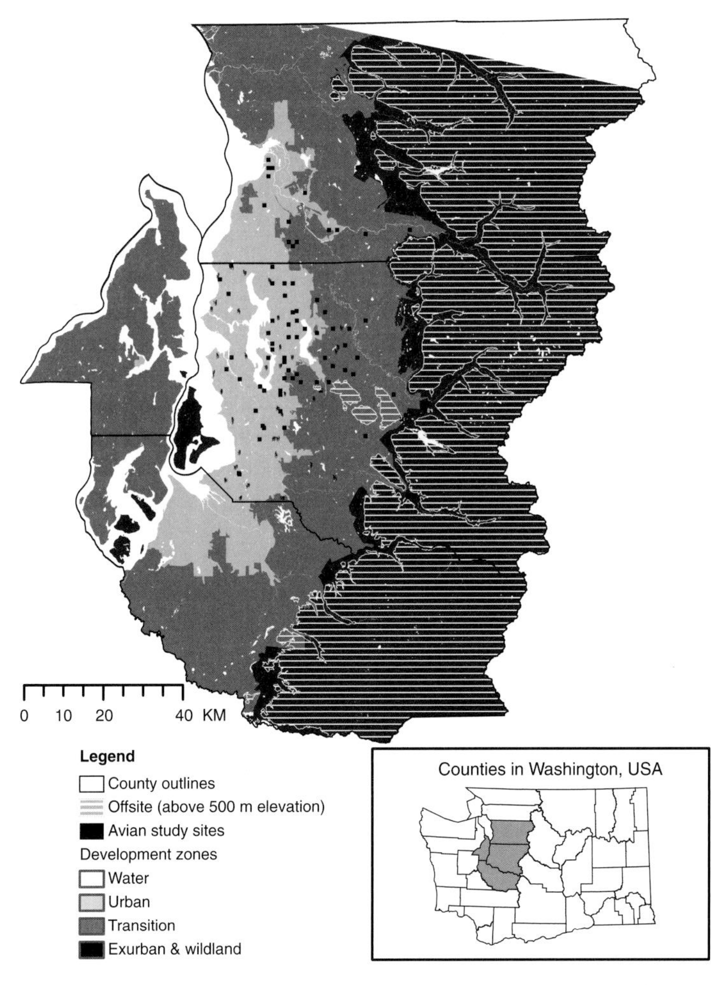

Figure 15.1. Four county study area of western Washington, USA, with 1-km^2 bird study sites (n = 139), 500-m elevation contour, and three zones of urban, transition, and exurban (defined by distance to urban center, population density, and elevation above sea level).

people by 2025 (Office of Financial Management, State of Washington: www.ofm.wa.gov/pop/gma/projections.asp). Currently, remnant forests exist in a variety of sizes and settings, from small urban parks or undeveloped parcels to large blocks of contiguous forest (Donnelly and Marzluff 2004b, 2006). Songbird diversity is greatest in landscapes with 50–60% forest cover, because such areas gain more synanthropic (e.g., human-associated species) and early successional species than the native forest species they lose (Marzluff 2005).

We present a habitat approach to estimate species richness for three development-sensitive guilds (native forest, early successional, and synanthropic) and four subguilds of the early successional guild. Insights from subguilds are compared with predictions from models of individual species relative abundance. We focus on the early successional guild because half of the total regional bird diversity is included in this group and previous efforts to model the responses of the entire guild were unsatisfactory (Hepinstall et al. 2009).

METHODS

Study Area

Our study area encompassed the lowland (<500 m above sea level) portions (8,800 km^2) of temperate, moist forest in four counties of the State of Washington (Fig. 15.1), including Seattle,

Bellevue, Tacoma, and Everett metropolitan areas. Forests were mostly coniferous, including western hemlock (*Tsuga heterophylla*), Douglas-fir (*Pseudotsuga menziesii*), and western red cedar (*Thuja plicata*), with a few red alder (*Alnus rubra*), bigleaf maple (*Acer macrophyllum*), black cotton wood (*Populus trichocarpa*), and Oregon ash (*Fraxinus latifolia*) occurring near riparian and disturbed areas (Franklin and Dyrness 1988).

Avian Study Sites

Across the urban-exurban gradient, we chose 139 study sites of 1 km^2, characterized as single-family residential (SFR) sites embedded in a forested matrix (n = 119), mixed use/commercial/industrial sites (n = 13), and fully forested ("control") sites with minimal development (n = 7; Fig. 15.1). We used 2002 land cover and 2000 parcel maps to randomly select SFR and control sites (n = 126) along four axes of urbanization (all calculated per km^2): (1) urbanization intensity (suburban and exurban based on percentage urban land cover); (2) mean patch size of urban land cover; (3) the probability that two randomly chosen adjacent pixels belong to the same class (contagion); and (4) mean development age (i.e., housing age 5–15 years [young], 40–50 years [middle-age], and >70 years [old]). Sites defined as suburban contained >70% urban land cover (per km^2) and <15% forest land cover. Exurban sites contained <50% urban land cover and >40% forest land cover. All SFR sites also contained patches of remnant forest. The mixed use/commercial/industrial sites were randomly selected using 2002 land use maps; we included this land use category to improve our understanding of bird communities in more intensely developed areas. Further details of site selection and site characteristics are presented in Donnelly and Marzluff (2004a, 2004b, 2006), Blewett and Marzluff (2005), and Hepinstall et al. (2009).

Avian Surveys

Trained observers conducted 6,437 fixed-radius (50 m) point-count surveys of breeding birds at 992 locations (302 forested, 690 human-developed) within the 139 1-km^2 study sites during the springs and summers of 1998–2005. Only one person surveyed any location within a given year, and individual differences in survey ability were minimized by having seven observers conduct all counts. Each observer was trained for two weeks prior to their first count in logistics, distance estimation, and species identification. All observers counted for multiple years (2–5 years per observer).

Individual sites were sampled from 1 to 7 years, dependent on study goals. Annual variation within a site was substantially less than variation among sites, so varying lengths of survey history were unlikely to bias our results (Donnelly and Marzluff 2006). Locations within sites were visited 3–5 times per year (late March–late August) with late-season surveys necessitated by our long breeding season. Ten-minute point counts were conducted between 30 minutes prior to and approximately 6 hours after sunrise (details in Donnelly and Marzluff 2004a, 2004b, 2006). Within most sites, eight point counts were conducted: six located in developed areas and two in remnant patches of forest. Greater effort was allocated to sampling birds in developed areas than in forests because a previous study of forest reserves in the same region indicated that birds and vegetation were more variable in developed areas (Donnelly and Marzluff 2006). All points were >150 m apart, which in our experience reduced double counting, with the exception of a small number where the separation was maximized within the only forest fragment that existed.

Species richness was calculated as the mean number of species occurring on a site over the years the site was surveyed. Relative abundance was calculated for each site as the mean number of individuals detected per point per survey. This method was preferable to others, such as maximum abundance per survey, because it limited the influence of young of the year and migrating individuals on abundance estimates.

Previous work had identified three development-sensitive habitat guilds using a subset of 57 common species (Appendix 15.1; Hepinstall et al. 2009). The guild approach centers on different colonization and extinction probabilities of native forest birds, synanthropic birds, and early successional birds in urbanizing landscapes (Marzluff 2005). "Native forest" birds (n = 19) are typically found in intact, second growth (60–100-year-old) coniferous forest, "early successional" species (n = 29) exploit the heterogeneous vegetation of fragmented landscapes following some type of physical disturbance, and "synanthropic" species (n = 9) are either native or nonnative birds that thrive in human dominated landscapes (i.e., "urban exploiters" and "urban adapters"; González-Oreja et al., online material). To better understand the response of species within the large early successional guild, we separated it into four subguilds: young and deciduous forest (n = 15), suburban (n = 6), open/grassy/brushy (n = 6), and aquatic (n = 2) (Appendix 15.1).

Modeling Framework

We used land use output from the UrbanSim urban development model (Waddell 2002, Waddell et al. 2003; www.urbansim.org) and land cover output from LCCM (Hepinstall et al. 2008, 2009) as input into simple linear regression models derived from avian field data described above to make spatially explicit predictions of avian species richness and individual species relative abundance 28 years into the future. Outputs from UrbanSim and LCCM were used as inputs into our avian community models. For a description of the models, inputs, and outputs, see Hepinstall et al. (2008, 2009).

Avian Models of Species Richness and Relative Abundance

We developed linear and nonlinear regression models to explain bird species richness and relative abundance as a function of land use and land cover composition and configuration. Bird point-count data (discussed above) were used to develop models of species richness for: (1) total species richness, (2) the three habitat guilds, and (3) the four early successional subguilds (Appendix 15.1). We also developed models of relative abundance for individual species within the early successional guild. We used land cover data from 2002 to calculate the percentage forest, percentage urban, the aggregation index of forest (Fragstats, ver. 3.3, Amherst, MA; McGarigal et al. 2002), the number of unique patches of forest, and the number and mean size of unique urban patches within each 1-km^2 bird study area. Land use data derived from 2002 parcel maps were used to calculate the percentage, patch density, and aggregation index of residential parcels and mean development age of parcels. We eliminated those landscape variables that were highly correlated (Pearson correlation coefficient >0.70). Errors were not adjusted for spatial autocorrelation because sampling units (1-km^2 study sites) were separated by at least several kilometers.

Two *a priori* models of species richness and relative abundance were defined based on previous studies (Donnelly and Marzluff 2004a, 2004b, 2006; Hepinstall et al. 2009), landscape measures relevant to urban planners, and variables available as output from UrbanSim and LCCM. Our purpose was not to select the "best" model from a series of models, but rather to use all available information to best account for variation in bird diversity and species abundance. Therefore, we fully specified only two models which differ in practicality and complexity rather than evaluating a full family of models using formal model selection procedures (Burnham and Anderson 1998). The simple model included: (1) percentage of forest (in linear and quadratic form); (2) aggregation of residential land use; and (3) mean development age of parcels, all within a 1-km^2 window. These independent variables were consistently related to avian diversity and abundance in our previous studies. The complex model included the aforementioned variables and added the following ones:

1. percentage of grass and agriculture;
2. forest aggregation index;
3. the number of unique patches of forest land cover;
4. number of unique patches of urban land cover;
5. the mean patch size of unique patches of urban land cover;
6. the percentage residential land use; and
7. the patch density of patches of residential land use, all within a 1-km^2 window.

Coupling Predicted Land Use and Land Cover with Avian Models

Parameter estimates from avian models were applied to the future landscapes generated by UrbanSim and LCCM to predict total, guild, and subguild species richness as well as relative abundance for all species within the early successional guild that had significant model equations. In addition, we summed the individual relative abundance estimates within each subguild to derive a total subguild relative abundance. We calculated the landscape variables in our avian models for each 30-m pixel in our study area using ArcGIS and Fragstats 3.3 (for aggregation index) with a 1-km^2 moving window to match the

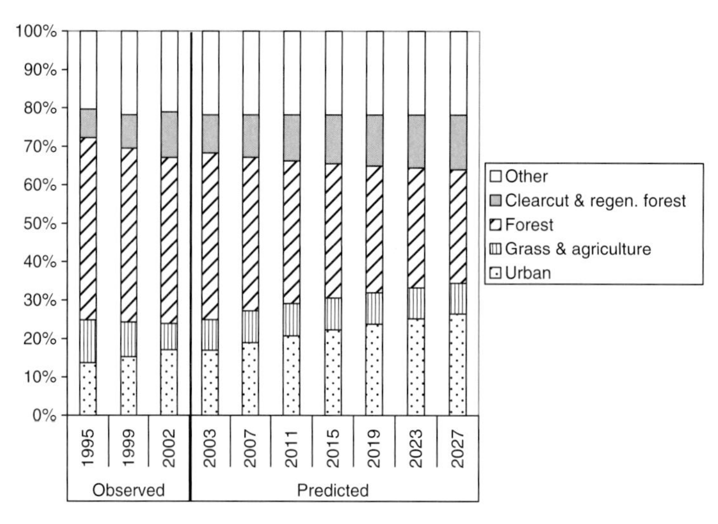

Figure 15.2. Land cover percentages for the study area observed (1995, 1999, 2002) and predicted (2003, 2015, 2027) derived from 1995–1999 observed transitions and 1999 base map. Classes included: urban (heavy urban, >80% impervious surfaces; medium urban, 50–80% impervious; light urban, 20–50% impervious); grass and agriculture; forest (deciduous and mixed forest, 100–25% deciduous species; coniferous forest, >75% conifers); clearcut and regenerating forest (clearcut, recent forest harvest with no regrowth yet visible on satellite imagery; regenerating forest, recent forestland harvest with regrowth visible).

study site design of the avian surveys. Regression coefficients were applied to 2003, 2015, and 2027 LCCM and UrbanSim outputs for our study area and standardized to yield estimates of species richness per km^2 or relative abundance per hectare. To allow for comparisons of specific locations across time and because our output maps contained predictions for each 30-m pixel (approximately 9 million), we randomly sampled 100,000 non-water pixels and summarized predicted species richness and relative abundance for these locations in the results section.

Avian model results for areas below 500 m elevation were compared by three zones of urbanization: urban, transition, and exurban (Fig. 15.1; Hepinstall et al. 2008). Zone comparisons allowed us to examine changes in avian richness and relative abundance as a function of proximity to and proportion of development. The urban zone was dominated by urban development (residential, commercial, industrial, institutional, and office land uses) and contained all the major cities of the study area. The transition zone contained a more heterogeneous mix of impervious areas, agricultural lands, and remnant woodlands. The exurban zone extended from >300 m elevation to the Cascade Mountains ridge and contains development primarily along river valleys. Areas above 500 m elevation were removed from our analysis because they were not sampled by our avian study sites (Fig. 15.1), contain little urban development, and were not the focus of the land use or land cover models.

RESULTS

Predictions of Future Land Cover

Predictions of land cover from 2003 to 2027 showed decreased forest land (from 43% of the study area to 29%) and increased developed land (heavy, medium, and low urban classes; 17% to 27%; Fig. 15.2). The percentage of grass and agriculture were predicted to be essentially stable (~8%), although the specific location of these classes may have changed. Clearcutting and regenerating forest increased from 9% in 2003 to 14% in 2027.

Avian Surveys

Across the 139 sites, we identified 114 species of birds. Each site contained a mean of 31.5 (±11.6 SD) species. By guild, mean species richness of a site was 11.2 (±4.6 SD) for native forest species, 11.3 (±4.6 SD) for early successional species, and 5.2 (±2.2 SD) for synanthropic species.

TABLE 15.1

Linear regression model derived from simple models of species richness (species/point/survey) for total species richness and for guilds and early successional subguilds as a function of landscape metrics for 139 suburban landscapes in Puget Sound, Washington, 1998–2005.

				Unstandardized coefficients and *P*-values				
Sp. Richness	Adj. r^2	Model *P*-value		Constant	%Forest	%Forest^2	RAI	MDA
Total	0.178	<0.001	β	28.722	0.086		0.097	−0.237
			$P\leq$	0.001	0.020		0.038	0.001
Native forest	0.440	<0.001	β	6.358	0.094		0.052	−0.068
			$P\leq$	0.001	0.001		0.001	0.002
Synanthropic	0.225	<0.001	β	5.239	−0.037		0.019	−0.014
			$P\leq$	0.001	0.001		0.025	0.250
Early successional guild								
	0.054	0.022	β	11.566	1.984	−2.972	2.155	−2.913
			$P\leq$	0.001	0.520	0.230	0.100	0.030
Early successional subguild								
Y&D Forest	0.212	0.001	β	3.205	3.730	−1.425	1.228	−1.490
			$P\leq$	0.008	0.018	0.250	0.07	0.028
Suburban	0.198	0.001	β	4.266	0.239	−1.745	0.606	−0.038
			$P\leq$	0.001	0.860	0.100	0.280	0.950
O/G/B	0.131	0.001	β	3.122	−2.068	0.541	0.294	−0.915
			$P\leq$	0.001	0.027	0.460	0.460	0.023
Aquatic	0.082	0.004	β	1.485	−0.273	−0.186	0.014	−0.669
			$P\leq$	0.001	0.550	0.610	0.940	0.001

NOTE: df = 3 regression, 135 error; 4 regression, 134 error for models with %forest^2.

RAI = residential aggregation index, MDA = mean development age, Y&D forest = young and deciduous forest, O/G/B = open/grassy/brushy.

Models of Avian Species Richness and Relative Abundance

Total Species Richness

Approximately 18% (simple model: Table 15.1; adjusted r^2) and 21% (complex model: Table 15.2; adjusted r^2) of the variation in total species richness was accounted for in our two *a priori* models. For both the simple and complex models, richness increased with increased percentage of forest and aggregation of residential development in the 1-km^2 study site, and decreased with mean age of residential development.

Habitat Guild

Within habitat guilds, we explained more (44%, simple model; 51%, complex model) of the variation in native forest species richness than in the richness of early successional (5% and 14%) or synanthropic species (23% and 28%; Tables 15.1, 15.2). Native forest species richness mirrored total species richness and increased with increasing percentage of forest and aggregation of residential development and decreased with mean age of residential development (Tables 15.1, 15.2). Native forest species richness also was positively correlated with forest aggregation in our complex

TABLE 15.2

Linear regression model results derived from complex models of species richness (species/point/survey) for total species richness and for guilds and early successional subguilds as a function of landscape metrics for 139 suburban landscapes in Puget Sound, Washington, 1998–2005.

Model	Adj. r^2	Model P-value		Constant	%F	%F²	RAI	MDA	%GR	FAI	NPF	NPU	MPSU	%RES	PDR
				Unstandardized coefficients and P-values											
Total	0.208	0.000	β	22.551	0.060		0.149	−0.270	0.459	0.072	0.117	−0.173	0.137	−0.091	0.097
			$P\leq$	0.001	0.373		0.021	0.001	0.039	0.230	0.580	0.440	0.720	0.080	0.510
Native Forest	0.508	<0.001	β	1.937	0.059		0.047	−0.052	0.020	0.076	0.120	−0.055	−0.057	−0.009	0.051
			$P\leq$	0.340	0.006		0.021	0.041	0.770	0.001	0.070	0.440	0.630	0.590	0.267
Synanthropic	0.281	<0.001	β	4.941	−0.031		0.034	−0.030	0.114	−0.004	−0.064	−0.055	0.068	−0.012	0.007
			$P\leq$	<0.001	0.012		0.003	0.038	0.004	0.730	0.100	0.180	0.320	0.210	0.790
Early successional guild															
	0.144	0.001	β	9.570	−2.294	−0.232	3.881	−3.514	11.555	2.238	−0.105	−0.140	0.066	−1.116	−0.025
			$P\leq$	0.001	0.650	0.950	0.024	0.015	0.001	0.240	0.160	0.070	0.610	0.460	0.560
Early successional subguild: Y&D forest															
	0.223	0.000	β	2.596	2.458	−0.587	1.302	−1.681	3.984	0.853	−0.053	−0.063	−0.008	0.623	0.003
			$P\leq$	0.080	0.360	0.740	0.150	0.027	0.026	0.400	0.180	0.120	0.910	0.430	0.890

Early successional subguild: Suburban														
0.260	<0.001	β	2.681	−1.822	0.011	1.219	−0.362	4.366	1.008	−0.018	−0.065	0.058	−0.059	0.007
		$P\leq$	0.027	0.400	0.990	0.100	0.550	0.003	0.220	0.570	0.050	0.300	0.930	0.700
Early successional subguild: O/G/B														
0.265	<0.001	β	3.325	−2.717	0.518	1.180	−0.943	2.035	0.307	−0.024	−0.005	0.011	−1.482	−0.031
		$P\leq$	0.001	0.070	0.600	0.018	0.024	0.038	0.580	0.270	0.810	0.780	0.001	0.017
Early successional subguild: Aquatic														
0.251	<0.001	β	1.648	−0.855	0.062	0.322	−0.753	1.959	0.068	−0.013	−0.002	0.006	−0.551	−0.012
		$P\leq$	0.001	0.230	0.900	0.180	<0.001	<0.001	0.800	0.220	0.860	0.730	0.011	0.048

NOTE: df = 10 regression, 128 error; 11 regression, 127 error for models with $\%forest^2$.

F = forest, RAI = residential aggregation index, MDA = mean development age, GR = grass and agriculture, FAI = forest aggregation index, NPF = number patches of forest, NPU = number of patches urban, MPSU = mean urban patch size, RES = residential land use, PDR = residential patch density.

model. Synanthropic species richness was significantly correlated with percentage forest (−) and residential aggregation (+) in the simple model and with mean development age (−) and percentage grass (+) in the complex model. Early successional species richness was significantly correlated only with mean development age (−) in the simple model and residential aggregation (+), mean development age (−), and percentage grass (+) in the complex model. However, within the complex model, only 14% of the observed variation in early successional richness was explained by our guild-level models (Table 15.2).

Early Successional Subguilds

Mean development age (−) and percentage grass (+) were the only significant variables in the complex model for the young and deciduous forest subguild (Table 15.2). The complex model for the aquatic subguild included these same variables as well as percentage residential. The complex open/grassy/brush subguild model had significant parameter estimates for all variables associated with residential land use (i.e., residential aggregation index [−], percentage residential [−], and residential patch density [−]) in addition to mean development age (−) and percentage grass (+). The complex subguild models each explained more of the observed variation (22–27%) than the complex guild or simple guild or subguild models. None of the subguild models or the early successional guild model had a significant curvilinear (quadratic) response of species richness to percentage forest (Tables 15.1, 15.2).

Simple (Table 15.3) and complex (Table 15.4) regression models of relative abundance of 29 early successional species were significant for 18 and 22 species, respectively. Eight species had curvilinear responses to percentage forest in the simple model, four in the complex model. Percentage forest, percentage forest squared, number of patches of forest, percentage residential, residential aggregation index, and residential patch density were the most consistent predictors of relative abundance for early successional species.

Comparison of Simple versus Complex Models

In general, our complex models explained little variation beyond that explained by the simple models. However, percentage grass was an important contributor to explaining variation in richness of those species that utilize developed landscapes, and forest configuration was important to native forest species.

Predictions of Future Bird Species Richness and Relative Abundance

Our models predicted declining species richness during the next two decades as native forest species respond to the loss of forest cover and early successional species respond to the aging of current and future developments (Fig. 15.3). We predicted decreased mean total species richness using our simple and complex models from 34 and 36 species in 2003 to 30 or 31 species in 2027, respectively (Fig. 15.3). Using our complex model, we predicted a decline in mean total species richness from 33 to 24 species in the urban zone and 37 to 31 species in the transition zone (Fig. 15.3b). The guild-specific results indicated that the loss of early successional (~3) and native forest (~2–3) species will be most noticeable in the urban and transition zones. Simple models predicted the diversity of the synanthropic guild will increase slightly (2–4 species) in the transition zone and remain relatively stable in the other zones (Fig. 15.3a). The complex model for the synanthropic guild predicted a decrease in species richness in the transition, exurban, and overall study area (Fig. 15.3b). The complex models, which included more landscape variables, consistently reduced mean predictions of species richness for the three habitat guilds by 1–3 species. Predicted changes in species richness were concentrated in those regions of the study area where land cover change was predicted to be most dramatic, primarily in the transition zone surrounding the present urban core where forest was lost to new development.

The young and deciduous forest early successional subguild was predicted to lose species across the study area and for each development zone (Figs. 15.4, 15.5). The mean species richness by development zone (Fig. 15.4) does not tell the entire story. Slight losses in the mean richness of the suburban subguild predicted in the transition zone and larger losses in the exurban zone were actually predicted gains of 2–3 species along the development front and losses of 1–4 species at higher elevations (Fig. 15.5). We predicted overall and transition zone losses for the open/grassy/

TABLE 15.3

Linear regression model results for simple models of relative abundance for species in the early successional guild as a function of landscape metrics for 139 suburban landscapes in Puget Sound, Washington, 1998–2005.

				Unstandardized coefficients and *P*-values				
	Adj. r^2	Model *P*-value		Constant	%Forest	%Forest^2	RAI	MDA
Young and deciduous forest subguild								
Black-headed Grosbeak	0.30	0.00	β	−0.14	0.08	0.41	0.12	0.04
			P≤	0.48	0.76	0.05	0.30	0.70
Cassin's Vireo	0.06	0.01	β	0.00	0.02		−0.01	0.00
			P≤	1.00	0.01		0.37	0.84
MacGillivray's Warbler	0.05	0.02	β	0.08	0.04		−0.08	0.01
			P≤	0.24	0.15		0.01	0.85
Pine Siskin	0.07	0.01	β	0.15	2.16	−1.58	0.30	−0.22
			P≤	0.79	0.01	0.01	0.37	0.52
Purple Finch	0.08	0.00	β	0.15	0.10		0.20	−0.21
			P≤	0.38	0.13		0.02	0.03
Rufous Hummingbird	0.13	0.00	β	−0.07	0.16		0.07	−0.01
			P≤	0.45	0.01		0.11	0.86
Song Sparrow	0.04	0.04	β	0.87	3.76	−2.50	−0.57	0.10
			P≤	0.37	0.01	0.01	0.29	0.85
Tree Swallow	0.04	0.03	β	0.02	0.00		−0.01	0.00
			P≤	0.03	0.45		0.01	0.72
Warbling Vireo	0.27	0.00	β	0.08	0.05		−0.11	0.02
			P≤	0.02	0.01		0.01	0.33
Yellow-rumped Warbler	0.36	0.00	β	0.54	0.71	−0.56	−0.59	0.03
			P≤	0.01	0.01	0.01	0.01	0.70
Suburban subguild								
Black-capped Chickadee	0.15	0.00	β	0.07	−0.89		0.38	1.30
			P≤	0.94	0.02		0.38	0.01
Bewick's Wren	0.12	0.00	β	−0.20	2.05	−1.91	−0.22	0.70
			P≤	0.73	0.01	0.01	0.50	0.04
Bushtit	0.12	0.00	β	0.94	−1.17		−0.22	0.57
			P≤	0.25	0.01		0.57	0.22
Northern Flicker	0.07	0.01	β	0.33	0.87	−0.95	−0.08	0.08
			P≤	0.35	0.06	0.01	0.67	0.70
Open/grassy/brushy subguild								
American Goldfinch	0.05	0.02	β	0.97	−0.35		−0.31	−0.11
			P≤	0.01	0.01		0.05	0.57
White-crowned Sparrow	0.10	0.00	β	1.33	−2.73	1.54	0.27	−0.26
			P≤	0.03	0.01	0.02	0.44	0.46
Aquatic subguild								
Common Yellowthroat	0.10	0.00	β	0.22	−0.06		−0.14	−0.01
			P≤	0.01	0.06		0.01	0.81
Red-winged Blackbird	0.07	0.01	β	0.88	0.52	−0.59	−0.50	−0.16
			P≤	0.01	0.14	0.04	0.01	0.29

RAI = residential aggregation index, MDA = mean development age.

TABLE 15.4

Linear regression model results for complex models of relative abundance for species in the early successional guild as a function of landscape metrics for 139 suburban landscapes in Puget Sound, Washington, 1998–2005.

	Adj. r^2	Model P-value		Unstandardized coefficients and P-values											
				Constant	%F	%F^2	%RAI	MDA	%GR	FAI	NPF	NPU	MPSU	%RES	PDR
Young and deciduous forest subguild															
Black-headed Grosbeak	0.29	0.00	β	−0.06	0.70		0.04	0.39	0.13	−0.07	−0.01	−0.02	−0.01	0.00	0.04
			$P\leq$	0.78	0.01		0.75	0.19	0.37	0.64	0.25	0.00	0.24	0.91	0.77
Cassin's Vireo	0.09	0.01	β	−0.01	0.02		−0.03	−0.03	0.01	0.00	0.00	0.00	0.00	0.00	0.00
			$P\leq$	0.59	0.08		0.01	0.30	0.24	0.95	0.37	0.59	0.15	0.53	0.67
MacGillivray's Warbler	0.15	0.00	β	0.25	0.24	−0.21	0.02	0.01	−0.14	−0.04	0.00	0.00	−0.01	0.00	0.04
			$P\leq$	0.00	0.08	0.02	0.66	0.90	0.01	0.41	0.42	0.62	0.01	0.14	0.25
Pine Siskin	0.12	0.00	β	−0.52	1.36		0.82	1.46	−0.08	−0.67	0.02	−0.02	0.06	0.00	−0.16
			$P\leq$	0.42	0.01		0.04	0.09	0.85	0.12	0.22	0.39	0.07	0.72	0.67
Purple Finch	0.08	0.03	β	0.17	0.20		0.18	−0.06	0.05	−0.11	0.00	0.01	0.01	0.00	−0.25
			$P\leq$	0.36	0.15		0.11	0.80	0.70	0.39	0.40	0.19	0.18	0.91	0.02
Rufous Hummingbird	0.10	0.01	β	−0.01	0.17		0.01	−0.01	0.09	−0.03	0.00	0.00	−0.01	0.00	−0.01
			$P\leq$	0.92	0.04		0.90	0.93	0.20	0.66	0.83	0.75	0.31	0.66	0.88
Song Sparrow	0.07	0.04	β	−0.04	−0.13		−0.58	3.54	−0.14	1.03	0.00	0.02	0.06	−0.01	−0.04
			$P\leq$	0.97	0.87		0.37	0.01	0.84	0.15	0.95	0.56	0.25	0.48	0.94
Tree Swallow	0.06	0.05	β	0.02	−0.01		0.00	−0.01	−0.02	0.01	0.00	0.00	0.00	0.00	0.00
			$P\leq$	0.13	0.09		0.57	0.57	0.02	0.19	0.81	0.03	0.12	0.23	0.78
Warbling Vireo	0.36	0.00	β	0.04	−0.08	0.14	0.07	0.05	−0.15	−0.02	0.00	0.00	0.00	0.00	0.04
			$P\leq$	0.41	0.27	0.01	0.00	0.32	0.00	0.56	0.94	0.70	0.03	0.01	0.06
Willow Flycatcher	0.23	0.00	β	0.36	−0.06		−0.09	0.68	0.12	−0.10	−0.01	0.00	−0.01	0.00	−0.10
			$P\leq$	0.00	0.47		0.18	0.01	0.10	0.20	0.11	0.90	0.02	0.35	0.11
Yellow-rumped Warbler	0.38	0.00	β	0.71	0.90	−0.73	0.04	−0.14	−0.69	−0.11	−0.01	0.00	0.00	−0.01	0.08
			$P\leq$	0.01	0.01	0.01	0.64	0.43	0.01	0.29	0.15	0.32	0.80	0.01	0.27

	Adj. r^2	Model P-value													
							Suburban subguild								
Black-capped Chickadee	0.24	0.00	β	−1.02	−1.65		1.02	1.30	−0.73	1.42	0.01	−0.03	−0.02	−0.02	1.78
Bewick's Wren	0.17	0.00	$P\leq$	0.29	0.02		0.08	0.30	0.25	0.03	0.81	0.27	0.74	0.30	0.01
			β	−1.62	−0.84		0.37	−0.05	−0.41	1.22	0.02	0.01	0.05	−0.01	0.89
			$P\leq$	0.01	0.08		0.35	0.96	0.33	0.01	0.36	0.51	0.12	0.62	0.02
Bushtit	0.11	0.01	β	0.20	−1.69		−0.44	0.30	0.18	1.00	0.01	−0.02	−0.01	0.02	0.44
			$P\leq$	0.83	0.02		0.43	0.81	0.77	0.11	0.56	0.49	0.87	0.26	0.40
Northern Flicker	0.19	0.00	β	−0.35	−0.31		0.69	0.72	−0.41	0.45	−0.01	−0.02	0.02	−0.01	0.25
			$P\leq$	0.35	0.27		0.01	0.14	0.10	0.07	0.29	0.11	0.31	0.13	0.24
							Open/grassy/bushy subguild								
American Goldfinch	0.16	0.00	β	0.97	−0.67		−0.42	0.88	−0.19	0.18	−0.01	0.01	0.03	−0.01	−0.15
			$P\leq$	0.01	0.01		0.05	0.05	0.40	0.44	0.12	0.29	0.12	0.02	0.45
Killdeer	0.19	0.00	β	1.02	0.03		−0.10	1.36	0.03	−0.57	−0.01	−0.01	−0.02	−0.01	−0.12
			$P\leq$	0.01	0.86		0.50	0.00	0.88	0.01	0.06	0.08	0.19	0.03	0.43
Orange-crowned Warbler	0.09	0.01	β	0.09	0.05		0.00	0.12	0.04	−0.09	0.00	0.00	0.00	0.00	0.01
			$P\leq$	0.06	0.13		1.00	0.05	0.25	0.00	0.08	0.07	0.09	0.04	0.82
Savannah Sparrow	0.13	0.00	β	0.02	−0.41	0.18	−0.08	0.16	0.04	0.15	0.00	0.00	0.00	0.00	−0.01
			$P\leq$	0.76	0.01	0.02	0.03	0.03	0.23	0.01	0.61	0.05	0.40	0.97	0.67
White-crowned Sparrow	0.23	0.00	β	1.92	−0.96		−1.30	−1.63	0.78	−0.43	−0.02	0.02	0.03	−0.03	−0.16
			$P\leq$	0.01	0.04		0.01	0.05	0.06	0.31	0.34	0.29	0.39	0.00	0.65
							Aquatic subguild								
Common Yellowthroat	0.25	0.00	β	0.01	0.00		0.00	0.00	0.00	0.00	0.00	0.00	0.00	0.00	0.00
			$P\leq$	0.01	0.17		0.27	0.46	0.08	0.36	0.44	0.07	0.67	0.34	0.00
Red-winged Blackbird	0.11	0.01	β	0.81	−0.61		−0.02	0.55	−0.58	0.31	0.00	0.01	0.00	−0.01	−0.07
			$P\leq$	0.01	0.01		0.90	0.15	0.00	0.12	1.00	0.13	0.99	0.09	0.66

F = forest, RAI = residential aggregation index, MDA = mean development age, GR = grass and agriculture, FAI = forest aggregation index, NPF = number patches of forest, NPU = number of patches urban, MPSU = mean urban patch size, RES = residential land use, PDR = residential patch diversity.

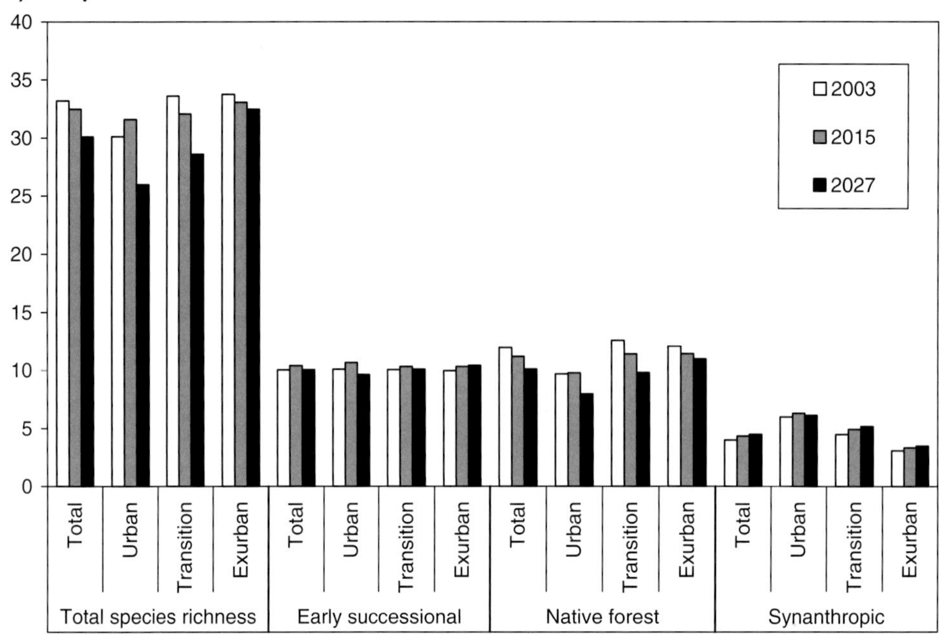

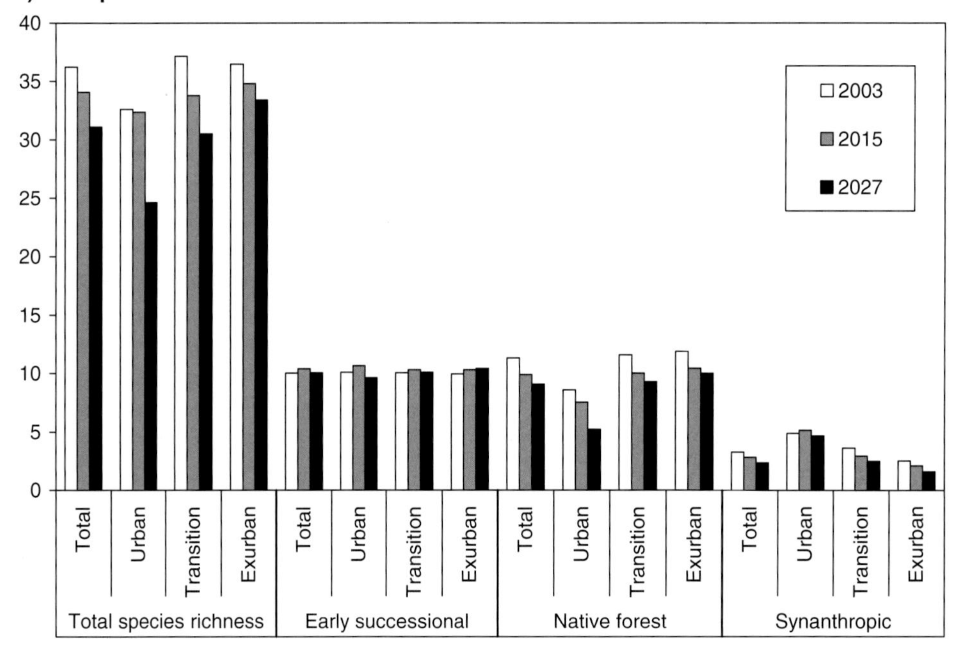

Figure 15.3. Comparison of mean predicted total and guild-specific species richness for study area and by land development zones (Fig. 15.1) for (a) the simple models and (b) the complex models.

brushy subguild, and losses in all but the exurban zone for aquatic species (Figs. 15.4, 15.5).

When we summed relative abundance estimates for the individual species comprising each subguild (Figs. 15.6, 15.7), we saw a pattern generally opposite to the pattern derived from the species richness models (Figs. 15.4, 15.5). The exceptions were for relative abundance of young and deciduous forest species and aquatic species, which were predicted to decrease in the urban

A) Young and deciduous forest species

B) Suburban species

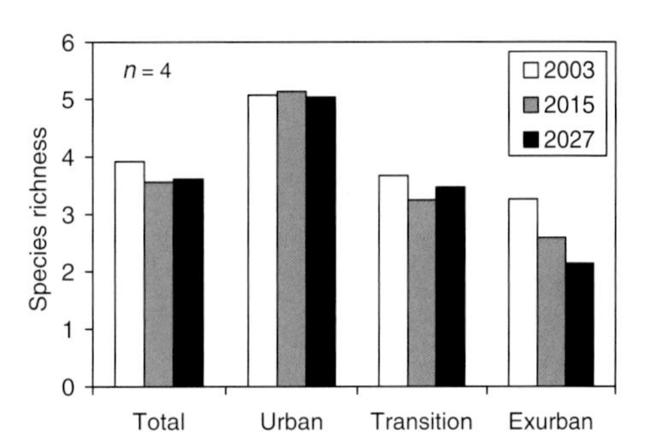

C) Open/grassy/brushy species

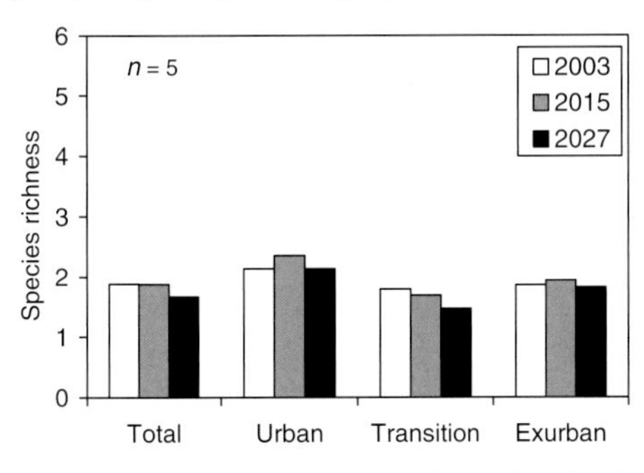

D) Aquatic species

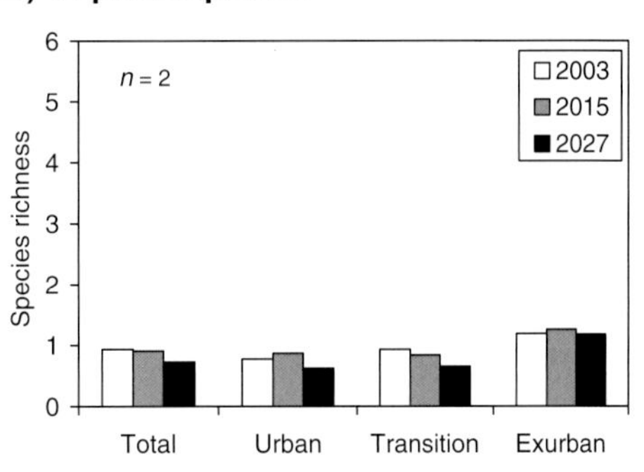

Figure 15.4. Mean predicted species richness for all early successional species for four early successional subguilds for study area and by land development zones derived from the complex models (Fig. 15.1).

zone. The greatest increase in numbers was predicted to occur in the transition zone for suburban species, going from less than 20 individuals per ha to almost 40 per ha, nearly approaching the mean relative abundance of this subguild in the urban zone (Fig. 15.6). Maps of changes in relative abundance of species in each sub-guild clearly show predicted losses of species in the young and deciduous forest and open/grassy/brushy subguilds in the agricultural lowlands and gains in the higher elevations (Fig. 15.7). Suburban species were predicted to increase the most in the transition zone (Fig. 15.7).

DISCUSSION

We predicted how avian communities will change as a result of predicted changes in land use and land cover. Land use predictions from UrbanSim, a microsimulation urban development model, and land cover predictions from LCCM were used as inputs into avian models to predict changes in avian species richness and individual species' relative abundance over time. Our development of an integrated model within a spatially explicit modeling framework included the effects of spatial interaction and neighborhoods, two priorities for research within the field of land use change modeling (Verberg et al. 2004).

The coupling of economic, biophysical, and wildlife habitat models allows policymakers to explore how policy options may alter landscapes and therefore assess the implications of different policies on land development, land cover, and biodiversity. Agencies charged with regional metropolitan planning (e.g., Puget Sound Regional Council in the Seattle, Washington, metropolitan area) require such integrated modeling systems to assess the impacts of different scenario alternatives and make effective planning decisions and investment choices (Waddell 2002, Waddell et al. 2003, Borning et al. 2007). Natural resource and wildlife conservation agencies and nongovernmental organizations also require tools to plan for landscape change.

There are several published studies that model future land use change and urban development and the potential impacts of these changes on

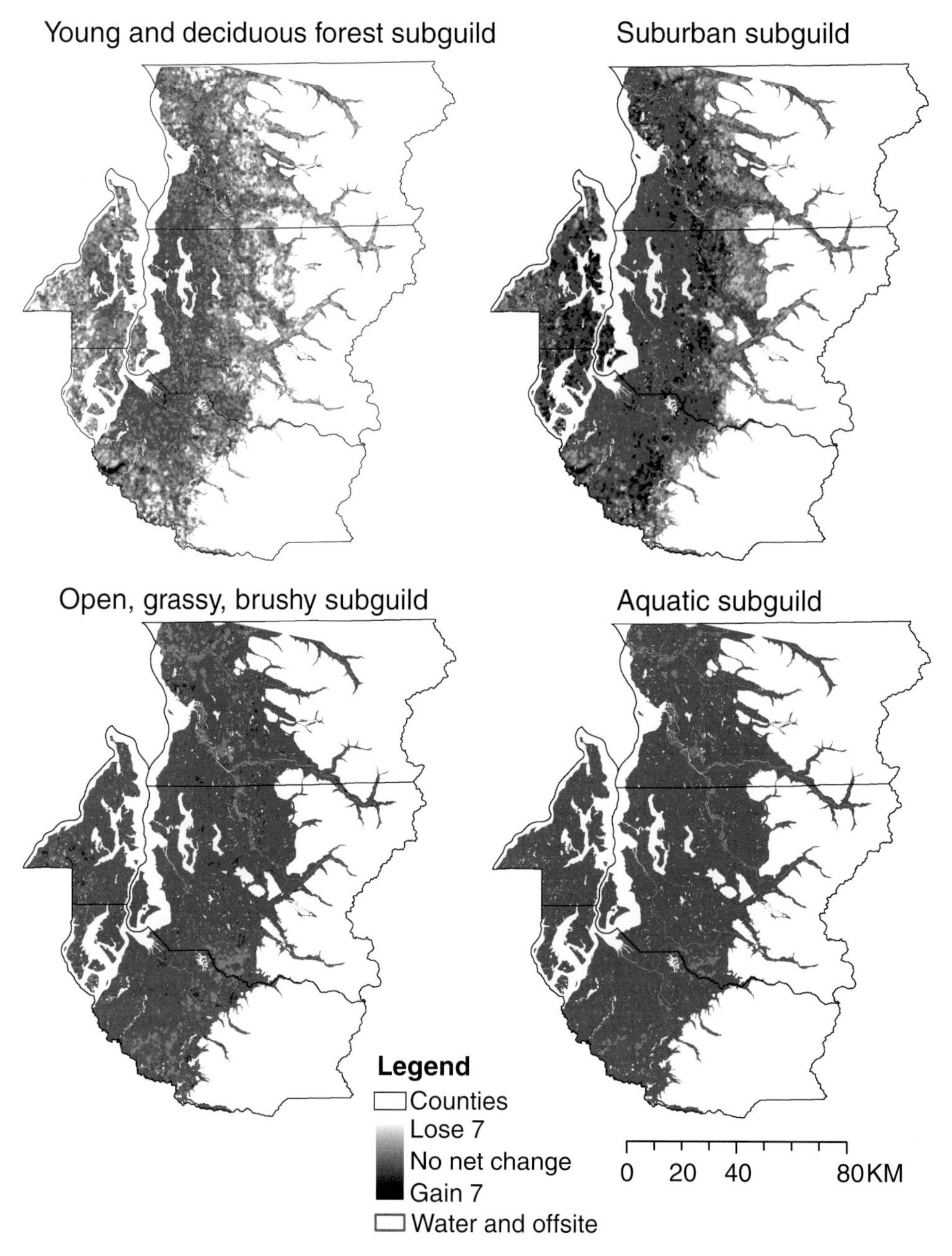

Figure 15.5. Mapped differences in the number of species predicted for the early successional subguilds between 2003 and 2027 over the study area <500 m elevation as predicted using complex models. (Note: A color version of this figure is available online.)

natural systems. For example, in the Willamette River basin in western Oregon, possible future scenarios were envisioned using guidance from citizen groups and generalized models of landscape change were created to depict how the landscape might change over 50 years (Baker et al. 2004, Hulse et al. 2004). These future landscapes were used as input in wildlife habitat (Schumaker et al. 2004) and fish and aquatic community (Van Sickle et al. 2004) models to determine potential changes in future biotic communities. Outputs of these modeling exercises have included future scenarios for use in salmon recovery (Hulse et al. 2004) and basin-wide restoration (Baker et al.

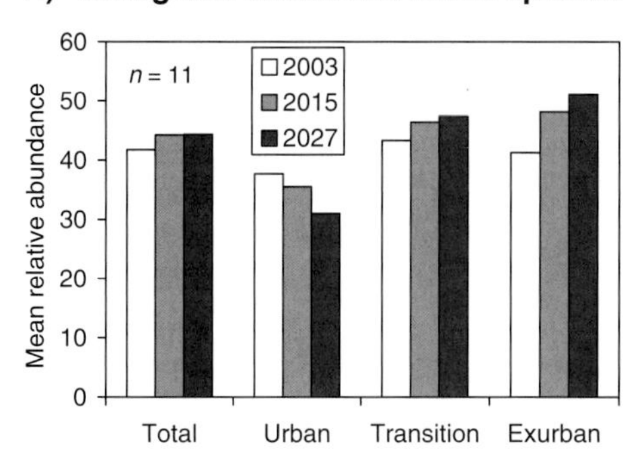

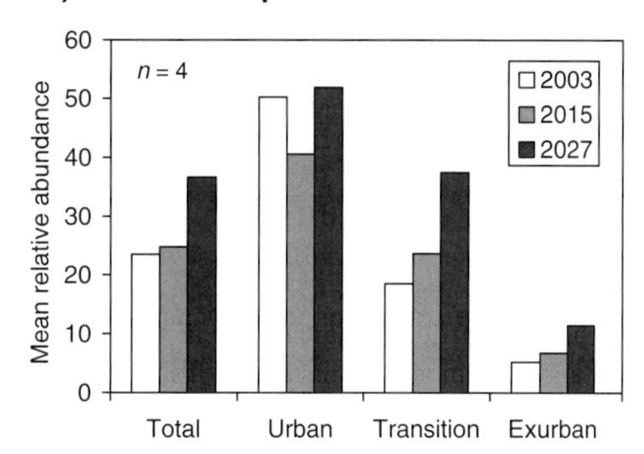

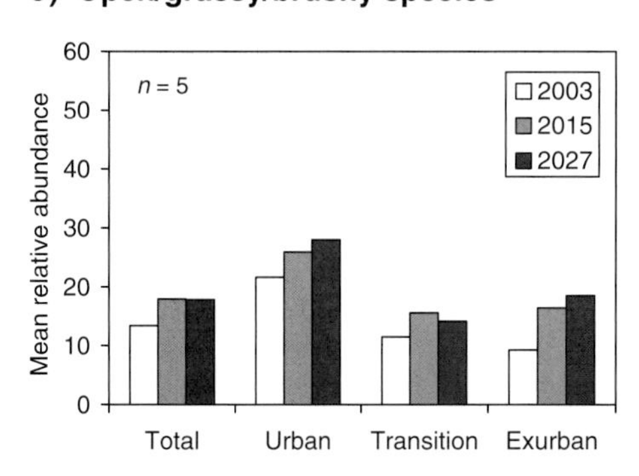

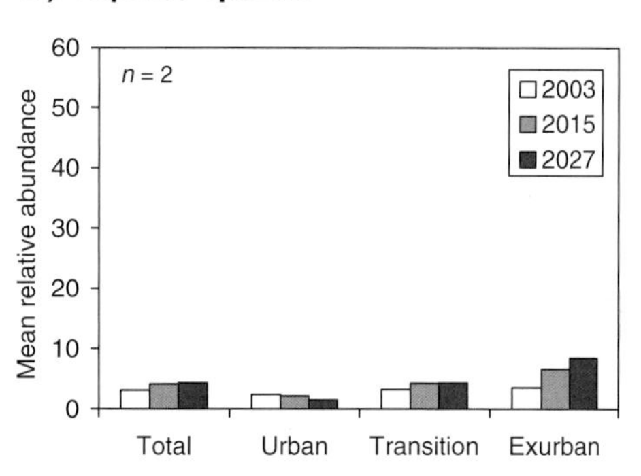

Figure 15.6. Early successional summed predictions of relative abundance per hectare for four early successional subguilds for the study area and by land development zones derived from the complex models (Fig. 15.1).

2004). Our work builds on these and other studies by including output from spatially explicit, sophisticated economic and biophysical models as input into our biodiversity models.

Previous field work examined the effects of habitat fragmentation (Bolger et al. 1997), spatial patterns of habitat (Hennings and Edge 2003, Blair 2004), neighboring habitats (Clergeau et al. 2001), and scale of analysis (Hostetler 1999) on avian communities across urban land uses. The substantial reduction in forest cover and increase in developed land predicted for the Seattle area in the next few decades is similar to observed and predicted changes in other regions of the country (Turner et al. 2003) and is expected to challenge the region's avifauna. The decline in overall diversity and changes in avian community composition that we predict have been observed in field studies (Blair 2004) and simulation studies (Schumaker et al. 2004). Blair (1996) observed a shift from native species dominating in undeveloped areas to exotic and invasive species in more developed sites. Clergeau et al. (2006) found that urbanization had the effect of homogenizing the avifauna in urban areas in three European countries. Whittaker and Marzluff (chapter 12, this volume) found that more impervious area was correlated with negative effects on juvenile mobility and percent forest or urban were correlated with different ranges of mobility for the species they studied.

The peak in avian diversity at intermediate levels of forest cover (35% and 50%) that we predicted is similar to results seen by others (Blair 2004, Marzluff 2005, Pennington and Blair, chapter 2, this volume). Initial urbanization generates more habitat heterogeneity, allowing more species to colonize, but as urbanization intensity increases, habitat becomes more homogeneous and dominated by built structures and impervious surfaces. Because urbanization represents a more permanent disturbance than other forms of disturbance such as logging and fire, the effects of urbanization likely will be longer-lasting and cumulative (McKinney 2002). We predicted that total avian diversity will decline substantially as forest cover is reduced below 40% and native forest and some early successional species decline.

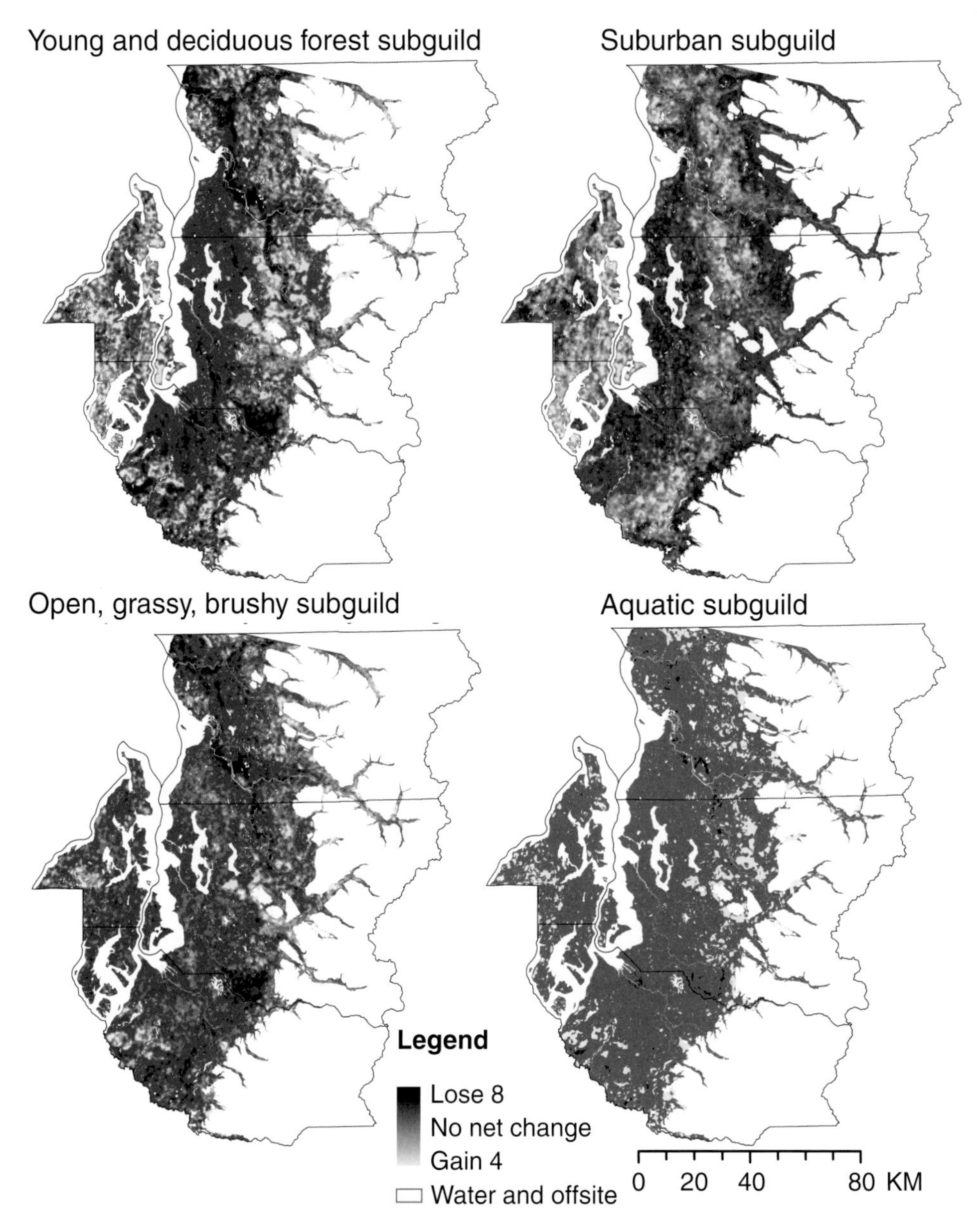

Figure 15.7. Mapped changes in total relative abundance (relative gain or loss of individuals) per hectare, summed for each species in the four early successional subguilds between 2003 and 2027 for study area predicted by the complex models. (Note: A color version of this figure is available online.)

Empirical studies in forested systems have seen a similar decrease in species diversity when forest cover is reduced below a threshold (Villard et al. 1999, Smith and Hellmann 2002, Fahrig 2003, Pennington and Blair, chapter 2, this volume). A trend for declines will be compensated for somewhat by increased relative abundance of synanthropic and early successional species in new developments, but as developments age, these species may decrease in abundance as vegetation regrows (Donnelly and Marzluff 2006).

Individual species as well as guild-level responses to the processes of forest conversion is obvious to even the casual observer; we captured these responses in our models by including measures of the amount and pattern of forest and development and age of development. While the mean diversity of birds was only predicted to decline by an average of 3–5 species, the region's bird diversity will be significantly more vulnerable to further loss of forest than it is at present. Increased vulnerability is expected because: (1) our land cover change model predicts more of the landscape will have <40% forest cover; and (2) avian diversity in our forested landscape exhibits a slight peak, with 35–50% forest cover. If more of our landscape contains <40% forest and cover continues to drop, we will see a continued decline in species richness. In contrast, avian diversity declines only slightly as forest cover exceeds 60% and bird communities come to be composed of mainly native forest species (Marzluff 2005). Thus, the reduced forest cover (from 43% in 2003 to 29% in 2027) that we predict is expected to substantially lower regional avian diversity, especially in those areas where forest currently occupies >40% of the landscape but is predicted to drop substantially <40% by 2027. Previous population studies have seen a drop in population viability when the patches of habitat decrease below 30–50% of the landscape, thereby limiting the movement potential between habitat patches (Smith and Hellmann 2002).

The spatial pattern of forest loss has additional consequences for future avian communities. Forest species increased in landscapes with increased amounts of aggregated forest patches, as well as aggregated housing developments, because in such landscapes substantial contiguous forests remained. Synanthropic species also increased in landscapes of aggregated development, but where built lands were extensive. Currently, avian diversity is greatest in areas with a diversity of land cover types characteristic of areas with little to medium levels of land development (Hansen et al. 2005). However, our land cover simulations indicated that avian diversity will decrease in coming years in areas closer to developed lands. Such a decrease in avian diversity and increase in the dominance of urban land cover will push the avian community toward a more homogeneous community composed of synanthropic species (Blair 1996; Donnelly and Marzluff 2004a, 2006; Marzluff 2005). Future bird community richness was predicted to increase gradually with distance from development and as the transition zone was transformed into denser development. The region will likely be characterized by spatially partitioned bird communities dominated by either adaptable, synanthropic species (in dense developments) or resilient native forest birds (in the exurban zone). Currently, forest cover is limited in the areas close to Seattle. Much of the transition zone east of Seattle is a 50:50 mixture of developed and forested lands with high avian diversity. Further east, in the forested higher-elevation areas, large patches of contiguous forest likely contain viable populations of native forest birds. By 2027, we predicted that the rich 50:50 landscape of the transition zone will be mostly lost and replaced with development of increased density, resulting in lowered bird diversity. Further east in the forested Cascade Mountain foothills, bird communities were predicted to remain more diverse. Because the LCCM does not model forest regrowth into mature forest, predicted total and native forest species richness may have underestimated richness in the higher-elevation production forests east of Seattle. Both UrbanSim and LCCM were designed primarily for application in urbanizing regions, and do not capture the forest landscape change as accurately as a model designed specifically for such lands (Hepinstall et al. 2008).

Our synanthropic guild predictions using the complex model clearly contained errors (e.g., species richness was predicted to decline in the transition zone, where an increase in developed land cover was predicted). The complex models included more variables that measured landscape configuration; however, species present within each guild or subguild may not be responding in the same direction or magnitude to these measures, thereby confounding our results. Schumaker et al. (2004) also observed complex and often contradictory results when projecting the response of wildlife populations to future landscape change. They attributed their results to the fact that individual species respond in different ways to incremental changes in the amount and configuration of habitat on the landscape. Further exploration of species responses to specific measures of landscape configuration is warranted. In addition, our results comparing predicted trends in species richness versus relative abundance indicates that community dynamics may not equate with population dynamics. Summing predictions for individual species abundance should, in theory, mirror the predicted total species richness; however, this is rarely the case due to confounding factors such as individualistic species responses to different aspects of urbanization (Rodewald, chapter 5, this volume) or differential species detection affecting our model parameter estimates (Dorazio et al. 2006).

Understanding the dynamics of early successional species is vital to understanding the overall response of birds to urbanization in our region (Marzluff 2005), because half of the total bird diversity is included in this group (Appendix 15.1). Our expanded efforts in predicting early successional species responses to changing landscapes produced conflicting results, with predictions of early successional species richness having counter-intuitive results of lower species richness in future landscapes. Our models of single species' relative abundance were better at capturing the diversity of observed responses than guild predictions of species richness. Individual species relative abundance predictions appeared more realistic, producing predictions of increased abundance which we expected to see as land is converted from remnant forest to residential areas, lawns, and other landscaped areas.

CONCLUSIONS AND FUTURE DIRECTIONS

In summary, we used data derived from years of fieldwork to develop models predicting biodiversity that allowed us to explore how land use and land cover change may influence future avian species richness and relative abundance in the central Puget Sound of western Washington. Current and future predictions clearly indicated that both landscape composition and configuration are important in determining avian community composition.

We predicted that future bird communities in urbanizing areas will be slightly less diverse and more vulnerable to future losses than they are at present. Predicted declines in some early successional species in urban areas and native forest species in areas undergoing land cover conversion will drive these losses: in urbanizing areas, early successional communities will shift to more individuals associated with suburban landscapes and fewer associated with young and deciduous forested or aquatic areas; native forest birds increasingly will become reliant on higher-elevation forests because most low-elevation forests will be converted to development too dense to support large populations. Bird populations at high elevations may be less sustainable due to harsher winters and shorter growing seasons that may limit survival and reproduction.

Our study has advanced urban avian studies by directly linking economic-derived land development models with land cover change models to predict changes in the landscape relevant to birds and predict how these changes will affect future species richness and relative abundance of individual species. Work continues on increasing the accuracy, flexibility, and applicability of each model. For instance, the LCCM's accuracy for exurban areas could be increased by using readily available forest growth models and modeling systems (e.g., ORGANON, LMS). Also, coupling LCCM with biophysical process models (e.g., climate) is critical when predicting land cover change and avian communities on a longer time scale.

Avian models of relative abundance in their current form have not been corrected for the effects of species detectability (Boulinier et al. 1998), which is something that may aid the development of community richness models from predictions of individual species abundance (Royle 2004, Kéry et al. 2005). Additional exploration into and inclusion of the mechanisms (e.g., dispersal, territorial size, source-sink dynamics) behind observed bird population responses to landscape change may also improve abundance and richness predictions.

ACKNOWLEDGMENTS

We thank R. Donnelly, T. Blewett, C. Ianni, K. Whittaker, J. DeLap, S. Rullman, T. Unfried, and D. Oleyar for their help collecting data. We thank S. Coe for development of land cover maps used in LCCM and the Center for Urban Simulation and Policy Analysis at the University of Washington. We thank anonymous reviewers, C. A. Lepczyk, P. S. Warren, and K. Merry, who provided helpful and constructive comments on previous drafts. Many private landowners graciously gave permission to conduct research on their property. Research was supported by the University of Washington, the UW's School of Forest Resources, particularly its Rachel Woods Endowed Graduate Program and the Denman Sustainable Resource Sciences Professorship.

LITERATURE CITED

Baker, J. P., D. W. Hulse, S. V. Gregory, D. White, J. V. Sickle, P. A. Berger, D. Dole, and N. H. Schumaker. 2004. Alternative futures for the Willamette River basin, Oregon. Ecological Applications 14: 313–324.

Blair, R. B. 1996. Land use and avian species diversity along an urban gradient. Ecological Applications 6:506–519.

Blair, R. B. 2001a. Birds and butterflies along urban gradients in two ecoregions of the United States: is urbanization creating a homogeneous fauna? Pp. 33–56 *in* J. L. Lockwood and M. L. McKinney (editors). Biotic homogenization. Kluwer Academic/ Plenum, New York, NY.

Blair, R. B. 2001b. Creating a homogeneous avifauna. Pp. 459–486 *in* J. M. Marzluff, R. Bowman, and R. Donnelly, editors. Avian ecology and conservation in an urbanizing world. Kluwer Academic, Norwell, MA.

Blair, R. B. 2004. The effects of urban sprawl on birds at multiple levels of biological organization. Ecology and Society 9:2.

Blewett, C. M., and J. M. Marzluff. 2005. Effects of urban sprawl on snags and the abundance and productivity of cavity-nesting birds. Condor 107:677–692.

Bolger, D. T., T. A. Scott, and J. T. Rotenberry. 1997. Breeding bird abundance in an urbanizing landscape in coastal southern California. Conservation Biology 11:406–421.

Borning, A., P. Waddell, and R. Förster. 2007. UrbanSim: Using simulation to inform public deliberation and decision-making. Pp. 439–466 *in* H. Chen, L. Brandt, V. Gregg, R. Traunmüller, S. Dawes, E. Hovy, A. Macintosh, and C. A. Larson (editors), E-government research, case studies, and implementation series: integrated series in information systems, Vol. 17. Springer-Verlag, New York, NY.

Boulinier, T., J. D. Nichols, J. R. Saur, J. E. Hines, and K. H. Pollock. 1998. Estimating species richness: the importance of heterogeneity in species detectability. Ecology 79:1018–1028.

Burnham, K. P., and D. R. Anderson. 1998. Model selection and inference: a practical information-theoretic approach. Springer-Verlag, New York, NY.

Clergeau, P., S. Croci, J. Jokimäki, M.-J. Kaisanlahti-Jokimäki, and M. Dinetti. 2006. Avifauna homogenization by urbanization: analysis at different European latitudes. Biological Conservation 127:336–344.

Clergeau, P., J. Jokimäki, and J.-P. L. Savard. 2001. Are urban bird communities influenced by the bird diversity of adjacent landscapes? Journal of Applied Ecology 38:1122–1134.

Doak, D. F., and L. S. Mills. 1994. A useful role for theory in conservation. Ecology 75:615–626.

Donnelly, R. E., and J. M. Marzluff. 2004a. Importance of reserve size and landscape context to urban bird conservation. Conservation Biology 18:733–745.

Donnelly, R. E., and J. M. Marzluff. 2004b. Designing research to advance the management of birds in urbanizing areas. Pp. 114–122 *in* W. W. Shaw, L. K. Harris, and L. Vandruff (editors), Proceedings of the 4th International Symposium on Urban Wildlife Conservation, 1–5 May 1999. University of Arizona Press, Tuscon, AZ.

Donnelly, R. E., and J. M. Marzluff. 2006. Relative importance of habitat quantity, structure, and spatial pattern to birds in urbanizing environments. Urban Ecosystems 9:99–117.

Dorazio, R. M., J. A. Royle, B. Soderstrom, and A. Glimskar. 2006. Estimating species richness and accumulation by modeling species occurrence and detectability. Ecology 87:842–854.

Dunning, J. B., D. J. Stewart, B. J. Danielson, B. R. Noon, T. L. Root, R. H. Lamberson, and E. E. Stevens. 1995. Spatially explicit population models: current forms and future uses. Ecological Applications 5:3–11.

Elvidge, C. D., K. E. Baugh, E. A. Kihn, H. W. Kroehl, and E. R. Davis. 1997. Mapping city lights with nighttime data from the DMSP operational linescan system. Photogrammatic Engineering and Remote Sensing 63:727–734.

Ewing, R. H. 1994. Characteristics, causes, and effects of sprawl: a literature review. Environmental and Urban Studies 21:1–17.

Fahrig, L. 2003. Effect of habitat fragmentation on biodiversity. Annual Review of Ecology, Evolution, and Systematics 34:487–515.

Fahrig, L., and G. Merriam. 1994. Conservation of fragmented populations. Conservation Biology 8:50–59.

Foley, J. A., R. DeFries, G. P. Asner, C. Barford, G. Bonan, S. R. Carpenter, F. S. Chapin, M. T. Coe, G. C. Daily, H. K. Gibbs, J. H. Helkowski, T. Holloway, E. A. Howard, C. J. Kucharik, C. Monfreda, J. A. Patz, I. C. Pretice, N. Ramankutty, and P. K. Snyder. 2005. Global consequences of land use. Science 309:570–574.

Franklin, J. F., and C. T. Dyrness. 1988. Natural vegetation of Oregon and Washington. Oregon State University Press, Corvallis, OR.

Hansen, A. J., R. L. Knight, J. M. Marzluff, S. Powell, K. Brown, P. H. Gude, and K. Jones. 2005. Effects of exurban development on biodiversity: patterns, mechanisms, and research needs. Ecological Applications 15:1893–1905.

Hennings, L. A., and W. D. Edge. 2003. Riparian bird community structure in Portland, Oregon: habitat, urbanization, and spatial scale patterns. Condor 105:288–302.

Hepinstall, J. A., M. Alberti., and J. M. Marzluff. 2008. Predicting land cover change and avian community responses in rapidly urbanizing environments. Landscape Ecology 28:1257–1276.

Hepinstall, J. A., J. M. Marzluff, and M. Alberti. 2009. Modeling the responses of birds to predicted changes in land cover in an urbanizing region. Pp. 625–659 *In* J. Millspaugh and F. R. Thompson III (editors), Models for planning wildlife conservation in large landscapes. Elsevier Science, Amsterdam, The Netherlands.

Hostetler, M. 1999. Scale, birds, and human decisions: a potential for integrative research in urban ecosystems. Landscape and Urban Planning 45:15–19.

Houghton, R. A. 1994. The worldwide extend of land-use change. BioScience 44:305–313.

Hulse, D. W., A. Branscomb, S. G. Payne. 2004. Envisioning alternatives: using citizen guidance to map future land and water use. Ecological Applications 14:325–341.

Imhoff, M. L., L. Bounoua, R. DeFries, W. T. Lawrence, D. Stutzer, C. J. Tucker, and T. Rickets. 2004. The consequences of urban land transformation on net primary productivity in the United States. Remote Sensing of Environment 89:434–443.

Kéry, M., J. A. Royle, and H. Schmid. 2005. Modeling avian abundance from replicated counts using binomial mixture models. Ecological Applications 15:1450–1461.

Kowarik, I. 1995. On the role of alien species in urban flora and vegetation. Pp. 85–103 *in* P. K. Pysek, M. Prach, M. Rejmanek, and P. M. Wade, (editors), Plant invasions: general aspects and special problems. SPB Academic, Amsterdam, The Netherlands.

Lawrence, W. T., M. L. Imhoff, N. Kerle, and D. Stutzer. 2002. Quantifying urban land use and impact on soils in Egypt using diurnal satellite imagery of the earth surface. International Journal of Remote Sensing 23:3921–3937.

Marzluff, J. M. 2001 Worldwide urbanization and its affects on birds. Pp. 19–47 *in* J. M. Marzluff, R. Bowman, and R. Donnelly (editors), Avian ecology and conservation in an urbanizing world. Kluwer Academic Publishers, Norwell, MA.

Marzluff, J. M. 2005. Island biogeography for an urbanizing world: how extinction and colonization may determine biological diversity in human-dominated landscapes. Urban Ecosystems 8:157–177.

McGarigal, K., S. A. Cushman, M. C. Neel, and E. Ene. 2002. FRAGSTATS: spatial pattern analysis program for categorical maps. University of Massachusetts, Amherst, MA.

McGarigal, K., and W. C. McComb. 1995. Relationships between landscape structure and breeding birds in the Oregon Coast Range. Ecological Monographs 65:235–260.

McKinney, M. L. 2002. Urbanization, biodiversity, and conservation. BioScience 52:883–890.

Meyer, W. B., and B. L. Turner II. 1992. Human population growth and global land-use/cover change. Annual Review of Ecology and Systematics 23:39–61.

Mladenoff, D. J. 2004. LANDIS and forest landscape models. Ecological Modelling 180:7–19.

Mörtberg, U. M., B. Balfors, and W. C. Knol. 2007. Landscape ecological assessment: a tool for integrating biodiversity issues in strategic environmental assessment and planning. Journal of Environmental Management 82:457–470.

Pickett, S. T. A., M. L. Cadenasso, J. M. Grove, C. H. Nilon, R. V. Pouyat, W. C. Zipperer, and R. Costanza. 2001. Urban ecological systems: linking terrestrial ecological, physical, and socioeconomic components of metropolitan areas. Annual Review of Ecology and Systematics 32:127–157.

Robinson, L., J. P. Newell, and J. M. Marzluff. 2005. Twenty-five years of sprawl in the Seattle region: growth management responses and implications for conservation. Landscape and Urban Planning 71:51–72.

Royle, J. A. 2004. N-mixture models for estimating population size from spatially replicated counts. Biometrics 60:108–115.

Ruckelshaus, M., C. Hartway, and P. Kareiva. 1997. Assessing the data requirements of spatially explicit dispersal models. Conservation Biology 11:1298–1306.

Sala, O. E., F. S. Chapin III, J. J. Armesto, E. Berlow, J. Bloomfield, R. Dirzo, E. Huber-Sanwald, L. F. Huenneke, R. B. Jackson, A. Kinzig, R. Leemans, D. M. Lodge, H. A. Mooney, M. Oesterheld, N. L. Poff, M. T. Sykes, B. H. Walker, M. Walker, and D. H. Wall. 2000. Global biodiversity scenarios for the year 2100. Science 287:1770–1774.

Schumaker, N. H., T. Ernst, D. White, J. Baker, and P. Haggerty. 2004. Projecting wildlife responses to alternative future landscapes in Oregon's Willamette basin. Ecological Applications 14:381–400.

Scott, J. M., F. Davis, B. Csuti, R. Noss, B. Butterfield, S. Caicco, C. Groves, T. C. Edwards Jr., J. Ulliman, H. Anderson, F. D'Erchia, and R. G. Wright. 1993. Gap analysis: a geographic approach to protection of biological diversity. Wildlife Monographs 123:3–41.

Smith, J. N. M., and J. J. Hellmann. 2002. Population persistence in fragmented landscapes. Trends in Ecology and Evolution 17:397–399.

Turner, M. G., S. M. Pearson, P. Bolstad, and D. N. Wear. 2003. Effects of land-cover change on spatial pattern of forest communities in the Southern Appalachian Mountains (USA). Landscape Ecology 18:449–464.

Van Sickle, J., J. Baker, A. Herlihy, P. Bayley, S. Gregory, P. Haggerty, L. Ashkenas, and J. Li. 2004. Projecting

the biological condition of streams under alternative scenarios of human land use. Ecological Applications 14:368–380.
Verburg, P. H., P. P. Schot, M. J. Dijst, and A. Veldkamp. 2004. Land use change modelling: current practice and research priorities. GeoJournal 61:309–324.
Villard, M., M. K. Trzcinski, and G. Merriam. 1999. Fragmentation effects on forest birds: relative influence of woodland cover and configuration on landscape occupancy. Conservation Biology 13:774–783.
Waddell, P. 2002. UrbanSim: modeling urban development for land use, transportation and environmental planning. Journal of the American Planning Association 68:297–314.
Waddell, P., A. Borning, M. Noth, N. Freier, M. Becke, and G. Ulfarsson. 2003. UrbanSim: a simulation system for land use and transportation. Networks and Spatial Economics 3:43–67.
Wear, D. N., and P. Bolstad. 1998. Land-use changes in southern Appalachian landscapes: spatial analysis and forecast evaluation. Ecosystems 1:575–594.
White, D., P. G. Minotti, M. J. Barczak, J. C. Sifneos, K. E. Freemark, M. V. Santelmann, C. F. Steinitz, A. R. Kiester, and E. M. Preston. 1997. Assessing risks to biodiversity from future landscape change. Conservation Biology 11:349–360.

APPENDIX 15.1

List of common bird species by guild membership, as detected in point count surveys of 139 study sites in Puget Sound, Washington, 1998–2005.

Common name	Scientific name
Native forest	
American Robin	*Turdus migratorius*
Black-throated Gray Warbler	*Setophaga nigrescens*
Brown Creeper	*Certhia americana*
Chestnut-backed Chickadee	*Poecile rufescens*
Dark-eyed Junco	*Junco hyemalis*
Downy Woodpecker	*Picoides pubescens*
Golden-crowned Kinglet	*Regulus satrapa*
Hairy Woodpecker	*Picoides villosus*
Hammond's Flycatcher	*Empidonax hammondii*
Hermit Thrush	*Catharus guttatus*
Hutton's Vireo	*Vireo huttoni*
Pacific-slope Flycatcher	*Empidonax difficilis*
Red-breasted Nuthatch	*Sitta canadensis*
Spotted Towhee	*Pipilo maculatus*
Steller's Jay	*Cyanocitta stelleri*
Swainson's Thrush	*Catharus ustulatus*
Western Tanager	*Piranga ludoviciana*
Wilson's Warbler	*Wilsonia pusilla*
Winter Wren	*Troglodytes troglodytes*

Common name	Scientific name
Synanthropic	
American Crow	*Corvus brachyrhynchos*
Anna's Hummingbird	*Calypte anna*
Barn Swallow	*Hirundo rustica*
Brewer's Blackbird	*Euphagus cyanocephalus*
Brown-headed Cowbird	*Molothrus ater*
European Starling	*Sturnus vulgaris*
House Finch	*Carpodacus mexicanus*
House Sparrow	*Passer domesticus*
Rock Pigeon	*Columba livia*
Early successional subguild: Young and deciduous forest	
Black-headed Grosbeak	*Pheucticus melanocephalus*
Band-tailed Pigeon	*Columba fasciata*
Cassin's Vireo	*Vireo cassinii*
MacGillivray's Warbler	*Geothlypis tolmiei*
Olive-sided Flycatcher	*Contopus cooperi*
Pine Siskin	*Carduelis pinus*
Purple Finch	*Carpodacus purpureus*
Red Crossbill	*Loxia curvirostra*
Rufous Hummingbird	*Selasphorus rufus*
Song Sparrow	*Melospiza melodia*
Tree Swallow	*Tachycineta bicolor*
Warbling Vireo	*Vireo gilvus*
Willow Flycatcher	*Empidonax traillii*
Western Wood-Pewee	*Contopus sordidulus*
Yellow-rumped Warbler	*Setophaga coronata*
Early successional subguild: Suburban	
Black-capped Chickadee	*Poecile atricapillus*
Bewick's Wren	*Thryomanes bewickii*
Bushtit	*Psaltriparus minimus*
Cedar Waxwing	*Bombycilla cedrorum*
Northern Flicker	*Colaptes auratus*
Violet-green Swallow	*Tachycineta thalassina*

Common name	Scientific name
Early successional subguild: Open/grassy/bushy	
American Goldfinch	*Carduelis tristis*
Killdeer	*Charadrius vociferus*
Northern Rough-winged Swallow	*Stelgidopteryx serripennis*
Orange-crowned Warbler	*Oreothlypis celata*
Savannah Sparrow	*Passerculus sandwichensis*
White-crowned Sparrow	*Zonotrichia leucophrys*
Early successional subguild: Aquatic	
Common Yellowthroat	*Geothlypis trichas*
Red-winged Blackbird	*Agelaius phoeniceus*

CHAPTER SIXTEEN

Interactions between People and Birds in Urban Landscapes

Richard A. Fuller, Katherine N. Irvine, Zoe G. Davies, Paul R. Armsworth, and Kevin J. Gaston

Abstract. A large body of work over the past few decades has revealed the manifestly dramatic impacts of urbanization on species' distributions and ecologies, many of which result from gross changes in land use and configuration. Less well understood are the rather more direct interactions between people and biodiversity in the urban arena. While there is a general concern that urbanization impoverishes human contact with nature, daily interaction with biodiversity in urban greenspaces and the widespread provision of food and nesting resources for wildlife form a part of many city-dwellers' experience. Using data from the UK, we show that supplementary resource provision aimed explicitly at enhancing avian populations can result in high levels of additional foraging and nesting opportunities, particularly in urban areas. However, our data also indicate that levels of such resource provision are strongly positively correlated with human population density at a regional scale, and within a large city. The proportion of households participating in bird feeding depends on social and economic features of the human population, suggesting that strong covariation between human and ecological communities will result. Indeed, we demonstrate that the abundances of some urban-adapted bird species are positively related to the density of feeding stations across the urban landscape, although such relationships were not apparent for other species that commonly use garden feeding stations. It has been suggested that interactions with nature, such as feeding birds, could have beneficial consequences for human health. A better understanding of this potential feedback is required.

Key Words: bird feeding, housing density, private gardens, socioeconomics, urban ecology.

The provision of feeding and nesting resources for birds is a popular activity across much of the world, particularly in industrialized nations. Between one-fifth and one-third of households in Europe, North America, and Australia provide supplementary food for wild birds (Clergeau et al. 1997, Rollinson et al. 2003, Lepczyk et al. 2004), and in the United States alone, 52 million people frequently feed garden birds (U.S. Fish and Wildlife Service

Fuller, R. A., K. N. Irvine, Z. G. Davies, P. R. Armsworth, and K. J. Gaston. 2012. Interactions between people and birds in urban landscapes. Pp. 249–266 *in* C. A. Lepczyk and P. S. Warren (editors). Urban bird ecology and conservation. Studies in Avian Biology (no. 45), University of California Press, Berkeley, CA.

2001). Surprisingly few studies have considered the role of gardens in supporting biodiversity (but see Savard et al. 2000; Beebee 2001; Thompson et al. 2003; Gaston et al. 2005a, 2005b; Daniels and Kirkpatrick 2006; Smith et al. 2006) or the impact of supplementary resources provisioned within gardens.

SPATIAL PATTERNS IN PROVISION OF RESOURCES FOR BIRDS

Significant sums of money are spent annually on deliberate resource provision for wild birds, and a supply industry has emerged, frequently importing feed from those tropical countries where much of it is grown. CJ WildBird Foods, Europe's largest wild bird food supplier, currently employs more than 150 staff and generated sales worth £20 million (US$39 million) in 2005/06 (http://www.birdfood.co.uk). Total annual expenditure on outdoor feeding of birds in the UK has recently been estimated at £200 million (US$390 million; British Trust for Ornithology 2006), while in the U.S. $3.5 billion is spent annually on bird food and feeding equipment (U.S. Fish and Wildlife Service 2001). International trade supplies a global market in specialty bird seed. For example, niger (*Guizotia abyssinica*) is grown mainly in India [22,609 tons (t) imported to the U.S. for bird food in 2003], Ethiopia (18,290 t), and Myanmar (7,043 t; Lin 2005).

The results of this deliberate resource provision to birds depend on its distribution across the landscape. A recent study in southeastern Michigan, USA, found that while the proportion of landowners providing food for birds did not vary among rural, suburban, and urban landscapes, the density of bird feeders per land parcel was significantly higher in urban than in rural and suburban areas (Lepczyk et al. 2004), presumably driven by the smaller size of plots in urban landscapes. If the density of human settlement predicts the density of bird feeders across the landscape, we might expect resource provision to occur disproportionately in (1) more densely populated regions and (2) more densely populated neighborhoods within cities. However, this will depend on how the popularity of bird feeding varies in relation to human population density and socioeconomic factors. Here, we use data on bird feeding across England at a regional scale, as well as information on small-scale variation in the activity within a large city, to describe spatial variation in the proportion of people engaging in supplementary feeding, and how this variation translates into patterns in the spatial density of resource provision.

DRIVERS OF BIRD-FEEDING ACTIVITY

While providing a significant resource base, the provision of food and nesting sites for birds also represents an opportunity for interaction between people and nature, and it occurs close to where people live and work on a daily basis (Miller and Hobbs 2002). Experiences of nature lead to a variety of measurable benefits, at both individual and societal levels (Vandruff et al. 1995, Mabey 1999, Irvine and Warber 2002, de Vries et al. 2003, Maller et al. 2005). For example, the presence of urban open spaces with trees and grass increased social interaction among neighbors, promoted a sense of community, and reduced crime in inner-city low-income housing areas of Chicago (Kuo et al. 1998, Kuo and Sullivan 2001), and the psychological benefits derived by visitors to urban greenspaces in Sheffield, UK, increased with plant species richness at the sites (Fuller et al. 2007). Given that a large proportion of the human population lives in cities, most of these human-nature interactions will inevitably focus on those species occurring in urban environments. However, surprisingly little is known about the drivers and consequences of these interactions in cities.

Levels of bird feeding and other forms of wildlife gardening vary enormously across the human population (Lepczyk et al. 2002, 2004; Gaston et al. 2007). Landowners participating in bird-feeding activity in southeastern Michigan tended to be older, were more likely to be women, and had achieved higher educational qualifications than those not participating (Lepczyk et al. 2004). Bird feeding was not related to the number of dwelling occupants, their occupation, or dwelling size as measured by floor area. Additional factors that might influence the likelihood of engaging in bird-feeding activity include economic and perceptual considerations, social context and garden size, interest in and knowledge about wildlife, and the amount of time that household members have available. As such, the level of participation in bird feeding is likely to vary consistently among different kinds of human communities, which themselves show complex patterns of spatial organization across urban landscapes (Harris et al. 2005).

Here, we investigate three possible socioeconomic drivers of bird-feeding activity: household income, age of dwelling occupants, and number of people composing the household, at both national and citywide scales.

ARE BIRD DENSITIES ASSOCIATED WITH LEVELS OF BIRD FEEDING?

Supplementary feeding clearly has the potential to improve the condition and increase the probability of survival of individual birds. Black-capped Chickadees (*Parus atricapillus*) with access to supplementary food during the winter months had greater body mass and higher overwinter survival rates than birds without such access (Brittingham and Temple 1988), and supplementary feeding improved nutritional condition as measured by feather growth rates in four North American bark-foraging species (Grubb and Cimprich 1990). Other studies have identified positive associations between urbanization and population density of supplementary feeding species (Jokimäki and Suhonen 1998) and positive effects of supplementary feeding on winter survival (van Balen 1980, Orell 1989).

Despite these specific examples, whether provision of food for birds in gardens can translate into higher population densities in general remains an open question. Given the popularity of bird feeding across much of the developed world, and the fact that bird feeders can reach very high densities in the landscape, one might expect population densities of those species best able to exploit the supplementary food to be positively correlated with levels of resource input at a landscape scale.

In this study we test this idea by relating the population density of six urban-adapted species to levels of supplementary bird-feeding activity across a large city. The six species were identified in a recent study as the most highly urbanized of the British avifauna (Cannon 2005), and they vary in their dietary requirements, specifically in the proportion of grains included in the diet. For comparison, we also present data for the Winter Wren (*Troglodytes troglodytes*), an insectivore that does not commonly take supplementary food in urban gardens, yet is found at reasonably high densities within urban environments. There is no *a priori* reason to assume that the density of this species will depend directly on provision of supplementary food.

In sum, the aims of this paper are threefold. First, we describe spatial patterns in bird-feeding activity and how this translates into resource availability on the ground, at both national and citywide scales. Second, we investigate variation in bird-feeding activity and resource availability in relation to housing density and human socioeconomic drivers, also at national and citywide scales. Third, we assess whether densities of selected bird species are associated with levels of supplementary resource provision across the urban landscape.

METHODS

This study was carried out at two spatial scales, first using data at the resolution of counties across the whole of England to characterize regional variation in bird-feeding activity, and second via a grid-based analysis of bird-feeding activity and the distributions of birds across the city of Sheffield, a large inland city in northern England. With a human population of ca. 513,000, Sheffield is the fifth largest municipality in the UK, and the ninth largest urban area (Office for National Statistics 2001, Beer 2005). The urban area of Sheffield was defined as the set of 1 km $\times$ 1 km squares within the administrative boundary of the city in which coverage by urban development exceeded 25% (Gaston et al. 2005b). This resulted in a Sheffield study area of 160 km^2.

Bird Feeding and Socioeconomic Variables

To investigate nationwide variation in bird feeding activity, we used data from the Survey of English Housing (SEH), an annual government-funded survey of ca. 30,000 households across England. The 2001/02 survey (NCSR and DETR 2004) included a small set of questions investigating participation in wildlife gardening. Respondents were asked whether they encourage wildlife in their garden, patio, yard, balcony, or roof terrace by (1) feeding the birds/providing bird feeders, bird tables, or birdbaths and/or (2) putting up nest boxes. Respondents were also asked to give annual gross total household income, age of the household reference person (the person in whose name the house is registered, or with highest income if the house is jointly registered, or the eldest occupant if incomes are equal), and the number of people living in the house.

For reasons of confidentiality, questionnaire data were only available aggregated at the scale of the local authority. Because this resulted in small sample sizes within some local authorities, for the present analysis we used data aggregated at county scale. There were 46 counties recognized in England in 2001, although the Isle of Wight, with only 50 respondents, was excluded from all analyses. We calculated *Proportion Feeding* (the number of households at which birds were fed in each county divided by the number of households in that county included in the survey) and *Feeder Density* (*Proportion Feeding* multiplied by the number of households in the county, derived from the 2001 UK Census; Office for National Statistics 2001). Household density in each county was calculated by dividing the number of households in each county by county area. Household income, household age (age of household reference person), and household size (number of people comprising the household) were expressed as mean values for all responding households in each county.

We sent a postal questionnaire to 2,421 randomly chosen residential addresses in three ca. 1-km^2 study sites in Sheffield, selected to capture a variety of urban forms and neighborhood types: a city center area, a low-density outer suburban area, and a high-density residential area situated between the center and suburbs (see Gaston et al. 2007 for further details). Of the questionnaires sent, 47.3% were returned (32.7%, 49%, and 61% in the inner, middle, and outer study areas, respectively). Respondents were asked to indicate whether they provide (1) food and/or (2) nest boxes for birds in their garden. The questionnaire contained 50 questions relating to a wider project on urban sustainability (Jones 2002) and thus the questions on bird feeding and nest box provision formed only a small part, a structure that minimized bias arising from the level of interest of people in wildlife and/or gardening influencing the likelihood of returning the form.

We used a national commercial classification of neighborhood types (Mosaic UK) developed by Experian's Business Strategies Division (see www.business-strategies.co.uk) to classify each household into one of 61 neighborhood types. This classification is based on a hierarchical cluster analysis across more than 400 social, economic, and demographic variables (Farr and Webber 2001, Harris et al. 2005). The cluster analysis identified 61 distinct neighborhood types, of which 47 occurred within urban Sheffield. Each household within the study area was assigned to one of these of neighborhood types (note that neighborhood types are not explicit spatial units—although the types tend to cluster spatially, adjoining houses can be assigned to different neighborhood types). As part of a national questionnaire sent to over 500,000 households by Experian, respondents were asked to indicate whether they provide food for birds on a regular basis. For each household in Sheffield, we used its neighborhood type to assign a probability that bird feeding was occurring at the household. For each 250 m × 250 m grid cell across the city, *Proportion Feeding* was the average of this probability across all houses in the grid cell, and thus depended on the relative numbers of households of different neighborhood types. *Feeder Density* was calculated for each grid cell by multiplying *Proportion Feeding* by the number of households. Grid cells with no houses ($n = 498$) were excluded from all analyses, apart from figures that report Sheffield-wide *Feeder Density*, because zero values need to be included in that instance.

Data associated with Experian's classification of neighborhood types were similarly used to characterize variation by grid cell in socioeconomic variables. Household income was expressed as the percentage of households where gross income exceeds £50,000 (US$97,000), household age was expressed as the percentage of householders over the age of 55 years, and household size was expressed as the percentage of households in each grid cell comprising more than two people.

Abundance of Selected Bird Species

Each 1-km square in the urban area of Sheffield (the set of 160 such squares within the urban area; Gaston et al. 2005b) was split into four 500 × 500 m cells, and a sampling point was randomly located within each, resulting in 640 points. Between 24 May and 1 July 2005, a point transect (a transect of zero length; Buckland et al. 2001) of 5 min duration was conducted at each survey point, or at the nearest accessible location within the same habitat type. In 318 (49.7%) of the 640 cases, the exact randomly chosen point location was accessible. Where it was not, the observer stood at the nearest accessible point in the same habitat type. The identity and distance from the observer of each detected bird were noted. Birds in flight were excluded from all analyses.

Distances were estimated in the field in 14 bands (0–4.9 m, 5–9.9 m, 10–14.9 m, 15–19.9 m, 20–24.9 m, 25–29.9 m, 30–39.9 m, 40–49.9 m, 50–59.9 m, 60–69.9 m, 70–79.9 m, 80–89.9 m, 90–99.9 m, 100 m+). Because the probability of detecting birds declined with increasing distance from the observer, data were analyzed using the Program Distance software (ver. 5, St. Andrews, Scotland; Thomas et al. 2005). Detection functions were calculated separately by species. Pointwise density estimates were calculated by applying the detection function for each species to the distance data from each survey point. Land cover characteristics within a 100-m buffer around each survey point were determined in a GIS, based on the classification of surface cover polygons by Ordnance Survey within the MasterMap digital cartographic data set at a 1:1,250 scale (Murray and Shiell 2003). Cover by greenspace in each 100-m buffer was determined by summing the area of all polygons classified as natural surface or garden in the MasterMap data.

Density data were extracted for the six species identified in a recent analysis as having the strongest positive association between distribution and urbanization in the UK (Blackbird, *Turdus merula*; Blue Tit, *Cyanistes caeruleus*; Great Tit, *Parus major*; House Sparrow, *Passer domesticus*; Starling, *Sturnus vulgaris*; and Common Wood-Pigeon, *Columba palumbus*; Cannon 2005). All species except Common Wood-Pigeon regularly take supplementary food from garden feeding stations (Cramp et al. 1977–1994). The Winter Wren was included as a comparator, as it is well adapted to urban conditions but is a strict insectivore, rarely taking artificially provided food. Given that bird-feeding data were available at the level of neighborhood type, relationships between bird feeding and bird abundance were investigated at this level. Bird density was the average from the survey points falling within each of the neighborhood types. Neighborhood types containing fewer than three bird survey points were excluded from analyses, resulting in a set of 35 neighborhood types for assessing relationships between *Feeder Density* and bird density.

Statistical Approach

We constructed mixed models in SAS (ver. 9.1, SAS Institute, Cary, NC). All variables were normally distributed or $\log_{10}$-transformed to achieve normality. No spatial autocorrelation was apparent in the England-wide data set assessed using GeoDA (release 0.95i, Spatial Analysis Laboratory, University of Illinois) with first order queen contiguity-based spatial weights for counties sharing a common boundary. However, there was a strong spatial signal in the Sheffield data set, so analyses at the citywide scale implement spatial correlation models that fit a spatial covariance matrix to the data and use this to adjust test statistics accordingly (Littell et al. 1996). The choice of the exponential over other spatial covariance structures was based on inspection of semi-variograms of independent error model residuals. In all cases, backward stepwise model-building procedures were employed to determine minimum adequate models. In the case of using socioeconomic variables to predict bird feeding, separate models were constructed for each predictor, together with its square term (household income, household age, household size, and density of households). In models predicting bird density, greenspace (the proportion of vegetated surface within 100 m of the bird survey point) and *Feeder Density* were entered initially into the model. Estimates of variance explained (i.e., r^2 values) cannot be derived from spatial models, but are provided for independent error models. The fit of alternative spatial models was compared using Akaike's Information Criterion (AIC). To show the relationships graphically, data points at county scale are presented individually, and grid cell data from Sheffield are split into equal interval groups, based on transformed values where necessary.

RESULTS

Spatial Patterns in Bird Feeding and Nest Box Provision

Household density showed strong spatial variation, both at nationwide (Fig. 16.1a) and citywide (Fig. 16.2a) scales, respectively reflecting the general pattern of urbanization across the country and variation in the intensity of urbanization within the city limits of Sheffield. Across England, 39.1% of respondents reported the presence of bird-feeding equipment and/or provided supplementary food for birds in the outside space associated with their property, and 18.1% provided one or more nest boxes. Questionnaire data from the three study areas in Sheffield indicated that 51.7% of respondents provided food for birds and 16.3% provided at least one nest box.

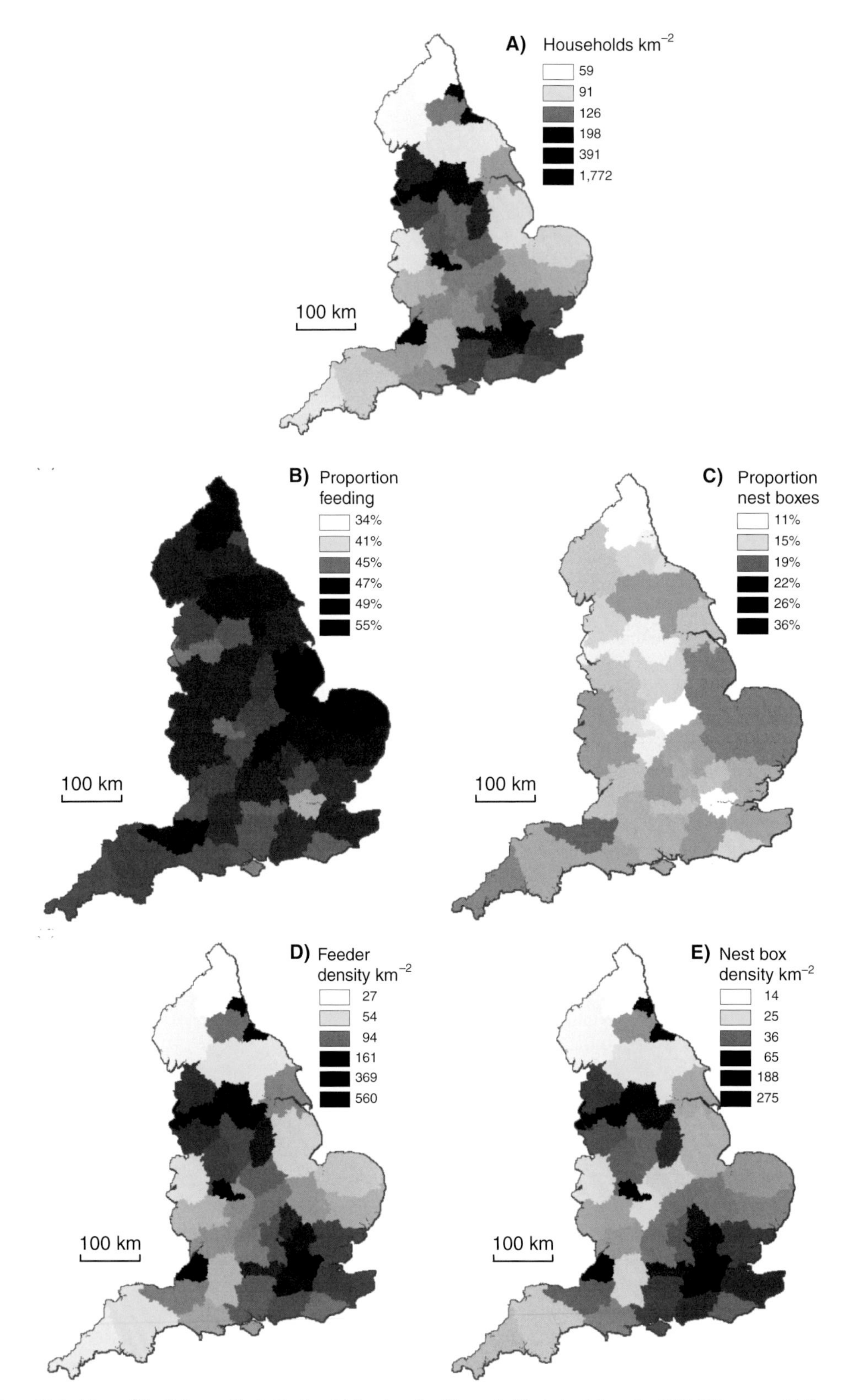

Figure 16.1. Maps of English counties indicating (a) the density of households derived from the 2001 UK census, (b) the proportion of respondents to the Survey of English Housing (SEH) reporting the presence of bird-feeding equipment and/or providing supplementary food for birds in the outside space associated with their property, (c) the proportion of SEH respondents reporting the presence of one or more nest boxes in the outside space associated with their property, and the density of (d) locations at which birds are fed, and (e) locations at which nest boxes are provided. Legends indicate the top of the range of values associated with each of six shades determined using the Jenks natural break classification.

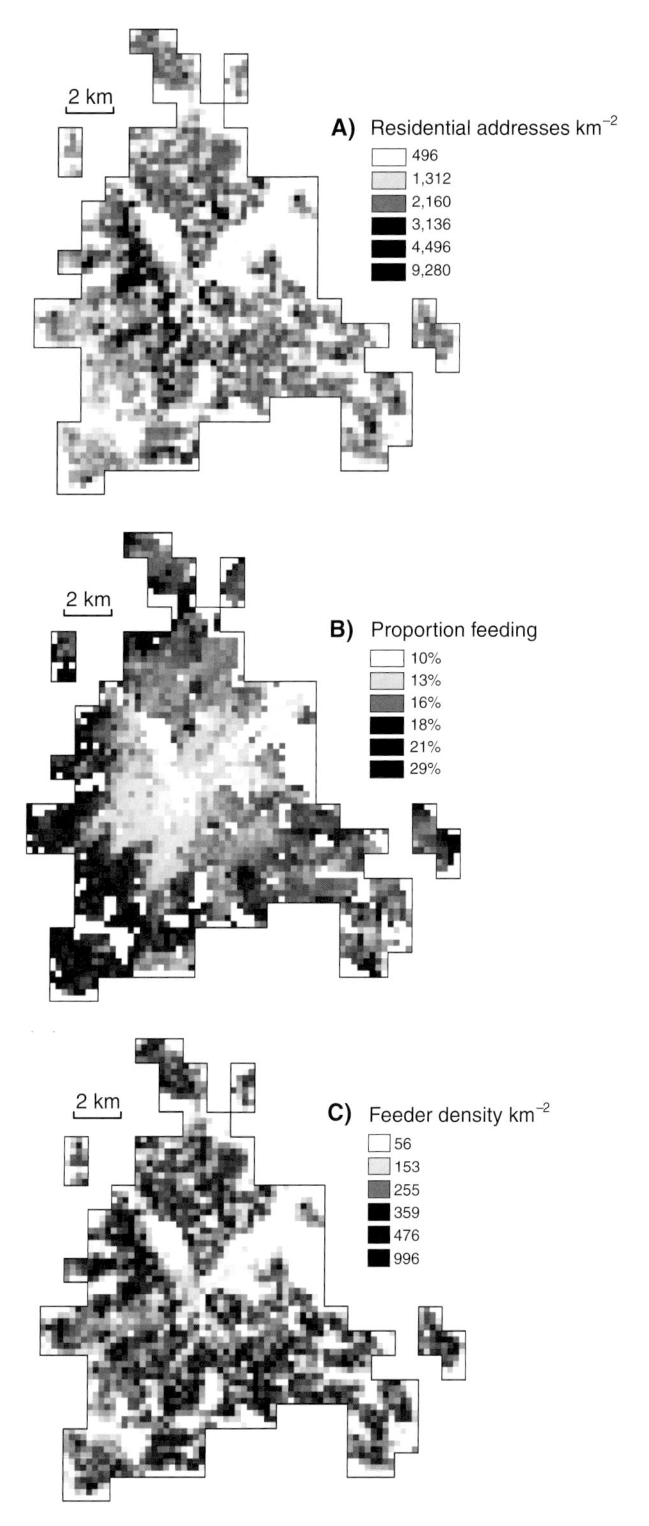

Figure 16.2. Map of urban Sheffield divided into 250 × 250-m grid cells indicating (a) the density of households derived from a count of residential addresses, (b) the estimated proportion of householders who feed birds, based on the numbers of households from different socioeconomic groups and questionnaire data stratified by socioeconomic group, (c) the density of locations at which birds are fed, obtained by multiplying (a) and (b). Legends indicate mid-point of the range of values associated with each of six shades determined using Jenks natural break classification.

The proportions of households providing food and nest boxes for birds showed patterns approximately inverse to that of household density across England as a whole, being higher in less densely populated counties (Fig. 16.1b, c). For bird feeding a similar pattern was evident within the city of Sheffield, where the activity clearly declined in prevalence toward the densely populated inner suburbs and the city center (Fig. 16.2b).

Average feeder density by county across England varied between 14.8 and 559.7 km^{-2} (mean = 107.5), while in the 250 × 250 m grid cells across Sheffield, feeder density varied between zero and 995.9 km^{-2} (mean = 197.7). Average nest box density by county across England varied between 2.6 and 274.8 km^{-2} (mean = 47.3). The spatial patterns of the density of bird-feeding stations depended almost entirely on household density, being more or less independent of *Proportion Feeding*, such that *Feeder Density* tended to be higher in more densely populated areas both at nationwide (Fig. 16.1d) and citywide (Fig. 16.2c) scales. Patterns in *Proportion Feeding* and *Feeder Density* in relation to household density were, therefore, strikingly similar at nationwide and citywide scales. A similar pattern for nest box density was apparent at the nationwide scale (Fig. 16.1e).

Household Density, Socioeconomics, and Bird-feeding Activity

Taking the national data first, household density was by far the strongest predictor of both *Proportion Feeding* and *Feeder Density* (Fig. 16.3a, b; Table 16.1). *Proportion Feeding* declined as household density increased. However, the strong positive relationship between *Feeder Density* and the density of households at the scale of the county, and the absence of a significant squared household density term from this model (Table 16.1), indicate that housing density was much more important in determining the overall density of feeders across the landscape than the popularity of bird feeding. While the relationship between *Feeder Density* and household density is constrained to be at least positive triangular (*Feeder Density* cannot exceed household density; see Fig. 16.3b), the relationship is linear, with no cases of obviously low feeder densities in counties with high household densities, has a slope significantly lower than 1 (b = 0.88, 95% CI = 0.85–0.93), and does not decelerate. None of the socioeconomic variables were an important predictor of *Proportion Feeding* or *Feeder Density* (Fig. 16.3c–h; Table 16.1). Household age was negatively related to *Feeder Density* (Fig. 16.3f), and household size was related negatively to *Proportion Feeding* and positively to *Feeder Density* (Fig. 16.3g, h), although the explanatory power of all three relationships was low (Table 16.1).

At the citywide scale across Sheffield, spatial models revealed a hump-shaped relationship between housing density and *Proportion Feeding*, with *Proportion Feeding* declining sharply at high housing densities (Fig. 16.4a; Table 16.1). The model resulted in a strong positive relationship between household density and *Feeder Density* (Fig. 16.4b; Table 16.1). The form of the relationship is notable, with *Feeder Density* initially increasing rapidly with household density (slope of the linear household density term > 1; b = 1.3, 95% CI = 1.24–1.35). The squared household density term was negative, indicating a decelerating effect of household density on *Feeder Density*, such that at high household densities little change in *Feeder Density* was apparent (Fig. 16.4b; Table 16.1), presumably due to a combination of a smaller proportion of households having access to a garden and a decline in the popularity of bird feeding at high household densities.

In contrast to the England-wide data, all three socioeconomic variables were strong predictors of *Proportion Feeding* and *Feeder Density* across Sheffield. Household income had a positive accelerating relationship with *Proportion Feeding* (Fig. 16.4c; Table 16.1), but a negative accelerating relationship with *Feeder Density* (Fig. 16.4d; Table 16.1), presumably reflecting the tendency for higher income groups to live in lower-density neighborhoods. The age of householders and household size showed hump-shaped relationships with *Proportion Feeding* and *Feeder Density*, with both the popularity of bird feeding and the spatial density of the resource peaking at intermediate levels of household age and size (Fig. 16.4e–h, Table 16.1).

Bird Feeding and the Abundance of Selected Bird Species

Densities of three of the seven urban-adapted bird species in each of the 35 neighborhood types across urban Sheffield were positively related to the density of feeders (Fig. 16.5). These relationships remained significant when greenspace

TABLE 16.1

Results of single-predictor regression models using housing density and three socioeconomic variables to predict the proportion of households at which food for birds is provided and the spatial density of feeding sites (a) across 45 English counties and (b) across 250 × 250 m grid cells comprising urbanized Sheffield (using spatial models).

	Density of households			Household income			Household age			Household size		
	Linear	Square	r^2	Linear	Square	r^2	Linear	Square	r^2	Linear	Square	r^2
England												
Prop. feeding	33.2^{---}		0.44							6.5^{-}		0.13
Feeder density	$2{,}043^{+++}$		0.98				7.4^{--}		0.15	5.58^{+}		0.12
Sheffield												
Prop. feeding	96.36^{+++}	122.3^{---}		89.31^{+++}	31.33^{+++}		607^{+++}	273.1^{---}		277.8^{+++}	220^{---}	
Feeder density	$2{,}236.4^{+++}$	130.7^{---}		56.3^{---}	58.56^{---}		38.91^{+++}	22.74^{---}		172.6^{+++}	186^{---}	

NOTE: Each model includes the linear and square term for the predictor. Backward stepwise selection was used to remove nonsignificant terms from each model, and only terms significant in the final model are shown. r^2 values are provided for each final model for the national data, but cannot be computed for spatial models. Superscript symbols after F values indicate effect direction and significance level ($P < 0.05$, $P < 0.01$, $P < 0.001$ for one, two, and three symbols, respectively).

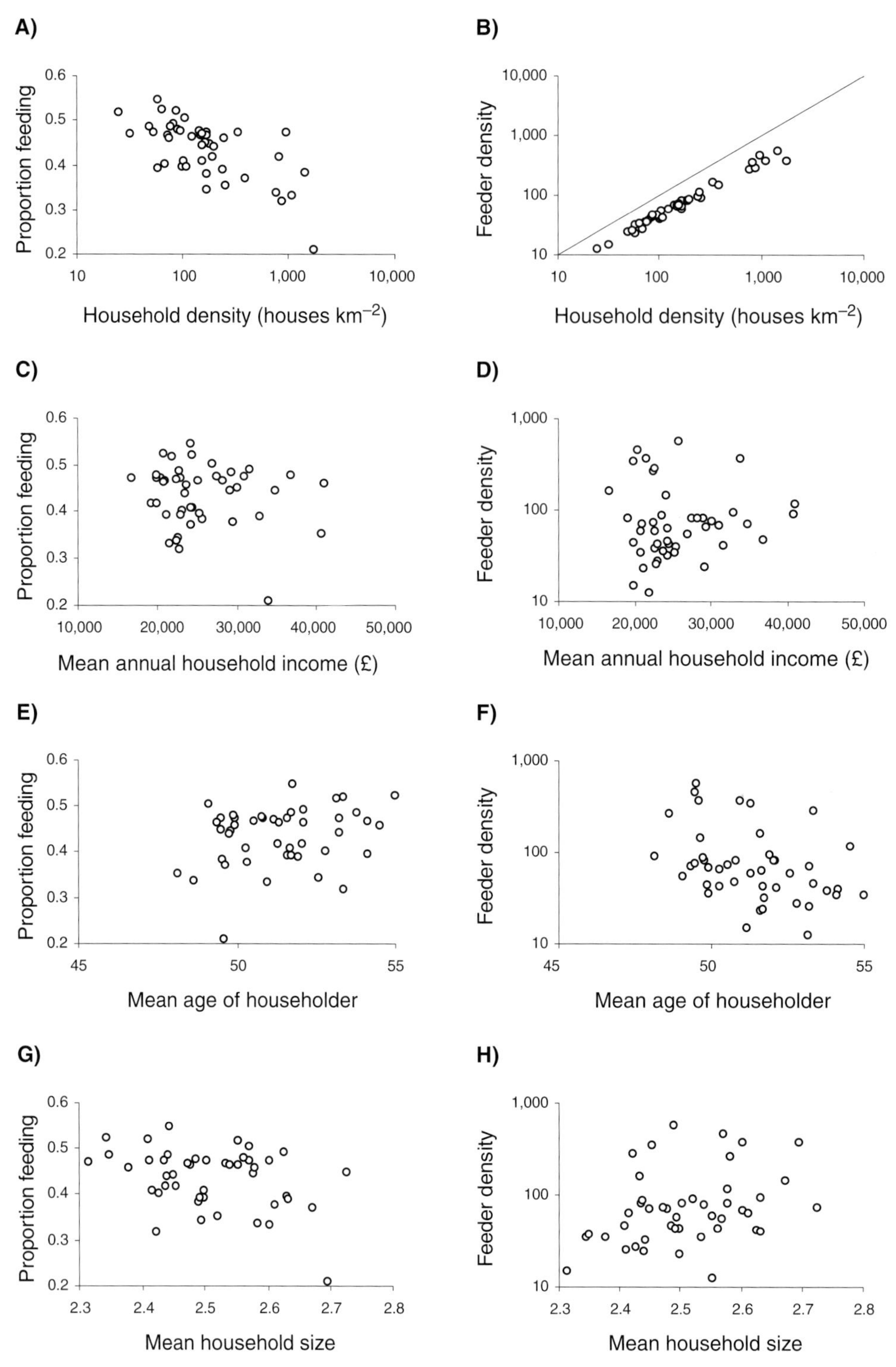

Figure 16.3. Relationships at county scale across England between household density derived from the 2001 UK census and (a) the proportion of respondents to the Survey of English Housing (SEH) reporting the presence of bird-feeding equipment and/or providing supplementary food for birds in the outside space associated with their property (*Proportion Feeding*), and (b) the density of such locations at which birds are fed (*Feeder Density*). Solid line indicates $y = x$. Also, the relationships between three socioeconomic variables derived from SEH responses and *Proportion Feeding* and *Feeder Density*—(c, d) mean gross annual household income, (e, f) mean age of the household reference person, and (g, h) household size, expressed as the mean number of occupants per household. Note that *Feeder Density* and household density are plotted on a log scale.

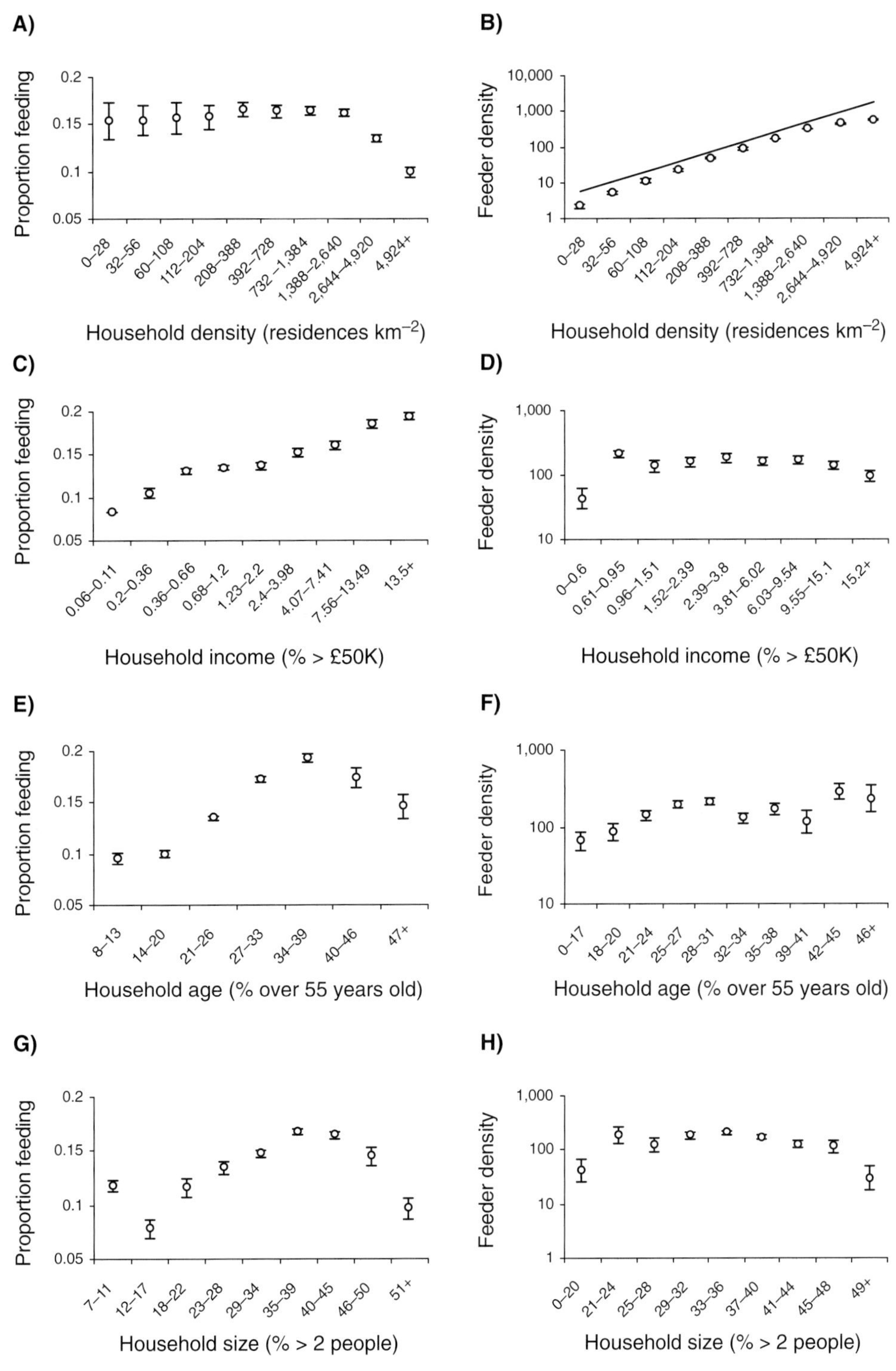

Figure 16.4. Relationships within Sheffield city at 250 × 250-m grid cell resolution between household density derived from counts of residential addresses and (a) the proportion of households in each grid cell providing supplementary food for birds in the outside space associated with their property (*Proportion Feeding*), and (b) the density of such locations at which birds are fed (*Feeder Density*). Solid line indicates $y = x$. Also, the relationships between three socioeconomic variables calculated using statistics by neighborhood type in the MOSAIC classification (see text) and *Proportion Feeding* and *Feeder Density*—(c, d) household income expressed as the percentage of households where gross income exceeds £50,000, (e, f) household age expressed as the percentage of householders over the age of 55 years, and (g, h) household size expressed as the percentage of households comprising more than two people. Note that *Feeder Density* and household density are plotted on a log scale.

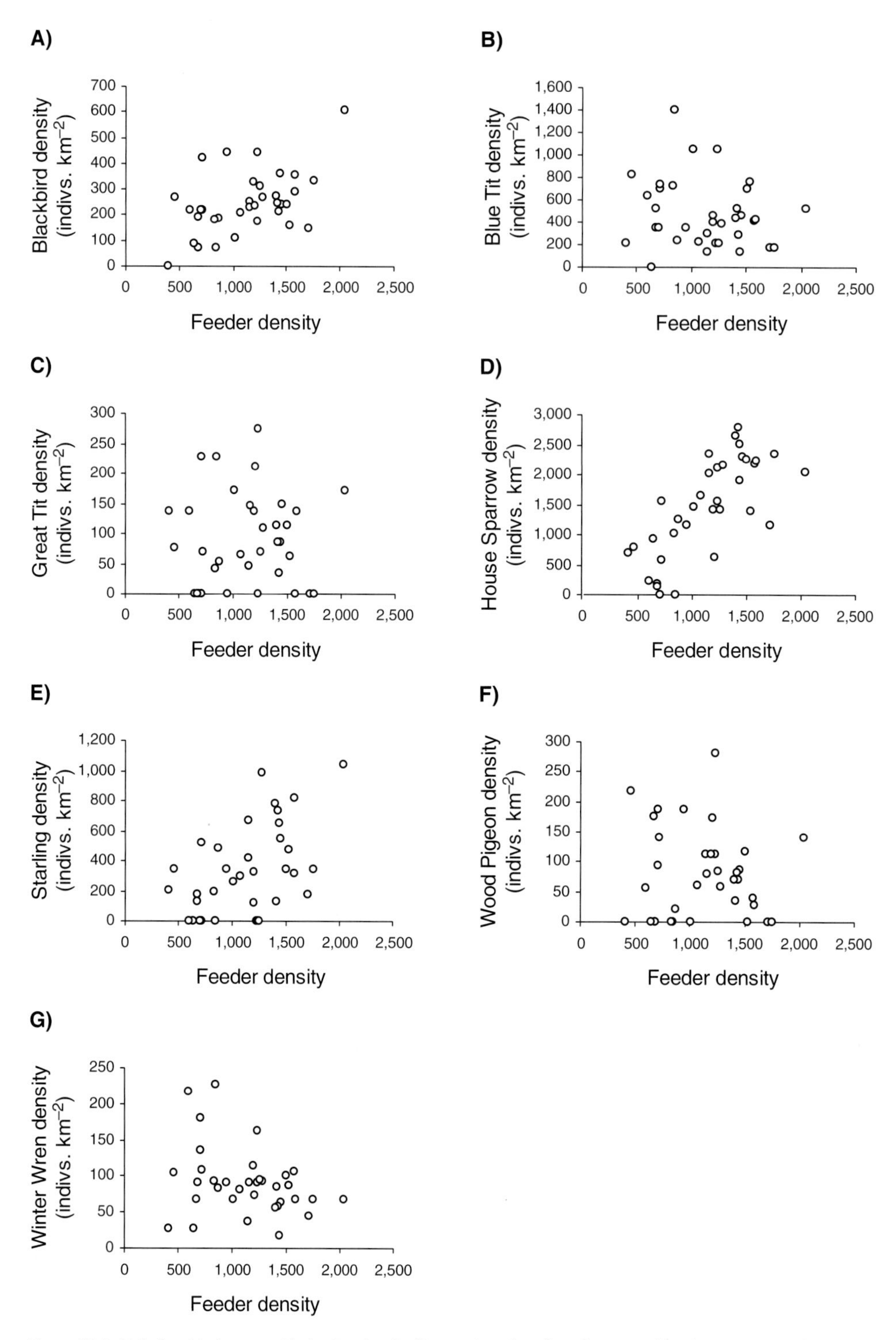

Figure 16.5. Relationship between *Feeder Density* (feeding stations km^{-2}) and species abundance in each of 35 neighborhood types across Sheffield. Species are (a) Blackbird, (b) Blue Tit, (c) Great Tit, (d) House Sparrow, (e) Starling, (f) Wood-Pigeon, and (g) a habitat-generalist insectivore, the Winter Wren. The relationships for Blackbird and Starling remain significant upon removal of the right-hand data point.

TABLE 16.2

Results of regression models using greenspace coverage and the density of feeding stations in 35 neighborhood types to predict the density of seven highly urbanized species within the city of Sheffield.

	Greenspace			Feeder density		
	β	F	r^2	β	F	r^2
Blackbird	0.14	10.07^{**}	0.23			
Blue Tit						
Great Tit						
House Sparrow	1,916.23	7.52^{*}	0.19	1.43	42.34^{***}	0.57
Starling				0.37	11.45^{**}	0.26
Common Wood Pigeon						
Winter Wren						

NOTE: Backward stepwise selection was used to remove nonsignificant terms from the full model, and only terms significant in the final model are shown. r^2 values are for the whole final model where there is only one predictor, and partial r^2 statistics where both predictors were retained in the final model. Superscript symbols after F values indicate significance level ($P < 0.05$, $P < 0.01$, $P < 0.001$ for one, two, and three symbols, respectively). β = model slope.

coverage was included in the model to account for variation in gross urban form (Table 16.2). The House Sparrow showed the strongest pattern, with a positive relationship with *Feeder Density*, explaining 57% of the variation in its abundance (Fig. 16.5d; Table 16.2). The abundances of Blackbird and Starling were also positively related to *Feeder Density*, which explained 23% and 26% of the variation in their numbers, respectively (Fig. 16.5a, e; Table 16.2). Densities of the remaining four species (Blue Tit, Great Tit, Common Wood-Pigeon, and Winter Wren) showed no significant relationship with greenspace or feeder density at the scale of the neighborhood (Fig. 16.5b, c, f, g; Table 16.2).

DISCUSSION

Extent of the Resource and Spatial Patterns of Bird Feeding

Our results confirm that the provision of food for birds is popular in the UK, with 39% of households across England engaging in the activity. We estimate an average feeder density across England of about 100 km^{-2}, and within Sheffield of about 200 km^{-2}. We are aware of no published estimates of the average amount of food for wild birds put out in individual gardens, and in particular how stocking scales up to a standing crop, so estimating the size of the resource base that this provision generates is not straightforward. Clearly, however, these densities of bird-feeding stations represent a large potential resource for birds, and one that is concentrated in more densely populated areas.

The resource base of supplementary food is unlikely to be static over time. Lepczyk et al. (2004) showed that the highest proportion of landowners feeding birds in southeastern Michigan occurred between December and March, with a decline through summer to autumn. Historically in the UK, feeding was mainly carried out in winter, in the belief that typical supplementary food types are unsuitable for fledglings and that adults could find all the natural food they needed during summer (Moss and Cottridge 1998). More recently, advice from the British Trust for Ornithology recommends that feeding be carried out year-round, with the additional provision of live food suggested during the summer months (Toms 2003), and a carryover effect has been demonstrated whereby winter-fed birds show increased productivity in the following breeding season (Robb et al. 2008). Much more work is required to document variation in the amount and types of food put out for wild birds in gardens, and how this varies temporally, both in the short term and over seasons (Jones and Reynolds 2008). A significant proportion of people feeding birds only do so on an infrequent basis (Gaston et al. 2007), so care is needed when interpreting data on bird feeding in terms of the amount of the resource available.

The provision of supplementary food and nesting sites for birds showed distinct spatial patterning, both at a regional scale across England and within Sheffield. The proportion of people providing resources for wild birds was generally negatively related to population density, both at the county and neighborhood scale, although in Sheffield the proportion feeding only declined noticeably at high household densities (Figs. 16.3a, 16.4a). However, because of marked variation in household density, the density of feeding stations depended much more closely on the density of human settlement than on the proportion of people engaged in supplementary resource provision. In practice, the link means that supplementary resources are being provided disproportionately in (1) more densely populated regions of the country and (2) more densely populated neighborhoods within cities. The pattern has obvious implications for the kinds of avian assemblages that will receive resource inputs and the kinds of environments in which these human-nature interactions are taking place.

The relationship between household density and the intensity of resource provision for birds indicates that the majority of these interactions between people and nature are occurring in highly urbanized areas. Urban sites are precisely the environments where enhanced contact with nature through gardens is likely to result in significant psychological, physical, and social benefits to the human population. The garden has long been considered an integral part of health and well-being (Gerlach-Spriggs et al. 1998). Access to a garden has been shown to reduce self-reported sensitivity to stress (Stigsdotter and Grahn 2004), while lack of access is associated with increased self-reported levels of depression and anxiety (Macintyre et al. 2003). While we are not aware of any studies that directly explore the contribution of wildlife to quality of life, a few studies do include insight into this question (Vandruff et al. 1995, Clergeau et al. 2001). The presence of wildlife has been cited as a part of planting and water gardening that make them enjoyable activities (Catanzaro and Ekanem 2004), and observing and feeding wildlife were found to predict neighborhood satisfaction (Frey 1981). However, any benefits to human well-being associated specifically with feeding wild birds remain unknown.

Socioeconomic Correlates of Bird Feeding

Three socioeconomic variables (household income, age of householders, and number of people comprising the household) were poor predictors of both the prevalence of bird-feeding activity across the human population and the resulting spatial density of bird-feeding stations at the national scale. Conversely, within Sheffield, the three variables were strongly related to both the prevalence of bird feeding and the spatial density of bird-feeding stations (Table 16.1; Fig. 16.3). The proportion of households feeding birds increased with household income, and showed hump-shaped relationships with both age of householders and number of people comprising the household. The density of bird-feeding stations across the urban landscape was negatively related to household income, but the hump-shaped relationships with household age and household size were retained (Table 16.1; Fig. 16.4). The differences in these relationships at the two scales suggest that local variation in socioeconomic status is much more important than regional differences in, for example, household income in determining the likelihood that a given household provides food for birds. By aggregating data at a county scale, important socioeconomic effects were averaged away. This contrast is important because it demonstrates the utility of fine-scale studies such as this one. Data at county scale are relatively easy to obtain, but they may be uninformative, as demonstrated here. However, it is important to note that such socioeconomic patterns might not be universal; our findings contrast with those of Lepczyk et al. (2004), who detected no influence of household size or wealth on levels of bird feeding in Michigan.

A positive local relationship between the proportion of households feeding birds and household income makes intuitive sense, yet it suggests that human socioeconomic deprivation is directly related to the quality of the experience that people have of nature. Although bird feeding can exclusively comprise throwing out kitchen scraps, a large proportion of people provide specialist feeds grown and purchased specifically for provision to wild birds and use specific equipment such as bird feeders and bird tables (Cowie and Hinsley

1988a, Moss and Cottridge 1998). In Cardiff, UK, 56% of questionnaire respondents fed birds daily or several times per week (Cowie and Hinsley 1988a). Such frequent replenishment of bird food carries a significant financial commitment, given that a standard birdseed mix currently retails at between approximately £1.00 (US$1.95) kg^{-1} and £1.50 (US$2.93) kg^{-1} depending on the quantity purchased (www.birdfood.co.uk).

The relationships between the prevalence of bird feeding and household age and size reveal additional variation across human society in the popularity of the activity. Given that public policy, at least in the UK, explicitly encourages wildlife-friendly garden management practices (DEFRA 2002), significant potential exists for directed interventions via public campaigns. The resultant levels of resource provision across the landscape imply that the status of avian populations, at least in urban areas, might be managed via such programs (Ilyichev et al. 1990, Fuller et al. 2008). Clergeau et al. (2001) showed that avian diversity was positively perceived by city dwellers in Rennes, France, and suggested that successful conservation of urban avifaunas will enhance human quality of life. Further work could profitably focus on how such socioeconomic variation relates to opportunity and motivation for, as well as the benefits of, bird feeding.

Bird Feeding and Bird Abundance across the Urban Landscape

We found significant relationships, all of which were positive, between the density of feeding stations and bird abundance in urban environments for Blackbird, House Sparrow, and Starling. All three species regularly take supplementary food in gardens (Cramp et al. 1977–1994). The relationship was strongest for the House Sparrow, a species native in the UK and in severe decline across the country (Robinson et al. 2005). Steep declines in urban House Sparrows in the past have been associated with declines in winter food supply, when the replacement of horse-drawn vehicles with motor vehicles led to a sharp drop in grain availability (Bergtold 1921). More recently, building on brownfield sites leading to loss of sites supporting ruderal plants has been proposed as a factor reducing food availability (Crick et al. 2002). All this suggests that urban House Sparrow populations may be food limited, and the positive relationship with the density of bird-feeding stations raises the intriguing possibility that the population may be responding to the availability of supplementary food at the neighborhood scale within Sheffield.

The Winter Wren is insectivorous (Cramp et al. 1977–1994) and, as such, there is no *a priori* expectation that it will respond directly to bird feeding. The absence of a relationship between Winter Wren density and feeder density suggests that the relationships for Blackbird, House Sparrow, and Starling were not being driven simply by variation in some other component of the urban landscape. Somewhat surprisingly, though, feeder density did not predict the abundance of Blue Tit and Great Tit, both species that feed commonly on garden bird feeders, including during the breeding season (Cowie and Hinsley 1988b). However, populations of both of these hole-nesting species are relatively stable in urban environments (Cannon et al. 2005), and there is no reason to suppose that their populations are especially food limited within UK cities. Large-scale experiments that can generate meaningful variation in supplementary food availability within urban areas are needed to isolate the effects of garden bird feeding on avian abundance.

Only 10 of 76 nest boxes provided for Great Tits and Blue Tits in a recent study in Sheffield were occupied, suggesting that nest sites were unlikely to be a limiting resource for those species (Cannon 2005). Conversely, for House Sparrows in the UK, older houses appear to be important for nesting sites, suggesting some nest site availability limitation and that newer developments may be less suitable (Wotton et al. 2002, Mason 2006). Individual species will doubtless respond to food and nest site availability in different ways, and detailed studies will be required to understand how these factors interact to determine population densities.

CONCLUSION

Our data indicate that garden bird feeding is a popular activity in England and within a typical large city. The level of resource provision across the landscape is strongly influenced by human population density, being higher in more densely populated areas. Garden bird feeding has strong socioeconomic predictors that are scale dependent, and is positively associated with the densities of some urban-adapted bird species. Garden bird feeding therefore represents a large resource input that, coupled with other forms of management, might be used as a conservation option in urban environments. Promoting interaction with garden birds might also lead to increased human engagement with nature and potential for a positive effect on quality of life, particularly within urban environments, where arguably such engagement is most needed.

ACKNOWLEDGMENTS

This work was funded by the Engineering and Physical Sciences Research Council (through the CityForm research consortium), and the Countryside Council for Wales, the Department for Environment, Food and Rural Affairs, the Environment and Heritage Service, Natural England, and the Scotland and Northern Ireland Forum for Environmental Research (SNIFFER, through the BUGS II project). Experian's Business Strategies Division kindly provided access to the Mosaic UK data set, and MasterMap data were supplied by Ordnance Survey, by license through the CityForm Consortium. We also wish to acknowledge the National Centre for Social Research (data collectors), the Department for Transport, Local Government and the Regions (sponsor), and the UK Data Archive (data distributor) for providing the Survey of English Housing (SEH) 2001/02 data set. Crown Copyright material is reproduced with the permission of the controller of Her Majesty's Stationery Office and the Queen's Printer for Scotland. These organizations bear no responsibility for the data analysis or interpretation of findings stated within this publication. We are grateful to members of the CityForm consortium who helped in collecting the questionnaire data; to O. Barbosa, J. Tratalos, and P. S. Warren for discussion; and to two anonymous referees for their helpful comments.

LITERATURE CITED

Beebee, T. J. 2001. British wildlife and human numbers: the ultimate conservation issue? British Wildlife 13:1–8.

Beer, A. 2005. The green structure of Sheffield. Pp. 40–51 *in* A. C. Werquin, B. Duhem, G. Lindholm, B. Oppermann, S. Pauleit, and S. Tjallingii (editors), Green structure and urban planning: final report of COST Action C11. Brussels, Belgium.

Bergtold, W. H. 1921. The English Sparrow (*Passer domesticus*) and the motor vehicle. Auk 38:244–250.

British Trust for Ornithology. 2006. We spend £200 million a year on wild bird food. BTO Press release No. 2006/12/76, December 2006. Thetford, UK.

Brittingham, M. C., and S. A. Temple. 1988. Impacts of supplemental feeding on survival rates of Black-capped Chickadees. Ecology 69:581–589.

Buckland, S. T., D. R. Anderson, K. P. Burnham, J. L. Laake, D. L. Borchers, and L. Thomas. 2001. Introduction to distance sampling: estimating abundance of biological populations. Oxford University Press, Oxford, UK.

Cannon, A. R. 2005. Wild birds in urban gardens: opportunity or constraint? Ph.D. dissertation, University of Sheffield, Sheffield, UK.

Cannon, A. R., D. E. Chamberlain, M. P. Toms, B. J. Hatchwell, and K. J. Gaston. 2005. Trends in the use of private gardens by wild birds in Great Britain 1995–2002. Journal of Applied Ecology 42:659–671.

Catanzaro, C., and E. Ekanem. 2004. Home gardeners value stress reduction and interaction with nature. Acta Horticulturae 639:269–276.

Clergeau, P., A. Sauvage, A. Lemoine, J.-P. Marchand, F. Dubs, and G. Mennechez. 1997. Quels oiseaux dans la ville? Les Annales de la Reserche Urbaine 74:119–130.

Clergeau, P., G. Mennechez, A. Sauvage, and A. Lemoine. 2001. Human perception and appreciation of birds: a motivation for wildlife conservation in urban environments of France. Pp. 69–88 *in* J. M. Marzluff, R. Bowman, and R. Donnelly (editors), avian ecology in an urbanizing world. Kluwer Academic Publishers, Norwell, MA.

Cowie, R. J., and S. A. Hinsley. 1988a. The provision of food and the use of bird feeders in suburban gardens. Bird Study 35:163–168.

Cowie, R. J., and S. A. Hinsley. 1988b. Feeding ecology of Great Tits (*Parus major*) and Blue Tits (*Parus caeruleus*), breeding in suburban gardens. Journal of Animal Ecology 57:611–626.

Cramp, S., K. E. L. Simmons, and C. M. Perrins. (1977–1994). Handbook of the birds of Europe, the Middle East and North Africa, Vols 1–9. Oxford University Press, Oxford, UK.

Crick, H. Q. P., R. A. Robinson, G. F. Appleton, N. A. Clark, and A. D. Rickard. 2002. Investigation into the causes of the decline of starlings and house sparrows in Great Britain. BTO Research Report No. 290. British Trust for Ornithology, Thetford, UK.

Daniels, G. D., and J. B. Kirkpatrick. 2006. Does variation in garden characteristics influence the conservation of birds in suburbia? Biological Conservation 133:326–335.

DEFRA. 2002. Working with the grain of nature. Department of the Environment, Food and Rural Affairs. Defra Publications, London, UK.

De Vries, S., R. A. Verheij, P. P. Groenewegen, and P. Spreeuwenberg. 2003. Natural environments–

healthy environments? An exploratory analysis of the relationship between greenspace and health. Environment and Planning A 35:1717–1731.

Farr, M., and Webber, R. 2001. MOSAIC: From an area classification system to individual classification. Journal of Targeting, Measurement and Analysis for Marketing 10:55–65.

Frey, J. E. 1981. Preferences, satisfactions, and the physical environment of urban neighborhoods. Ph.D. dissertation, University of Michigan, Ann Arbor, MI.

Fuller, R. A., K. N. Irvine, P. Devine-Wright, P. H. Warren, and K. J. Gaston. 2007. Psychological benefits of greenspace increase with biodiversity. Biology Letters 3:390–394.

Fuller, R. A., P. H. Warren, P. R. Armsworth, O. Barbosa, and K. J. Gaston. 2008. Garden bird feeding predicts the structure of urban avian assemblages. Diversity and Distributions 14:131–137.

Gaston, K. J., R. M. Smith, K. Thompson, and P. H. Warren. 2005a. Urban domestic gardens (II): experimental tests of methods for increasing biodiversity. Biodiversity and Conservation 14:395–413.

Gaston, K. J., P. H. Warren, K. Thompson, and R. M. Smith. 2005b. Urban domestic gardens (IV): the extent of the resource and its associated features. Biodiversity and Conservation 14:3327–3349.

Gaston, K. J., R. A. Fuller, A. Loram, C. MacDonald, S. Power, and N. Dempsey. 2007. Urban domestic gardens (XI): variation in urban wildlife gardening in the UK. Biodiversity and Conservation 16: 3227–3238.

Gerlach-Spriggs, N., R. E. Kaufman, and S. B. Warner, Jr. 1998. Restorative gardens: the healing landscape. Yale University Press, New Haven, CT.

Grubb, T. C., and D. A. Cimprich. 1990. Supplementary food improves the nutritional condition of wintering woodland birds: evidence from ptilochronology. Ornis Scandinavica 21:277–281.

Harris, R., Sleight, P., and Webber, R. 2005. Geodemographics, GIS and neighbourhood targeting. John Wiley and Sons Ltd., Chichester, UK.

Ilyichev, V. D., V. M. Konstantinov, and B. M. Zvonov. 1990. The urbanized landscape as an arena for mutual relations between man and birds. Pp. 122–130 *in* M. Luniak (editor), Urban ecological studies in central and eastern Europe. Ossolineum, Wroclav, Poland.

Irvine, K. N., and S. L. Warber. 2002. Greening healthcare: practicing as if the natural environment really mattered. Alternative Therapies in Health and Medicine 8:76–83.

Jokimäki, J., and J. Suhonen. 1998. Distribution and habitat selection of wintering birds in urban environments. Landscape and Urban Planning 39:253–263.

Jones, C. A. 2002. EPSRC Sustainable Urban Form Consortium. Planning, Practice & Research 18:231–233.

Jones, D. N., and S. J. Reynolds. 2008. Feeding birds in our towns and cities: a global research opportunity. Journal of Avian Biology 39:265–271.

Kuo, F. E., and W. C. Sullivan. 2001. Environment and crime in the inner city: does vegetation reduce crime? Environment and Behavior 33:343–367.

Kuo, F. E., W. C. Sullivan, R. L. Coley, and L. Brunson. 1998. Fertile ground for community: inner-city neighborhood common spaces. American Journal of Community Psychology 26:823–851.

Lepczyk, C. A., A. G. Mertig, and J. Liu. 2002. Landowner perceptions and activities related to birds across rural-to-urban landscapes. Pp. 251–259 *in* D. Chamberlain and E. Wilson (editors), Avian landscape ecology. Proceedings of the 2002 Annual UK-IALE Conference. Colin Cross Printers Ltd., UK.

Lepczyk, C. A., A. G. Mertig, and J. Liu. 2004. Assessing landowner activities related to birds across rural-to-urban landscapes. Environmental Management 33:110–125.

Lin, E. 2005. Production and processing of small seeds for birds. Agricultural and Food Engineering Technical Report. Food and Agriculture Organisation of the United Nations, Rome, Italy.

Littell, R. C., G. A. Milliken, W. W. Stroup, and R. D. Wolfinger. 1996. SAS system for mixed models. SAS Institute Inc., Cary, NC.

Mabey, R. 1999. The unofficial countryside. Pimlico, London, UK.

Macintyre, S., A. Ellaway, R. Hiscock, A. Kearns, G. Der, and L. McKay. 2003. What features of the home and the area might help to explain observed relationships between housing tenure and health: evidence from the west of Scotland. Health and Place 9:207–218.

Maller, C., M. Townsend, A. Pryor, P. Brown, and L. St. Leger. 2005. Healthy nature healthy people: "contact with nature" as an upstream health promotion intervention for populations. Health Promotion International 21:45–54.

Mason, C. F. 2006. Avian species richness and numbers in the built environment: can new housing developments be good for birds? Biodiversity and Conservation 15:2365–2378.

Miller, J. R., and R. J. Hobbs. 2002. Conservation where people live and work. Conservation Biology 16:330–337.

Moss, S., and Cottridge, D. 1998. Attracting birds to your garden. New Holland, London, UK.

Murray, K. J., and D. Shiell. 2003. A new geographic information framework for Great Britain. Photogrammetric Engineering and Remote Sensing 69:1175–1182.

NCSR and DETR. 2004. Survey of English Housing, 2001–2002 (computer file). SN: 5021. National Centre for Social Research and Department for Transport, Local Government and the Regions, UK Data Archive, Colchester, UK.

Office for National Statistics. 2001. Census: standard area statistics (England and Wales). ESRC/JISC Census Programme, London, UK.

Orell, M. 1989. Population fluctuations and survival of Great Tits (*Parus major*) dependent on food supplied by man in winter. Ibis 131:112–127.

Robb, G. N., R. A. McDonald, D. E. Chamberlain, S. J. Reynolds, T. J. E. Harrison, and S. Bearhop. 2008. Winter feeding of birds increases productivity in the subsequent breeding season. Biology Letters 4:220–223.

Robinson, R. A., G. M. Siriwardena, and H. Q. P. Crick. 2005. Size and trends of the House Sparrow (*Passer domesticus*) population in Great Britain. Ibis 147:552–562.

Rollinson, D. J., R. O'Leary, and D. N. Jones. 2003. The practice of wildlife feeding in suburban Brisbane. Corella 27:52–58.

Savard, J. L., P. Clergeau, and G. Mennechez. 2000. Biodiversity concepts and urban ecosystems. Landscape and Urban Planning 48:131–142.

Smith, R. M., K. J. Gaston, P. H. Warren, and K. Thompson. 2006. Urban domestic gardens (VIII): environmental correlates of invertebrate species abundance. Biodiversity and Conservation 15:2515–2545.

Stigsdotter, U. A., and Grahn, P. 2004. A garden at your doorstep may reduce stress: private gardens as restorative environments in the city. Paper 00015. Proceedings of the Open Space-People Space Conference, 27–29 October 2004, Edinburgh, Scotland.

Thomas, L., J. L. Laake, S. Strindberg, F. F. C. Marques, S. T. Buckland, D. L. Borchers, D. R. Anderson, K. P. Burnham, S. L. Hedley, J. H. Pollard, J. R. B. Bishop, and T. A. Marques. 2005. Distance 5.0, release beta 5. Research Unit for Wildlife Population Assessment, University of St. Andrews, UK.

Thompson, K., K. C. Austin, R. M. Smith, P. H. Warren, P. G. Angold, and K. J. Gaston. 2003. Urban domestic gardens (I): putting small-scale plant diversity in context. Journal of Vegetation Science 14:71–78.

Toms, M. 2003. The BTO/CJ Garden Birdwatch book. British Trust for Ornithology, Thetford, UK.

U.S. Fish and Wildlife Service. 2001. 2001 National survey of fishing, hunting and wildlife associated recreation. <www.fws.gov>.

van Balen, J. H. 1980. Population fluctuations of the Great Tit and and feeding conditions in winter. Ardea 68:143–164.

Vandruff, L. W., D. L. Leedy, and F. W. Stearns. 1995. Urban wildlife and human well-being. Pp. 203–211 *in* H. Sukopp, M. Numata, and A. Huber (editors), Urban ecology as the basis of urban planning. SPB Academic Publishing, Amsterdam, The Netherlands.

Wotton, S. R., R. Field, R. H. W. Langston, and D. W. Gibbons. 2002. Homes for birds: the use of houses for nesting by birds in the UK. British Birds 95:586–592.

CHAPTER SEVENTEEN

Who Feeds the Birds?

A COMPARISON ACROSS REGIONS

Christopher A. Lepczyk, Paige S. Warren, Louis Machabée, Ann P. Kinzig, and Angela G. Mertig

Abstract. Humans often engage in activities on their property that can directly or indirectly influence population dynamics and ecological processes, including gardening to attract birds and butterflies, installing ponds or bird baths, keeping pets indoors, and perhaps most commonly, feeding birds. Yet we know little about the spatial or temporal distribution of these human resource subsidies or the factors that motivate people to engage in them. As a first step to understanding the spatial distribution of bird-related activities, we sought to assess how often and in what ways people elect to participate in three activities that influence birds. In parallel studies in southeastern Michigan and Phoenix, Arizona, we surveyed ~3,800 people regarding their participation in bird feeding, planting vegetation for birds, and allowing cats outdoors. In both regions, a large proportion of respondents fed birds (66% in Michigan and 43% in Arizona) and planted vegetation (55% in Michigan and 42% in Arizona), whereas fewer allowed cats outdoors (26% in Michigan and 18% in Arizona). The predominant food types were commercial seed mixtures in both locations, with regional differences in specific food types. Striking differences between the two regions came in the levels of engagement in each activity, with Michigan respondents much more likely to feed birds and participating in more bird-related activities than Arizona respondents. Although many factors varied across rural-urban gradient, the urban landscape in Michigan was nearly identical in all ways to the Arizona sample, all of which was urban. This suggests the possibility of general characteristics in the ways that urban residents restructure local ecosystems. Altogether, our findings highlight how human activities at local scales may alter ecological processes at larger scales and demonstrate the importance of using a social survey approach for understanding ecological questions.

Key Words: bird feeding, Breeding Bird Survey, human dimensions, landowners, rural-to-urban gradients, social surveys, socioeconomic.

Lepczyk, C. A., P. S. Warren, L. Machabée, A. P. Kinzig, and A. G. Mertig. 2012. Who feeds the birds: a comparison across regions. Pp. 267–284 *in* C. A. Lepczyk and P. S. Warren (editors). Urban bird ecology and conservation. Studies in Avian Biology (no. 45), University of California Press, Berkeley, CA.

Research on human-environment interactions often focuses on human activities with negative consequences for wildlife, such as habitat destruction (Forester and Machlis 1996, Vitousek et al. 1997, Marzluff 2001). But humans also engage in activities that either create new environments or subsidize existing ones in ways that may ameliorate some of the impacts of habitat loss. Examples range from gardening to attract birds and butterflies to building bird or bat houses for nesting and roosting, installing ponds or bird baths, and perhaps most commonly, hanging bird feeders and filling them with food. People engage in these activities with the explicit intention to benefit the environment and improve their property for wildlife habitat (Stokes and Stokes 1987, Sargent and Carter 1999). These resource subsidies and habitat modifications may provide opportunities for native species in urbanizing lands (Emlen 1974, Atchison and Rodewald 2006), but may also fundamentally alter ecological processes such as interspecific competition and trophic dynamics (Shochat 2004, Faeth et al. 2005, Shochat et al. 2006, Warren et al. 2006). Yet ecologists have for the most part done little to address the systemic effects of bird feeding, gardening, and other private landowner activities (but see Martinson and Flaspohler 2003; Cannon et al. 2005; Gaston et al. 2005; Smith et al. 2005; Robb et al. 2008a, 2008b). A first step toward greater understanding of human resource subsidies for birds is documenting the extent to which people engage in bird-related activities and the factors that influence people to do so.

Human activities that influence birds are largely carried out on private lands. Thus, while the processes influencing birds operate at larger scales, such as neighborhoods or landscapes (Hostetler 1999, Hostetler and Holling 2000), decisions are usually made at smaller scales, such as the household or parcel (Liu et al. 2003, Kinzig et al. 2005, Grove et al. 2006). Considering that private lands encompass >65% of the area of the U.S. (Dale et al. 2000), such household or parcel decisions are thus quite important. Local decisions are especially important in urban areas, where, the majority of urban greenspace is found on private lots (Grove et al. 2006). Many environmental and governmental organizations have recognized the ecological importance of landowners and encourage people to manage their property in ways that are beneficial to wildlife, plants, and the larger ecosystem (Sargent and Carter 1999, Mizejewski 2004). Similarly, citizen science initiatives such as Project FeederWatch (Hochachka et al. 1999, Lepage and Francis 2002) and Garden BirdWatch (Toms 2003) engage people in monitoring birds at their homes and on their lands. But many more people likely engage in bird-related activities than join these initiatives.

Ultimately, better management of human-dominated landscapes requires knowing what motivates people to act. Survey research provides a long-established means for directly assessing how people engage in bird-related activities (Cowie and Hinsley 1988, Brittingham and Temple 1989, Dunn and Tessaglia 1994, Cannon 1999) and is increasingly being used in applied ecological studies (White et al. 2005). In the case of ornithological studies, previous surveys have either been focused on amateur birders and ornithologists and people who participated in these activities (Brittingham and Temple 1989, Wiedner and Kerlinger 1990) or have sought volunteers (Project FeederWatch; Wells et al. 1998), but see Clergeau et al. (2001) for an exception. As a result, many previous ornithological surveys may be biased toward those people who are interested in birds (Lepczyk 2005). By targeting all people in a specific geographical location, without regard to their demonstrated interest in birds, we can achieve a more complete understanding of the distribution of bird-related activities and resource subsidies (Lepczyk et al. 2004a, 2004b).

A growing body of social science research indicates that people differ in their propensity to engage in nature-oriented activities and that these differences can be associated with a wide variety of social, cultural, and economic factors (Kaplan and Talbot 1988, Carr and Williams 1993, Fraser and Kenney 2000, Gobster 2002, Pullis La Rouche 2003). Classic sociological variables such as education, income, and ethnicity are often cited as associated with differing levels of engagement in nature-oriented activities. For instance, the users of wildland parks are more commonly white/Anglo-Americans than African Americans and tend on average to be wealthier (Gobster 2002). Similarly, bird watchers also tend to be white/Anglo-Americans, with higher incomes, higher education levels, and in their middle years (Pullis La Rouche 2003). By contrast, our previous study (Lepczyk et al. 2004a, 2004b) only found slight correspondence between bird-related activities on private lands and sociological variables such as age,

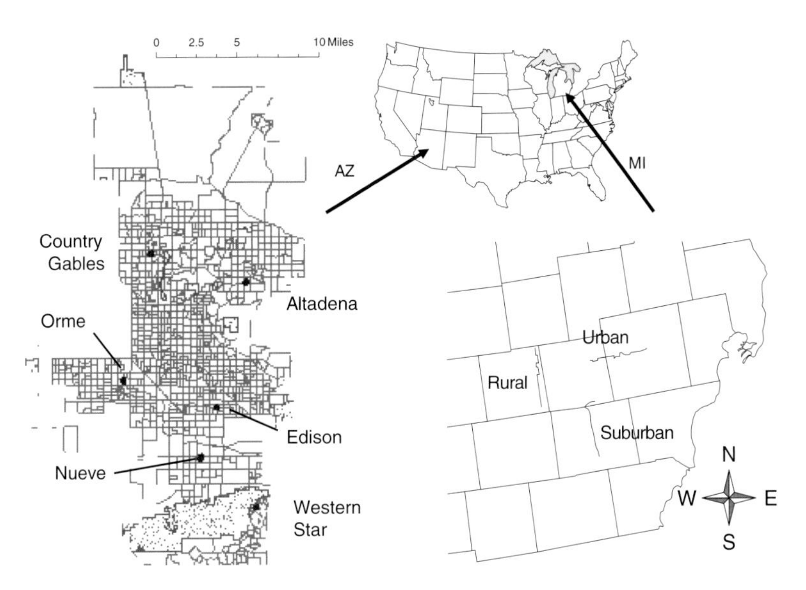

Figure 17.1. The study regions of southeastern Michigan and Phoenix, Arizona, indicating locations surveyed within each region. In Michigan, landowners were surveyed along transects in three landscape types: urban, suburban, and rural. Lines on the southeastern Michigan map indicate county boundaries. In Phoenix, neighborhoods surrounding six parks in Phoenix were surveyed with a random sample of 350 residents within a mile radius drawn from postal route data. Lines in Phoenix indicate census block groups, and speckled areas are public lands, typically large desert parks. The six parks fall on a socioeconomic gradient, lying within two upper-income (Altadena and Western Star), two middle-income (Country Gables and Orme), and two lower-income (Edison and Nueve) neighborhoods.

gender, and occupation. Instead, the Michigan survey found that position along a rural–urban gradient was a stronger predictor of people's participation in bird-related activities than sociological makeup. For instance, urban landowners had a greater density of bird feeders and houses and owned more cats, but planted or maintained vegetation at a lower frequency compared to rural landowners (Lepczyk et al. 2004a, 2004b). A greater density of bird feeders and houses in the urban landscape is consistent with the broadly reported observation that bird densities are greater in urban than in rural or wildlands areas (Marzluff 2001, McKinney 2002, Shochat et al. 2006). Before we can extrapolate more broadly from these rural–urban gradients to predict distributions of human subsidies, we need to test whether the patterns documented in Michigan are representative.

Comparative studies across regions are necessary to identify more comprehensively the factors that influence people's participation in bird-related activities. Human populations in different regions of the U.S. differ demographically and socioeconomically, and thus may also differ in their motivations to engage in activities such as bird feeding. For example, people in areas with high avian diversity might find it more rewarding to engage in bird feeding than those in areas of lower avian diversity (Fuller et al., chapter 16, this volume). Alternatively, the presence of cultural groups historically involved in bird watching might increase participation in a community more generally. Thus, social and demographic influences need to be investigated across multiple regions if we are to determine frequencies and motivations of people to engage in bird feeding and other activities with profound influences on bird communities. As a result, a parallel survey conducted in Phoenix, Arizona, in 2002 provides an opportune comparison to the Michigan survey. Within the U.S., these two regions are almost maximally distinct from one another, both ecologically and sociologically. Therefore, a comparison between the findings of the two studies provides opportunities to identify both general trends as well as factors underlying differences in people's engagement in bird-related activities.

TABLE 17.1

Survey questions from Michigan and Arizona used in the study.

Numbers in parentheses indicate question numbers.

Michigan	Arizona
(2) Does anyone in your household feed birds on your property? (Yes, No) (7) What type(s) of food do you feed the birds? (Check all that apply) Suet, thistle, corn, sunflower seeds, millet, sugar water/hummingbird nectar, bread, oranges, commercial seed mixture, other (Please specify)	(24) Do you regularly (at least once a week) feed birds in your yard using any of the following kinds of food? (Circle all that apply) 1. Commercial seed mixture 2. Thistle 3. Sunflower seeds 4. Millet 5. Sugar water 6. Bread 7. Orange slices 8. Mealworms 9. Other __________ 10. No, I do not regularly feed wild birds in my yard
—	—
(12) Have you planted vegetation or maintained landscaping on your property in order to benefit or encourage use by birds? (Yes, No)	(25) Do you plant trees, shrubs, or flowers in order to attract birds? (Yes, No)
(25) How many cats does your household own that are allowed access to the outside? _____Cats	(26) How many cats do you own that regularly go outside? __________
(30) About how long have you owned or lived on this parcel of land? __________Years	(1) How long have you been living at your current address? _______ (months or years)
(53) In what year were you born? 19____	(27) What is your present age? __________
(55) How many people currently live in your household? ________People	(28) How many OTHER people live in your household? __________
(54) Are you: 1. Male 2. Female	(30) Are you: 1. Male 2. Female

Given the opportunity to identify factors associated to participation in bird-related activities, our primary goal was to address the following questions: (1) Who engages in activities perceived as beneficial and detrimental for birds? (2) Are activities carried out in conjunction with one another? and (3) Does participation in these activities vary between the regions? In addressing these three questions, we predicted that traditional social status variables would be poor predictors of whether people engage in bird-oriented activities or not. We base this prediction on our previous work (Lepczyk et al. 2004a, 2004b) and on the view that bird-related activities more likely reflect lifestyle or cultural background than social status. Instead, we predict that age is a better general predictor of engaging in bird-related activities, while income is a better indicator of the extent to which people engage in costly activities. Last, we suggest that people who engage in one bird-oriented activity typically engage in others, even in negative activities such as allowing their house cats outdoors. Our comparative analyses represent a first step toward understanding the factors that motivate people to act to influence the presence of birds on private lands and in urban systems. Ultimately, knowing these motivations will aid in predicting the dynamics of resource availability for birds in human-dominated ecosystems as well as in managing these places for birds specifically and for wildlife in general.

Michigan	Arizona
(51) Do you have a house or residential structure on your land? (Yes, No)	—
	(34) What was your approximate family income from all sources, before taxes, in 2002? 1. <$15,000 2. $15,000–29,999 3. $30,000–49,999 4. $50,000–79,999 5. >$80,000
(52) If Yes, what is the approximate size of the house or residence? (Check one) 1. <1,000 square feet 2. 1,000–1,499 square feet 3. 1,500–1,999 square feet 4. 2,000–2,499 square feet 5. 2,500–2,999 square feet 6. 3,000–3,499 square feet 7. 3,500 square feet or larger 8. Unsure	—
(57) What is the highest level of school completed or degree you have received? (Check one) 1. Some school completed, but no high school diploma 2. High school graduate or GED 3. Some college, but no degree 4. Associates degree 5. Bachelor's degree 6. Master's, professional, or doctoral degree	(35) Which is the highest level of education that you have had the chance to complete? 1. Sixth grade or less 2. <12 yr 3. High school graduate 4. Some college 5. Technical or vocational school 6. College graduate 7. Courses beyond college, but no graduate degree 8. Graduate degree

METHODS

Study Sites

To address our main questions we compared two contrasting study regions of the U.S. with one another. The first study region was southeastern Michigan, near the Detroit metropolitan area (Fig. 17.1), where >90% of the land is privately owned and the landscape is undergoing rapid urbanization (Rutledge and Lepczyk 2002). Within the southeastern Michigan region, we selected three specific study landscapes that represent a continuum from rural to urban, based on their geographic locations, average land parcel sizes, and socio-demographic compositions (see Lepczyk et al. 2004a, 2004b for further details). The second study region was the Central Arizona–Phoenix Long-Term Ecological Research (CAP LTER) site, in which neighborhoods surrounding six small urban parks were selected (Fig. 17.1). Each neighborhood was selected to represent a range of income levels, and racial and ethnic compositions. However, for the purposes of our current study we consider all six parks as urban and have thus pooled the data for analysis. Ecologically the two study regions are quite different, as Phoenix lies in the desert biome compared to southeastern Michigan, which lies in the temperate deciduous forest. In addition, Phoenix is a much newer and economically booming city, while southeastern Michigan encompasses the economically depressed region near Detroit and has been settled far longer. Thus, while acknowledging the given restraints in any study, the combined results from the two regions may be applicable to similar urban areas within the U.S.

Surveys

To determine how people influence birds on their property in both study regions, we employed mail survey instruments. Specifically, we identified all private, domestic (non-business) landowners along three southeastern Michigan study landscapes and six Phoenix neighborhoods using addresses obtained from local tax assessor databases and county equalization offices, with verification by plat maps and visual inspection of all parcels. The identification process yielded a total of 1,694 private landowners (331 on rural, 390 on suburban, and 973 on urban Breeding Bird Survey [BBS] survey routes) in southeastern Michigan and 2,100 households in Phoenix, with 350 selected randomly from each of six neighborhoods. Surveys were administered following the total/tailored design method (Dillman 2000), which uses personal letters and multiple mailings to elicit high response rates.

The southeastern Michigan (MI) region was surveyed first, between October and December 2000, using a 58-question survey instrument (see Lepczyk 2002 for complete survey instrument). Based on Lepczyk's (2002) initial survey, a similar landowner survey was developed for Phoenix (AZ) that contained 37 questions and was administered between April and July 2003. Notably, while each survey differed in content, there were nine questions from the Michigan survey that directly corresponded to the Arizona survey (Table 17.1).

Linking Surveys and Statistical Analyses

While several of the questions were identical on both surveys, other questions needed to be recoded to make responses comparable between the surveys. We matched the survey questions and responses between the two study regions (Table 17.1) in the following manner. First, in the case of matching bird-feeding responses, we considered that any respondent answering AZ question 24 (except category 10) corresponded to a Yes, and hence corresponded to a Yes answer for question 2 in the MI survey. Second, we matched question 7 in the MI survey to the corresponding categories in question 24 pertaining to food types. Third, because question 12 in MI was followed by specific subcategories of plant types, we coded the response as Yes/No and matched it to question 25 in the AZ survey. Fourth, the only difference in question 30 for MI and question 1 in AZ was that MI landowners could be absentee and thus have no buildings on the property. Fifth, because during survey design of the MI instrument we considered income to be a potentially offensive question, we chose to use size of home as a proxy for income (Zaremba 1961, U.S. Census Bureau 2006). To match the number of categories of house size and income, we lumped categories 5, 6, and 7 together into category 5, because they all are >2,400 square feet, which is considered to be a large home (Anthony 2006). Finally, to make the eight categories of question 35 in AZ compatible with the six categories of question 57 in MI, the eight categories were collapsed into six by lumping together categories 1 and 2, and 7 and 8.

We conducted all statistical analyses using SYSTAT 10 (ver. 10, Systat Software, San Jose, CA). Response rate, the proportion of respondents engaging in an activity, and gender of respondents were compared using two-way and multi-way contingency tables with a Pearson chi-square test statistic. Comparisons between regions were carried out using *t*-tests, while comparisons across the four locations (i.e., three landscapes and AZ) were carried out with ANOVA for each activity and socio-demographic factors of age, number of people in the household, education, and income. Location differences were compared using a Bonferroni post hoc test (Zar 1996). In comparing respondents who participated in a given activity versus respondents who did not participate in activities across the four locations based upon socio-demographic factors, we used a general linear model framework. Because general linear models do not allow for post hoc comparisons, we only report whether or not locations were different, but not which ones, for participants versus non-participants. Data are reported as means ±SE (because 100% of the population was sampled, but only ~59% and 30% responded in MI and AZ, respectively), unless otherwise noted, with a P-value of ≤ 0.05 considered significant.

Survey Response Rate

After removing ineligible responses (e.g., from a business rather than from a domestic landowner, or from an individual who owned land outside the sampling region), nondeliverables, and multiple responses, the final population size of landowners in MI was 1,654. From among these 1,654 we received 968 completed surveys, yielding a 58.5% response rate. Response rates in different landscapes were 64.8% for Rural (212 of 327), 61.5% for Suburban (233 of 379), and 55.2% for Urban (523 of 948), which were significantly different ($\chi^2 = 11.11$; df = 2; $P = 0.0039$). In AZ a total of 638 surveys were received, yielding an overall 30% response rate. Response rates out of 350 surveys administered to each neighborhood were 41.4% (145) and 40.3% (141) in the two upper-income neighborhoods, 33.7% (118) and 25.4% (89) in the two middle-income neighborhoods, and 24.6% (86) and 16.6% (58) in the two lower-income neighborhoods.

RESULTS

Comparison of Respondent Activities

Michigan respondents were significantly more likely than Arizona respondents to engage in bird-related activities, across all activities surveyed. A total of 634 (of 968 = 65.5%) respondents fed birds in MI compared to 272 (out of 638 = 42.6%) in AZ, resulting in a greater proportion of MI respondents engaging in bird feeding compared to AZ (Pearson $\chi^2 = 81.7$, df = 1, $P < 0.0001$). Because the proportion of respondents engaging in bird feeding did not differ across the three landscapes of MI (Table 17.2; also see Lepczyk et al. 2004a), we did not consider the effect of location. Respondents in MI fed an average of 3.0 ± 0.08 types of food compared to 1.93 ± 0.06 in AZ ($t_{904} = -10.02$, $P < 0.0001$). In both regions, commercial seed mixtures were the most commonly used bird food (Table 17.2). However, other food types differed drastically in their relative importance, as exemplified by thistle, which was the third most common in MI compared to the least used feed in AZ. Of the respondents who indicated they fed other types of food, 55 from MI and 29 from AZ gave a detailed list (Appendix 17.1). Notably, this list includes both common foods for wildlife as well as unusual items, such as garbage and meat.

In the case of planting vegetation for birds, 527 (of 954 = 55.2%) respondents in MI engaged in this activity compared to 262 (of 628 = 41.7%) in AZ, indicating a greater proportion of participants in MI (Pearson $\chi^2 = 27.70$, df = 1, $P < 0.0001$). Notably, while participation in planting vegetation differed significantly within MI across the three landscapes (60.5% for rural, 58.3% for suburban, and 51.7% for urban; Lepczyk et al. 2004a), they were all markedly greater than participation in AZ (Pearson $\chi^2 = 32.62$, df = 3, $P < 0.0001$; Table 17.2).

Allowing cats outdoors differed between the two regions (Pearson $\chi^2 = 14.91$, df = 1, $P = 0.0001$), with 26.1% (253 of 968) of MI respondents engaging in this activity versus 17.6% (114 of 638) of AZ respondents. Furthermore, the proportion of respondents allowing cats outdoors differed across the three MI landscapes and AZ, with all MI respondents engaging in the activity to a greater degree than the AZ respondents (Table 17.2; Pearson $\chi^2 = 36.82$, df = 3, $P < 0.0001$). Of the respondents who allowed cats outdoors, the average MI respondent allowed 2.6 ± 0.20 cats outdoors, which was significantly greater than the 1.9 ± 0.19 allowed outdoors in AZ ($t_{364} = -2.24$, $P = 0.025$). However, comparing the three MI landscapes to AZ indicates that while the differences remain significant ($F_{3,362} = 6.97$, $P < 0.001$), the main regional difference is between AZ and the Rural MI landscape (Table 17.2), suggesting that the regional differences in outdoor cat ownership may actually reflect urban–rural gradients in pet ownership.

Of the three landowner activities investigated, MI respondents engaged in an average of 1.46 ± 0.03 activities compared to AZ respondents, who engaged in 1.02 ± 0.04 ($t_{1604} = -9.19$, $P < 0.0001$). Similarly, when considering the number of activities engaged in on each landscape, there were significant differences in participation between all locations ($F_{3,1599} = 32.2$, $P < 0.001$; Table 17.2). To compare if activities were carried out in conjunction with one another, pairwise correlations were conducted between the three activities in both regions. In MI, 43.7% of respondents engaged in both feeding and planting vegetation, 16.3% engaged in both planting vegetation

TABLE 17.2

Summary of the number of respondents engaged in each activity and food type as well as mean ±SE number of activities engaged in.

Numbers in parentheses are the percent of the total population engaged in the activity, except for specific food types which are the percent of respondents engaged among just those feeding birds.

	MI rural	MI suburban	MI urban	AZ
Respondents feeding birds	135 (65.5)	157 (68.0)	340 (66.0)	272 (42.6)
No. of types of food provided	3.1 ± 0.1	3.1 ± 0.1	2.9 ± 0.1	1.9 ± 0.1
Commercial seed	93 (68.8)	115 (73.2)	255 (75.0)	141 (51.8)
Thistle	70 (51.9)	86 (54.8)	163 (47.9)	13 (4.8)
Sunflower	91 (67.4)	107 (68.2)	217 (63.8)	59 (21.7)
Millet	27 (20.0)	34 (21.7)	49 (14.4)	17 (6.3)
Nectar	91 (67.4)	79 (50.3)	136 (40.0)	113 (41.5)
Bread	31 (23.0)	37 (23.6)	108 (31.8)	127 (46.7)
Orange	9 (6.7)	9 (4.7)	39 (11.5)	22 (8.1)
Other	9 (6.7)	15 (9.6)	29 (8.5)	33 (12.1)
Respondents planting vegetation for birds	124 (59.3)	134 (57.5)	267 (51.1)	262 (41.1)
Respondents allowing cats outdoors	70 (34.0)	75 (32.2)	107 (20.5)	114 (17.9)
No. of outdoor cats[c,f]	3.66 ± 0.34	2.56 ± 0.33	1.93 ± 0.27	1.87 ± 0.27
Total no. of activities[a,b,c,d,e,f]	1.58 ± 0.06	1.57 ± 0.06	1.37 ± 0.04	1.02 ± 0.04

NOTE: Superscript letters represent significant differences between locations based on Bonferroni *post-hoc* test as follows:
[a] AZ differed from MI suburban
[b] AZ differed from MI urban
[c] AZ differed from MI rural
[d] Suburban differed from urban
[e] Suburban differed from rural
[f] Urban differed from rural

for birds and allowing cats outdoors, and 19.9% engaged in both feeding birds and allowing cats outdoors. On the other hand, in AZ 25.5% of respondents engaged in both feeding and planting vegetation, 9.1% engaged in both planting vegetation for birds and allowing cats outdoors, and 8.9% engaged in both feeding birds and allowing cats outdoors. The people who said they allowed cats outside were a minority of respondents in both regions, but these respondents were as likely, if not more likely, to feed and plant to attract birds as were the people without outdoor cats. Thus, in both regions, the two activities people were most likely to engage in simultaneously were feeding and planting plants for birds, but outdoor cat owners did not avoid attracting birds to their homes and property.

Comparison of Demographics

Educationally, we found no difference overall between MI and AZ respondents (3.5 ± 0.05 vs. 3.5 ± 0.07; $t_{1554} = -0.21$, $P = 0.84$). However, separating the MI landscapes and comparing them with AZ indicates a significant difference ($F_{3,1549} = 17.19$, $P < 0.001$) in all locations except for the urban landscape versus AZ (Table 17.3). Similarly, age was the same between the two regions (50.6 ± 0.4 vs. 49.6 ± 0.7 for MI and AZ, respectively; $t_{1566} = -1.33$, $P = 0.18$), but

TABLE 17.3

Summary demographics of all survey respondents presented as means ±SE, except for gender.

Numbers in parentheses are the sample size for each question.

Demographic variable	MI rural	MI suburban	MI urban	AZ
Age[a,d]	51.7 ± 1.1 (203)	52.8 ± 1.0 (221)	49.2 ± 0.7 (503)	49.6 ± 0.6 (638)
Education level[a,c,d,e,f]	3.03 ± 0.11 (201)	4.09 ± 0.10 (223)	3.46 ± 0.07 (505)	3.50 ± 0.06 (624)
No. of people in house	2.8 ± 0.1 (204)	2.9 ± 0.1 (220)	2.8 ± 0.1 (508)	3.1 ± 0.1 (619)
Gender (M:F)	110:93	110:112	276:230	221:398
House size/income[a,d,e]	3.01 ± 0.09 (178)	3.58 ± 0.08 (208)	2.96 ± 0.05 (485)	3.01 ± 0.05 (543)

NOTE: Superscript letters represent significant differences between locations as described in Table 17.2.

differed when broken down by the landscapes ($F_{3,1561}$ = 921.18, P = 0.008; Table 17.3). In the case of number of people in the household, AZ respondents had significantly more individuals (3.08 ± 0.07 vs. 2.85 ± 0.05; t_{1552} = 2.85, P = 0.004). Notably, when compared among landscapes, the model was significant ($F_{3,1547}$ = 96.99, P = 0.034; Table 17.3), but none of the post hoc comparisons were. In terms of gender, MI had slightly more males who responded (46.7% females vs. 53.3% males) compared to AZ, which had a greater proportion of females responding (64.3% females vs. 35.7% males; Pearson χ^2 = 46.47, df = 1, $P < 0.001$). Finally, MI and AZ respondents had similar incomes (3.11 ± 0.04 vs. 3.01 ± 0.06 for MI and AZ, respectively; $t_{1409} = -1.58$, P = 0.12). However, when considered across the landscapes, income was significantly greater in the Suburban landscape compared to all other locations ($F_{3,1405}$ = 14.14, $P < 0.001$; Table 17.3).

Relationships between Demographics and Activities

We compared the socio-demographic characteristics of respondents who engaged in a given activity with those who did not engage in an activity (i.e., participants vs. non-participants) across the landscapes for age, education, number of people in the house, income, and gender. In the case of age, there was a difference between participants who fed birds and those who did not ($F_{1,1557}$ = 8.05, P = 0.005; Table 17.4), which differed consistently across the four landscapes ($F_{3,1557}$ = 3.92, P = 0.008; Table 17.4; Fig. 17.2). On the other hand, age did not differ between respondents engaged in planting vegetation for birds ($F_{1,1557}$ = 3.45, P = 0.06) or allowing cats outdoors ($F_{1,1557}$ = 0.09, P = 0.76), but did differ across the four landscapes for both activities (vegetation: $F_{3,1557}$ = 3.56, P = 0.014; cats: $F_{3,1557}$ = 3.18, P = 0.023; Table 17.4).

Educationally, we found differences between respondents who fed birds compared to those who did not ($F_{1,1545}$ = 10.90, $P < 0.001$), with those participating having a lower educational level. However, no differences were found in education between respondents who planted vegetation ($F_{1,1545}$ = 1.12, P = 0.29) or allowed cats outdoors ($F_{1,1545}$ = 2.97, P = 0.09; Table 17.4) and those who did not. Location did result in significant differences in education between participants and non-participants for all three activities (feeding birds: $F_{3,1545}$ = 17.04, $P < 0.0001$; vegetation: $F_{3,1545}$ = 15.86, $P < 0.001$; cats: $F_{3,1545}$ = 15.56, $P < 0.0001$; Table 17.4).

The number of people living in a respondent's home only differed for those respondents who allowed their cats outdoors compared to respondents who did not ($F_{1,1543}$ = 6.74, P = 0.01; Table 17.4). In the case of respondents who fed birds ($F_{1,1543}$ = 0.22, P = 0.64) or planted vegetation for birds ($F_{1,1543}$ = 0.01, P = 0.94), no differences were found in household size. However, household size was consistently different across the four landscapes for respondents engaged in feeding birds ($F_{3,1543}$ = 7.10, P = 0.03) and planting vegetation for birds ($F_{3,1543}$ = 7.59, P = 0.02), but not for allowing cats outdoors ($F_{3,1543}$ = 0.89, P = 0.62; Table 17.4).

TABLE 17.4

Relationship between demographic variables and activities for respondents engaged in an activity (participants) and those not engaged in an activity (non-participants).

All variables are means ±SE, except for gender. Statistically significant comparisons are summarized in Table 17.5.

	Participants				Non-participants			
Activity and demographic variable	MI rural	MI suburban	MI urban	AZ	MI rural	MI suburban	MI urban	AZ
Feeding birds								
Age	52.9 ± 1.3	53.0 ± 1.2	50.6 ± 0.8	50.8 ± 0.9	49.5 ± 1.8	52.4 ± 1.8	46.4 ± 1.1	48.7 ± 0.8
Education	2.9 ± 0.1	3.9 ± 0.1	3.4 ± 0.1	3.3 ± 0.1	3.2 ± 0.2	4.4 ± 0.2	3.6 ± 0.1	3.6 ± 0.1
No. people in house	2.7 ± 0.1	3.0 ± 0.1	2.8 ± 0.1	3.2 ± 0.1	2.9 ± 0.2	2.8 ± 0.2	2.8 ± 0.1	3.0 ± 0.1
Gender (M:F)	70:61	69:81	170:157	95:173	40:32	41:31	06:73	26:225
House size/income	3.0 ± 0.1	3.5 ± 0.1	3.0 ± 0.1	3.0 ± 0.1	3.0 ± 0.2	3.9 ± 0.2	2.8 ± 0.1	3.0 ± 0.1
Planting vegetation								
Age	51.6 ± 1.4	53.5 ± 1.3	50.9 ± 0.9	50.5 ± 0.9	51.8 ± 1.7	51.8 ± 1.6	47.4 ± 1.0	48.9 ± 0.8
Education	3.0 ± 0.1	4.2 ± 0.1	3.5 ± 0.1	3.7 ± 0.1	3.1 ± 0.2	4.0 ± 0.2	3.5 ± 0.1	3.4 ± 0.1
No. people in house	2.7 ± 0.1	2.9 ± 0.1	2.8 ± 0.1	3.2 ± 0.1	2.9 ± 0.2	2.9 ± 0.2	2.9 ± 0.1	3.0 ± 0.1
Gender (M:F)	65:56	58:73	128:130	19:178	45:37	52:39	148:100	142:220
House size/income	3.0 ± 0.1	3.6 ± 0.1	3.0 ± 0.1	3.2 ± 0.1	3.0 ± 0.1	3.6 ± 0.1	3.0 ± 0.1	2.9 ± 0.1
Allowing cats outdoors								
Age	51.3 ± 1.9	52.6 ± 1.8	50.3 ± 0.3	48.3 ± 1.4	51.9 ± 1.3	52.9 ± 1.2	48.9 ± 0.8	49.8 ± 0.7
Education	3.1 ± 0.2	4.2 ± 0.2	3.6 ± 0.2	3.7 ± 0.1	3.0 ± 0.1	4.0 ± 0.1	3.6 ± 0.2	3.5 ± 0.7
No. people in house	3.2 ± 0.2	3.1 ± 0.2	3.0 ± 0.2	3.0 ± 0.1	2.6 ± 0.1	2.8 ± 0.1	2.8 ± 0.1	3.1 ± 0.1
Gender (M:F)	34:33	38:35	52:50	28:84	76:60	72:77	224:180	193:314
House size/income	3.0 ± 0.2	3.6 ± 0.1	3.2 ± 0.1	3.3 ± 0.1	3.0 ± 0.1	3.5 ± 0.1	2.9 ± 0.1	2.9 ± 0.1

In the case of gender, a lower proportion of respondents were women in MI compared to AZ for all three activities (Table 17.4). The differences remained when investigated across the landscapes, but were only significant in the case of planting vegetation (Table 17.4; feeding birds: Mantel–Haenszel $\chi^2 = 2.52$, $P = 0.11$; planting vegetation: Mantel–Haenszel $\chi^2 = 11.49$, $P < 0.001$; cats: Mantel–Haenszel $\chi^2 = 3.75$, $P = 0.05$).

Income levels differed significantly between respondents who allowed cats outdoors ($F_{1,1401} = 4.78$, $P = 0.03$) compared to respondents who did not allow cats outdoors. But in the case of respondents who fed birds ($F_{1,1401} = 1.14$, $P = 0.29$) or planted vegetation for birds ($F_{1,1401} = 0.61$, $P = 0.44$), we found no differences in income. Income levels were significantly different across the four landscapes between participants and non-participants for all three activities (feeding birds: $F_{3,1401} = 16.60$, $P < 0.001$; vegetation: $F_{3,1401} = 13.51$, $P < 0.001$; cats: $F_{3,1401} = 10.20$, $P < 0.0001$; Table 17.4). Interestingly, among respondents who fed birds in AZ, income levels were significantly different for the seven food types (Fig. 17.3). Specifically, as income increased, a greater proportion of respondents fed thistle ($\chi^2 = 12.31$, df = 4, $P = 0.02$) and nectar ($\chi^2 = 15.04$, df = 4, $P = 0.005$), whereas fewer respondents fed bread ($\chi^2 = 27.51$, df = 4, $P < 0.001$). On the other hand, the proportion of respondents who fed different food types did not differ based on income in MI or in the Urban MI landscape ($P > 0.05$; e.g., thistle: $\chi^2 = 2.69$, df = 4, $P = 0.61$).

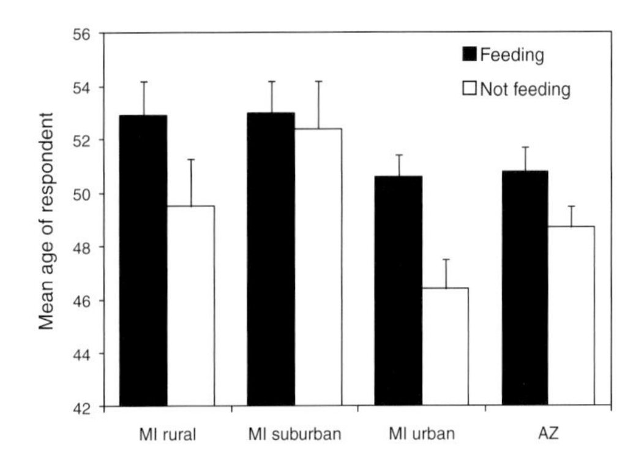

Figure 17.2. Differences in ages of respondents who feed birds versus do not feed birds across regions and landscapes. Significant differences occurred between participants who fed birds and those who did not ($F_{1,1557} = 8.05$, $P = 0.005$), as well as across the four landscapes ($F_{3,1557} = 3.92$, $P = 0.008$).

DISCUSSION

In both regions of our study, many people engaged in food subsidies for birds, suggesting that interest in birds is high (Table 17.2). In fact, 76% of respondents in MI and 82% in AZ engaged in at least one of the two activities oriented toward attracting birds to their home or property that we measured. However, we also detected significant differences between the two study regions in the level of engagement in bird-related activities. Interestingly, in all of the factors measured, southeastern MI residents were more likely than AZ residents to engage in both positive and negative bird-related activities, and engaged in a greater number of activities (Table 17.2). One consequence of this regional difference is that the availability of human supplemental food and other resources for birds in privately owned greenspaces is lower in AZ than MI. Since these lands account for the majority of lands in and around cities, the differences we observed in human subsidies likely have large impacts at the landscape scale.

Regional Differences

The central finding of the regional comparison is that more MI respondents engaged in activities that influence birds and engage in them to a greater degree than AZ respondents. The 2001 US Fish and Wildlife Service survey of wildlife-associated recreation reports more bird watchers in the northern states than in southern states (Pullis La Rouche 2003), suggesting that the differences between MI and AZ that we found do indeed reflect broader, regional trends. Because demographic factors did not explain these differences well, this suggests that the regional difference is due to factors beyond the scope of our measurements. Two explanations in particular deserve consideration. First, MI lies in a region that experiences severe winters, during which birds face harsh environmental conditions. The greater use of sunflower seed and thistle by MI respondents (Table 17.2) may also be a consequence of winter feeding practices, since these foods, particularly sunflower, are often marketed as important sources of calories for winter birds and have been shown to increase winter survival rates (Brittingham and Temple 1988). Also, as we noted in our MI study

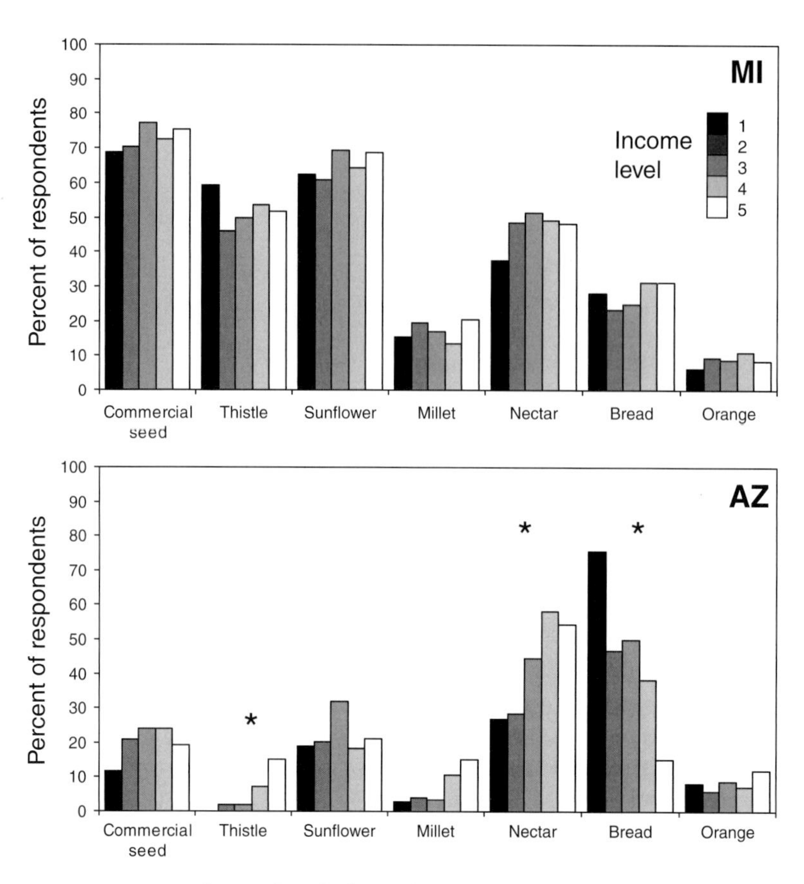

Figure 17.3. Percent of respondents feeding each type of bird food based on income in each region. Higher numbers reflect higher income levels. See Table 17.1 and methods in the text for definitions of income categories. Asterisks indicate cases where variation across income categories was significant by chi-square test ($P < 0.05$).

(Lepczyk et al. 2004a), more people feed birds during the winter months than during the summer months. Still, even the lowest rates of participation in MI were higher than in AZ. This raises a second, related possibility that regional patterns of bird feeding and other activities are culturally or historically contingent. We cannot distinguish between these two possibilities, but each suggests a useful line of inquiry for future research.

Some explanations for regional differences in bird-related activities can be ruled out on the basis of the direction of the differences, that is, greater engagement in MI than AZ. Because bird diversity is greater in AZ than in MI, the likelihood of seeing a variety of bird species is not the primary motivation for feeding birds, at least at these broad regional scales. Interestingly, however, Fuller et al. (chapter 16, this volume) provide tantalizing evidence that at local scales in cities in Britain, people may be motivated to provide more bird feeders in areas with higher bird diversity. Furthermore, people may be more motivated to see a few brightly colored birds or neighborhood resident birds at their feeder than in the number of species. In any case, both Fuller et al.'s and our findings suggest novel mechanisms, more psychological than sociological, underlying people's motivation to engage in bird-related activities on private lands.

Urban Similarities

Although levels of participation in bird-related activities differed between MI and AZ, the patterns in activity were remarkably similar, especially when considering only the urban landscape of MI and AZ. In both of these urban

TABLE 17.5

Comparison of respondents involved in a given activity (participants) versus those not involved (non-participants) based upon the socio-demographic factors.

	Socio-demographic factors				
Activity	Age	Education	No. people in house	Gender	Income
Feeding birds	X	X		X	
Planting vegetation				X	
Allowing cats outdoors			X	X	X

NOTE: An X indicates a significant difference was observed between the two groups.

environments, the dominant activity was feeding birds, followed by planting vegetation to attract birds and allowing cats outdoors (Table 17.2). In addition, similar proportions of residents provided nectar or sugar water, presumably to attract hummingbirds, though several other food types differed between the regions, notably thistle and sunflower seeds. Sociologically, the two urban locations exhibited more similarities than differences (Tables 17.3, 17.4). In fact, there were more differences between the urban, rural, and suburban MI respondents than between the urban MI and AZ respondents. Furthermore, while some statistical differences between the regions (e.g., number of people in household) were observed, they were generally too small to be ecologically or sociologically relevant. Also, some expected differences between MI and AZ, such as an older population in AZ, were not observed. The latter result likely a function of sampling within Phoenix, which is not as dominated by retirement communities as are other municipalities in the region. Ongoing surveys of the broader Phoenix metropolitan region will address these issues. Nevertheless, the similarity in behavior between the two randomly designed urban samples from such different regions suggests the possibility that there are general characteristics in the ways that urban residents restructure local ecosystems.

Relatively fewer urban respondents engage in practices that support a diversity of bird species, such as providing food for specialists or planting plants to attract birds. Instead, commercial seed mixtures, which appeal primarily to generalist bird species, are the dominant form of food provided by urban respondents (Table 17.2; Appendix 17.1). Commercial foods raise the concern that activities like bird feeding, which is generally thought to benefit birds, might instead contribute to processes that diminish urban biodiversity, such as increased dominance of exotic species (Shochat 2004).

Lack of Socio-demographic Associations

As predicted, traditional socioeconomic variables for the most part were not strongly associated with engagement in bird-related activities (Table 17.5), with a few revealing exceptions. Age proved to be the most consistent explanatory socio-demographic characteristic across the regions and landscapes for bird feeding, with older respondents being more likely to feed birds in all locations (Fig. 17.2). Our results are consistent with the US Fish and Wildlife Service survey findings showing greater proportions of bird watchers in their middle years or older, particularly those who primarily watched birds close to home (Pullis La Rouche 2003). Age is often included among socio-demographic indicators of social status, that is, social status increasing with age (Grove et al. 2006). But age is also a lifestyle characteristic (Grove et al. 2006), and we suggest that the differences we detect here in bird feeding are likely a consequence of the lifestyle changes associated with age, such as retirement and reduced physical activity. Regardless of mechan-ism, the findings suggest an important medium for communication about bird conservation to the general public might be to make use of media outlets aimed at older people.

We found some intriguing relationships between income variables and feeding certain types of food (Fig. 17.3). In particular, while income level—with house size as a proxy for income—was not strongly associated with bird-related activities (Table 17.4), it was associated with feeding particular food types in AZ. No such patterns were detected for MI respondents, though a different indicator of income level was used (house size) and the MI study did not aim to survey a cross-section of income levels as the AZ study did. In AZ, feeding of two costly food items (nectar and thistle seeds) increased with income level, while feeding bread, a less costly food item, declined sharply with income level (Fig. 17.3). If these household-level patterns aggregate to larger scales, then the distribution of certain specialized resources, like nectar for hummingbirds and thistle seed for finches, may contribute to patterns of increasing bird diversity with neighborhood income level reported for Phoenix and other cities (Kinzig et al. 2005, Melles 2005, Fuller et al., chapter 16, this volume).

Intriguingly, feeding nectar peaked in the moderate income levels, as did several other food types in both MI and AZ, though only the pattern for nectar was significant (Fig. 17.3). Our finding corresponds with other ongoing work in Phoenix and Baltimore, Maryland, in which the highest densities of bird feeders were in moderate-income neighborhoods (P. S. Warren, unpubl. data). The non-linear, hump-shaped relationships suggest a more complex set of underlying mechanisms than a simple association of feeding with wherewithal to purchase food. Instead, lifestyle behaviors, such as life stage or family size (Grove et al. 2006) may better account for people's interactions with birds.

A Note of Caution

While our study found a number of similarities between urban systems, we caution against extrapolating them widely as encompassing patterns at the regional level or in all urban systems. In particular, our Arizona survey focuses on only one municipality of 24 in a vast metropolitan area. Ongoing survey work is aimed at sampling more broadly from the Phoenix metropolitan area. In addition, we expect that a broader survey of Arizona residents would find significantly greater engagement in bird-oriented activities. Arizona is a major international tourist destination for birders, particularly the southeastern portion of the state, an area not covered by our survey. Another important caveat to bear in mind is that the methodological approach undertaken here compares two surveys that were targeted toward different groups of people and utilized two differing survey instruments. Thus, while the survey questions analyzed herein are similar, they were not always identical. Nevertheless, the differences in the study areas are worth considering, particularly the differences in proportions of people who feed birds. The proportion of people feeding birds in Arizona was smaller than the proportions in all three landscapes in Michigan. In other words, even taking only the most urban portions of each study area, we find significant differences in the number of people feeding birds in Michigan and Arizona.

CONCLUSIONS AND IMPLICATIONS

Many people in two different regions of the U.S. engage in activities that can positively or negatively influence birds. Moreover, people living in different urban areas are behaving in similar ways toward birds. But fewer people are providing food that supports a diversity of species, which poses a dilemma: Is it better to encourage citizens to feed birds, regardless of the food they provide, thereby encouraging people to interact with local environments? Or should we discourage citizens from feeding at all if they cannot afford to provide high-quality foods, due to either time or financial constraints? We are exploring the possibility of using marketing analysis to identify the best avenues of communication with people about birds. The large number of people we found engaging in activities that influence birds in two distinctly different regions of the U.S. highlights the importance for the ecologist of understanding the factors that motivate people to engage in these activities and to document their spatial and temporal distribution.

ACKNOWLEDGMENTS

We would like to thank everyone who helped make this study possible, including D. Stuart, D. Hinnebusch, P. Tarrant, K. Baker, J. Egeler, J. Liu, M. Mascarenhas, the CAP LTER bird technicians, and the Phoenix Parks, Recreation and Libraries Department, the Maricopa County Tax Assessor's office, and the Ingham, Livingston, Oakland, and Washtenaw County Equalization Offices. W. Porter provided invaluable assistance with data management. Finally, we appreciate the encouraging comments of two anonymous reviewers. The work was funded by NSF DEB grant #9714833, the NSF IGERT program at Arizona State University, a Michigan Agricultural Experiment

Station grant, and U.S. EPA STAR Fellowship grant #U-91580101-0.

LITERATURE CITED

Anthony, J. 2006. State growth management and housing prices. Social Science Quarterly 87: 122–141.

Atchison, K. A., and A. D. Rodewald. 2006. The value of urban forests to wintering birds. Natural Areas Journal 26:280–288.

Brittingham, M. C., and S. A. Temple. 1988. Impacts of supplemental feeding on survival rates of Black-capped Chickadees. Ecology 69:581–589.

Brittingham, M. C., and S. A. Temple. 1989. Patterns of feeder use by Wisconsin birds: a survey of WSO members. Passenger Pigeon 51:321–324.

Cannon, A. 1999. The significance of private gardens for bird conservation. Bird Conservation International 9:287–297.

Cannon, A. R., D. E. Chamberlain, M. P. Toms, B. J. Hatchwell, and K. J. Gaston. 2005. Trends in the use of private gardens by wild birds in Great Britain 1995–2002. Journal of Applied Ecology 42:659–671.

Carr, D. S., and D. R. Williams. 1993. Understanding the role of ethnicity in outdoor recreation experiences. Journal of Leisure Research 25:22–38.

Clergeau, P., G. Mennechez, A. Sauvage, and A. Lemoine. 2001. Human perception and appreciation of birds: a motivation for wildlife conservation in urban environments of France. Pp. 69–88 *in* J. M. Marzluff, R. Bowman, and R. Donnelly (editors), Avian ecology in an urbanizing world. Kluwer Academic Press, Boston, MA.

Cowie, R. J., and S. A. Hinsley. 1988. The provision of food and the use of bird feeders in suburban gardens. Bird Study 35:163–168.

Dale, V. H., S. Brown, R. A. Haeuber, N.T. Hobbs, N. Huntley, R. J. Naiman, W. E. Riebsame, M. G. Turner, and T. J. Valone. 2000. Ecological principles and guidelines for managing the use of land. Ecological Applications 10:639–670.

Dillman, D. A. 2000. Mail and Internet surveys: the tailored design method. 2nd ed. John Wiley & Sons, Inc. New York.

Dunn, E. H., and D. L. Tessaglia. 1994. Predation of birds at feeders in winter. Journal of Field Ornithology 65:8–16.

Emlen, J. T. 1974. An urban bird community in Tucson, Arizona: derivation, structure, regulation. Condor 76:184–197.

Faeth, S. H., P. S. Warren, E. Shochat, and W. A. Marussich. 2005. Trophic dynamics in urban communities. Bioscience 55:399–407.

Forester, D. J., and G. E. Machlis. 1996. Modeling human factors that affect the loss of biodiversity. Conservation Biology 10:1253–1263.

Fraser, E. D. G., and W. A. Kenney. 2000. Cultural background and landscape history as factors affecting perceptions of the urban forest. Journal of Arboriculture 26:106–112.

Gaston, K. J., P. H. Warren, K. Thompson, and R. M. Smith. 2005. Urban domestic gardens (IV): the extent of the resource and its associated features. Biodiversity and Conservation 14:3327–3349.

Gobster, P. H. 2002. Managing urban parks for a racially and ethnically diverse clientele. Leisure Sciences 24:143–159.

Grove, J. M., A. R. Troy, J. P. M. O'Neil-Dunne, W. R. Burch, M. L. Cadenasso, and S. T. A. Pickett. 2006. Characterization of households and its implications for the vegetation of urban ecosystems. Ecosystems 9:578–597.

Hochachka, W. M., J. V. Wells, K. V. Rosenberg, D. L. Tessaglia-Hymes, and A. A. Dhondt. 1999. Irruptive migration of Common Redpolls. Condor 101: 195–204.

Hostetler, M. 1999. Scale, birds, and human decisions: a potential for integrative research in urban ecosystems. Landscape and Urban Planning 45:15–19.

Hostetler, M., and C. S. Holling. 2000. Detecting the scales at which birds respond to structure in urban landscapes. Urban Ecosystems 4:25–54.

Kaplan, R., and J. F. Talbot. 1988. Ethnicity and preference for natural settings: a review and recent findings. Landscape and Urban Planning 15: 107–117.

Kinzig, A. P., P. S. Warren, C. Martin, D. Hope, and M. Katti. 2005. The effects of human socioeconomic status and cultural characteristics on urban patterns of biodiversity. Ecology and Society 10:23.

Lepage, D., and C. M. Francis. 2002. Do feeder counts reliably indicate bird population changes? 21 years of winter bird counts in Ontario, Canada. Condor 104:255–270.

Lepczyk, C. A. 2002. Effects of human activities on birds across landscapes of the Midwest. Ph.D dissertation, Michigan State University, East Lansing, MI.

Lepczyk, C. A. 2005. Integrating published data and citizen science to describe bird diversity across a landscape. Journal of Applied Ecology 42:672–677.

Lepczyk, C. A., A. G. Mertig, and J. Liu. 2004a. Assessing landowner activities that influence birds across rural-to-urban landscapes. Environmental Management 33:110–125.

Lepczyk, C. A., A. G. Mertig, and J. Liu. 2004b. Landowners and cat predation across rural-to-urban landscapes. Biological Conservation 115:191–201.

Liu, J., G. C. Daily, P. R. Ehrlich, and G. W. Luck. 2003. Effects of household dynamics on resource consumption and biodiversity. Nature 421:530–533.

Martinson, T. J., and D. J. Flaspohler. 2003. Winter bird feeding and localized predation on simulated bark-dwelling arthropods. Wildlife Society Bulletin 31:510–516.

Marzluff, J. M. 2001. Worldwide urbanization and its effects on birds. Pp. 19–47 *in* J. M. Marzluff, R. Bowman, and R. Donnelly (editors), Avian ecology in an urbanizing world. Kluwer Academic Press, Boston, MA.

McKinney, M. L. 2002. Urbanization, biodiversity, and conservation. Bioscience 52:883–890.

Melles, S. 2005. Urban bird diversity as an indicator of social diversity and economic inequality in Vancouver, British Columbia. Urban Habitats 3:25–48.

Mizejewski, D. 2004. Attracting birds, butterflies, and other backyard wildlife. Creative Homeowner, Upper Saddle River, NJ.

Pullis La Rouche, G. 2003. Birding in the United States: a demographic and economic analysis. Addendum to the 2001 National Survey of Fishing, Hunting and Wildlife-Associated Recreation. Report 2001-1. U.S. Fish and Wildlife Service, Arlington, VA.

Robb, G. N., R. A. McDonald, D. E. Chamberlain, and S. Bearhop. 2008a. Food for thought: supplementary feeding as a driver of ecological change in avian populations. Frontiers in Ecology and the Environment 6:476–484.

Robb, G. N., R. A. McDonald, D. E. Chamberlain, S. J. Reynolds, T. J. E. Harrison, and S. Bearhop. 2008b. Winter feeding of birds increases productivity in the subsequent breeding season. Biology Letters 4:220–223.

Rutledge, D. T., and C. A. Lepczyk. 2002. Landscape change: patterns, effects, and implications for adaptive management of wildlife resources. Pp. 312–333 *in* J. Liu and W. W. Taylor (editors), Integrating landscape ecology into natural resources management. Cambridge University Press, Cambridge, MA.

Sargent, M., and K. S. Carter. 1999. Managing Michigan's wildlife: a landowner's guide. Michigan Department of Natural Resources and Michigan United Conservation Club, Lansing, MI.

Shochat, E. 2004. Credit or debit? Resource input changes population dynamics of city-slicker birds. Oikos 106:622–626.

Shochat, E., P. S. Warren, S. H. Faeth, N. E. McIntyre, and D. Hope. 2006. From patterns to emerging processes in mechanistic urban ecology. Trends in Ecology and Evolution 21:186–191.

Smith, R. M., K. J. Gaston, P. H. Warren, and K. Thompson. 2005. Urban domestic gardens (V): relationships between landcover composition, housing and landscape. Landscape Ecology 20:235–253.

Stokes, D., and L. Stokes. 1987. The bird feeder book. Little, Brown and Co., New York.

Toms, M. 2003. The BTO/CJ Garden Birdwatch book. British Trust for Ornithology, Thetford, UK.

U.S. Census Bureau. 2006. Current housing reports, series H150/05. American Housing Survey for the United States: 2005.U.S. Government Printing Office, Washington, DC.

Vitousek, P. M., H. A. Mooney, J. Lubchenco, and J. M. Melillo. 1997. Human domination of Earth's ecosystems. Science 277:494–499.

Warren, P. S., C. Tripler, D. T. Bolger, S. H. Faeth, N. Huntly, C. A. Lepczyk, J. L. Meyer, T. Parker, E. Shochat, and J. Walker. 2006. Urban food webs: predators, prey, and the people who feed them. Bulletin of the Ecological Society of America 87: 387–394.

Wells, J. V., K. V. Rosenberg, E. H. Dunn, D. L. Tessaglia-Hymes, and A. A. Dhondt. 1998. Feeder counts as indicators of spatial and temporal variation in winter abundance of resident birds. Journal of Field Ornithology 69:577–586.

White, P. C. L., N. V. Jennings, A. R. Renwick, and N. H. L. Barker. 2005. Questionnaires in ecology: a review of past use and recommendations for best practice. Journal of Applied Ecology 42:421–430.

Wiedner, D., and P. Kerlinger. 1990. Economics of birding: a national survey of active birders. American Birds 44:209–213.

Zar, J. H. 1996. Biostatistical analysis. 3rd ed. Prentice Hall, Upper Saddle River, NJ.

Zaremba, J. 1961. Family income and wood consumption. Journal of Forestry 59:443–448.

APPENDIX 17.1

Other types of food provided to birds in the two study regions as indicated by number of respondents.

Food type	Michigan	Arizona	Total
Apple	5	1	6
Berries	1	1	2
Bread	3		3
Cat food		2	2
Cereal	3	2	5
Cheese	1		1
Chips		1	1
Corn	1	1	2
Crackers	1		1
Dog food	2	1	3
Dried fruit	1		1
Dried mango		1	1
Flower seeds	1		1
Food scraps	1		1
French fries		1	1
Fruit	2	1	3
Garbage	1		1
Leftovers	2		2
Melon	1		1
Nuts	2		2

APPENDIX 17.1 (*continued*)

APPENDIX 17.1 (CONTINUED)

Food type	Michigan	Arizona	Total
Peanut butter	5		5
Peanuts	7	2	9
Pears	1		1
Pineapple	1		1
Popcorn	5	2	7
Raisins	2		2
Rice		7	7
Safflower seed	8		8
Table scraps		1	1
Tortillas		3	3
Vegetable scraps	1	2	3
Wheat	1		1

PART FOUR

Future Directions

CHAPTER EIGHTEEN

Urban Evolutionary Ecology

John M. Marzluff

Abstract. Understanding the evolutionary responses of birds to urbanization has lagged behind understanding ecological responses. Therefore, I provide a conceptual framework for understanding evolutionary processes in urban environments and distill key features of birds that enable them to evolve with the novel features of urban environments. I start by reviewing the growing field of contemporary evolution, or current micro-evolution, that occurs in less than a few hundred years. I then broaden the genetic focus of contemporary evolution by also considering the evolution of cultural traits. This effort builds on the extensive European research on synurbization, or the adjustment of animals to urban environments. Contemporary evolution of cultural and genetic traits is well documented in urban environments. Changes in predation, industrial pollution, food sources, noise, and climate have caused the rapid evolution of bird life histories, coloration, morphology, migratory behavior, singing, and foraging behavior. The wide range of genetically and culturally heritable traits suggests that many other changes have evolved in response to urbanization. The uniqueness of the urban environment and the size and isolation of urban bird populations affects genotypic and phenotypic variability and the interactions between genotypes and phenotypes. Responses to selection (or random changes in phenotypes) that are inherited or learned socially define what evolves in urban bird populations, but population size is also important because stochastic events may extinguish small populations before they adapt to the novel urban environment. Because of the close association between people and birds in urban environments, coevolutionary relationships are possible. Coevolution may involve genetic and cultural traits. For example, humans and corvids appear to be culturally coevolving in American, European, and Asian cities. A variety of characteristics enable contemporary evolution by urban birds: large population size, sociality and innovation, exploitation of people, phenotypic plasticity, high annual reproductive rate, short generation time, environmental uniqueness, and population isolation. Studying contemporary evolution in urban environments requires correlating changes in heritable genetic and cultural traits with environmental conditions, demonstrating that selection is working on these traits, and making a causal connection between the environment and the species' response. Investigations require careful study and ideally a combination of longitudinal and experimental approaches. But it is possible, and showing the public that evolution occurs in their backyards may provide a unique way to engage citizens in science.

Key Words: adaptation, contemporary evolution, cultural evolution, extinction, natural selection, urbanization.

Marzluff, J. M. 2012. Urban evolutionary ecology. Pp. 287–308 *in* C. A. Lepczyk and P. S. Warren (editors). Urban bird ecology and conservation. Studies in Avian Biology (no. 45), University of California Press, Berkeley, CA.

Our understanding of the ecology of birds in an urbanizing world is blossoming. Correlational connections between land cover change and bird community diversity or population density are serving as strong platforms from which to study the mechanisms causing bird responses to urbanization (Shochat et al. 2006). We are learning the subtle details of how birds respond to urban sources of food (Shochat et al. 2004) and that decoupling from variation in natural prey may allow some species to attain spectacular population growth rates (Withey and Marzluff 2005), while others face extirpation (Schoech and Bowman 2003). We know more about the complexities of nest parasitism and predation in the urban context (Chace et al. 2003, Sinclair et al. 2005, Marzluff et al. 2007), and how it may interact with planted vegetation to influence population viability (Schmidt and Wheland 1999, Rodewald, chapter 5, this volume). The mysteries of post-fledging movements, survival, and dispersal within the urban landscape are being revealed (Whittaker and Marzluff, chapter 12, this volume). Scenarios of regional avian response to urbanization are becoming available to urban planners and policymakers (Hepinstall et al., chapter 15, this volume).

An understanding has enabled our science to begin moving from description to prediction. For example, I suggested that the response of avian community diversity to urbanization could be predicted by considering the colonization and extinction of guilds of species that were differentially sensitive to land cover change (Marzluff 2005). In this model, richness reflects the balance between the loss of species sensitive to the conversion and degradation of native land cover and the colonization of species able to exploit newly created land covers and human subsidies. A wide range of regional patterns of avian diversity measurable along gradients from urban to rural, exurban, or natural lands are predicted depending on factors that influence local extirpations such as the intensity of land use and configuration of reserved lands, characteristics of the urban source populations affecting colonization by synanthropic species, and the contrast between developed areas and the natural vegetation of the region that affects the attractiveness of built areas to early successional species.

Many advances in avian urban ecology have been motivated by a desire to reduce the loss of native species and avoid geographic homogenization of avifaunas (Blair 2001, 2004; Clergeau et al. 2006). This conservation focus is especially obvious in the many recent studies from rapidly urbanizing regions of North America (Kluza et al. 2000, Melles et al. 2002, Donnelly and Marzluff 2004, Blewett and Marzluff 2005, Er et al. 2005, Fraterrigo and Wiens 2005, Gehlbach 2005), South America (Reynaud and Thioulouse 2000), Asia (Lim and Sodhi 2004, Lee et al. 2005, Posa and Sodhi 2006), and Australia (Recher and Serventy 1991, Daniels and Kirkpatrick 2006, Hodgson et al. 2006, Parsons et al. 2006). Our conservation focus is productive and relevant, but it is also incomplete. Rarely do we consider the details of successful urban species. Yet increasingly it is apparent that such species are rapidly evolving to meet our challenges (Johnston 2001, Palumbi 2001).

European urban ecologists have concerned themselves with the adaptation of birds to urbanization. The term, "synurbization," refers to the general adjustment of animal populations to urban environments (Luniak 2004). It is not surprising that scientists in historic cities would focus on adaptation, while those in more recently established cities would focus on extirpation. The relative importance of extirpation versus adaptation may vary predictably with a city's age. Nonetheless, a holistic urban bird ecology should include a thorough understanding of changes in ecological processes over evolutionary time. Understanding how species adapt to a human-dominated world may allow us to better understand the dynamics of colonization and extinction and more accurately predict what future avifaunas will look like.

This chapter aims to stimulate an evolutionary view of urban ecology. I review and synthesize a growing literature on "contemporary" or rapid evolution (Endler 1986, Thompson 1998, Hendry and Kinnison 1999, Stockwell et al. 2003). By "contemporary evolution," researchers generally mean current change in heritable genetic traits within a species or population (micro-evolution) that occur in less than a few hundred years (Hendry and Kinnison 1999, Stockwell et al. 2003). I illustrate the sorts of traits in birds that we know to be heritable and capable of rapid change. I also review the aspects of urban environments that may select for directional adaptive change. I broaden the genetic focus of traditional contemporary evolution by considering the evolution of cultural, as well as genetic, traits. Inclusion of these traits allows a more complete consideration of the many behavioral adjustments that European colleagues have addressed in their investigations of synurbization and brings recent thoughts on cultural interactions between people and animals into the mix (Marzluff and Angell 2005a, 2005b). My objectives are to: (1) provide a conceptual framework for understanding evolutionary processes in urban environments, (2) distill key features of birds that enable them to evolve in urban systems, and (3) encourage researchers to engage more in the study of evolutionary responses of birds to urbanization.

CONTEMPORARY EVOLUTION IN URBAN ECOSYSTEMS

The most obvious pattern in empirical studies documenting adaptive evolution is that most of them document the response to anthropogenic changes in the environment. . . .

Reznick and Ghalambor 2001

As the above quote emphasizes, contemporary evolution is conspicuous in response to anthropogenic change. Overharvest, habitat degradation and fragmentation, toxic pollution, and the introduction of species into novel environments often cause plants and animals to adapt or go extinct (Endler 1986, Palumbi 2001). Adaptation by one species may further affect coevolved interspecific interactions between symbionts, predators and prey, parasites and hosts, diseases, vectors and hosts, and competitors (Thompson 1998). Urban environments are characterized by novel mixes of predators, competitors, diseases, food sources, vegetation, land cover, and land use. They also induce local climatic anomalies (e.g., heat islands and heat waves; Endlicher et al. 2006), provide abnormal light patterns (Rich and Longcore 2006), expose inhabitants to altered acoustical environments (Patricelli and Blickley 2006, Warren et al. 2006), introduce synthetic pollutants, and place animals and people in close, daily contact. Therefore, evolution should be conspicuous and rapid in cities (Johnston 2001). Indeed Martin and Clobert (1996) suggested that a suite of life-history differences between European and North American birds evolved in response to different urban environments. Reduction in nest predation characteristic of long-standing persecution of predators in European cities was thought to explain reduced clutch size, increased iteroparity, and increased adult survival in European birds, relative to North American birds.

In addition to the responsiveness of life-history traits to changed predator communities, I found 19 detailed studies of contemporary evolution in urban ecosystems (Table 18.1). This sample is not an exhaustive search of the literature, and it is reflective of my ornithophilic bias. The studies show that diverse human activities in cities around the world during the last two centuries have exposed native and introduced species to novel climatic regimes, ameliorated breeding seasons, new foods, and persistent low-frequency noise, and altered the very backdrop to life. Animal populations have responded genetically and culturally in a few generations with changed morphology and behavior.

The literature I reviewed consistently identifies the novel abiotic conditions of cities as important selective forces. Our introductions of species and industrial lifestyles often place urban birds, fish, and insects into environments that cause rapid genetic responses. Europeans seeking to acclimate to new lands, for example, enriched native avifaunas with birds from their homeland and aesthetically pleasing songsters. As a result, House Sparrows (*Passer domesticus*), House Finches (*Carpodacus mexicanus*), and Common Mynas (*Acridotheres tristis*) were moved across and between continents (Elliot and Arbib 1953, Johnston and Selander 1964, Baker and Moeed 1979, Baker 1980). Since introduction, their morphology, coloration, and basic migratory habits (House Finch) evolved in response to novel climatic conditions. Western mosquitofish (*Gambusia affinis*) introduced from Texas to settlements in California and Nevada to control mosquitoes similarly evolved morphologies and life histories to match new thermal environments. As Europe became increasingly reliant on coal to fuel 18th- and 19th-century industrialization, soot from urban factories darkened the substrates animals lived among and allowed predators to drive the evolution of coloration in many insects and at least one bird, the Rock Dove (*Columba livia*). Initially, heavy soot provided an advantage to dark (melanic) forms of moths, spiders, beetles, and pigeons whose better match to the dark background reduced predation and provided a selective

TABLE 18.1

Contemporary genetic and cultural evolution by animals in response to specific human activities in urban ecosystems.

Species	Location	Human activity	Population size	Years	Generations
Red Fox	Zurich, Switzerland	Food subsidy	>500	1985–2000	5–7
Dark-eyed Junco	San Diego, CA, USA	Heat island, land cover change	160	1983–2002	8
House Sparrow	North America	Introduction	Large	1850–1965	<111
House Sparrow	New Zealand	Introduction	Large	1862–1977	~100
Common Myna	New Zealand	Introduction	Large N of 40° S	1870–1978	50–100
House Finch	Eastern USA	Introduction	Large	1940–1966	~25
Rock Pigeon	Manchester, UK	Industry	Large	1800–1970	>100
European Blackbird	Continental Europe	Artificial light, heat island, reduced persecution, supplemental food	Large	Early 1800s–2000	~50[a]
Peppered Moth	Wales-Liverpool, UK	Industry	Large	1952–1984	15
Western Mosquitofish	Nevada, USA	Mosquito control		1934–1995	110–165
Rock Pigeon	Montréal, Canada	Experiment	~100	May–June 1985	<1
Titmice	United Kingdom	Milk delivery	Large	1921–1947	<20
Titmice	Ireland	Milk delivery	Large	1937–1947	<5
Magpie	Chepstow, Gwent, UK	Milk delivery	Large	1991–1993	1
Carrion Crow	Sendai, Japan	Planting trees	Large	1960–1975	~5
House Finch	Milwaukee, WI, USA	Introduction	<500	1973–1992	<20

Selective force	Trait	Mode	Evolutionary processes			Reference
			Plasticity	Drift	Selection	
New food	Microsatellite allelic diversity	Genetic	–	+	–	Wandeler et al. 2003
Climate, sex, long breeding season	% white in tail	Genetic	–	–	+	Yeh 2004
Climate	Plumage, morphology	Genetic	–	–	+	Johnston and Selander 1964
Climate	Morphology	Genetic	–	+	+	Baker 1980
Precipitation, temperature	Morphology	Genetic	–	+	+	Baker and Moeed 1979
Temperature, food	Migration	Genetic	+	–	+	Able and Belthoff 1998
Pollution, predation	Melanism	Genetic	–	–	+	Bishop and Cook 1980
Increased population density, long breeding season,	Tameness, reduced migration, timing of reproduction, stress response	Genetic	+	–	+	Walasz 1990; Partecke 2003; Partecke et al. 2004, 2005, 2006a, 2006b
Pollution, predation	Melanism	Genetic	–	–	+	Cook et al. 1986
Temperature	Breeding size, fat content, embryo size	Genetic	+	+	+	Stockwell and Weeks 1999
New food	Foraging trait	Cultural	+	–	+	Lefebvre 1986
New food	Removing lid from milk bottle	Cultural	+	–	+	Lefebvre 1995
New food	Removing lid from milk bottle	Cultural	+	–	+	Lefebvre 1995
New food	Removing lid from milk bottle	Cultural	+	–	+	Vernon 1993
New food	Cracking walnuts with automobiles	Cultural	+	–	+	Nihei and Higuchi 2001
? Sex	Song syntax, repertoire size, and sharing	Cultural	+	–	?	Pytte 1997

TABLE 18.1 (*continued*)

TABLE 18.1 (continued)

Species	Location	Human activity	Population size	Years	Generations
Great Tit	11 European cities[b]	Traffic, large buildings, industry	Large	Early 1900s–2000	~20[c]
House Finch	Los Angeles, CA, USA	Same as Great Tit			
Song Sparrow	Portland, OR, USA	Same as Great Tit			

[a] Exact timing of these adaptations to urban life is not known, but was likely soon after Blackbirds invaded cities in the 1800s.

[b] Great Tits were studied in London, Leiden, Prague, Amsterdam, Paris, Brussels, Antwerp, Rotterdam, Berlin, Nottingham, and Luxembourg.

[c] Exact timing of acoustic divergence between urban and nearby rural populations is unknown, but at least in western USA, urban noise was negligible prior to the early to mid-1900s.

[d] Possible genetic divergence has not been confirmed, Slabbekoorn and Smith 2002.

advantage over lighter morphs (Kettlewell 1973, Bishop and Cook 1980). This classic demonstration of natural selection leading to "industrial melanism" is continuing to respond to urban policies. With increased pollution control in urban areas in the latter half of the 19th century, the lighter-colored trees, rocks, and buildings provided safe resting sites for paler morphs, and natural selection has adjusted downward the relative frequency of melanic forms (Cook et al. 1986).

Birds in urban environments have learned to eat a wide variety of exotic and prepared foods (Goodwin 1978, Marzluff and Angell 2005a). Some of these innovations have spread through local populations as culturally evolved traditions. The cultural evolution of novel feeding behavior has been experimentally demonstrated within urban Rock Doves (Lefebvre 1986). It has been implicated in the removal of milk bottle lids by tits and corvids in Britain (Martínez del Rio 1993, Vernon 1993, Lefebvre 1995) and in the use of cars as nutcrackers by Carrion Crows (*Corvus corone*) in Japan (Nihei and Higuchi 2001).

Traffic, industry, and built surfaces that characterize urban areas have profoundly altered the acoustical environments that many birds rely on. Low-frequency noise, disruption of sound waves by large, flat surfaces, and altered sound channels have selected for shorter, higher-frequency, louder, faster, and novel songs (Patricelli and Blickley 2006, Warren et al. 2006). Such adjustments benefit singing birds because they enable potential mates and territorial intruders to hear vocal advertisements in noisy environments (Slabbekoorn and Smith 2002). In oscine songbirds, social learning is allowing such changes to proceed rapidly by cultural evolution, which may initially constrain but later promote speciation (Slabbekoorn and Smith 2002).

The Great Tit (*Parus major*) is a model species illustrating adaptive, contemporary, cultural evolution of song in urban environments. Loud, low-frequency noise in major European cities favors short, fast, and high-frequency song (Slabbekoorn and Peet 2003, Slabbekoorn and den Boer-Visser 2006). Phenotypic plasticity in singing enables tits to vary song characters. Learning may rapidly fit songs to the acoustic environment because songs that can be heard can also be copied and those that do not elicit responses can be dropped (Slabbekoorn and den Boer-Visser 2006). Because tits learn in part through interactions with neighbors on their breeding territory, urban tit songs are consistent and different from rural tit songs (Slabbekoorn and den Boer-Visser 2006). Assortative mating for habitat-dependent song characteristics and their genetic basis may lead to reproductive isolation and possible speciation of urban tits (Slabbekoorn and Smith 2002, Ellers

Selective force	Trait	Mode	Evolutionary processes: Plasticity	Drift	Selection	Reference
Low-frequency noise, large flat surfaces, altered sound channels	Song frequency, amplitude, duration,	Cultural and genetic[d]	+	−	+	Slabbekoorn and Peet 2003, Slabbekoorn and den Boer-Visser 2006
						Fernández-Juricic et al. 2005
						Wood and Yezerinac 2006

and Slabbekoorn 2003). But currently, phenotypic plasticity and cultural evolution are simply allowing tits of various genotypes to persist in urban environments, thereby softening the force of natural selection on genetic variation.

Some evolution in urban populations is neither directional nor clearly an adaptive response to natural selection. Changes in the microsatellite allelic diversity in foxes (Wandeler et al. 2003) and the culturally derived unique songs of eastern House Finches (Pytte 1997), for example, may reflect the random action of genetic drift or the chance occurrence of unique individuals in small founding populations.

The detailed work needed to understand the relative contributions of phenotypic plasticity, genetic drift, and adaptive responses to natural selection is exemplified in Yeh's (2004, Yeh and Price 2004) study of plumage evolution in urban Dark-eyed Juncos (*Junco hyemalis*). By raising juncos from urban and wildland nests in a common environment, determining the effective population size, and quantifying the rate of evolution, Yeh concluded that the observed reduction in the proportion of white in the tails of urban (San Diego) juncos was an adaptive response to natural selection. A variety of aspects of the San Diego environment may have favored less-conspicuous tails, including a lengthened breeding season that favors greater parental investment over aggressive defense of a territory, increased risk of predation, and novel climate, food, and vegetation.

Similar experimentation is also documenting the interplay between phenotypic plasticity and genetic differentiation in urban Blackbirds (*Turdus merula*). In European cities, lack of persecution by people, artificial light, and supplemental food have enabled Blackbirds to attain unprecedented densities (Erz 1964). Urban Blackbirds are tamer, less migratory, breed and molt earlier, and suppress stress responses relative to rural Blackbirds (Walasz 1990; Partecke 2003; Partecke et al. 2004, 2006). Changes in stress response, migratory behavior, and timing of reproduction are mainly a result of phenotypic plasticity with some genetic change in adaptive characters (Partecke et al. 2004, 2006). Despite adaptive genetic microevolution, divergence in neutral alleles has not been detected (Partecke et al. 2006). Experiments on Blackbirds and Dark-eyed Juncos show how phenotypic plasticity, behavioral innovation, and the effects of small founding populations may set the initial range of responses that birds exhibit in novel urban environments, and how heritable traits quickly evolve in adaptive responses to natural selection (Price et al. 2003).

I suspect that increased scrutiny of urban birds will reveal many more cases of genetic and cultural evolution. Many traits are at least weakly heritable (Table 18.2). These suggest that in addition to changes in morphology, migration, song, and feeding behavior (Tables 18.1, 18.2), we should expect urban birds to respond to selection for changes in egg size, clutch size, laying rhythm, nest size, nest spacing, personality, fear, dispersal distance, and immunity (Table 18.2). The response of birds to stresses (e.g., diseases) and subsidies (e.g., food) in urban systems may

TABLE 18.2

Examples of traits known to be genetically heritable in wild bird populations.

Trait	Species	Reference
Body size and weight	Collared Flycatcher	Merila et al. 2001
	Japanese Quail	Akbas et al. 2004
	Eurasian Magpie	De Neve et al. 2004
	Snow Petrel	Barbraud 2000
	Pinyon Jay	Marzluff and Balda 1988
Body condition	Blue Tit	Merila et al. 1999
Clutch size	Domestic goose	Rosinski et al. 2006
	Numerous	Postma and van Noordwijk 2005
	Collared Flycatcher	Sheldon et al. 2003
Colony size and spacing	Barn Swallow	Møller 2002
Coloration	Blue Tit	Hadfield et al. 2006
	Collared Flycatcher	Garant et al. 2004
Dispersal distance	Great Reed Warbler	Hansson et al. 2003
	Red-cockaded Woodpecker	Pasinelli et al. 2004
Egg size	Pied Flycatcher	Potti 1993
Fear	Domestic Chicken	Altan et al. 2005
Fledgling weight	Great Tit	Vantienderen and De Jong 1994
Head length	Common Gull	Larsson et al. 1997
Laying date	Collared Flycatcher	Sheldon et al. 2003
Laying rhythm	Domestic goose	Rosinski et al. 2006
Migratory activity	Blackcap	Pulido et al. 2001, Berthold 2003
	Barn Swallow	Møller 2001
Parental care	Long-tailed Tit	MacColl and Hatchwell 2003
Personality	Great Tit	Drent et al. 2003, Carere et al. 2005, van Oers et al. 2004
	Great Tit	Dingemanse et al. 2004
Tail length and nest size	Barn Swallow	Møller 2006
Tumor growth	Domestic Chicken	Praharaj et al. 2004
Volume of song control nuclei	Zebra Finch	Airey et al. 2000

constrain heritability (Charmantier et al. 2004, De Neve et al. 2004) and hence evolutionary potential (Christe et al. 2000). But there is ample evidence from wild populations of birds that beak morphology (Grant and Grant 1995, Smith et al. 1995), plumage coloration (Johnston and Selander 1964) and shape (Møller and Szep 2005), migratory behaviors (Berthold and Pulido 1994), rejection of brood parasites' eggs (Cruz and Wiley 1989), and song types (Baker et al. 2003) evolve in 10–200 years (a few to several hundred generations).

KEY PROCESSES IN CONTEMPORARY EVOLUTION

Our basic understanding of the evolutionary process has changed little since Darwin's and Wallace's time. The environment poses challenges to organisms. Organisms vary in their ability to solve these challenges, which affects individual reproduction and survival. Within a population, through time the frequency of traits possessed by successful individuals increases while those characteristic of less successful individuals decrease. Phenotypic characteristics enabling success diverge from their ancestral condition to become recognized adaptations of a local race, and possibly distinguishing features of unique species.

While understanding the basic process of phenotypic change in response to natural selection has stood the test of time, the key processes within adaptive evolution have been revealed with increasing complexity and mechanistic clarity. I have illustrated some of these new insights within the context of birds living in urban environments (Fig. 18.1). I emphasize four processes: (1) the influence of the environment on the genotypes and phenotypes in urban bird populations; (2) the complex interplay between genotypes and phenotypes that determines the variation within a population upon which selection, drift, and gene flow work; (3) the interactions between heritable (genetic and learned) changes, population size, and environmental stochasticity; and (4) the outcomes of phenotypic and genotypic change through time.

The Urban Environment as a Selective Force

Urban environments affect genetic and phenotypic diversity of a population by posing novel challenges (selective pressure) and perhaps by exposing individuals to mutagenic pollutants and toxins. There are many novel challenges to birds in urban environments, including new mixes of predators, competitors, and plants; abundant, unique, and often changing foods; buildings, cars, and people; and altered light, climate, noise, and disturbance regimes (Fig. 18.1). The hazards of these novelties can drive evolution because of the negative consequences on reproduction and survival of native birds in urban environments (Marzluff 2001). But novelty also provides opportunities that some native and introduced birds exploit with beneficial consequences to their fitness (Luniak 2004). Birds that survive in urban areas are under selection to expand the range of foods they eat, adjust their response to human activity, avoid dangerous anthropogenic settings (e.g., roads and buildings), breed and migrate in synchrony with new foods, temperatures, and photo-regimes, recognize new predators, avoid the masking effects of low-frequency noise on their vocalizations, and nest in unique places. These are just a few of the obvious ways urban environments may drive the evolution of birds.

Interacting Genotypes and Phenotypes

The genotypic and phenotypic adjustments to urban life are interactive. The shaded circle in Figure 18.1 shows some of the interactive processes that affect the variation in genotypes and phenotypes within and between generations. The phenotypic plasticity, or range of behaviors, actions, and morphologies among individuals, may affect the genotypic frequencies in a population through survival and reproduction of the fittest, genetic assimilation (Waddington 1961), and gene-culture coevolution (Durham 1991). Birds introduced into urban environments or those living in areas as they are converted to cities may exhibit phenotypic plasticity, usually by adjusting existing behaviors (e.g., timing of breeding; Price et al. 1988, Partecke et al. 2004) or learning new behaviors (e.g., placing nuts on roads to be opened by passing cars; Nihei and Higuchi 2001, singing higher-pitched songs; Slabbekoorn and den Boer-Visser 2006).

Phenotypic plasticity may be a key to initial survival in a new environment (Yeh and Price 2004), and it may facilitate genotypic change at the population level if extreme phenotypes are favored (Price et al. 2003). Such "genetic assimilation," for example, may account for the rapid evolution of genetically determined migratory behavior from initially plastic behavioral responses to climate and food shortage, and other cases of genetically determined behaviors. Plasticity enables genetic assimilation because extreme forms of a plastic response and the genes underlying them enjoy a fitness benefit and increase in frequency. Extreme cultural traits may also increase in a population through selective advantage (e.g., responsive and easily heard songs; Slabbekoorn and den Boer-Visser 2006), but phenotypic traditions with selective advantages can also drive genetic change across generations (Durham 1991). An example of the latter case, known as "gene-culture coevolution," is the evolution of lactase production in adult humans who evolved in dairying cultures (Durham 1991). Innovative behaviors are not only

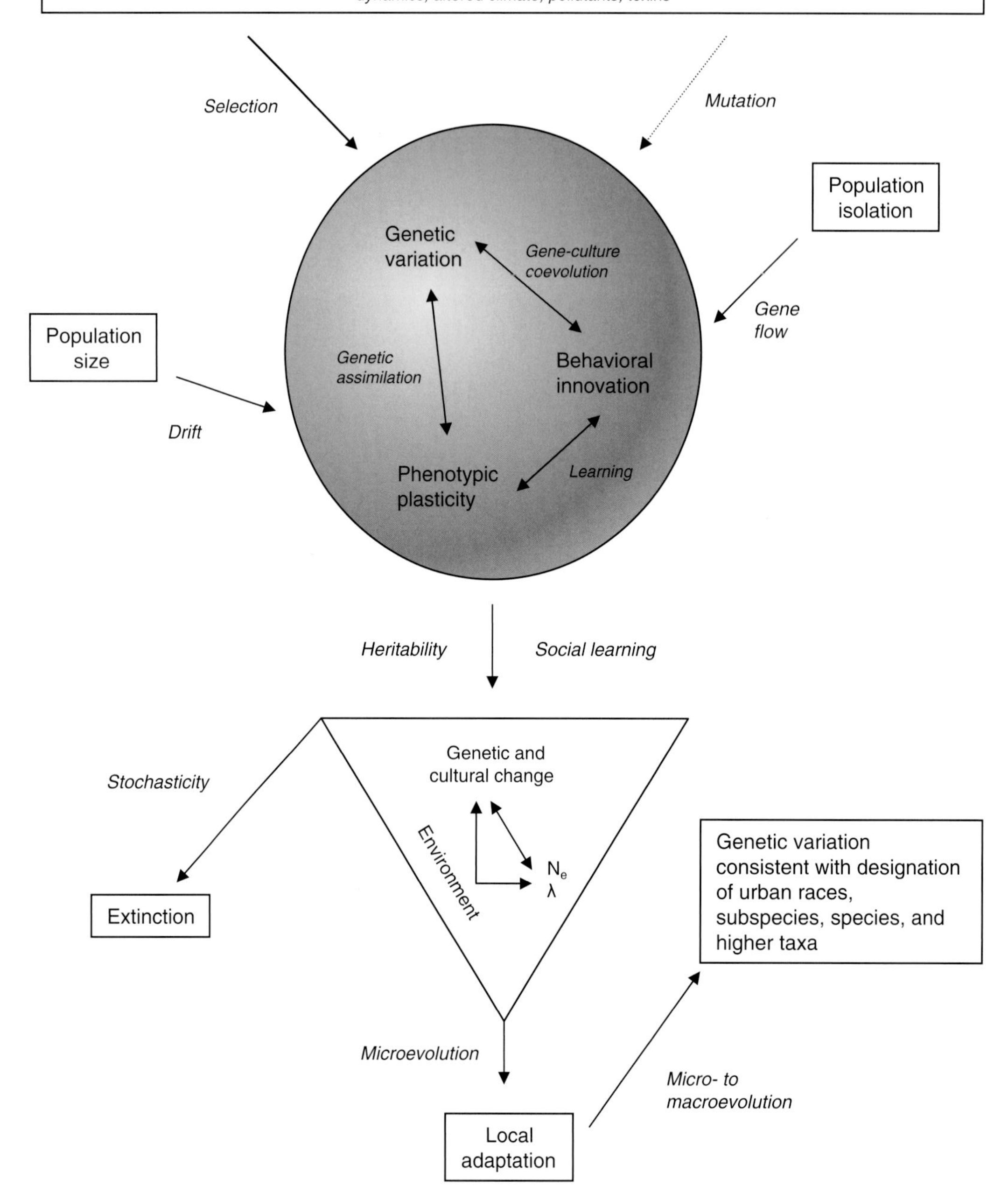

Figure 18.1. Key processes and interactions of contemporary evolution in urban ecosystems. Many novel and challenging aspects of urban environments affect phenotypic and genotypic variability (gray circle) of an urban animal population, but they do so in two fundamental ways: as agents of natural selection and as mutagenic toxins and pollutants. Genotypes (lighter portion of circle) and phenotypes (darker portion of circle) interact in complex ways within and between generations to determine phenotypic variation that is exposed to selection. Responses of a population to selection may be mediated by small population size (drift) and gene flow, but genetic traits are inherited and cultural traits are passed on by social learning to change the genetic and cultural composition of future generations. This change, or evolutionary adjustment to the environment, interacts with environmental and demographic stochasticity (triangle) in such a way that small populations are at risk of extinction as they are adapting to the novel urban environment. Larger or rapidly growing populations are at less risk of stochastic extinction and evolve local adaptations that may be deemed sufficient to warrant taxonomic designations of urban populations as races, subspecies, or full species.

enabled by phenotypic plasticity, they may expand the range of phenotypes within a population (Slabbekoorn and Smith 2002). By their exposure to the novel aspects of an urban environment, the variety of phenotypes within a bird population are selected among, which causes the associated genes and memes (as well as those linked to them by physical proximity or coevolved mutual dependence) to change in frequency within and between generations (genetic and cultural evolution). I am using "memes" here to mean the units of information transferred among individuals by social learning (e.g., rules, songs, religions, or even specific behaviors like handshakes or dietary choices; Dawkins 1976, Blackmore 1999). For example, it appears that higher-pitched, faster acoustical memes of Great Tits (and to a lesser extent their genetic bases) have increased in urban populations because they are effective at attracting mates and expelling rivals (Slabbekoorn and den Boer-Visser 2006).

Population Size and Isolation

Two important aspects of the population, its size and isolation, initially affect how the interacting genotypes and phenotypes respond to selection (Fig. 18.1). Population size influences evolution in many ways. The size of the founding population affects its genetic variation, and in small populations gene frequencies may change substantially over generations as a simple result of taking small, random samples of a larger whole (genetic drift). Drift may affect the course of evolution (Wright 1931, 1932), disrupt directional selection, and even steer directional changes in phenotypes in large populations if selection is weak (Lande 1976). The isolation of a population also affects how it responds to selection. Isolated urban populations may be able to respond quickly to selection, but populations connected by gene flow to non-urban populations may respond more slowly because new individuals, carrying phenotypes and genotypes unmodified by urban life, interbreed with urban individuals or influence the spread of traditions, thus reducing the uniqueness of the next generation.

Response to Selection

Urban genotypes and phenotypes change through time as they are genetically inherited or socially learned. The heritability of genetic traits and reliability of social learning are thus key processes that affect how urban bird populations respond to novel aspects of their environment. However, the result of genetic and cultural change is a complex interplay between the environment, population size and growth rate, and the amount of change that occurs each generation (triangle in Fig. 18.1). Populations that are large and poorly adapted to the current environment can change drastically and rapidly as differences in reproduction and survival quickly sort among phenotypes. Well-adapted populations may change little, but attain high growth rates. These are key features of a population because small and slow growing populations are prone to extinction from environmental and demographic stochasticity (Lande 1993). Obtaining large population size may be necessary for urban populations to persist long enough to evolve genetic adaptations to local conditions (Gomulkiewicz and Holt 1995). Cultural adaptation may evolve more quickly and enable small populations to track rapid environmental change and survive stochastic events. The resulting microevolution of local adaptations creates distinguishing features of races or local populations that may accumulate to produce species or higher taxa (macroevolution; Arnold et al. 2001). The connections between microevolution and macroevolution have intrigued evolutionary biologists for decades (Simpson 1953, Mayr 1963, Arnold et al. 2001).

Conceptualizing evolution in urban environments highlights two important points (Fig. 18.1). First, small populations may not be able to evolve adaptations to novel environments. Our focus has been on the mechanisms that reduce fitness in urban birds without clear recognition of the importance of small population size and stochastic variation. Second, if extinction is averted, novel environments become engines of evolutionary change. Rarely do we study the creative aspects of urban life, yet because of the pace of evolution in urban ecosystems, we may be ideally suited to document contemporary evolution and observe if and how the similarities and differences between microevolution and macroevolution play out.

EVOLVING WITH PEOPLE

A unique aspect of contemporary evolution in urban ecosystems is the potential coevolutionary relationship between people and birds. Here,

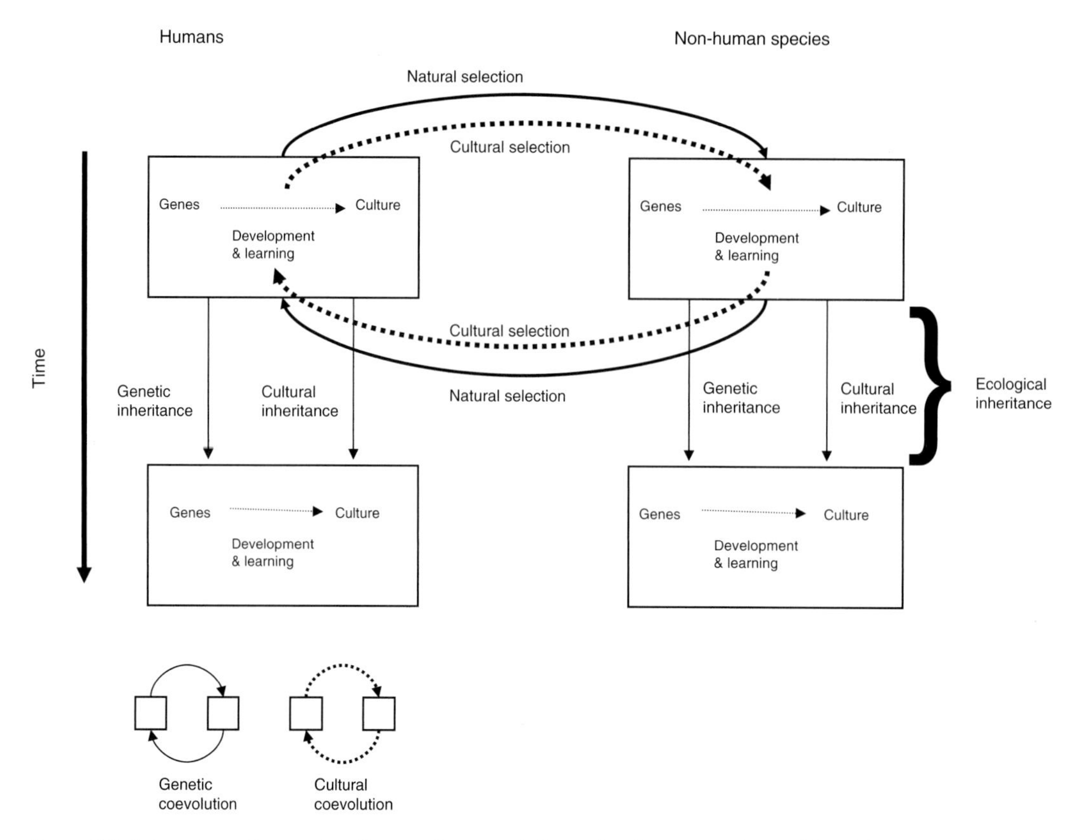

Figure 18.2. Integration and expansion of gene-culture coevolution and niche construction models to illustrate genetic and cultural coevolution between two interacting species (here humans and another social animal in their environment, after Laland et al. 2000). The collection of phenotypes within a population (boxes) emerge as a joint product of genetic, individually aquired, and socially learned (culture) information. Populations evolve as genetic and learned information is transferred through time by genetic and cultural inheritance. As people interact with their environment they cause changes in other organisms' inclusive fitness (natural selection) or the cultural fitness of their memes (cultural selection). Reciprocal changes in humans caused by organisms in their environment can lead to coevolved genes (bottom panel, solid lines) or cultures (bottom panel, dotted lines). Redrawn from Marzluff and Angell (2005b).

I explore some of the processes in this intriguing relationship by extending the theory of niche construction (Odling-Smee et al. 2003) to explicitly consider how people influence and in turn can be influenced by aspects of their environment. I suggest that when humans interact with other social species, who themselves have the ability to evolve culture, then simple feedbacks from a culturally evolving "environment" can stimulate rapid cultural evolution in humans. Exploring this "cultural coevolution" (Marzluff and Angell 2005a, 2005b) may expand our understanding of evolution in social urban birds and increase our awareness of the important cultural services people obtain from nature.

Humans and other social animals have a dual inheritance system whereby gene frequencies change through time in response to mutation, natural and cultural selection, and drift (genetic inheritance), and meme frequencies change through time in response to innovation, natural and cultural selection, learning, and drift (cultural inheritance; Fig. 18.2). At any point in time, human culture is composed of memes that reflect genetic, individually learned, and socially transmitted information (boxes in Fig. 18.2; Durham 1991, Laland et al. 2000). I follow the more mechanistic definition of culture—variation acquired and maintained by indirect and direct social learning (Boyd and Richerson 1985, Rendell and Whitehead 2001, Castro and Toro 2004).

As described in the previous section (Fig. 18.1), humans are potent agents of natural selection, affecting the microevolution of organisms and modifying the configuration and functioning of the physical environment. Environmental responses to these effects can force cultural and natural selection on humans, affecting an individual's inclusive fitness and the cultural fitness of their memes, as originally postulated by gene-culture and niche construction theory. Niche construction theory recognizes that the environment changes in response to human natural and cultural selection so that humans "inherit" a change in ecology as well as a change in gene and meme frequency (Laland et al. 2000). I suggest that this "ecological inheritance" is not only the physical and ecological change wrought by people, but also the cultural change in response to human activity by animals capable of social learning. Many animals, especially social, long-lived, and intelligent ones, acquire information through social learning and develop traditions that meet the social learning-based definition of culture (Avital and Jablonka 2000, Rendell and Whitehead 2001, Danchin et al. 2004). Where human activity results in differential cultural fitness of another animal's memes (Fig. 18.2; cultural selection from humans to the environment) and the resulting cultural evolution in the animal affects the cultural fitness of human memes (cultural selection from the environment to humans), then human and animal memes may become coevolved. Cultural coevolution is analogous to traditional genetic coevolution, where reciprocal natural selection between organisms drives mutual change in genes (e.g., crossbill bill depth and conifer cone structure; Benkman 2003).

Corvids, a common component of urban bird communities throughout the world, provide many examples of cultural coevolution with people (Marzluff and Angell 2005a). Interacting with people favors cultural adjustment by corvids because human attitudes and important resources change regularly and rapidly. Three aspects of human culture appear especially important to corvid culture: persecution, provision of new food, and creation of new opportunities. The power of persecution is evident in the nest defense culture of crows and ravens. American Crows (*Corvus brachyrhynchos*) and Common Ravens (*Corvus corax*) in the western USA aggressively defend their nests in cities and towns where shooting is outlawed and in general where persecution is frowned upon, but quietly retreat out of gun range in rural areas where an aggressive bird would be wounded or killed (Knight 1984, Knight et al. 1987). We do not know if these cultural changes in crows and ravens affect human culture, but annoyance with aggressive city crows is common and responded to with control efforts in some cities (e.g., Lancaster, PA), which may diminish future aggressive tendencies of city crows. In other settings, people bond with crows almost as closely as they bond with their pets, feeding them daily. Feeding favors tameness and solicitation by crows who recognize the individuals who provide for them (Marzluff and Angell 2005a). The responses of crows to people, be it aggressive or solicitous, has stimulated a radiation of popular culture from the naming of sports teams and rock bands to the myriad trappings and tales traditional in American Halloween celebrations. Other examples of coupled interactions between crows and people in cities include Jungle Crow (*Corvus macrorhynchus*) scavenging in Tokyo, which prompted improved trash disposal methods by residents and subsequent novel foraging behavior by crows (Higuchi 2003), use of imported walnuts by Carrion Crows in Sendai that stimulated placement of nuts in roads for cracking and driving behavior to swerve and hit the nuts (Nihei and Higuchi 2001, H. Higuchi pers. comm.), and milk drinking in Britain where concerns about disease transmission to people by tits and Eurasian Magpies (*Pica pica*) that open milk bottle lids may change milk delivery methods (Vernon 1993). The observed spread in nutcracking and milk-drinking behaviors are consistent with social learning and cultural evolution (Lefebvre 1995, Nihei and Higuchi 2001), suggesting that human and bird cultures have coevolved.

The cultural responses of people to nature often depend on the frequency and effect of the interaction. This may be an important factor in the likelihood of coevolution. When birds like corvids are rare, people often ignore or revere them, but once they become competitors they are viewed as pests and despised, harassed, and perhaps controlled. Our interaction with urban nature thus has a built in negative feedback mechanism that favors novel cultures in people and birds as the frequency and type of their interaction changes. When fewer in numbers and less a rival for resources, ravens, for example, were birds of the gods, even gods themselves, and useful guardians, navigators for mortals, and efficient sanitation engineers. When their abundance was thought to reduce valuable game or their consumption of human flesh evoked horror, guns and control policies were used to reduce their numbers and change their culture. Persecuted ravens became rare and shy around people. A contemporary rareness brought mystery, wonder, and concern from enough people that a culture of restoration developed (Glandt 2003). Understanding our role in such cyclical cultural phenomena may be important to conservation and restoration efforts in urban settings.

WHAT CHARACTERS ENABLE ADAPTATION RATHER THAN EXTINCTION?

We are increasingly familiar with the characteristics of birds threatened by urbanization (Marzluff et al. 2001, Chace and Walsh 2006), but what sorts of species might we expect to evolve, rather than go extinct, in response to the novel and strong selective challenges posed by our increasingly urban lifestyles? Our mechanistic understanding of genetic and cultural evolution and coevolution provides critical insights into differences that may separate urban creations from urban casualties (Figs. 18.1, 18.2).

Here I discuss some of the traits that have caused and should continue to cause contemporary evolution by birds in urban environments. The proposed traits are not necessarily the same list one might generate to characterize birds that survive or even thrive in urban settings (McKinney 2002). My focus is on the characteristics that enable genetic and cultural changes in response to urbanization. Some of these characteristics may also enable birds to exploit urban environments. But, as argued above, while the ability to evolve may be sufficient to avert extinction, it may or may not be necessary to evolve in order to live in urban environments.

Population Size

Species that have large populations in urban environments will be less responsive to drift (Lynch 1996) and less vulnerable to stochastic extinction (Lande 1993, Gomulkiewicz and Holt 1995) and therefore more responsive to selection. High reproductive rates and large genetic and phenotypic variation may be corollaries of large population size, which would also increase the ability of populations to change in response to selection. Cause and effect may be confused over time as population size enables adaptive evolution and adaptation enables population growth. Species that should have an evolutionary advantage in urban environments because of large population size include those that directly benefit from anthropogenic subsidies, actions, and structures. Species that nest in our buildings, eat our food, and use our planted landscapes should be able to build large populations that are responsive to novel challenges such as exotic predators, reduced nest site availabilities, variable food supplies, traffic hazards, pollution, and localized persecution.

Sociality and Innovation

Social species have a dual inheritance system that allows cultural and genetic evolution (Fig. 18.2). Innovative species, often those with large brains relative to their body sizes (Sol et al. 2005), may have increased phenotypic plasticity and variability, which can enhance survival during colonization (Sol et al. 2002), and respond to selection within a generation by the mechanism of social learning. Innovations are often public information in social species (Danchin et al. 2004). Advantageous innovations are thus visible and available for rapid incorporation by a population able to observe and learn or imitate. Long-lived, large-brained, social species should be especially capable of innovation and adaptive cultural evolution. Corvids, psitticines, larids, and parids may survive and rapidly evolve in urban areas because social learning allows them to develop traditions and avoid stochastic extinction. Oscine songbirds may also rapidly evolve cultural dialects in urban areas. As discussed, these may or may not be adaptive, but increasingly the selective advantages of adjusting vocal communication relative to a city's noise is causing the cultural evolution of vocal behavior (Leader et al. 2005, Slabbekoorn and den Boer-Visser 2006, Warren et al. 2006). Assortative mate choice for learned songs may initially limit population divergence (Slabbekoorn and Smith 2002), but if it later results in reduced gene flow between social neighborhoods, this form of cultural evolution may also increase the rate of genetic evolution (Ellers and Slabbekoorn 2003, Lachlan and Servedio 2004).

Exploitation of People

Human activity is a root cause of the close general association between anthropogenic change and contemporary evolution (Reznick and Ghalambor 2001). But living in daily contact with people may further select for rapid change. Individual learning and innovation may be needed to respond to erratic human activities, but more lasting changes and those with high associated risk may be efficiently tracked by social learning (Feldman et al. 1996). Social and innovative species that interact routinely with people may thus evolve culturally especially rapidly. Cultural evolution may explain how some corvid populations traditionally avoid or seek particular humans that they live among (J. M. Marzluff et al. unpubl. data). Rapidly evolved cultural traditions may also drive rapid genetic evolution (Durham 1991).

Phenotypic Plasticity

Behaviorally variable species are expected to rapidly evolve in response to selection (Price et al. 2003). Their plasticity may enable them to attain large populations resistant to stochastic events (e.g., many generalist species). But plasticity may also expose extreme phenotypes with fitness consequences to selection, thereby driving rapid responses to novel selective forces and speeding the genetic assimilation of behavioral variation.

High Annual Reproductive Rate

In addition to its effect on population size, a high reproductive rate may further accelerate evolution by increasing the amount of random phenotypic variation within a generation. Species with large broods or multiple broods may increase the production of extreme phenotypes and thereby the chance of rapid adaptive genetic evolution. Eurasian Collared-Doves (*Streptopelia decaoto*) have been able to colonize and spread throughout European cities (and following introduction to North America, throughout warm U.S. cities as well) in part because of their ability to produce multiple broods (Hengeveld and Vandenbosch 1991).

Generation Time

Increased genetic variability (Simpson 1953), high population growth rate (Cole 1954), and an increased ability to more quickly respond to selection all enable species with short generation times to evolve more rapidly (genetically) than those with long generation times (Marzluff and Dial 1991). Generation time scales negatively with body size (Peters 1983), so small birds should evolve especially rapidly in urban environments. Evolution might also be generally more rapid in warmer cities where breeding occurs year-round and some species have generation times that are shorter than one year.

Environmental Uniqueness

The strength of selection is a fundamental driver of the degree and speed of evolution (Van Tienderen and De Jong 1994), so species in particularly extreme environments, fundamentally different environments, and perhaps under the influence of sexual as well as natural selection should evolve rapidly (Arnold et al. 2001). Species introduced to urban environments are often under strong selection and evolve rapidly. Species in newly urbanizing areas or in the center versus the periphery of a city may be under stronger selection and therefore evolve most rapidly. Sexual selection in novel environments may increase the rate of evolution in secondary sexual characteristics, as is apparently the case for junco plumage in San Diego (Yeh 2004).

Isolation

Population isolation often leads to stochastic extinction, but where large populations are sufficiently isolated from homogenizing gene and culture flow, evolution should be especially rapid. This was the case for song repertoires of Dupont's Lark (*Chersophilus duponti*) in fragmented shrublands (Laiolo and Tella 2005). Even over small spatial scales, non-random dispersal can reduce gene flow and promote evolutionary diversification (Garant et al. 2005). Species with constrained dispersal or patchy population structures imposed by urbanization may evolve rapidly. Rapid evolution may be the case for species with many of the previous characteristics that live in megacities with little contact with the nearest natural area. It may also be more relevant for resident or introduced species than for migratory species that wander widely. Some gene and culture flow may be necessary for rapid macroevolution and may speed microevolution by increasing the range of variation available for selection to act upon (Wyles et al. 1983). Careful study of the variation in evolutionary rates in response to varying gene and culture flow would be a productive research area for urban evolutionary ecologists.

EXPLORING COMTEMPORARY EVOLUTION IN URBAN SYSTEMS

> Of all the biological questions, those about evolution are the toughest to answer. The problem is one of time. For one thing, the events we want to understand happened long ago.
>
> Kroodsma 2005

Kroodsma obviously was not thinking about urban ecosystems and microevolution when he commented on the time frame of evolution. In contrast, urban ecologists may be especially well situated to study evolution because selection pressures are strong, large populations of social birds often live in these environments, and previous studies suggest evolutionary responses can occur in a few decades (Table 18.1). Rapid evolutionary adjustment to novel selective pressures may quickly stabilize (Yeh 2004), so newly or rapidly changing urban areas may be especially fruitful places to study contemporary evolution in action. The tracks of contemporary evolution, however, should be readily apparent in population comparisons along gradients of urbanization worldwide.

The study of evolution is difficult, as Kroodsma rightly points out. It is especially difficult to distinguish phenotypic plasticity or behavioral adjustments (synurbization) to a novel environment from evolved adjustments with genetic or cultural underpinnings. Endler (1986) lays a clear foundation for the study of natural selection, and Yeh (2004) and Partecke et al. (2004) provide superb examples of how to distinguish among phenotypic plasticity, drift, and adaptive evolution in an urban bird. Here I briefly summarize their suggestions for the approach to evolutionary study in urban ecosystems.

1. Describe your system. Find heritable (Table 18.2) or socially learned traits that should respond to specific challenges posed by your urban environment. Correlate change in trait values with change along a gradient of urbanization.
2. Demonstrate selection is acting on the traits you have identified. Evidence requires experimental perturbation, correlations between genetic trait values and demography, or comparisons of trait values among cohorts. Studying responses to perturbations may be especially feasible in urban areas because large-scale changes are frequent, replicated, and sometimes reversed (Cook et al. 1986). Standard experiments may be necessary to rule out phenotypically plastic responses to urbanization.
3. Demonstrate the causal connection between the environment and the species' response that produces cultural or genetic change. Fitness differences in genetic traits should be tied to environmental conditions to explain observed frequency shifts in traits through time. Social learning of specific cultural traits should be tied to the occurrence and spread of the traits in a natural setting (Lefebvre 1995).

Urban environments are among the best places in the world to study contemporary evolution. The city's strong and novel selection pressures and the pervasiveness of genetic and cultural evolution in challenging environments make evolution in cities a certainty. However, as explained above, extra care must be taken to distinguish evolution from plasticity or environmental response. The revelation that evolutionary changes can occur after a few generations of selection, drift, or cultural spread means that ecologists can observe and measure evolution. No longer must we concern ourselves with only the extinction of species in urban settings. We can also enjoy the creative, evolutionary responses of species to the challenges we pose. The fact that evolution can be demonstrated in modern species where people live is an educational opportunity that biologists should capitalize on. What better way to engage the public in science than to show them evolution in their own backyards?

ACKNOWLEDGMENTS

I would like to thank R. Bowman for suggesting the importance of evolution in urban ecosystems and P. Price for kindling within me (and many others) a keen interest in evolution. D. Oleyar, S. Rullman, T. Unfried, B. Webb, K. Whittaker, and P. S. Warren took the time to improve my thoughts with valuable comments on this manuscript. C. Stockwell and J. Allen supplied key research, critical thinking, and new information helpful to my understanding of contemporary evolution. H. Slabbekoorn opened my eyes to the cultural evolution of song in urban ecosystems and provided a thoughtful and useful critique of my ideas. Funding was provided by the College of Forest Resources' Denman Professorship in Sustainable Resources Science at the University of Washington.

LITERATURE CITED

Able, K. P., and J. R. Belthoff. 1998. Rapid "evolution" of migratory behaviour in the introduced House Finch of eastern North America. Proceedings of the Royal Society (London) B 265:2063–2071.

Airey, D. C., H. Castillo-Juarez, G. Casella, E. J. Pollak, and T. J. DeVoogd. 2000. Variation in the volume of Zebra Finch song control nuclei is heritable: developmental and evolutionary implications. Proceedings of the Royal Society (London) B 267: 2099–2104.

Akbas, Y., C. Takma, and E. Yaylak. 2004. Genetic parameters for quail body weights using a random regression model. South African Journal of Animal Science 34:104–109.

Altan, O., P. Settar, Y. Unver, and M. Cabuk. 2005. Heritabilities of tonic immobility and leucocytic response in sire and dam layer lines. Turkish Journal of Veterinary and Animal Sciences 29:3–8.

Arnold, S. J., M. E. Pfrender, and A. G. Jones. 2001. The adaptive landscape as a conceptual bridge between micro- and macroevolution. Genetic 112–113: 9–32.

Avital, E., and E. Jablonka. 2000. Animal traditions: behavioural inheritance in evolution. Cambridge University Press, Cambridge, UK.

Baker, A. J. 1980. Morphometric differentiation in New Zealand populations of the House Sparrow (*Passer domesticus*). Evolution 34:638–653.

Baker, A. J., and A. Moeed. 1979. Evolution in the introduced New Zealand population of the Common Myna, *Acidotheres tristis* (Aves: Sturnidae). Canadian Journal of Zoology 57:570–584.

Baker, M. C., M. S. A. Baker, and E. M. Baker. 2003. Rapid evolution of a novel song and an increase in repertoire size in an island population of an Australian songbird. Ibis 145:465–471.

Barbraud, C. 2000. Natural selection on body size traits in a long-lived bird, the Snow Petrel, *Pagodroma nivea*. Journal of Evolutionary Biology 13: 81–88.

Benkman, C. W. 2003. Divergent selection drives the adaptive radiation of crossbills. Evolution 57: 1176–1181.

Berthold, P. 2003. Genetic basis and evolutionary aspects of bird migration. Advances in the Study of Behavior 33:175–229.

Berthold, P., and F. Pulido. 1994. Heritability of migratory activity in a natural bird population. Proceedings of the Royal Society (London) B 257:311–315.

Bishop, J. A., and L. M. Cook. 1980. Industrial melanism and the urban environment. Advances in Ecological Research 11:373–404.

Blackmore, S. 1999. The meme machine. Oxford University Press, Oxford, UK.

Blair, R. B. 2001. Birds and butterflies along urban gradients in two ecoregions of the United States: is urbanization creating a homogeneous fauna? Pp. 33–56. *in* J. L. Lockwood and M. L. McKinney (editors), Biotic homogenization. Kluwer Academic/Plenum, New York, NY.

Blair, R. B. 2004. The effects of urban sprawl on birds at multiple levels of biological organization. Ecology and Society 9(2). <http://www.ecologyandsociety.org/vol9/iss5/art2>.

Blewett, C. M., and J. M. Marzluff. 2005. Effects of urban sprawl on snags and the abundance and productivity of cavity-nesting birds. Condor 107: 677–692.

Boyd, R., and P. J. Richerson. 1985. Culture and the evolutionary process. University of Chicago Press, Chicago.

Carere, C., P. J. Drent, J. M. Koolhaas, and T. G. G. Groothuis. 2005. Epigenetic effects on personality traits: early food provisioning and sibling competition. Behaviour 142:1329–1355.

Castro, L., and M. A. Toro. 2004. The evolution of culture: from primate social learning to human culture. Proceedings of the National Academy of Sciences USA 101:10235–10240.

Chace, J. F., and J. J. Walsh. 2006. Urban effects on native avifauna: a review. Landscape and Urban Planning 74:46–69.

Chace, J. F., J. J. Walsh, A. Cruz, J. W. Prather, and H. M. Swanson. 2003. Spatial and temporal activity patterns of the brood parasitic Brown-headed Cowbird at an urban/wildland interface. Landscape and Urban Planning 64:179–190.

Charmantier, A., L. E. B. Kruuk, and M. M. Lambrechts. 2004. Parasitism reduces the potential for evolution in a wild bird population. Evolution 58: 203–206.

Christe, P., A. P. Møller, N. Saino, and F. de Lope. 2000. Genetic and environmental components of phenotypic variation in immune response and body size of a colonial bird, *Delichon urbica* (the House Martin). Heredity 85:75–83.

Clergeau, P., S. Croci, J. Jokimäki, M.-L. Kaisanlahti-Jokimäki, and M. Dinetti. 2006. Avifauna homogeniszation by urbanization: analysis at different European latitudes. Biological Conservation 127:336–344.

Cole, L. C. 1954. The population consequences of life history phenomena. Quarterly Review of Biology 29:103–137.
Cook, L. M., G. S. Mani, and M. E. Varley. 1986. Post-industrial melanism in the peppered moth. Science 231:611–613.
Cruz, A., and J. W. Wiley. 1989. The decline of an adaptation in the absence of a presumed selection pressure. Evolution 43:55–62.
Danchin, É., L.-A. Giraldeau, T. J. Valone, and R. H. Wagner. 2004. Public information: from nosy neighbors to cultural evolution. Science 305:487–491.
Daniels, G. D., and J. B. Kirkpatrick. 2006. Does variation in garden characteristics influence the conservation of birds in suburbia? Biological Conservation 133:326–335.
Dawkins, R. 1976. The selfish gene. Oxford University Press, Oxford, UK.
De Neve, L., J. J. Soler, T. Perez-Contreras, and M. Soler. 2004. Genetic, environmental and maternal effects on magpie nestling-fitness traits under different nutritional conditions: a new experimental approach. Evolutionary Ecology Research 6:415–431.
Dingemanse, N. J., C. Both, P. J. Drent, and J. M. Tinbergen. 2004. Fitness consequences of avian personalities in a fluctuating environment. Proceedings of the Royal Society (London) B 271:847–852.
Donnelly, R., and J. M. Marzluff. 2004. Importance of reserve size and landscape context to urban bird conservation. Conservation Biology 18:733–745.
Donnelly, R. E., and J. M. Marzluff. 2006. Relative importance of habitat quantity, structure, and spatial pattern to birds in urbanizing environments. Urban Ecosystems 9:99–117.
Drent, P. J., K. van Oers, and A. J. van Noordwijk. 2003. Realized heritability of personalities in the Great Tit (*Parus major*). Proceedings of the Royal Society (London) B 270:45–51.
Durham, W. H. 1991. Coevolution: genes, culture, and human diversity. Stanford University Press, Stanford, CA.
Ellers, J., and H. Slabbekoorn. 2003. Song divergence and male dispersal among bird populations: a spatially explicit model testing the role of vocal learning. Animal Behaviour 65:671–681.
Elliott, J. J., and R. S. Arbib, Jr. 1953. Origin and status of the House Finch in the eastern United States. Auk 70:31–37.
Endler, J. A. 1986. Natural selection in the wild. Princeton University Press, Princeton, NJ.
Endlicher, W., G. Jendritzky, J. Fischer, and J.-P. Redlich. 2006. Heat waves, urban climate and human health. Pp. 103–114 *in* W. Wuyi, T. Krafft, and F. Kraas (editors), Global change, urbanization and health. China Meteorological Press, Beijing, China.
Er, K. B. H., J. L. Innes, K. Martin, and B. Klinkenberg. 2005. Forest loss with urbanization predicts bird extirpations in Vancouver. Biological Conservation 126:410–419.
Erz, W. 1964. Populationsökologische Untersuchungen an der Avifauna zweier norddeutscher Großstädte. Zeitschrift für wissenschaftliche Zoologie 170:1–111.
Feldman, M. W., K. Aoki, and J. Kumm. 1996. Individual versus social learning: evolutionary analysis in a fluctuating environment. Anthropological Science 104:209–231.
Fernández-Juricic, E., R. Poston, K. Decollibus, T. Morgan, B. Bastain, C. Marin, K. Jones, and R. Treminio. 2005. Microhabitat selection and singing behavior patterns of male House Finches (*Carpodacus mexicanus*) in urban parks in a heavily urbanized landscape in the western U.S. Urban Habitats 3:49–69.
Fraterrigo, J. M., and J. A. Wiens. 2005. Bird communities of the Colorado Rocky Mountains along a gradient of exurban development. Landscape and Urban Planning 71:263–275.
Garant, D., L. E. B. Kruuk, T. A. Wilkin, R. H. McCleery, and B. C. Sheldon. 2005. Evolution driven by differential dispersal within a wild bird population. Nature 433:60–65.
Garant, D., B. C. Sheldon, and L. Gustafsson. 2004. Climatic and temporal effects on the expression of secondary sexual characters: genetic and environmental components. Evolution 58:634–644.
Gehlbach, F. R. 2005. Native Texas avifauna altered by suburban entrapment and a method for easily assessing natural avifaunal value. Bulletin of the Texas Ornithological Society 38:35–47.
Glandt, D. 2003. Der Kolkrabe. AULA-Verlag, Wiebelsheim, Germany.
Goodwin, D. 1978. Birds of man's world. British Museum of Natural History, London, UK.
Gomulkiewicz, R., and R. D. Holt. 1995. When does evolution by natural selection prevent extinction? Evolution 49:201–207.
Grant, P. R., and B. R. Grant. 1995. Predicting microevolutionary responses to direction selection on heritable variation. Evolution 49:241–251.
Hadfield, J. D., M. D. Burgess, A. Lord, A. B. Phillimore, S. M. Clegg, and I. P. F. Owens. 2006. Direct versus indirect sexual selection: genetic basis of colour, size, and recruitment in a wild bird. Proceedings of the Royal Society (London) B 273:1347–1353.
Hansson, B., S. Bensch, and D. Hasselquist. 2003. Heritability of dispersal in the Great Reed Warbler. Ecology Letters 6:290–294.
Hegyi, G., J. Torok, L. Toth, L. Z. Garamszegi, and B. Rosivall. 2006. Rapid temporal change in the

expression and age-related information content of a sexually selected trait. Journal of Evolutionary Biology 19:228–238.
Hendry, A. P., and M. T. Kinnison. 1999. Perspective: the pace of modern life: measuring rates of contemporary microevolution. Evolution 53:1637–1653.
Hendry, A. P., and M. T. Kinnison. 2001. An introduction to microevolution: rate, pattern, process. Genetica 112–113:1–8.
Hengeveld, R., and F. Vandenbosch. 1991. The expansion of the Collared Dove (*Streptopelia decaocto*) population in Europe. Ardea 79:67–72.
Higuchi, H. (editor). 2003. Conflict between crows and humans in urban areas. Global Environmental Research 7:129–205.
Hodgson, P., K. French, and R. E. Major. 2006. Comparison of foraging behaviour of small, urban-sensitive insectivores in continuous woodland and woodland remnants in a suburban landscape. Wildlife Research 33:591–603.
Johnston, R. F. 2001. Synanthropic birds of North America. Pp. 49–68 *in* J. M. Marzluff, R. Bowman, and R. Donnelly (editors), Avian ecology and conservation in an urbanizing world. Kluwer Academic Publishers, Norwell, MA.
Johnston, R. F., and R. K. Selander. 1964. House Sparrows: rapid evolution of races in North America. Science 144:548–550.
Kettlewell, B. 1973. The evolution of melanism: the study of a recurring necessity, with special reference to industrial melanism in the Lepidoptera. Clarendon Press, Oxford, UK.
Kinnison, M. T., and A. P. Hendry. 2001. The pace of modern life II: from rates of contemporary microevolution to pattern and process. Genetic 112:145–164.
Kluza, D. A., C. R. Griffin, and R. M. DeGraaf. 2000. Housing developments in rural New England: effects on forest birds. Animal Conservation 3:15–26.
Knight, R. L. 1984. Responses of nesting ravens to people in areas of different human densities. Condor 86:345–346.
Knight, R. L., D. J. Grout, and S. A. Temple. 1987. Nest-defense behavior of the American Crow in urban and rural areas. Condor 89:175–177.
Kroodsma, D. E. 2005. The singing life of birds: the art and science of listening to birdsong. Houghton Mifflin, Boston, MA.
Lachlan, R. F., and M. R. Servedio. 2004. Song learning accelerates allopatric speciation. Evolution 58:2049–2063.
Laiolo, P., and J. L. Tella. 2005. Habitat fragmentation affects culture transmission: patterns of song matching in Dupont's Lark. Journal of Applied Ecology 42:1183–1193.
Laland, K. N., J. Odling-Smee, and M. W. Feldman. 2000. Niche construction, biological evolution, and cultural change. Behavioral and Brain Sciences 23:131–175.
Lande, R. 1976. Natural selection and random genetic drift in phenotypic evolution. Evolution 30:314–334.
Lande, R. 1993. Risks of population extinction from demographic and environmental stochasticity and random catastrophes. American Naturalist 142: 911–927.
Larsson, K., K. Rattiste, and V. Lilleleht. 1997. Heritability of head size in the Common Gull, *Larus canus*, in relation to environmental conditions during offspring growth. Heredity 79:201–207.
Leader, N., J. Wright, and Y. Yom-Tov. 2005. Acoustic properties of two urban song dialects in the Orange-tufted Sunbird (*Nectarinia osea*). Auk 122:231–245.
Lee, T. M., M. C. K. Soh, N. Sodhi, L. P. Koh, and S. L. H. Kim. 2005. The effects of habitat disturbance on mixed species bird flocks in a tropical sub-montane rainforest. Biological Conservation 122:193–204.
Lefebvre, L. 1986. Cultural diffusion of a novel food-finding behaviour in urban pigeons: an experimental field test. Ethology 71:295–304.
Lefebvre, L. 1995. The opening of milk bottles by birds: evidence for accelerating learning rates, but against the wave-of-advance model of cultural transmission. Behavioural Processes 34:43–54.
Lim, H. C., and N. S. Sodhi. 2004. Responses of avian guilds to urbanization in a tropical city. Landscape and Urban Planning 66:199–215.
Luniak, M. 2004. Synurbization—adaptation of animal wildlife to urban development. Pp. 50–55 *in* W. W. Shaw, L. K. Harris, and L. Vandruff (editors), Proceedings of the 4th International Symposium on Urban Wildlife Conservation. University of Arizona Press, Tucson, AZ.
Lynch, M. 1996. A quantitative-genetic perspective on conservation issues. Pp. 471–501 *in* J. C. Avise and J. L. Hamrick (editors), Conservation genetics, case histories from nature. Chapman and Hall, New York.
MacColl, A. D. C., and B. J. Hatchwell. 2003. Heritability of parental effort in a passerine bird. Evolution 57:2191–2195.
Martin, T. E., and J. Clobert. 1996. Nest predation and avian life-history evolution in Europe versus North America: a possible role of humans? American Naturalist 147:1028–1046.
Martínez del Rio, C. 1993. Do British tits drink milk or just skim the cream? British Birds 86:321–322.
Marzluff, J. M. 2001. Worldwide urbanization and its effects on birds. Pp. 19–47 *in* J. M. Marzluff, R. Bowman, and R. Donnelly (editors), Avian conservation and ecology in an urbanizing world. Kluwer Academic Publishers, Norwell, MA.

Marzluff, J. M. 2005. Island biogeography for an urbanizing world: how extinction and colonization may determine biological diversity in human-dominated landscapes. Urban Ecosystems 8:155–175.

Marzluff, J. M., and T. Angell. 2005a. In the company of crows and ravens. Yale University Press, New Haven, CT.

Marzluff, J. M., and T. Angell. 2005b. Cultural coevolution: how the human bond with crows and ravens extends theory and raises new questions. Journal of Ecological Anthropology 9:67–73.

Marzluff, J. M., and R. P. Balda. 1988. Pairing patterns and fitness in free-ranging Pinyon Jays: what do they reveal about mate choice? Condor 90:201–213.

Marzluff, J. M., R. Bowman, and R. Donnelly (editors). 2001. Avian ecology and conservation in an urbanizing world. Kluwer Academic Publishers, Norwell, MA.

Marzluff, J. M., and K. P. Dial. 1991. Life history correlates of taxonomic diversity. Ecology 72:428–439.

Marzluff, J. M., J. C. Withey, K. A. Whittaker, M. D. Oleyar, T. M. Unfried, S. Rullman, and J. DeLap. 2007. Consequences of habitat utilization by nest predators and breeding songbirds across multiple scales in an urbanizing landscape. Condor 109:516–534.

Mayr, E. 1963. Animal species and evolution. Harvard University Press, Cambridge, MA.

McKinney, M. L. 2002. Urbanization, biodiversity, and conservation. BioScience 52:883–890.

Melles, S., S. Glenn, and K. Martin. 2002. Urban bird diversity and landscape complexity: species-environment associations along a multi-scale habitat gradient. Conservation Ecology 7(1):5. <http://www.consecol.org/vol7/iss1/art5/>.

Merila, J., R. Przybylo, and B. C. Sheldon. 1999. Genetic variation and natural selection on Blue Tit body condition in different environments. Genetical Research 73:165–176.

Merila, J., L. E. B. Kruuk, and B. C. Sheldon. 2001. Natural selection on the genetical component of variance in body condition in a wild bird population. Journal of Evolutionary Biology 14:918–929.

Møller, A. P. 2001. Heritability of arrival date in a migratory bird. Proceedings of the Royal Society (London) B 268:203–206.

Møller, A. P. 2002. Parent-offspring resemblance in degree of sociality in a passerine bird. Behavioral Ecology and Sociobiology 51:276–281.

Møller, A. P. 2006. Rapid change in nest size of a bird related to change in a secondary sexual character. Behavioral Ecology 17:108–116.

Møller, A. P., and T. Szep. 2005. Rapid evolutionary change in a secondary sexual character linked to climatic change. Journal of Evolutionary Biology 18:481–495.

Murton, R. K., R. J. P. Thearle, and J. Thompson. 1972. Ecological studies of the feral pigeon, *Columba livia*, var. I, II. Journal of Applied Ecology 9:835–889.

Nihei, Y., and H. Higuchi. 2001. When and where did crows learn to use automobiles as nutcrackers? Tohoku Psychologica Folia 60:93–97.

Odling-Smee, F. J., K. N. LaLand, and M. W. Feldman. 2003. Niche construction: the neglected process in evolution. Princeton University Press, Princeton, NJ.

Palumbi, S. R. 2001. Humans as the world's greatest evolutionary force. Science 293:1786–1790.

Parsons, H., R. E. Major, and K. French. 2006. Species interactions and habitat associations of birds inhabiting urban areas of Sydney, Australia. Austral Ecology 31:217–227.

Partecke, J. 2003. Annual cycles of urban and forest-living European Blackbirds (*Turdus merula*): genetic differences or phenotypic plasticity? Ph.D. Dissertation, Ludwig-Maximilians University, Munich, Germany.

Partecke, J., E. Gwinner, and S. Bensch. 2006a. Is urbanization of European Blackbirds (*Turdus merula*) associated with genetic differentiation? Journal für Ornithologie 147:549–552.

Partecke, J., I. Schwabl, and E. Gwinner. 2006b. Stress and the city: urbanization and its effects on the stress physiology in European blackbirds. Ecology 87:1945–1952.

Partecke, J., T. Van't Hof, and E. Gwinner. 2004. Differences in the timing of reproduction between urban and forest European Blackbirds (*Turdus merula*): result of phenotypic flexibility or genetic differences? Proceedings of the Royal Society of London (B) 271:1995–2001.

Pasinelli, G., K. Schiegg, and J. R. Walters. 2004. Genetic and environmental influences on natal dispersal distance in a resident bird species. American Naturalist 164:660–669.

Patricelli, G. L., and J. L. Blickley. 2006. Avian communication in urban noise: causes and consequences of vocal adjustment. Auk 123:639–649.

Peters, R. H. 1983. The ecological implications of body size. Cambridge University Press, Cambridge, UK.

Posa, M. R. C., and N. S. Sodhi. 2006. Effects of anthropogenic land use on forest birds and butterflies in Subic Bay, Philippines. Biological Conservation 129:256–270.

Postma, E., and A. J. van Noordwijk. 2005. Genetic variation for clutch size in natural populations of birds from a reaction norm perspective. Ecology 86:2344–2357.

Potti, J. 1993. Environmental, ontogenic, and genetic-variation in egg size of Pied Flycatchers. Canadian Journal of Zoology 71:1534–1542.

Praharaj, N., C. Beaumong, G. Damgrine, D. Soubieux, L. Merat, D. Bouret, G. Luneau, J. M. Alletru, M. H.

Pinard-van der Laan, P. Thoraval, and S. Mignon-Grasteau. 2004. Genetic analysis of the growth curve of Rous sarcoma virus-induced tumors in chickens. Poultry Science 83:1479–1488.

Price, T. D., M. Kirkpatrick, and S. J. Arnold. 1988. Directional selection and the evolution of breeding date in birds. Science 240:798–799.

Price, T. D., A. Qvarnström, and D. E. Irwin. 2003. The role of phenotypic plasticity in driving genetic evolution. Proceedings of the Royal Society (London) B 270:1433–1440.

Pulido, F., P. Berthold, G. Mohr, and U. Querner. 2001. Heritability of the timing of autumn migration in a natural bird population. Proceedings of the Royal Society (London) B 268:953–959.

Pytte, C. L. 1997. Song organization of House Finches at the edge of an expanding range. Condor 99:942–954.

Recher, H. F., and D. L. Serventy. 1991. Long term changes in the relative abundances of birds in Kings Park, Perth, Western Australia. Conservation Biology 5:90–102.

Rendell, L., and H. Whitehead. 2001. Cetacean culture: still afloat after the first naval engagement of the culture wars. Behavioral and Brain Sciences 24:360–382.

Reynaud, P. A., and J. Thioulouse. 2000. Identification of birds as biological markers along a neotropical urban-rural gradient (Cayenne, French Guiana), using co-inertia analysis. Journal of Environmental Management 59:121–140.

Reznick, D. N., and C. K. Ghalambor. 2001. The population ecology of contemporary adaptations: what empirical studies reveal about the conditions that promote adaptive evolution. Genetica 112–113: 183–198.

Rich, C., and T. Longcore (editors). 2006. Ecological consequences of artificial night lighting. Island Press, Washington, DC.

Rosinski, A., S. Mowaczewski, H. Kontecka, M. Bednarczyk, G. Eliminowska-Wenda, H. Bielinska, and A. Maczynska. 2006. Analysis of the laying rhythm and reproductive traits of geese. Folia Biologica-Krakow 54:145–152.

Schmidt, K. A., and C. J. Whelan. 1999. Effects of exotic *Lonicera* and *Rhamnus* on songbird nest predation. Conservation Biology 13:502–506.

Schoech, S., and R. Bowman. 2003. Does differential access to protein influence differences in timing of breeding of Florida Scrub-Jays in suburban and wildland habitats? Auk 120:1114–1127.

Sheldon, B. C., L. E. B. Kruuk, and J. Merila. 2003. Natural selection and inheritance of breeding time and clutch size in the Collared Flycatcher. Evolution 57:406–420.

Shochat, E., S. B. Lerman, M. Katti, and D. B. Lewis. 2004. Linking optimal foraging behavior to bird community structure in an urban-desert landscape: field experiments with artificial food patches. American Naturalist 164:232–243.

Shochat, E., P. S. Warren, S. H. Faeth, N. E. McIntyre, and D. Hope. 2006. From patterns to emerging processes in mechanistic urban ecology. Trends in Ecology and Evolution 21:186–191.

Simpson, G. G. 1953. The major features of evolution. Columbia University Press, New York, NY.

Sinclair, K. E., G. R. Hess, C. E. Moorman, and J. H. Mason. 2005. Mammalian nest predators respond to greenway width, landscape context and habitat structure. Landscape and Urban Planning 71:277–293.

Slabbekoorn, H., and A. den Boer-Visser. 2006. Cities change the songs of birds. Current Biology 16:2326–2331.

Slabbekoorn, H., and M. Peet. 2003. Birds sing at a higher pitch in urban noise. Nature 424:267.

Slabbekoorn, H., and T. B. Smith. 2002. Bird song, ecology and speciation. Philosophical Transactions of the Royal Society of London (B) 357:483–503.

Smith, T. B., L. A. Freed, J. K. Lepson, and J. H. Carothers. 1995. Evolutionary consequences of extinctions in populations of a Hawaiian honeycreeper. Conservation Biology 9:107–113.

Sol, D., S. Timmermans, and L. Lefebvre. 2002. Behavioural flexibility and invasion success in birds. Animal Behaviour 63:490–502.

Sol, D., R. P. Duncan, T. M. Blackburn, P. Cassey, and L. Lefebvre. 2005. Big brains, enhanced cognition, and response of birds to novel environments. Proceedings of the National Academy of Sciences USA 102:5460–5465.

Stockwell, C. A., and S. C. Weeks. 1999. Translocations and rapid evolutionary responses in recently established populations of western mosquitofish (*Gambusia affinis*). Animal Conservation 2:103–110.

Stockwell, C. A., A. P. Hendry, and M. T. Kinnison. 2003. Contemporary evolution meets conservation biology. Trends in Ecology and Evolution 2:94–101.

Thompson, J. N. 1998. Rapid evolution as an ecological process. Trends in Ecology and Evolution 13:329–332.

van Oers, K., P. J. Drent, P. de Goede, and A. J. van Noordwijk. 2004. Realized heritability and the repeatability of risk-taking behaviour in relation to avian personalities. Proceedings of the Royal Society (London) B 271:65–73.

Van Tienderen, P. H., and G. De Jong. 1994. A general model of the relation between phenotypic selection and genetic response. Journal of Evolutionary Biology 7:1–12.

Vernon, J. D. R. 1993. Magpies and milk bottles. British Birds 86:315.

Walasz, K. 1990. Experimental investigations on the behavioural differences between urban and forest Blackbirds. Acta Zoologica Cracoviensia 33: 235–271.

Wandeler, P., S. M. Funk, C.R. Largiadèr, S. Gloor, and U. Breitenmoser. 2003. The city-fox phenomenon: genetic consequences of a recent colonization of urban habitat. Molecular Ecology 12:647–656.

Warren, P. S., M. Katti, M. Ermann, and A. Brazel. 2006. Urban bioacoustics: it's not just noise. Animal Behaviour 71:491–502.

Withey, J. C., and J. M. Marzluff. 2005. Dispersal by juvenile American Crows influences population dynamics across a gradient of urbanization. Auk 122:206–222.

Wood, W. E., and S. M. Yezerinac. 2006. Song Sparrow (*Melospiza melodia*) song varies with urban noise. Auk 123:650–659.

Wright, S. 1931. Evolution in Mendelian populations. Genetics 16:97–159.

Wright, S. 1932. The roles of mutation, inbreeding, crossbreeding, and selection in evolution. Proceedings of the Sixth International Congress of Genetics 1:356–366.

Wyles, J. S., J. G. Kunkel, and A. C. Wilson. 1983. Birds, behavior, and anatomical evolution. Proceedings of the National Academy of Sciences USA 80:4394–4397.

Yeh, P. J. 2004. Rapid evolution of a sexually selected trait following population establishment in a novel habitat. Evolution 58:166–174.

Yeh, P. J., and T. D. Price. 2004. Adaptive phenotypic plasticity and the successful colonization of a novel environment. American Naturalist 164:531–542.

INDEX

Page numbers preceded by "a" or "b" denote pages in the online Special Topics A and B.

STUDIES IN AVIAN BIOLOGY

1. Kessel, B., and D. D. Gibson. 1978.
   Status and Distribution of Alaska Birds.
2. Pitelka, F. A., editor. 1979.
   Shorebirds in Marine Environments.
3. Szaro, R. C., and R. P. Balda. 1979.
   Bird Community Dynamics in a Ponderosa Pine Forest.
4. DeSante, D. F., and D. G. Ainley. 1980.
   The Avifauna of the South Farallon Islands, California.
5. Mugaas, J. N., and J. R. King. 1981.
   Annual Variation of Daily Energy Expenditure by the Black-billed Magpie: A Study of Thermal and Behavioral Energetics.
6. Ralph, C. J., and J. M. Scott, editors. 1981.
   Estimating Numbers of Terrestrial Birds.
7. Price, F. E., and C. E. Bock. 1983.
   Population Ecology of the Dipper (Cinclus mexicanus) *in the Front Range of Colorado.*
8. Schreiber, R. W., editor. 1984.
   Tropical Seabird Biology.
9. Scott, J. M., S. Mountainspring, F. L. Ramsey, and C. B. Kepler. 1986.
   Forest Bird Communities of the Hawaiian Islands: Their Dynamics, Ecology, and Conservation.
10. Hand, J. L., W. E. Southern, and K. Vermeer, editors. 1987.
    Ecology and Behavior of Gulls.
11. Briggs, K. T., W. B. Tyler, D. B. Lewis, and D. R. Carlson. 1987.
    Bird Communities at Sea off California: 1975 to 1983.
12. Jehl, J. R., Jr. 1988.
    Biology of the Eared Grebe and Wilson's Phalarope in the Nonbreeding Season: A Study of Adaptations to Saline Lakes.
13. Morrison, M. L., C. J. Ralph, J. Verner, and J. R. Jehl, Jr., editors. 1990.
    Avian Foraging: Theory, Methodology, and Applications.
14. Sealy, S. G., editor. 1990.
    Auks at Sea.
15. Jehl, J. R., Jr., and N. K. Johnson, editors. 1994.
    A Century of Avifaunal Change in Western North America.
16. Block, W. M., M. L. Morrison, and M. H. Reiser, editors. 1994.
    The Northern Goshawk: Ecology and Management.
17. Forsman, E. D., S. DeStefano, M. G. Raphael, and R. J. Gutiérrez, editors. 1996.
    Demography of the Northern Spotted Owl.
18. Morrison, M. L., L. S. Hall, S. K. Robinson, S. I. Rothstein, D. C. Hahn, and T. D. Rich, editors. 1999.
    Research and Management of the Brown-headed Cowbird in Western Landscapes.
19. Vickery, P. D., and J. R. Herkert, editors. 1999.
    Ecology and Conservation of Grassland Birds of the Western Hemisphere.
20. Moore, F. R., editor. 2000.
    Stopover Ecology of Nearctic–Neotropical Landbird Migrants: Habitat Relations and Conservation Implications.
21. Dunning, J. B., Jr., and J. C. Kilgo, editors. 2000.
    Avian Research at the Savannah River Site: A Model for Integrating Basic Research and Long-Term Management.
22. Scott, J. M., S. Conant, and C. van Riper, II, editors. 2001.
    Evolution, Ecology, Conservation, and Management of Hawaiian Birds: A Vanishing Avifauna.
23. Rising, J. D. 2001.
    Geographic Variation in Size and Shape of Savannah Sparrows (Passerculus sandwichensis).
24. Morton, M. L. 2002.
    The Mountain White-crowned Sparrow: Migration and Reproduction at High Altitude.
25. George, T. L., and D. S. Dobkin, editors. 2002.
    Effects of Habitat Fragmentation on Birds in Western Landscapes: Contrasts with Paradigms from the Eastern United States.

26. Sogge, M. K., B. E. Kus, S. J. Sferra, and M. J. Whitfield, editors. 2003. *Ecology and Conservation of the Willow Flycatcher.*

27. Shuford, W. D., and K. C. Molina, editors. 2004. *Ecology and Conservation of Birds of the Salton Sink: An Endangered Ecosystem.*

28. Carmen, W. J. 2004. *Noncooperative Breeding in the California Scrub-Jay.*

29. Ralph, C. J., and E. H. Dunn, editors. 2004. *Monitoring Bird Populations Using Mist Nets.*

30. Saab, V. A., and H. D. W. Powell, editors. 2005. *Fire and Avian Ecology in North America.*

31. Morrison, M. L., editor. 2006. *The Northern Goshawk: A Technical Assessment of Its Status, Ecology, and Management.*

32. Greenberg, R., J. E. Maldonado, S. Droege, and M. V. McDonald, editors. 2006. *Terrestrial Vertebrates of Tidal Marshes: Evolution, Ecology, and Conservation.*

33. Mason, J. W., G. J. McChesney, W. R. McIver, H. R. Carter, J. Y. Takekawa, R. T. Golightly, J. T. Ackerman, D. L. Orthmeyer, W. M. Perry, J. L. Yee, M. O. Pierson, and M. D. McCrary. 2007. *At-Sea Distribution and Abundance of Seabirds off Southern California: A 20-Year Comparison.*

34. Jones, S. L., and G. R. Geupel, editors. 2007. *Beyond Mayfield: Measurements of Nest-Survival Data.*

35. Spear, L. B., D. G. Ainley, and W. A. Walker. 2007. *Foraging Dynamics of Seabirds in the Eastern Tropical Pacific Ocean.*

36. Niles, L. J., H. P. Sitters, A. D. Dey, P. W. Atkinson, A. J. Baker, K. A. Bennett, R. Carmona, K. E. Clark, N. A. Clark, C. Espoz, P. M. González, B. A. Harrington, D. E. Hernández, K. S. Kalasz, R. G. Lathrop, R. N. Matus, C. D. T. Minton, R. I. G. Morrison, M. K. Peck, W. Pitts, R. A. Robinson, and I. L. Serrano. 2008. *Status of the Red Knot* (Calidris canutus rufa) *in the Western Hemisphere.*

37. Ruth, J. M., T. Brush, and D. J. Krueper, editors. 2008. *Birds of the US–Mexico Borderland: Distribution, Ecology, and Conservation.*

38. Knick, S. T., and J. W. Connelly, editors. 2011. *Greater Sage-Grouse: Ecology and Conservation of a Landscape Species and Its Habitats.*

39. Sandercock, B. K., K. Martin, and G. Segelbacher, editors. 2011. *Ecology, Conservation, and Management of Grouse.*

40. Forsman, E. D., et al. 2011. *Population Demography of Northern Spotted Owls.*

41. Wells, J. V., editor. 2011. *Boreal Birds of North America: A Hemispheric View of Their Conservation Links and Significance.*

42. Paul, E., editor. 2012. *Emerging Avian Disease.*

43. Ribic, C. A., F. R. Thompson, III, and P. J. Pietz, editors. 2012. *Video Surveillance of Nesting Birds.*

44. Bart, J. R., and V. H. Johnston, editors. 2012. *Arctic Shorebirds in North America: A Decade of Monitoring.*

45. Lepczyk, C. A., and P. S. Warren, editors. 2012. *Urban Bird Ecology and Conservation.*